Microorganisms for Sustainability

Volume 52

Series Editor
Naveen Kumar Arora, Environmental Microbiology, School for Environmental Science, Babasaheb Bhimrao Ambedkar University, Lucknow, India

Microorganisms have been in existence since the origin of life on earth and can survive the most extreme habitats or conditions on earth. Microorganisms are involved in regulating biogeochemical cycles, maintaining plant and animal health, and sustaining the global food chain. Moreover, they play crucial roles in addressing the challenges of climate change and achieving the targets of Sustainable Development Goals (SDGs).

This multidisciplinary book series captures the role of microbes towards building a sustainable world, while encompassing cutting-edge technologies and current needs across various fields such as agriculture sustainability, bioremediation, restoration of degraded habitats and wastelands, and food security. Additionally, this series explores microbial applications in industries, and building their utilization in clean and green energy solutions. Furthermore, themes like microbial secondary metabolites, extremophilic microbes and modern omics, including next generation sequencing and metagenomics, are also covered in this series.

With contributions from researchers across the globe, this series addresses the important call of 'One Planet-One Health-One Future'. It comprises a collection of diverse volumes that provides insights for scientists, young researchers, educators and decision-makers in the government, private sector, and non-governmental organizations, empowering their efforts to achieve the global goals.

The series invites, evaluates, and accepts book proposals to ensure a diverse, inclusive, and evolving program. The final decision regarding acceptance rests with the series editor.

Peer-review

This book series follows a stratified review process. Proposals for individual volumes are reviewed by the series editor and then the editorial board members. On a case-to-case basis, external reviewers are also invited for further evaluation of the book proposal. Review of the chapters is the responsibility of the volume editor(s). A manuscript submission platform has been recently made available to the authors, volume editors and the series editor.

Mukesh Meena • Andleeb Zehra
Prashant Swapnil • Chandra Shekhar Seth
Editors

Metabolic Sustainability of Endophytes

Current Status, Challenges and Potential

Springer

Editors
Mukesh Meena
Department of Botany
Mohanlal Sukhadia University
Udaipur, Rajasthan, India

Andleeb Zehra
Institute of Science, Department of Botany
Banaras Hindu University
Varanasi, Uttar Pradesh, India

Prashant Swapnil
Department of Botany, School of Basic Sciences
Central University of Punjab
Bathinda, Punjab, India

Chandra Shekhar Seth
Department of Botany
University of Delhi
New Delhi, Delhi, India

ISSN 2512-1901 ISSN 2512-1898 (electronic)
Microorganisms for Sustainability
ISBN 978-981-96-4003-4 ISBN 978-981-96-4004-1 (eBook)
https://doi.org/10.1007/978-981-96-4004-1

This Springer imprint is published by the registered company Springer Nature Singapore Pte Ltd.
The registered company address is: 152 Beach Road, #21-01/04 Gateway East, Singapore 189721, Singapore

Preface

Endophytes, the symbiotic microorganisms that reside within plant tissues without causing disease, have captivated the scientific community over the last few decades. Their ability to enhance plant growth, improve stress tolerance, and synthesize bioactive compounds places them at the forefront of research into sustainable agricultural and industrial practices. This emerging field of study presents a transformative opportunity for integrating microbial symbiosis into a broad spectrum of applications, ranging from agriculture to environmental management and biotechnology. However, the journey to fully understanding and utilizing the metabolic potential of endophytes is still in its nascent stages. With the global push toward sustainable practices, these microscopic allies may hold the key to addressing some of the most pressing challenges we face today—namely, food security, environmental sustainability, and the mitigation of climate change.

This book, *Metabolic Sustainability of Endophytes: Current Status, Challenges, and Their Potential*, seeks to present a holistic view of the role of endophytes in plant metabolism and sustainability, highlighting both their ecological importance and their potential in industrial applications. By focusing on the metabolic processes that endophytes influence within their plant hosts, this volume addresses several critical questions: How do endophytes modulate plant physiology to enhance resilience? What are the metabolic pathways involved in the biosynthesis of bioactive compounds? Can endophytes be engineered or applied on a large scale to meet the needs of modern agriculture?

Chapters will unfold the positions of fungal endophytes as reservoirs of bioactive metabolites with immense potential. The application of fungal endophytes in sustainable agriculture will emerge as a recurring theme, delving into their future role in plant health, anticipated abiotic stress management, and their contribution to the future production of extracellular enzymes. Moreover, the volume will explore the promising future applications of endophytes as biostimulant to offer sustainable solutions for anticipated crop production. This exploration will extend to their future in medicine and agriculture. Beyond agriculture, endophytes are also being explored for their ability to produce secondary metabolites with pharmaceutical value, such as antimicrobial, anticancer, and anti-inflammatory agents. This multifunctionality underlines the importance of deepening our understanding of these microorganisms and their metabolic capabilities.

This book brings together leading researchers and experts in the field to address key issues such as the variability of endophyte-plant interactions across different environments, the complexity of metabolic pathways, and the lack of standardized protocols for isolating and applying endophytes in real-world settings that pose significant hurdles. It provides a comprehensive overview of the current state of endophyte research while also exploring the potential of these microorganisms to contribute to sustainable practices. The chapters cover a wide range of topics, from the fundamental biology and diversity of endophytes to their metabolic interactions with plants, as well as their potential in biotechnological applications. Importantly, this volume also sheds light on the future directions of endophyte research. Advancements in genomics, metabolomics, and synthetic biology will undoubtedly accelerate the discovery of new endophyte strains with unique properties and the development of innovative applications for endophytes in agriculture, medicine, and beyond.

As editors, we are pleased to present *Metabolic Sustainability of Endophytes: Current Status, Challenges, and Their Potential* as a critical contribution to the field of endophyte research. We hope that this book serves as a valuable resource for researchers, academicians, and professionals alike, and that it will inspire future studies aimed at unlocking the full potential of endophytes for sustainable solutions in the face of global challenges. By deepening our understanding of the metabolic processes governed by endophytes, we can move closer to achieving a more sustainable and resilient world.

We extend our heartfelt thanks to all the contributing authors who have shared their expertise and insights, as well as to our reviewers for their critical evaluations. We are confident that this compilation of knowledge will be a timely and important addition to the scientific literature, inspiring further research and innovation in this exciting and rapidly evolving field.

Udaipur, Rajasthan, India — Mukesh Meena
Varanasi, Uttar Pradesh, India — Andleeb Zehra
Bathinda, Punjab, India — Prashant Swapnil
New Delhi, Delhi, India — Chandra Shekhar Seth

Acknowledgments

The successful completion of this book is the culmination of the collective efforts of many individuals and organizations. The editors extend their heartfelt appreciation to the contributing authors for their insightful chapters and unwavering dedication to this project. We are deeply grateful to the reviewers for their valuable feedback, which greatly enhanced the quality, coherence, and presentation of the chapters. Special thanks go to the members of the Editorial Advisory Board for their guidance and support throughout the compilation process. We also express our sincere gratitude to our respective institutions for their support and encouragement during this endeavor. In particular, we would like to thank the Vice-Chancellor, Director, Dean, and Head of the associated departments for their continuous backing and motivation. Finally, we are extremely appreciative of the publishing team at Springer Nature for their professional guidance and assistance in bringing this book to completion.

Mukesh Meena
Andleeb Zehra
Prashant Swapnil
Chandra Shekhar Seth

Contents

Editors and Contributors

About the Editors

Mukesh Meena completed his M.Sc. and Ph.D. in Botany from the Department of Botany, Banaras Hindu University, Varanasi, India. He is currently serving as an Assistant Professor in the Department of Botany at Mohanlal Sukhadia University, Udaipur, Rajasthan. Dr. Meena previously worked as a post-doctoral research fellow at the Department of Biotechnology, Faculty of Science, Jamia Hamdard, New Delhi, India. He is an active member of several prominent scientific societies, including the Mycological Society of India (MSI), the Indian Science Congress Association (ISCA), the Association of Microbiologists of India (AMI), and the Asian PGPR Society for Sustainable Agriculture. Among his accolades, he received the Best Paper Presentation Award at the Fourth International Conference on Advances in Applied Science and Environmental Technology (ASET) held in Bangkok, Thailand, in 2016. He has participated in numerous international and national conferences and is well-versed in a range of analytical tools relevant to mycopathological research. Dr. Meena has authored over 150 peer-reviewed publications, including research articles, review articles, and book chapters. Additionally, he has published six books with international publishers. He has been recognized in the "World Ranking of Top 2% Scientists" database, released by Stanford University, USA. This marks the third consecutive years—2022, 2023, and 2024—that he has been listed among the top 2% of scientists globally.

Andleeb Zehra has completed her Ph.D. from Centre of Advanced Study in Botany, Banaras Hindu University, Varanasi. She is a visiting lecturer in Mianz International College. She has extensive research experience in plant-microbe interaction, fungal biology, biological control, and plant growth promoting microbes. She is the recipient of one of the prestigious fellowships of UGC "Dr. D. S. Kothari Post-doctoral fellowship." She was awarded for her outstanding academic achievement by the governor of Uttar Pradesh, India. She has published more than 40 national and international articles and delivered numerous oral and poster presentations in several national and international conferences.

Prashant Swapnil graduated from Veer Bahadur Singh Purvanchal University and post-graduated from the Department of Botany, Banaras Hindu University, Varanasi, India. Dr. Swapnil did his Ph.D. from the Department of Botany, Banaras Hindu University, Varanasi. Currently, he is working as an Assistant Professor in the Department of Botany, Central University of Punjab, India. He has more than 20 peer-reviewed international publications and more than 12 book chapters published. He has participated in several International and National conferences. Dr. Swapnil has more than 7 years of teaching (from Banaras Hindu University, India, 2013–2015; Acharya Narendra Dev College, Delhi University, India, 2019–2020, Department of Botany, North Campus, Delhi University, India, 2020–2021) and 2 years of postdoctoral experience (from International Centre for Genetic Engineering and Biotechnology, New Delhi, India). Dr. Swapnil has expertise in Cyanobacterial Stress Physiology as well as Plant Pathogen Interaction.

Chandra Shekhar Seth is an Associate Professor at the Department of Botany, University of Delhi, New Delhi-110007, Delhi, India. He obtained his M.Sc. degree in Botany from Banaras Hindu University, Varanasi (U.P.) in 2003 and Ph.D. degree in Botany from University of Lucknow (U.P.) in 2008. His major research interest area includes how various phytohormones particularly Salicyclic acid (SA), Jasmonic acid (JA), and Nitric Oxide (NO) mitigate the heavy metals (HMs) and salinity induce toxicity in plants, photosynthesis and nitrogen metabolism under abiotic stress, and role of nanomaterials in mitigating the HMs and salt stresses in plants. Oxidative stress and tolerance mechanism via antioxidants and phytochelatins, phytoremediation of HMs, potential use of algae as biofuels, nutrients and pigments are other important interesting area of his work. He is the fellow of two prestigious societies, namely, Fellow of the Academy of Environmental Biology (FAEB), Lucknow-226020 (U.P.) India, and Fellow Society for Plant Research (FSPR) Meerut-250110, UP, India. He has been the Life members of ISCA (The Indian Science Congress Association) Kolkata-700017, India; ISEB (International Society of Environmental Botanist), Lucknow-226001 (U.P.); AEB (The Academy of Environmental Biology), Lucknow-226020 (U.P.), India; ISPM (International Society of Plant Morphologists), New Delhi-110007, India; and DUBS (Delhi University Botanical Society), New Delhi-110007, India. He has published many peer-reviewed, excellent research papers, review papers, and several book and book chapters related to this field. He is an internationally recognized reviewer for many journals. He has been serving as an editorial board member for Soil and Environmental Health.

Contributors

Jaya Arora Laboratory of Biomolecular Technology, Department of Botany, Mohanlal Sukhadia University, Udaipur, Rajasthan, India

Rosana Babu Department of Computational Biology, Oxford University, Oxford, UK

Kiran Bala Department of Biochemistry, School of Applied Sciences, Om Sterlling Global University, Hisar, India

Sneh Punia Banger Department of Food, Nutrition & Packaging Sciences, Clemson University, Clemson, SC, USA

Anupam Berwal Department of Microbiology, Kalpana Chawla Govt. Medical College, Karnal, India

Rupesh Bhardwaj Cyanobacterial and Plant Stress Biology Lab, School of Basic Science, Department of Botany, Central University of Punjab, Bathinda, Punjab, India

School of Basic Science, Department of Botany, Central University of Punjab, Bhatinda, Punjab, India

Basanti Brar Department of Biotechnology, School of Applied Sciences, Om Sterlling Global University, Hisar, India

Lokvendra Singh Budania Department of Anesthesiology, Kalpana Chawla Govt. Medical College, Karnal, India

Payal Chaturvedi Department of Microbiology and Biotechnology, IIS (Deemed to be University), Jaipur, Rajasthan, India

Sushree Pralipta Chhotaray School of Basic Science, Department of Botany, Central University of Punjab, Bhatinda, Punjab, India

Alen D' Rozario Department of Biotechnology, St. Xavier's College (Autonomous), Kolkata, West Bengal, India

Amadou Hamadoun Dicko Faculty of Agronomy and Animal Health, University of Segou, Campus of Sebougou, Segou, Mali

Purnamsree Gogoi Department of Forestry, North Eastern Regional Institute of Science and Technology, Nirjuli, Arunachal Pradesh, India

Applied Microbiology Laboratory, Department of Forestry, North Eastern Regional Institute of Science and Technology, Nirjuli, Arunachal Pradesh, India

Ritika Gupta Department of Biotechnology, St. Xavier's College (Autonomous), Kolkata, West Bengal, India

Sheikh Tajamul Islam Department of Bioresources, School of Biological Sciences, University of Kashmir, Srinagar, India

Vidhi Jain Department of Microbiology, Mohanlal Sukhadia University, Udaipur, Rajasthan, India

Applied Microbiology Laboratory, Department of Forestry, North Eastern Regional Institute of Science and Technology, Nirjuli, Arunachal Pradesh, India

Seema Joshi Department of Environment Science, School of Applied Sciences, Om Sterlling Global University, Hisar, India

Madhu Kamle Applied Microbiology Laboratory, Department of Forestry, North Eastern Regional Institute of Science and Technology, Nirjuli, Arunachal Pradesh, India

Department of Biochemistry, University of Lucknow, Lucknow, India

Department of Biotechnology, University of Lucknow, Lucknow, Uttar Pradesh, India

Pradeep Kumar Applied Microbiology Lab, Department of Botany, University of Lucknow, Lucknow, Uttar Pradesh, India

College of Life Science & Biotechnology, Korea University, Seoul, Republic of Korea

Priyanshu Kumar School of Basic Science, Department of Botany, Central University of Punjab, Bhatinda, Punjab, India

Sumit Kumar Department of Plant Pathology, College of Agriculture, Rajmata Vijayaraje Scindia Krishi Vishwa Vidyalaya, Gwalior, Madhya Pradesh, India

Pritee Kumari Cyanobacterial and Plant Stress Biology Lab, School of Basic Science, Department of Botany, Central University of Punjab, Bathinda, Punjab, India

School of Basic Science, Department of Botany, Central University of Punjab, Bhatinda, Punjab, India

Avinash Marwal Department of Biotechnology, Vigyan Bhawan-Block B, New Campus, Mohanlal Sukhadia University, Udaipur, Rajasthan, India

Mukesh Meena Laboratory of Phytopathology and Microbial Biotechnology, Department of Botany, Mohanlal Sukhadia University, Udaipur, Rajasthan, India

Indurekha Meher School of Basic Science, Department of Botany, Central University of Punjab, Bhatinda, Punjab, India

Mohamed Ahmed Mosa Nanotechnology & Advanced Nano-Materials Laboratory (NANML), Plant Pathology Research Institute, Agricultural Research Center, Giza, Egypt

Azha Ufaq Nabi Department of Bioresources, School of Biological Sciences, University of Kashmir, Srinagar, India

Nisha Parashar Department of Microbiology and Biotechnology, IIS (Deemed to be University), Jaipur, Rajasthan, India

Muskan Punjabi Department of Microbiology and Biotechnology, IIS (Deemed to be University), Jaipur, Rajasthan, India

Nitish Rai Department of Biotechnology, Mohanlal Sukhadia University, Udaipur, Rajasthan, India

Department of Zoology, University of Lucknow, Lucknow, Uttar Pradesh, India

Reiaz Ul Rehman Department of Bioresources, School of Biological Sciences, University of Kashmir, Srinagar, India

Aryadeep Roychoudhury Discipline of Life Sciences, School of Sciences, Indira Gandhi National Open University, New Delhi, India

Baljeet Singh Saharan Department of Microbiology, Chaudhary Charan Singh Haryana Agricultural University, Hisar, India

Abhishek Sahoo Laboratory of Phytopathology and Microbial Biotechnology, Department of Botany, Mohanlal Sukhadia University, Udaipur, Rajasthan, India

Priyanshu Sao Department of Biotechnology and Microbiology, Faculty of Sciences, ISBM University, Gariyaband, Chhattisgarh, India

Faamiya Shajar Department of Bioresources, School of Biological Sciences, University of Kashmir, Srinagar, India

Charu Sharma Department of Microbiology and Biotechnology, IIS (Deemed to be University), Jaipur, Rajasthan, India

Sana Sheikh Department of Botany, St Aloysius (Deemed to be University), Mangaluru, India

Sareen Sheikh Department of Chemistry, Yenepoya Institute of Arts, Science and Management, Yenepoya Deemed to be University, Mangaluru, India

Laishram Amarjit Singh Cyanobacterial and Plant Stress Biology Lab, School of Basic Science, Department of Botany, Central University of Punjab. Bathinda, Punjab, India

School of Basic Science, Department of Botany, Central University of Punjab, Bhatinda, Punjab, India

Namita Ashish Singh Department of Microbiology, Mohanlal Sukhadia University, Udaipur, Rajasthan, India

Applied Microbiology Laboratory, Department of Forestry, North Eastern Regional Institute of Science and Technology, Nirjuli, Arunachal Pradesh, India

Bhaskarjyoti Saikia Sonowal School of Basic Science, Department of Botany, Central University of Punjab, Bhatinda, PunjabIndia

Prashant Swapnil Cyanobacterial and Plant Stress Biology Lab, School of Basic Science, Department of Botany, Central University of Punjab, Bathinda, Punjab, India

School of Basic Science, Department of Botany, Central University of Punjab, Bhatinda, Punjab, India

V.S. Swaroop Kumar Department of Plant Science, Central University of Kerala, Kasaragod, India

Alapati Prema Siva Naga Teja Department of Microbiology, Chaudhary Charan Singh Haryana Agricultural University, Hisar, India

Aina Thakur School of Basic Science, Department of Botany, Central University of Punjab, Bhatinda, Punjab, India

Kanika Thakur School of Basic Science, Department of Botany, Central University of Punjab, Bhatinda, Punjab, India

A. Usha Raja Nanthini Department of Biotechnology, Mother Teresa Women's University, Kodaikanal, Tamilnadu, India

Poonam Verma Department of Biotechnology and Microbiology, Faculty of Sciences, ISBM University, Gariyaband, Chhattisgarh, India

Ankush Yadav Cyanobacterial and Plant Stress Biology Lab, School of Basic Science, Department of Botany, Central University of Punjab, Bathinda, Punjab, India

School of Basic Science, Department of Botany, Central University of Punjab, Bhatinda, Punjab, India

Andleeb Zehra MI College, Male, Maldives

Classification and Occurrence of Fungal Endophytes

1

Sana Sheikh, Sareen Sheikh, V. S. Swaroop Kumar, and Rosana Babu

Abstract

Endophytic fungi are the symbiotic-associated organisms in the plant host, assisting the plant with secondary metabolite production. These fungi working as factory of production of chemicals are widespread, enhancing plant growth and resilience against stress. These fungal communities are categorised based on the reproductive pattern and occurrence in various plant groups. This chapter elaborates on their classification and asymptomatic occurrence in various plant groups.

Keywords

Endophytic fungi · Plant · Classification · Occurrence

1.1 Introduction

"Microbiome" as quoted by Joshua Lederberg is the "ecological community of commensal microorganisms, symbionts or pathogens, that literally occupy a space in our body." In plants, the microbiome majorly comprises of bacteria, fungi (including arbuscular mycorrhizal fungi (AMF)), oomycetes, viruses, and archaea

S. Sheikh (✉)
Department of Botany, St Aloysius (Deemed to be University), Mangaluru, India

S. Sheikh
Department of Chemistry, Yenepoya Institute of Arts, Science and Management, Yenepoya Deemed to be University, Mangaluru, India

V. S. Swaroop Kumar
Department of Plant Science, Central University of Kerala, Kasaragod, India

R. Babu
Department of Computational Biology, Oxford University, Oxford, UK

M. Meena et al. (eds.), *Metabolic Sustainability of Endophytes*, Microorganisms for Sustainability 52, https://doi.org/10.1007/978-981-96-4004-1_1

interacting with plant tissues (Andreote et al. 2014). Although it is usually assumed that most of the plant interactions with other species are pathogenic or saprophytic in character, the surveys have changed our original belief and oriented the true nature of such interactions to be more diversified and not always pathogenic. (Gouda et al. 2016). Many of such interactions hold essential functions in maintaining the plant's health and development. The early evolution and diversification of land plants are very much attributable to fungal symbionts (Turner et al. 2013a, b; Lutzoni et al. 2018). The development of improved plant growth, stress tolerance, resistance to pathogens, herbivory, and desiccation were particularly associated with symbiotic interactions (Rashmi et al. 2019; Aamir et al. 2020). A single species of plant could contain thousands of microbes in the form of epiphytes or endophytes (Turner et al. 2013a, b; Andreote et al. 2014, Gouda et al. 2016). In contrast to "epiphytes," which refers to fungi that are found on the surface of plants, Anton de Bary developed the term "endophytes" in 1866 to characterise fungi that reside within the tissues of their host plants (Dutta et al. 2014). The term "endophyte" is the focus of this chapter and shall therefore shed light on its classification, occurrence, and its importance.

1.1.1 Fungal Endophytes

Fungi are a widespread category of creatures that can be found in almost any kind of ecosystem. Fungal diversity is believed to comprise at least 1.5 million species; however, only approximately 75,000 of those species—or 5% of the total—have been described so far (Hawksworth 2001). Less than 1% of all plausible fungal species on Earth may have inadequate knowledge of their physiology and biochemistry (Hawksworth 1991). A great deal of nature needs to be investigated, mostly microbiological habitats (Newmann et al. 2003). Apparently, Fungal endophytes are microorganisms that live in plant tissues without causing harm to the host (Nisa et al. 2015; Petrini 1986; Wilson 1995; Hardoim et al. 2015). Fungal endophytes refer to any organism that lives within plant tissues, including diseases and mycorrhizal root symbionts. Endophytes, according to most mycologists, are organisms that colonise inside plant tissues without causing harm to their hosts. These numerous microorganisms that live inside the tissues of living plants without generating any visible negative effects or symptoms on the outside are one such undiscovered and unexplored species. (Aly et al. 2011). However, when the host ages, it may become pathogenic (Rodriguez & Redman 2008). Plant endophytes consist of bacteria, fungus, actinomycetes, and mycoplasma (Arnold 2007). These various creatures can be found in a variety of plant types, including trees, fodder, vegetables, fruits, and medicinal plants.

Endophytes are regarded to be a large and unique source of bioactive compounds with a wide range of potential applications in the culinary, medical, and agricultural industries (Strobel et al. 2004). Approximately 80% of endophytic fungi produce biologically active compounds, such as antioxidants, nematicidals, herbicides, anticancer, and anti-inflammatory medicines (Schulz et al. 2002). Many studies have been undertaken on the role of endophytes in host plants, and the findings indicate that these microbes can promote plant growth, increase plant tolerance to environmental stressors, improve disease resistance, and recycle nutrients (Sturz and Nowak 2000). Certain endophytes that live within plants produce some of the useful natural substances that the plants create. For example, it has been shown that some endophytic fungi can create taxol, an anticancer chemical produced by specific types of yew trees (Taxus sp.), such as *Taxomyces Andreanae* (Strobel and Daisy 2003). As a result, the endophytes found inside plants have been appropriately dubbed chemical factories.

The name "endophyte" comes from the Greek words endon (inside) and phyte (plant). Since de Bary defined "Endophytes are the living organisms that colonise within plant tissues" (De Bary 1866). There are various changes to the original definition by other researchers (Caroll 1986; Petrini et al. 1991; Wilson 1995). Plant endophytes include a varied spectrum of bacteria, actinomycetes, mycoplasma, and fungi (Khiralla et al. 2016). Endophytes are further classified based on their need for a suitable host for survival as obligatory, which is entirely dependent on the host for metabolic activities and reproduction, and facultative, which exists freely outside during a particular point of their life cycle (Gouda et al. 2016). In contrast to mycorrhizal fungi, which dwell and colonise the root tissues and the rhizosphere, endophytic fungi solely occupy the interior plant tissues of the root, stem, and leaves, seldom appearing except for sporulation at the death of the host tissues (Rodriguez et al. 2009).

Almost every major plant lineage, including liverworts, mosses, pteridophytes, gymnosperms, and angiosperms, has a symbiotic connection with fungi as endophytes (Arnold and Lutzoni 2007). These fungal endophytes are a polyphyletic group of ascomycetous fungi that vary in their prevalence and functions (Arnold 2007). Several members are also found to belong to Basidiomycota (Rodriguez and Redman 2008). Endophytes have been found in a wide range of natural settings, from the arctic to the tropics, including grasslands, croplands, woods, and even deserts (Rashmi et al. 2019). Endophytic fungi are primarily divided into two families based on variations in their systematics, host-specificity, and ecological roles: clavicipitaceous endophytes and non-clavicipitaceous endophytes (Rodriguez et al. 2009; Gouda et al. 2016; Selim et al. 2017).

1.2 Classification of Fungal Endophytes

Fungal endophytes are categorised into several types depending on their colonisation patterns and interactions with host plants. For example, they can be classed as epiphytic or endophytic endophytes based on whether they colonise the exterior surface of plant tissues or live within them. Some fungal endophytes may create mutualistic relationships with host plants, offering nutrients or other benefits in exchange for resources. Endophytic fungal communities can be classified into two groups based on host occurrence and reproductive pattern: Clavicipitaceous/Balansiaceous (C-group) and non-Clavicipitaceous/non-Balansiaceous (NC-group).

C-group endophytic fungi infect host plant ovules and propagate vertically through host seeds from parents to offspring. Their intended host range is restricted to grass species (Poaceae), although they will colonise live rhizomes and shoots of host plants as target tissues. C-group species are typical obligatory endophytes (Carroll 1988), where they protect their hosts from herbivore assaults or allow the hosts to live under drought circumstances by secreting defensive or supporting bioactive compounds, respectively (Poveda et al. 2021). NC-group endophytic fungi, which are endophytic fungi that do not host grass, are widely dispersed and have a high biodiversity from tropical to polar regions. Their hosts include plant populations that are woody, vascular, or nonvascular. They spread asexually or sexually by producing spores or conidia, which aid in the induction of symbiosis or horizontal proliferation. These endophytic fungi (EF) can exist in a quiescent state until they detect chemical changes from host plants that have endured wounds, injuries, or other environmental stresses; thus, they are not intimately tied to the host plants (Carroll 1988). The NC-group usually colonises aerial organs in a localised, confined, limited, and primarily intercellular manner; however, it colonises roots or the rhizosphere in a widespread, organised, systematic, intercellular, and intracellular manner. Endophytic mycobiomes of the NC-group include Fusarium spp., *Piriformospora indica*, and dark septate mycobiota (Schulz et al. 2002).

1.2.1 Clavicipitaceous Endophytes

The majority of the members of the family Clavicipitaceae (Hypocreales; Ascomycota) make up the clavicipitaceous endophytes, often known as grass endophytes or Class 1 endophytes (Rodriguez et al. 2009). The clavicipitaceous endophytes are a group of endophytic fungi found primarily in the grass family (Poaceae) (Terhonen et al. 2019). These fungi benefit the grasses by increasing their ecological fitness. The fundamental shortcoming of grasses is their inability to manufacture secondary metabolites; however, this inability is totally offset by the development of symbiotic relationships, which are capable of producing protective chemicals through fungal symbionts (Bamisile et al. 2018). In the late nineteenth century, European researchers discovered clavicipitaceous endophytes in grass seeds of *Lolium temulentum*, *Lolium arvense*, *Lolium linicolum*, and *Lolium remotum* (Rodriguez et al. 2009).

The endophytes in this group are predominantly obligatory biotrophic intercellular symbiotic ascomycetes from the Clavicipitaceae family (Kuldau and Bacon 2008; Rodriguez et al. 2009). These endophytes exist only in three families: Poaceae, Cyperaceae, and Juncaceae. These fungi are primarily from the tribe Balansiae, which consists of five genera (Atkinsonella, Myriogenospora, Balansia, Balansiopsis, and Epichloe) and approximately 30 species (Clay 1989; Kuldau and Bacon 2008; Rodriguez et al. 2009). This categorization is based on the differences in conidial morphology and ascospore states among the taxa.

1.2.2 Non-Clavicipitaceous Endophytes

These are a large group that has not been well defined taxonomically, but the majority of the species belong to the phyla Ascomycota and Basidiomycota, represented by the genera *Alternaria*, *Arthrobotrys*, *Aspergillus*, *Cladosporium*, *Colletotrichum*, *Coprinellus*, *Curvularia*, *Fusarium*, *Paecilomyces*, *Penicillium*, *Phanerochaete*, *Phoma* and others (Schulz and Boyle 2005; Gutierrez et al. 2012). These extremely diverse fungi can be roughly categorised into three functional groups according to the variances in life cycle, ecological interactions, and fundamental features. Non-clavicipitaceous fungi are categorised as Class 2 endophytes, Class 3 endophytes, and Class 4 endophytes (Rodrigues et al. 2009).

Class 2 endophytes are a diverse group of species that are all members of the Dikarya (Ascomycota or Basidiomycota). The majority of this class's endophytic fungi constitute Ascomycota, with only a few Basidiomycota endophytes present. These endophytes are carried vertically via seed coats, seeds, or rhizomes and are capable of colonising an extensive spectrum of tissues in plant stems and leaves. These stimulate plant root and shoot biomass (Rodriguez et al. 2008; Macia et al. 2008).

Class 3 endophytes are categorised depending on their distribution primarily among above-ground tissues (plant shoots) via horizontal transmission. These endophytic fungi elicit highly localised infections and exhibit significant biodiversity in plants. Class 3 endophytes are the varied endophytic fungi associated with the leaves of tropical trees (Gamboa and Bayman 2001), as well as very diversified supporters in above-ground tissues of nonvascular plants, seedless vascular plants, woody and herbaceous angiosperms in tropical forest biomes (Higgins et al. 2007; Murali et al. 2007; Davis and Shaw 2008). Class 3 endophytes can be found in flowers, fruits, asymptomatic wood, and inner bark, as well as photosynthetic and herbaceous tissues (Kumar and Hyde 2004; Tejesvi et al. 2005). Asymptomatic lichens also contain Class 3 endophytes, which are known as "endolichenic" fungus. Class 3 endophytes are found in all of the major nonlichenised clades. They are common among the Pezizomycetes, Leotiomycetes, and Eurotiomycetes, and especially among the highly endophyte-rich Sordariomycetes and Dothideomycetes (Arnold 2007; Higgins et al. 2007). Class 3 endophytes reproduce through hyphal fragmentation and/or the formation of sexual or asexual spores on dead or senescent tissue (Herre et al. 2005).

Class 4 endophytes, known as dark septate endophytes, are connected with root tissues of angiosperms that include dark-pigmented fungus. Figure 1.1 depicts the classification of fungal endophytes. Various endophytic microbial species penetrate plants, particularly the ones that are considered perennial (Stone et al. 2004). Ascomycota fungal endophytes are horizontally spreading fungi that reside in plant roots and aerial tissues without creating symptoms that are apparent (Suryanarayanan et al. 2009).

Facultative endophytes can reside inside plants at certain stages of their life cycles, or they can live outside the plant and form a connection with the proximal rhizosphere soil of the host plant. In contrast, obligatory strains persist inside plants

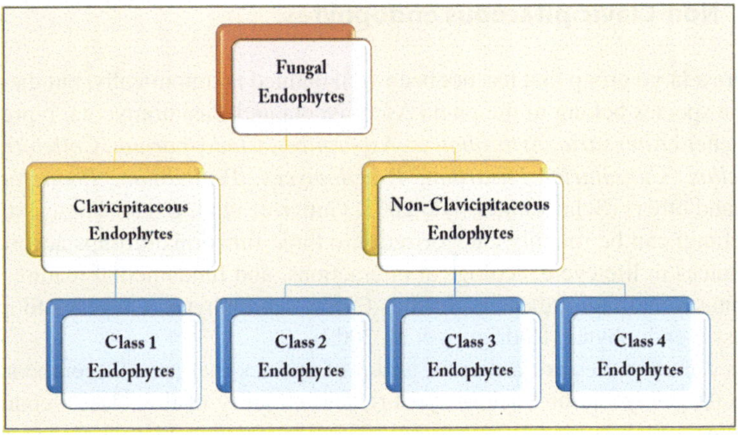

Fig. 1.1 Depicting the classification of fungal endophytes

throughout their life cycles. According to Gouda et al. (2016), they usually proliferate vertically through plant generations, exploiting or altering the metabolic functions and by-products of plants to secure their own existence.

According to T.N. Taylor and E.L. Taylor (2000), fossilized stem and leaf tissues contain evidence of fungi associated with plants. Carroll (1988) proposed that fungal endophytes may have evolved as diseases lost their virulence. Fungi and bacteria are the most prevalent microorganisms that live as endophytes. Other microbial species, such as mycoplasmas and archaebacteria, appear to occur as endophytes in plants, although no evidence has been produced to support this. Fungi are the most often isolated endophytes. It appears that the vast majority of plants have not been examined for their endophytes. As a result, vast prospects exist for the recovery of novel fungal forms, taxa, and biotypes (Strobel and Daisy 2003).

Endophytic fungi have been found to host algae (Tan and Zou 2001), liverworts, mosses, hornworts (Aly et al. 2011; Sun et al. 2012), grasses (Su et al. 2010), lycophytes, ferns, Equisetopsida (Aly et al. 2011), shrubs, deciduous and coniferous trees (Sun et al. 2011), gymnosperms, angiosperms, and annual/perennial herbaceous and broad-leaved plants. Endophytic fungi have also been spread in a large range of geographic habitats, such as tropical, temperate, arctic tundra, alpine, aquatic, and xerophytic ecosystems for more than 400 million years (Tan and Zou 2001; Sun et al. 2012). Endophytes are found in medicinal plants (Strobel 2002), and they protect the host from infectious pathogens while also allowing it to live in adverse environmental conditions. It is so critical to determine the endophyte diversity of therapeutic plants. They occupy distinct biological niches and thrive in a variety of odd situations (Strobel and Daisy 2003; Strobel et al. 2004). Marine algae-associated endophytic fungi have emerged as a major source of bioactive chemicals. The researchers described an endophytic fungus, *Chaetomium globosum*, associated with marine green algae (Kamat et al. 2020). Diaporthe sp. SC-J0138, an endophytic fungus, was isolated from the leaves of *Cyclosorus*

Table 1.1 Occurrence of fungal endophytes in various families of medicinal plants

Family	Host	Fungus	References
Celastraceae	*Maytenus hookeri*	*Alternaria alternata*	Tuan et al. (2010)
	Maytenus ilicifolia	*Pestalotiopsis* sp.	Gomes et al. (2007)
Fabaceae	*Piptadenia adiantoides*	*Cochliobolus* sp.	Fernanda et al. (2008)
	Bauhinia brevipes	*Phomopsis* sp.	Mariana et al. (2011)
Rubiaceae	*Alibertia macrophylla*	*Penicillium* sp.	Camila et al. (2009)
	Cinchona ledgeriana	*Phomopsis, Diaporthe, Schizophyllum, Penicillium, Fomitopsis*, and *Arthrinium* sp.	Shoji et al. (2010)

parasiticus, a pteridophyte. Several pteridophytes, bryophytes, and spermatophytes have been found to contain fungal endophytes (Gao et al. 2019). Xylariales are the main fungus in the rhizomes of *Acrostichum speciosum*, while Hypocreales are the major fungi in the petioles, roots, and rhizomes of *Acrostichum aureum* (Zhu et al. 2024).

According to Raviraja (2005), medicinal plants in India's Western Ghats (one of the world's biodiversity hotspots) feature a varied variety of endophytic fungi. Endophytes have been found in medicinal plants from the Western Ghats (Naik et al. 2014). Ruma et al. (2011) isolated 106 endophytic fungi from *Artocarpus hirsutus* Lam. and *Vateria indica* Linn., two endemic medicinal plants of the Western Ghats, and documented them using classic morphological methods. The most common genera were Coniothyrium sp. (96.5%), Trichoderma sp. (84.5%), Mortierella sp. (36.75%), Phyllosticta sp. (19%), and Acremonium sp. (21.5%). Meenavalli et al. (2013) investigated the distribution of Xylaria endophytes in the leaves of 22 tree species in a dry thorn forest and 27 tree species in a stunted montane evergreen forest in southern India's Western Ghats. *Muscodor kashayum*, a new species of endophyte, was recently isolated from the medicinal species *Aegle marmelos* (bael tree) in the Western Ghats of Muthanga, Wayanad Wildlife Sanctuary, Kerala, India (Meshram et al. 2013). Sixty-five endophytic fungi were isolated from healthy leaves of *Indigofera suffruticosa* Miller, a medicinal plant (Fabaceae) used in folk medicine to cure a variety of ailments (Santos et al. 2015). Table 1.1 shows the occurrence of fungal endophytes in several plant groups.

1.3 Conclusions

Endophytes exhibit a high level of biological diversity, particularly in temperate and tropical vegetation. They play a crucial role in the health and survival of host plants by providing various benefits. These benefits include enhancing the plant's ability to tolerate abiotic stresses such as drought and metal contamination, preventing the colonization of pathogenic organisms, and synthesizing novel organic compounds with various biological activities. Fungal endophytes can be found in diverse habitats, such as forests, grasslands, and agricultural fields. These microorganisms have a widespread occurrence globally and can be found in both natural and man-made environments. Their presence in plants can be attributed to various factors, including transmission through seeds, colonization through root systems, and horizontal transfer between plants. Fungal endophytes have been found to play a significant role in plant health and growth. They can enhance plant tolerance to environmental stressors, promote nutrient uptake, and provide defence against pathogens and herbivores. Furthermore, fungal endophytes have been reported to produce secondary metabolites with potential pharmaceutical and agricultural applications. These metabolites can have antimicrobial, antitumor, and insecticidal properties, making fungal endophytes a valuable resource for drug discovery and sustainable agriculture practices. Overall, fungal endophytes are highly diverse and can be found in a wide range of plant species. These fascinating group of microorganisms play a crucial role in plant health and survival by providing various potential benefits.

References

Aamir M, Rai KK, Zehra A, Kumar S, Yadav M, Shukla V et al (2020) Fungal endophytes: classification, diversity, ecological role, and their relevance in sustainable agriculture. In: Kumar A, Singh VK (eds) Microbial endophytes. Woodhead Publishing, Cambridge, MA, pp 291–323. https://doi.org/10.1016/b978-0-12-818734-0.00012-7

Aly AH, Debbab A, Proksch P (2011) Fungal endophytes: unique plant inhabitants with great promises. Appl Microbiol Biotechnol 90:1829–1845

Andreote FD, Gumiere T, Durrer A (2014) Exploring interactions of plant microbiomes. Sci Agric 71:528–539

Arnold AE (2007) Understanding the diversity of foliar endophytic fungi: progress, challenges, and frontiers. Fungal Biol Rev 21(2–3):51–66

Arnold AE, Lutzoni F (2007) Diversity and host range of foliar fungal endophytes: are tropical leaves biodiversity hotspots? Ecology 88(3):541–549

Bamisile BS, Dash CK, Akutse KS, Keppanan R, Wang L (2018) Fungal endophytes: beyond herbivore management. Front Microbiol 9:544

Camila MO, Geraldo HS, Luis OR, Lisinéia MZ, Alana HE, Maria CM, Youngc, Vanderlan SB, Angela RAZ (2009) Bioactive metabolites produced by *Penicillium* sp.1 and sp.2, two endophytes associated with *Alibertia macrophylla* (Rubiaceae). Naturforsch 64:824–830

Carroll GC (1986) The biology of endophytism in plants with particular reference to woody perennials. Microbiol Phyllosphere:203–222

Carroll GC (1988) Fungal endophytes in stems and leaves: from latent pathogen to mutualistic symbiont. Ecology 69(1):2–9

Clay K (1989) Clavicipitaceous endophytes of grasses: their potential as biocontrol agents. Mycol Res 92(1):1–12

Davis EC, Shaw AJ (2008) Biogeographic and phylogenetic patterns in diversity of liverwort—associated endophytes. Am J Bot 95:914–924

De Bary A (1866) Morphologie und physiologie der pilze, flechten und myxomyceten, vol 1. Engelmann

Dutta D, Puzari KC, Gogoi R, Dutta P (2014) Endophytes: exploitation as a tool in plant protection. Braz Arch Biol Technol 57:621–629

Fernanda FC, Luiz HR, Betania BC, Rachel BC, Ana LTR, Tânia MAA, Carlos AR, Carlos LZ (2008) Leishmanicidal metabolites from *Cochliobolus* sp., an endophytic fungus isolated from *Piptadenia adiantoides* (Fabaceae). PLoS Negl Trop Dis 2:12

Gamboa MA, Bayman P (2001) Communities of endophytic fungi in leaves of a tropical timber tree (Guarea Guidonia: Meliaceae). Biotropica 33:352–360

Gao Y-H, Bai W-X, Sun W-H, Zhou W-N, Wu G-L, Zhu Z-Q, Ma J-Y, Li H-Y (2019) Diversity of culturable endophytic fungi associated with bryophytes, Pteridophytes and spermatophytes from Dawei Mountain Nature Reserve, China. Chiang Mai J Sci 46:626–638

Gomes FJ, Pimentel IC, Vicente VA, Pie MR, Kava CV, Galli TL, Pereira JO, deSouza AQ, Glienke C (2007) Bioprospecting highly diverse endophytic *Pestalotiopsis* spp. with antibacterial properties from *Maytenus ilicifolia*, a medicinal plant from Brazil. Can J Microbiol 53(10):1123–1132

Gouda S, Das G, Sen SK, Shin HS, Patra JK (2016) Endophytes: a treasure house of bioactive compounds of medicinal importance. Front Microbiol 7:1538

Gutierrez MPR, Gonzalzez MNA, Ramirez MA (2012) Compounds derived from endophytes: A review of phytochemistry and pharmacology. Curr Med Chem 19(18):2992–3030

Hardoim PR, Van Overbeek LS, Berg G, Pirttilä AM, Compant S, Campisano A, Sessitsch A (2015) The hidden world within plants: ecological and evolutionary considerations for defining functioning of microbial endophytes. Microbiol Mol Biol Rev 79(3):293–320

Hawksworth DL (1991) The fungal dimension of biodiversity: magnitude, significance, and conservation. Mycol Res 95:641–655

Hawksworth DL (2001) The magnitude of fungal diversity: the 1.5 million species estimate revisited. Mycol Res 105:1422–1432

Herre EA, Bael SAV, Maynard Z, Robbins N, Bischoff J, Arnold AE, Rojas E, Mejía LC, Cordero RA, Woodward C, Kyllo DA (2005) Tropical plants as chimera: some implications of foliar endophytic fungi for the study of host plant defense, physiology, and genetics. In: Burslem DFRP, Pinard MA, Hartley SE (eds) Biotic interactions in the tropics. Cambridge University Press, Cambridge, pp 226–240

Higgins KL, Arnold AE, Miadlikowska J, Sarvate SD, Lutzoni F (2007) Phylogenetic relationships, host affinity, and geographic structure of boreal and arctic endophytes from three major plant lineages. Mol Phylogenet Evol 42:543–555

Kamat S, Kumari M, Sajna KV et al (2020) Endophytic fungus, *Chaetomium globosum*, associated with marine green alga, a new source of Chrysin. Sci Rep 10:18726. https://doi.org/10.1038/s41598-020-72497-3

Khiralla A, Spina R, Yagi S, Mohamed I, Laurain-Mattar D (2016) Endophytic fungi: occurrence, classification, function and natural products. Endophytic fungi: diversity, characterization and biocontrol, pp 1–19

Kuldau GA, Bacon CW (2008) Clavicipitaceous endophytes: their ability to enhance resistance of grasses to multiple stresses. Biol Control 46(1):57–71

Kumar DSS, Hyde KD (2004) Biodiversity and tissue recurrence of endophytic fungi in *Tripterygium wilfordii*. Fungal Divers 17:69–90

Lutzoni F, Nowak MD, Alfaro ME, Reeb V, Miadlikowska J, Krug M et al (2018) Contemporaneous radiations of fungi and plants linked to symbiosis. Nat Commun 9(1):5451

Macia-Vicente JG, Jansson HB, Abdullah SK, Descals E, Salinas J, Lopez-Llorca LV (2008) Fungal root endophytes from natural vegetation in Mediterranean environments with special reference to Fusarium spp. FEMS Microbiol Ecol 64:90–105

Mariana PAH, Fernando AOS, Yumi O, Leonardo R, Jean CS, Ary CJ, Geraldo WF, Carlos AR (2011) Distribution of the endophytic fungi community in leaves of *Bauhinia brevipes* (Fabaceae). Acta Botanica Brasilica Feira de Santana 25:4

Meenavalli B, Govinda R, Nagamani TA (2013) Endophytic Xylariaceae from the forests of Western Ghats, southern India: distribution and biological activities. Mycology 4(1):29–37

Meshram V, Kapoor N, Saxena S (2013) *Muscodor kashayum* sp. nov.—a new volatile anti-microbial producing endophytic fungus. Mycology 4(4):196–204. https://doi.org/10.1080/21501203.2013.877990

Murali TS, Suryanarayanan TS, Venkatesan G (2007) Fungal endophyte communities in two tropical forests of southern India: diversity and host affiliation. Mycol Prog 6:191–199

Naik BS, Krishnappa M, Krishnamurthy YL (2014) Biodiversity of endophytic fungi from seven herbaceous medicinal plants of Malnad region, Western Ghats, southern India. J Forest Res 25(3):707–711

Newmann JA, Abner ML, Dado RG, Gibson DJ, Brookings A, Parsons AJ (2003) Effects of elevated CO_2, nitrogen and fungal endophyte-infection on tall fescue: growth photosynthesis, chemical composition and digestibility. Glob Chang Biol 9:425–437

Nisa H, Kamili AN, Nawchoo IA, Shafi S, Shameem N, Bandh SA (2015) Fungal endophytes as prolific source of phytochemicals and other bioactive natural products: a review. Microb Pathog 82:50–59

Petrini O (1986) Taxonomy of endophytic fungi of aerial plant tissues. In: Fokkema NJ, Van Den Huevel J (eds) Microbiology of the phyllosphere. Cambridge University Press, Cambridge, pp 175–187

Petrini O, Andrews JH, Hirano SS (1991) Microbial ecology of leaves. Spring-Verlag, New York, pp 179–197

Poveda J, Eugui D, Abril-Urías P, Velasco P (2021) Endophytic fungi as direct plant growth promoters for sustainable agricultural production. Symbiosis 85:1–19. https://doi.org/10.1007/s13199-021-00789-x

Rashmi M, Kushveer JS, Sarma VV (2019) A worldwide list of endophytic fungi with notes on ecology and diversity. Mycosphere 10(1):798–1079

Raviraja NS (2005) Fungal endophytes in five medicinal plant species from Kudremukh range, Western Ghats of India. J Basic Microbiol 45(3):230–235

Rodriguez R, Redman R (2008) More than 400 million years of evolution and some plants still can't make it on their own: plant stress tolerance via fungal symbiosis. J Exp Bot 59(5):1109–1114

Rodriguez RJ, White JF, Arnold AE, Redman RS (2009) Fungal endophytes: diversity and functional roles. New Phytol 182(2):314–330

Ruma K, Sekhar S, Kumara KKS, Siddapura RN, Prakash HS (2011) Diversity of fungal endophytes from two endemic tree species *Artocarpus hirsutus* lam. And *Vateria indica* L. of Western Ghats, India. WJAS 7(5):577–582

Santos DIP, da Silva LC, da Silva MV, de Araújo JM, Cavalcanti MS, Lima VL (2015) Antibacterial activity of endophytic fungi from leaves of *Indigofera suffruticosa* Miller (Fabaceae). Front Microbiol 6:350

Schulz B, Boyle C (2005) The endophytic continuum. Mycol Res 109:661–686

Schulz B, Boyle C, Draeger S, Rommert A, Krohn K (2002) Endophytic fungi: a source of novel biologically active secondary metabolites. Mycol Res 106(9):996–1004

Selim KA, Nagia MM, Ghwas DEE (2017) Endophytic fungi are multifunctional biosynthesizers: ecological role and chemical diversity. In: Endophytic fungi: diversity, characterization and biocontrol, vol 39. Nova

Shoji M, Partomuan S, Kazuyoshi O, Hirotaka S (2010) Composition of endophytic fungi living in *Cinchona ledgeriana* (Rubiaceae). J Nat Med 64:227–230

Stone JK, Polishook JD, White JRJ (2004) Endophytic fungi. In: Mueller G, Bills GF, Foster MS (eds) Biodiversity of fungi: inventory and monitoring methods. Elsevier, Burlington, pp 241–270

Strobel GA (2002) Microbial gifts from rain forests. Can J Plant Pathol 24:14–20

Strobel GA, Daisy B (2003) Bioprospecting for microbial endophytes and their natural products. Microbiol Mol Biol Rev 67:491–502

Strobel G, Daisy B, Castillo U, Harper J (2004) Natural products from endophytic microorganisms. J Nat Prod 67:257–268

Sturz AV, Nowak J (2000) Endophytic communities of Rhizobacteria and the strategies required creating yield-enhancing associations with crops. Appl Soil Ecol 15:183

Su YY, Guo LD, Hyde KD (2010) Response of endophytic fungi of Stipa grandis to experimental plant function group removal in Inner Mongolia steppe, China. Fungal Diversity 431:93–101

Sun X, Guo LD, Hyde KD (2011) Community composition of endophytic fungi in *Acer truncatum* and their role in decomposition. Fungal Divers 47:85–95

Sun X, Ding Q, Hyde KD, Guo LD (2012) Community structure and preference of endophytic fungi of three woody plants in a mixed forest. Fungal Ecol 5:624. https://doi.org/10.1016/j.funeco.2012.04.001

Suryanarayanan T, Thirunavukkarasu N, Rajulu G, Sasse F, Jansen R, Murali TS (2009) Fungal endophytes and bioprospecting. Fungal Biol Rev 23:9–19. https://doi.org/10.1016/j.fbr.2009.07.001

Tan RX, Zou WX (2001) Endophytes: a rich source of functional metabolites. Nat Product Report 18:448–459

Taylor TN, Taylor EL (2000) The Rhynie chert ecosystem: a model for understanding fungal interactions. In: Bacon CW, White JF (eds) Microbial endophytes. Marcel Decker, New York, pp 31–48

Tejesvi MV, Mahesh B, Nalini MS, Prakash HS, Kini KR, Subbiah V, Shetty HS (2005) Endophytic fungal assemblages from inner bark and twig of *Terminalia arjuna* W. & A. (Combretaceae). World J Microbiol Biotechnol 21:1535–1540

Terhonen E, Blumenstein K, Kovalchuk A, Asiegbu FO (2019) Forest tree microbiomes and associated fungal endophytes: functional roles and impact on forest health. Forests 10(1):42

Tuan M, Li RQ, Wen QS, An L, Zhang, Jin MG (2010) Metabolites produced by an endophyte *Alternaria alternata* isolated from *Maytenus hookeri*. Chem Nat Compd 46:3

Turner TR, James EK, Poole PS (2013a) The plant microbiome. Genome Biol 14(6):209. https://doi.org/10.1186/gb-2013-14-6-209. PMID: 23805896; PMCID: PMC3706808

Turner TR, James EK, Poole PS (2013b) The plant microbiome. Genome Biol 14(6):1–10

Wilson D (1995) Endophyte: the evolution of a term, and clarification of its use and definition. Oikos 73:274–276

Zhu H, Zeng W, Chen M, He D, Cheng X, Jing Y, Liu Y, Yougen W, Yang D (2024) Endophytic fungal diversity of mangrove ferns *Acrostichum speciosum* and *A. Aureum* in China. Plan Theory 13, no. 5:685. https://doi.org/10.3390/plants13050685

Diversity and Functional Role of Fungal Endophytes

2

Kiran Bala, Seema Joshi, Basanti Brar, Anupam Berwal, Sneh Punia Banger, and Lokvendra Singh Budania

Abstract

In plant species, fungal endophytes are common and live in living tissues like leaves, stems, branches, and roots without causing any symptoms. Endophytes and their host plants have complicated connections that can range from mutualistic to harmful. Beneficial endophytes have the ability to improve plant fitness, which has raised attention in their possible uses in integrated plant health management. Fungal endophytes play important roles in plant defensive system, which is also elaborated in this chapter, with a focus on defense against non-insect pests like nematodes, bacteria, viruses, mites, and fungi. We look at the strategies endophytes use to combat these pests and emphasize the importance of these fungi in both present and future farming methods. We also offer a comprehensive list of endophytes that are beneficial to plants, including information on

K. Bala
Department of Biochemistry, School of Applied Sciences, Om Sterlling Global University, Hisar, India

S. Joshi
Department of Environment Science, School of Applied Sciences, Om Sterlling Global University, Hisar, India

B. Brar (✉)
Department of Biotechnology, School of Applied Sciences, Om Sterlling Global University, Hisar, India

A. Berwal
Department of Microbiology, Kalpana Chawla Govt. Medical College, Karnal, India

S. P. Banger
Department of Food, Nutrition & Packaging Sciences, Clemson University, Clemson, SC, USA

L. S. Budania
Department of Anesthesiology, Kalpana Chawla Govt. Medical College, Karnal, India

M. Meena et al. (eds.), *Metabolic Sustainability of Endophytes*, Microorganisms for Sustainability 52, https://doi.org/10.1007/978-981-96-4004-1_2

their typical habitats and uses in different research projects. This review is a thorough resource for learning about endophytes' potential in extensive agricultural applications. Fungal endophytes and plants have symbiotic connections in natural ecosystems, impacting plant populations through lower water consumption, higher biomass, and improved stress tolerance. Even after more than a century of study, little is known about the ecological significance of endophytes. Based on phylogenetic and life history characteristics, endophytes have traditionally been divided into two groups: clavicipitaceous (C) and nonclavicipitaceous (NC). According to our data, NC-endophytes are classified into three different functional groups that are distinguished by their capacity for host colonization, transmission, biodiversity, and fitness benefits. We suggest important areas for future endophyte biology research by comparing the life histories, host relationships, and ecophysiological roles of C- and NC-endophytes using this framework.

Keywords

Fungal taxonomy · Plant-fungal interactions · Symbiosis · Biocontrol; Antagonism; Defense activation

2.1 Introduction

Endophytes are mainly classified into four types as described in Fig. 2.1. Vertical transmission characterizes Class 1 endophytes, where maternal plants transfer fungi to offspring through seed infections, typically resulting in a single dominant fungal isolate or genotype within colonized plants (Ren et al. 2011). These endophytes often enhance plant growth, increase drought resilience, and produce animal-deterrent compounds, thereby reducing herbivory (Clay 1988). However, the

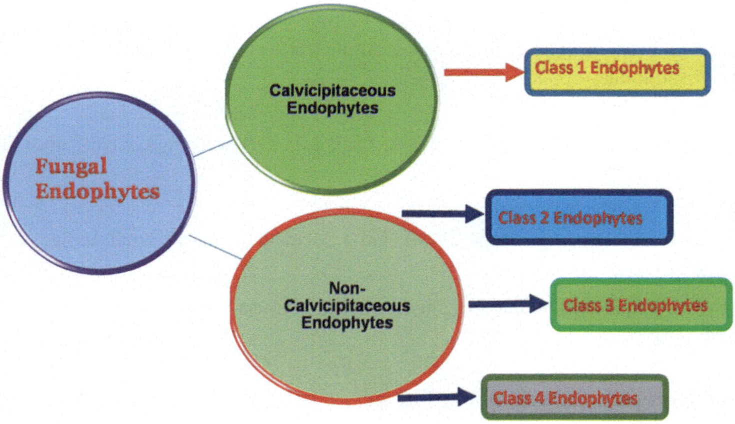

Fig. 2.1 Classification of fungal endophytes

benefits conferred by Class 1 endophytes are highly context-dependent, varying with genotypes, plant species, and environmental factors (Saikkonen et al. 2016). In contrast, *nonclavicipitaceous* (NC) endophytes comprise a vast and phylogenetically diverse group of primarily ascomycetous fungi, with complex and often undefined ecological roles. Research has recovered NC-endophytes from all major land plant lineages and terrestrial ecosystems, spanning agro-ecosystems and biomes from tropical to tundra regions (Baldrian et al. 2022). The sheer diversity of NC-endophytes, their ecological significance, evolutionary dynamics, and potential applications, as well as their ability to transition between endophytic and free-living lifestyles, have sparked significant interest among interdisciplinary scientists (Bard et al. 2024).

Nonclavicipitaceous (NC) endophytes are categorized into three classes based on their host transmission and colonization, plant biodiversity patterns, and ecological roles. These classes exhibit distinct characteristics: Class 2 endophytes colonize both below- and above -ground tissues, whereas class 3 endophytes are restricted to above-ground tissues and Class 4 endophytes to roots. Colonization patterns also vary: Class 3 endophytes exhibit localized infections, however, Class 3 and 4 exhibit systemic infections. Research reveals that individual host plants harbor limited diversity of Class 2 endophytes (Rodriguez et al. 2009), whereas Class 3 endophytes can exhibit remarkably high diversity, with up to 20 species identified in single leaf (Arnold et al. 2007). The Class 4 endophytes diversity remains understudied. Differences in plant biodiversity between Class 2 and Class 3 endophytes may stem from distinct host colonization and transmission patterns. While both classes are transmitted horizontally, Class 2 endophytes also employ vertical transmission via seeds, seed coats, as well as rhizomes (Caruso et al. 2022).

Class 2 endophytes possess a distinctive trait: imparting habitat-specific stress tolerance to their host plants, thereby enhancing their fitness in a particular environment. This adaptive benefit differs from the general benefits provided by other endophytes, which are not habitat-specific. Notably, Class 2 endophytes are the only ones demonstrated to confer habitat-adapted stress tolerance. Further investigation is necessary to comprehend the composition, ecological functions and potential advantages of Class 3 as well as Class 4 endophytes, as most remain unexplored in terms of fitness benefits, transmission modes, and rhizosphere competence. The global agricultural sector faces significant challenges, with plant pests and pathogens causing estimated yearly crop production losses of 30–50% (Bvenura et al. 2022). Traditionally, chemical pesticides have been employed to mitigate these losses and meet food demands (Popp et al. 2013). However, growing prohibition on chemical pesticides and increasing consumer concerns about residues in food and the environment have prompted governments and private agriculture industries to explore alternative, sustainable solutions (Choudhary et al. 2023).

One promising approach gaining attention is the utilization of beneficial endophytes as biological control agents for crop protection. Endophytes are microorganisms that inhabit plants for part or all of their life cycle, generally without causing harm (Barkodia et al. 2018). Research has shown that nearly all vascular plants harbor endophytes, which originate from the phyllosphere as well as rhizosphere and enters inside the plants through injury or wounds (Krishnaraj et al. 2017). Recent studies have investigated endophytic communities associated with various plant species, revealing an underestimated diversity of fungal endophytes (Riva et al. 2022). Furthermore, the distribution of certain endophytes has been found to be host- and/or environment-specific (Hou et al. 2022).

This chapter provides the multifaceted contributions of fungal endophytes in enhancing agricultural crop resilience, specifically their defensive mechanisms against a broad spectrum of pathogens and pests, including fungal, bacterial, nematode, viral, and mite threats. Furthermore, we investigate how environmental factors and host plant interactions influence the dynamics of fungal endophytes, and delve into the modes of endophyte transmission, encompassing both horizontal transfer between hosts and vertical inheritance from parent plants, to provide a comprehensive understanding of these beneficial microorganisms' complex relationships with their hosts and surroundings (Hou et al. 2022).

2.2 *Clavicipitaceous* Endophytes (Class 1)

The discovery of *Clavicipitaceous* endophytes in grasses dates back to the nineteenth century, with European researchers identifying them in seeds of *Lolium* species (Rai et al. 2014). Initially, scientists suspected a link between these endophytes and toxic syndromes in animals consuming infected plant tissues. This hypothesis gained traction when *Neotyphodium coenophialum* was linked to "summer syndrome" in grazing cattle *Festuca arundinacea.* Keith Clay's 1988 proposal that *Clavicipitaceous* endophytes serve as protective mutualists for host grasses sparked widespread research into their natural history, evolution, ecology, and physiology (Clay 1988). This led to the development of "endophyte-enhanced" turf grass cultivars, capitalizing on the beneficial effects of these endophytes.

Phylogenetically, *Clavicipitaceae* belongs to the *Hypocreales* order of *Ascomycota*, comprising both free-living and symbiotic species associated with insects, fungi, grasses, rushes, and sedges (Bills et al. 2016). Molecular sequence data suggest that soil-inhabiting species, which are geographically widespread, descended from saprotrophic ancestors. Ancestral state reconstructions indicate that *Clavicipitaceous* endophytes evolved from insect-parasitic ancestors, diversifying via inter-kingdom host flows (El Mansy et al. 2020). The arise of endophytic *Balansia* and *Epichloë* likely involved transitioning from free-living insect parasites to epibiotic plant biotrophs, eventually gaining direct access to grass hosts. Future genomic sequencing will enable comprehensive comparison of these within phylogenetic analysis (El Mansy et al. 2020).

The origin of *Clavicipitaceous endophytes* from insect pathogens can explain their proficiency in producing toxins targeting insects and other animals. Arsenal employed by free-living forms, including *Cordyceps,* likely shares precursors with endophytes, enabling their widespread presence in grasses (Uzma et al. 2019). *Epichloë* species, symbiotic partners of cool-season grasses, exhibit distinct characteristics. Their mycelium inhabits culms, leaf sheaths, and rhizome intracellular spaces, occasionally extending to leaf blades. Upon grass flowering, the fungus forms stromata, arresting inflorescence development and preventing seed head formation (choke disease) (Ross et al. 2016). Some *Epichloë* species display stromata on select tillers providing vertical transmission and partial seed production, whereas other Type 1 endophytes produce stomata on majority of tillers providing endophyte infection in escaping tillers. Additionally, *Epichloë* can produce auxins that modify plant tissue development. *Epichloë* fungi produce spermatia-bearing stromata, requiring mating type transfer for fertile perithecia and ascospore formation (Bacon et al. 2018). Botanophila flies, acting as symbiotic "pollinators," facilitate this process by vectoring spermatia. Although fertilized stromata generate infective ascospores, the initial infection mechanisms remain unclear. Interestingly, Epichloë's sexual cycle and stromata production are exclusive to Northern Hemisphere grasses (Patil et al. 2016). In contrast, Southern Hemisphere C-endophytes lack stromata formation and sexual reproduction. This disparity suggests alternative infection methods for Southern Hemisphere grasses, such as epiphyllous conidial production, non-stromata-dependent transmission, and loss of stromata production capacity in Southern Hemisphere endophytes. This geographical distinction highlights the complexity of *Epichloë's* reproductive strategies and warrants further investigation.

Certain *Epichloë* species have undergone evolutionary changes, resulting in the loss of sexual stage development. These asexual or type III endophytes, reclassified as *Neotyphodium* spp., exhibit no visible symptoms of infection throughout plant development (Tadych et al. 2014). They colonize leaves, culms, rhizomes, and inflorescence primordia, eventually infecting ovules, seeds, scutellum, and embryo axis. Phylogenetic analysis reveals multiple origins of Type III C-endophytes, distributed across both Northern and Southern Hemispheres. Vertical transmission occurs through seeds, while some retain epiphyllous mycelium, potentially facilitating horizontal transmission via water-released conidia. Study has revealed the tiller colonization as primary infection sites, limited growth in mature plant tissue, rapid growth in nutrient-rich meristematic tissues, systemic colonization of above-ground grass tissues, and exclusion from root meristems (EI Mansy et al. 2020).

Type III endophytes produce conidia in culture, indicating functional propagules. However, neighboring plant tissues resist endophytic growth, except in meristematic regions. The complex life cycle of Type III C-endophytes involves vertical transmission through seeds, potential horizontal transmission via epiphyllous conidia, systemic colonization of grass tissues, and limited growth in mature plant tissue. Despite significant research, the life histories and population dynamics of *Clavicipitaceous* endophytes (C-endophytes) remain poorly understood. Notably, phylogenetic analysis revealed that asexual C-endophytes may not be entirely clonal, suggesting complex evolutionary histories like interspecific hybridizations among asexual C-endophytes, phylogenetic evidence of recombination events, and potential mechanisms for maintaining genetic diversity. Two theories explain the significance of interspecies recombination in asexual C-endophytes: genetic variability and genome viability maintenance, and recombination may compensate for the lack of sexual reproduction, ensuring adaptability and survival (Duperron 2016).

Vertical transmission as a "rescue mechanism": Hybridized endophytes, unable to undergo sexual reproduction, may rely on vertical transmission to propagate. These findings highlight the intricate relationships between C-endophytes' reproductive strategies, genetic diversity, and ecological success (EI Mansy et al. 2020).

2.3 Class 2 Endophytes

These comprises a diverse group of fungal species within the *Dikarya* lineage, predominantly *Ascomycota* (specifically *Pezizomycotina*) and to a lesser extent *Basidiomycota* (mainly *Agaricomycotina* and *Pucciniomycotina*) (Naranjo-Ortiz and Gabaldón 2019). These endophytes are distinguished from nonclavicipitaceous endophytes by their widespread colonization of plant roots, stems, and leaves, extensive infection capabilities, transmission via seed coats and rhizomes, low rhizosphere abundance, and provision of habitat-specific adaptive benefits, with remarkably high infection rates (90–100%) in stressed plants. Although not all examples have undergone comprehensive analysis, they meet sufficient criteria for inclusion. The study of Class 2 endophytes was pioneered by research on a Phoma species inhabiting *Calluna vulgaris*, initially misidentified as mycorrhizal, but later found to extensively colonize the entire plant without forming characteristic mycorrhizal structures; subsequent research identified Phoma species as prevalent root endophytes enhancing plant fitness in Mediterranean flora (Naranjo-Ortiz and Gabaldón 2019).

Throughout the twentieth century, research periodically highlighted the benefits of Class 2 endophytes to their host plants, such as the symbiotic relationship between the brown alga *Ascophyllum nodosum* and *Mycophycia ascophylli*, which is essential for normal growth and development (EI Mansy et al. 2020). Class 2 endophytes have also been shown to enhance plant biomass production and confer resistance to various stressors, including disease, drought, desiccation, heat, and salinity. Despite their importance, studies on colonization patterns, host specificity, and fitness benefits remain limited. The infection process involves specialized structures like appressoria or direct hyphal penetration, followed by intercellular growth with minimal impact on host cell integrity. While healthy plants exhibit restricted sporulation, Class 2 endophytes proliferate and sporulate during host senescence. Ecological research is ongoing to understand their distribution and abundance in the rhizosphere, with certain species displaying high soil abundance, whereas others, such as *Fusarium culmorum* and *Colletotrichum magna*, are relatively rare; however, traditional soil analysis methods can obscure distinctions between endophyte propagation and spore deposition from senescing plants (Nelson et al. 2020).

Certain fungal species, such as Arthrobotrys, display versatile lifestyles, existing as both nematophagous and endophytic organisms, which enables them to attain high concentrations in the rhizosphere due to the abundance of nematodes (Schouten 2016). In contrast, Phoma species achieve high soil abundance through emergence, sporulation, and spore dissemination from senescing plant hosts. While horizontal transmission is evident, Class 2 endophytes frequently employ vertical transmission via seed coats inherited from maternal plants. Interestingly, these culturable endophytes grow well on various simple media, indicating adaptability. Their widespread presence in plant hosts and soils suggests varying competitive capabilities, with some species potentially exhibiting multifaceted lifestyles, seamlessly transitioning between symbiotic and saprophytic modes, highlighting their remarkable ecological flexibility (Frank et al. 2017).

2.4 Class 3 Endophytes

These are predominant in aerial plant tissues, facilitated by horizontal transmission pathways, forming localized and specialized infections that confer context-dependent benefits or drawbacks to their hosts, irrespective of specific habitats (Yang et al. 2023). This endophyte class boasts exceptionally high biodiversity within plant hosts, encompassing diverse fungal communities in tropical tree leaves, nonvascular plants, seedless vascular plants, conifers, and woody and herbaceous angiosperms across biomes ranging from tropical rainforests to boreal and Arctic/Antarctic ecosystems. Class 3 endophytes inhabit a broad range of plant tissues, including photosynthetic parts, flowers, fruits, asymptomatic wood, and inner bark, with analogous life strategies found in asymptomatic lichens as "endolichenic" fungi (Yang et al. 2023). Remarkably, individual tissues, plants, and populations display astonishing intra-host diversity, with multiple discrete infections within seemingly healthy leaves, contributing to a complex, high-density mosaic of diverse endophyte species. Studies in tropical forests reveal individual leaves can host up to one isolate per 2 mm^2 of tissue, often harboring dozens of species, with several leaves on the same plants supporting distinct endophyte assemblages and individual plants hosting more than a hundred of species, while plant species through their indigenous ranges can be inhabited by thousands of unique endophyte spp. (Lugtenberg et al. 2016; Hamzazai 2020; Ngubane 2022).

The remarkable diversity of Class 3 endophytes, evident in certain tropical species and locales, is not unique to these environments. Temperate and boreal plant communities also exhibit an impressive array of these endophytes (Oita 2020). Class 3 endophytes, transmitted horizontally, are generally varied from pathogens linked with same host species, as well as epiphyllous fungi inhabiting the same leaves. While their distinctiveness from saprotrophic communities is debated, emerging evidence suggests Class 3 endophytes include unique species not found among saprophytic fungi. Furthermore, these endophytes are rarely recovered from seeds. Class 3 endophytes vast diversity raises fundamental questions about its ecological functions (EI Mansy et al. 2020).

Class 3 endophytes have primarily focused on isolated experiments, neglecting the intricate interplay between multiple environmental and biotic stressors, as well as diverse endophyte-plant relationships (Sena et al. 2024). Despite evidence of enhanced disease resilience, herbivore resistance, and altered drought tolerance, the scarcity of comprehensive studies is striking, considering the vast diversity of Class 3 endophytes. To truly grasp the ecological significance of these fungi, future research must embrace the complexity of endophyte communities and explore multitrophic interactions. This emerging field mirrors the rapid expansion of mycorrhizal symbiosis research over the past few decades, presenting a thrilling opportunity for discovery (Sena et al. 2024).

Class 3 endophytes predominantly belong to the *Dikaryomycota* phylum, with a notable bias toward *Ascomycota*. Specifically, the *Pezizomycotina* subgroup is extensively represented, alongside a smaller presence of endophytic *Saccharomycotina*. Within *Pezizomycotina*, Class 3 endophytes are dispersed across all major nonlichenized clades, exhibiting a strong presence in *Pezizomycetes, Leotiomycetes*, and *Eurotiomycetes*. Notably, the *Sordariomycetes* and *Dothideomycetes* classes harbor an exceptionally high diversity of Class 3 endophytes. Interestingly, these same lineages have given rise to numerous ascomycetous plant pathogens and saprotrophic organisms. The relative abundance of each class within endophyte communities varies significantly across biomes and host lineages, with endophytic Leotiomycetes being particularly prevalent in conifers and *Sordariomycetes* dominating tropical plant communities (Hardoim et al. 2015).

Certain *Basidiomycota* lineages, specifically *Agaricomycotina, Pucciniomycotina*, and *Ustilaginomycotina*, also comprise Class 3 endophytes, albeit with relatively lower reported frequencies compared to their *ascomycetous* counterparts. Generally, *basidiomycetous* endophytes exhibit a predilection for woody tissues over foliage. However, recent studies suggest that endophytic *Basidiomycota* may be highly prevalent in foliage than previously thought, particularly when employing culture-independent methods. For instance, research on *Pinus taeda* revealed a fourfold increase in *basidiomycetous* species detection using environmental PCR technique on surface-sterilized leaves, compared to traditional culturing techniques (Gray 2016). Furthermore, *Ustilago maydis* has been found to be a significant component of the endophyte community in maize (*Zea mays*), highlighting the importance of considering *Basidiomycota* in endophyte diversity assessments.

Class 3 endophytes propagate through two primary methods: hyphal fragmentation and the production of asexual or sexual spores on dying or deceased plant tissues. These reproductive structures are released through various means, including passive liberation, herbivore-mediated transport, and environmental disruptions like wind and precipitation (Florea et al. 2015). Certain Class 3 endophytes, such as *Phyllosticta* sp., generate mucilaginous spores that rely partially on rainfall for dispersal. Conversely, Ingoldian fungi forms spores that require water for both infection and dispersal. While intense wind and rain facilitate optimal spore movement, gentle precipitation can still effectively disperse *Colletotrichum conidia*. Moreover, subtle wind patterns generated by daily temperature fluctuations significantly contribute to the dry propagules dispersal. Additionally, some Class 3 endophytes exploit insect vectors, such as moths, for long-distance dispersal by adhering to their bodies (Florea et al. 2015).

Seedlings raised in sterile conditions lack culturable Class 3 endophytes, distinguishing them from Class 2 endophytes, which can be vertically transmitted. Class 3 endophyte colonization occurs rapidly, facilitated by airborne inoculum, high relative humidity, and leaf surface wetting from rain, dew, or fog. Field experiments with endophyte-free *Theobroma cacao* seedlings revealed 80% leaf colonization within 2 weeks of emergence at the time of early wet season among tropical forests (Melnick 2010). Notably, leaf hardness and chemistry do not influence colonization, with young and mature leaves being colonized equally rapidly. Class 3 endophyte infection frequency, richness, and diversity generally follow a latitudinal gradient, peaking in tropical regions and decreasing in Arctic tundra and boreal forests. However, local abiotic conditions, land-use patterns, and host-specific factors, such as long-lived foliage in evergreens, can lead to unexpectedly high endophyte abundances and diversities in temperate zones, highlighting the complexity of Class 3 endophyte distribution and ecology (Melnick 2010).

2.5 Class 4 Endophytes

The discovery of dark septate endophytes (DSE), now classified as Class 4 endophytes, observation of a brown to blackish, pigmented fungus associated with terrestrial plant roots, initially termed "mycelium radicis atrovirens" (MRA) and later referred to as "pseudomycorrhizal" fungi and revealed 135 angiosperm species hosting these darkly pigmented fungi in root tissues (Mane and Vedamurthy 2018). Characterized by melanized septa and restricted to plant roots, Class 4 endophytes are primarily ascomycetous, conidial, or sterile fungi forming inter- and intracellular hyphae and microsclerotia. A comprehensive understanding of dark septate endophytes (DSE) diversity and host ranges requires extensive global studies of plant-DSE associations. Two latitudinal studies found DSE associated with 587 plant species across 320 genera and 114 families (Hidayat 2019). Colonization studies using five described DSE taxa (*Chloridium paucisporum*, *Leptodontidium orchidicola*, *Phialocephala dimorphospora*, *Phialocephala fortinii*, and *Phialophora finlandia*) revealed a broad host range, with *Phialocephala fortinii* colonizing 20 plant species. However, DSE classification requires caution due to ambiguous criteria, relying primarily on asexual, darkly pigmented, septate endophytes in plant roots. Misidentification can occur, as exemplified by *Curvularia protuberata*, a Class 2 endophyte that colonizes entire plants, not just roots. To avoid errors, researchers must analyze all plant tissues and verify fungal septa using specific dyes. The asexual designation of DSE may also require reevaluation, considering the potential overlap between soil fungi, saprophytic rhizoplane-inhabiting fungi, pathogenic fungi, mycorrhizal fungi, and fungal endophytes, highlighting the complexity of DSE ecology and taxonomy (Grünig et al. 2008). The widespread presence of dark septate endophytes (DSE) in soils and plant roots suggests horizontal transmission as the primary mode of dissemination. Laboratory studies have confirmed that mycelial fragmentation and conidial dispersal facilitate transmission. Although the anamorph-teleomorph connections remain largely unknown for most DSE species, the potential for sexual reproduction cannot be ruled out (Terhonen et al. 2019). The asexual reproductive structures and morphology of DSE-colonized roots, which exhibit similarities with *ectomycorrhizas*, *endomycorrhizas*, and *pseudomycorrhizas*, highlighting the complex interactions between DSE and their plant hosts (Sieber 2002).

2.6 Role and Importance of Fungal Endophytes

Endophytic microorganisms employ diverse mechanisms to promote plant growth and safeguard against pests and pathogens (Afzal et al. 2019). These beneficial microorganisms produce and secrete secondary metabolites, including volatile compounds, which inhibit pathogen growth and trigger plant defense mechanisms, such as systemic acquired resistance (SAR) and induced systemic resistance (ISR) (Rana et al. 2020; Mengistu et al. 2020; Chandran et al. 2021; Yadav and Meena 2022). Certain endophytes, like *Piriformospora indica*, induce host defense responses through jasmonic acid-dependent pathways. Others exhibit biocontrol potential by secreting antifungal and antibacterial compounds, engaging in mycoparasitic activity, or competing with pathogens for space and nutrients. For instance, an *Enterobacter* sp. strain isolated from finger millet suppresses *Fusarium graminearum* in host plant roots and produces antifungal compounds. Foliar application of endophytes to *Theobroma cacao* leaves protected against Phytophthora-induced necrosis and mortality, highlighting the complex interspecific interactions, including competition and mutual antagonism, that mediate host defense outcomes (Grabka et al. 2022). These findings underscore the multifaceted roles of endophytes in plant protection and growth promotion.

Beyond direct pathogen protection, numerous endophytes possess plant growth-promoting (PGP) properties that enhance host plant resilience. These PGP endophytes provide essential nutrients like nitrogen, phosphate, and iron and stimulate growth and development through various mechanisms (Olenska et al. 2020; Kumar et al. 2022; Meena et al. 2023). Associated with roots, PGP microbes produce chemical compounds influencing plant growth, including plant hormones such as indole-3-acetic acid (IAA), gibberellins, cytokinins, and 1-aminocyclopropane-1-carboxylic acid (ACC) deaminase, which promotes mycorrhization. Endophytes also modulate plant hormone levels, including auxin, cytokinin, ethylene, and gibberellin, and produce bioactive compounds (Kurepin et al. 2014). By enhancing host plant growth and overall health, PGP microbes indirectly contribute to protection against pathogens and pests, conferring a competitive advantage to colonized plants compared to non-colonized counterparts.

Fungal endophytes, asymptomatic inhabitants of plant tissue, are ubiquitous across all plant parts and species, whether wild or cultivated. A single plant can harbor multiple endophytic species, which may provide tissue-specific or systemic protection against disease, with some interactions exhibiting symbiotic and potentially mutualistic relationships (Grabka et al. 2022). The pervasive presence of endophytes has shifted research focus from traditional phytopathogenic and mycorrhizal fungi to plant-fungal interactions. Estimates suggest over one million endophytic species reside within 300,000 plant species, yet only a small fraction have been isolated and studied for their roles within their host plants, highlighting the vast unexplored potential of these microorganisms. Research has shown that endophytes, microorganisms living within plant tissues, can significantly enhance their hosts' resilience and productivity (Kamran et al. 2022). By mitigating the impacts of environmental stressors, such as drought, salinity, and extreme temperatures,

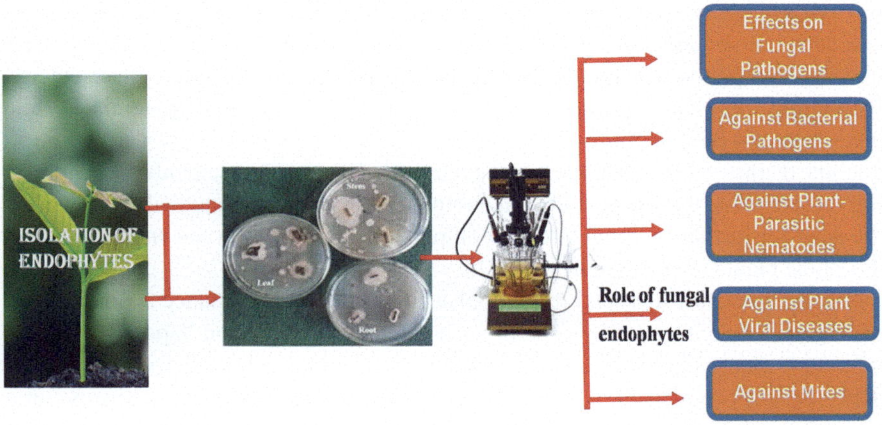

Fig. 2.2 Role of fungal endophytes in plant defense system

endophytes can improve plant fitness and overall well-being. Additionally, they provide various benefits, including protection against pathogens, insects, and herbivores, as well as promoting plant growth and development through stimulants (Fig. 2.2).

2.6.1 Fungal Endophytes Against Fungal Pathogens

Fungal pathogens inflict severe damage on crops, leading to plant death, reduced yields, and compromised quality, with some also producing harmful mycotoxins affecting human and livestock health (Anthony et al. 2021). While synthetic fungicides are widely used to control these pathogens, they can have detrimental environmental consequences, such as harming beneficial fungi crucial for crop health, like *arbuscular mycorrhizae*, and selectively targeting nontarget microorganisms. Fortunately, biocontrol endophytes, such as Ampelomyces, offer an eco-friendly alternative, effectively reducing pathogen prevalence while preserving mutualistic fungi. Integrating these biocontrol agents into pest management strategies can enhance agricultural sustainability, maintain soil health, and mitigate chemical pesticide resistance. By adopting diverse pest management approaches, the agricultural sector can minimize environmental harm and promote a healthier ecosystem, underscoring the importance of sustainable practices in crop protection (Anthony et al. 2021).

Research on secondary metabolites produced by endophytes is gaining momentum, focusing on harnessing natural products for agrochemical applications (Armengol et al. 2021). Various compounds have been extracted and isolated from specific fungal taxonomic orders using top-down approaches, with *Xylariales* exhibiting remarkable diversity in bioactive metabolites, including *glucosides*, *cytochalasans*, and *terpenoids* (Armengol et al. 2021). Alternatively, studies have employed a targeted approach to identify antifungal compounds from endophytes, involving characterization of fungal diversity within host plants, dual culture assays, and antagonism assessments. Endophytes demonstrating anti-pathogen activity undergo compound analysis using liquid chromatography and mass spectrometry, facilitating the identification of candidate species for biocontrol potential. While investigations have explored the antifungal efficacy of endophyte-derived compounds against various pathogens and their plant fitness enhancements, the underlying mechanisms of these benefits often remain unclear or understudied, highlighting the need for further research into the complex interactions between endophytes and their host plants (Armengol et al. 2021).

Endophytes play a crucial role in bolstering host plant resistance against fungal pathogens through systemic responses triggered by colonization. Upon endophytic infection, plants deploy defensive strategies, including cell wall fortification via deposits, to prevent pathogen penetration. Interestingly, endophytes possess specialized mechanisms, such as exoenzymes, allowing them to access strengthened cells, while blocking pathogen entry. Endophytes also prime plant defense responses through transcriptional reprogramming, influencing key defense-related genes involved in salicylic acid, jasmonic acid, and ethylene signaling pathways (Yan et al. 2019). Furthermore, endophyte colonization has been linked to enhanced photosynthesis rates, increased chlorophyll content, trichome and stomata density, antioxidant enzyme activity, callose deposition, cell lignification, and phytoalexin accumulation. Additionally, endophytes competitively exclude pathogenic fungi by occupying potential niches, providing protection even in the absence of other defense mechanisms, thereby safeguarding the host plant against fungal infections (Yan et al. 2019).

Fungal endophytes from the *Daldinia* genus have demonstrated potent inhibitory effects against plant pathogens *Sclerotium rolfsii* and *Colletotrichum acutatum*. Notably, *Daldinia eschscholtzii*, extracted from Stemona and ginger root, has been found to give a diverse array of compounds, with elemicin, benzaldehyde dimethyl acetal, and ethyl sorbate being the most prominent. Elemicin has been identified as a highly effective antifungal agent against various *Colletotrichum* species. Furthermore, *Daldinia* cf. *concentrica*, isolated from olive, has been shown to produce 27 volatile organic substances, consisting 3-methyl-1-butanol and isoamyl acetate, which exhibit broad-spectrum antifungal activities when combined. Additionally, *Daldinia* species have been shown to form antimicrobial substances, such as 11-diene, α-guaiene, guaia-1(10), and thujopsene, highlighting their potential as valuable sources of bioactive metabolites (Khruengsai et al. 2021).

The *Fusarium* genus encompasses a diverse range of species, some of which are notorious plant pathogens, while others exhibit endophytic properties, inhibiting the growth of other fungal pathogens. Although *Fusarium* metabolites have been extensively used for its pharmaceutical potential as antimicrobial agents, their antifungal properties and agricultural applications have received relatively less attention. Research has shown that extracts from *Fusarium proliferatum*, isolated from *Cissus quadrangularis*, demonstrate potent inhibitory effects against *Rhizoctonia solani* and *F. oxysporum*, with effective concentrations varying from 0.2–2.5 mg/mL. Chemical analysis revealed a rich composition of terpenoids, phenolics, and unsaturated alkenes. Moreover, the chitinases enzyme derived from *Fusarium chlamydosporum* has been found to effectively lyses the cell walls of *Puccinia arachidis*, preventing urediniospore germination and highlighting the potential of Fusarium-derived compounds as biocontrol agents in agricultural settings (Segaran and Sathiavelu 2019).

Research has explored the antifungal potential of secondary metabolites extracted from various endophyte species, including *Colletotrichum*, *Aspergillus*, *Gliocladium*, *Diaporthe*, *Phyllosticta*, *Lecanicillium*, and *Trichoderma*. Notably, *Trichoderma* species, such as *T. asperellum*, *T. longibrachiatum*, and *T. atroviride* isolated from soybeans, demonstrated significant antifungal activity against *Rhizoctonia solani*, reducing seed infection by 64%, 60%, and 55%, respectively (Sallam et al. 2021). These species synthesized hydrolytic enzymes like chitinase and pectinase, which degrade fungal cell walls, as well as siderophores that restrict iron availability to pathogens and indole acetic acid (IAA) that enhances plant growth. Furthermore, *Trichoderma erinaceum*, isolated from Stemona and ginger root, exhibited strong inhibitory effects against *Sclerotium rolfsii*, which is causative agent of southern stem rot disease, with 64% growth inhibition in double culture assays and 58% reduced infection in pot experiments. Chemical studies revealed that *T. erinaceum* produces bioactive compounds, including β-1,3 glucanase, 6-n-pentyl-2H-pyran-2-one (6PAP), and chitinase, contributing to its antifungal properties (Suebrasri et al. 2020).

Investigations into fungal endophytes from medicinal plants have unveiled promising antifungal and antimicrobial compounds (Caruso et al. 2022). For instance, *Aspergillus neoniger* extracts, isolated from *Ficus carica*, demonstrated potent inhibitory effects against *Penicillium* and *Aspergillus* species, with growth inhibition exceeding 80% (Alsenani et al. 2023). Subsequent analysis identified aurasperone A and D as key bioactive compounds, exhibiting minimum inhibitory concentrations of 76 and 67 μg/mL against *Fusarium oxysporum*. Similarly, an Aspergillus species from *Bethencourtia palmensis* produced antifungal compounds mellein and neoaspergillic acid, effectively inhibiting *Alternaria*, *Botrytis*, and *Fusarium* species. Other endophytes, including *Gliocladium catenulatum* and *Lecanicillium lecanii*, were found to produce chitinase, inhibiting growth and germination of *Rhizoctonia solani* and *Fusarium oxysporum*. Furthermore, *Colletotrichum coccodes* and *Phyllosticta capitalensis* from *Houttuynia cordata* displayed antifungal activity against *Candida albicans*, with bioactive compounds including geranylgeraniol, farnesol, squalene, and 1-octacosanol. Notably, *Diaporthe caatingaensis* from *Buchanania axillaris* produced camptothecin, a compound traditionally isolated from plants, exhibiting antibacterial, anticancer, and antifungal properties, warranting further investigation into its fungal-derived antifungal efficacy (Dhakshinamoorthy et al. 2021).

A comprehensive study isolated *T. lentiforme* and *Trichoderma harzianum* from watermelon (*Citrullus lanatus*), alongside 438 more endophytes. Seven selected species were screened for antagonistic properties against soil-borne pathogens, including *Fusarium oxysporum*, *Macrophomina phaseolina*, and *Monosporascus cannonballus*. Notably, *T. harzianum* and *T. lentiforme* exhibited exceptional inhibitory effects, suppressing pathogen growth by up to 93% in different assays. In vivo tests on watermelon plants further demonstrated disease decrease up to 67% (González et al. 2020). Investigation into their modes of action revealed that these *Trichoderma* species employ multiple strategies to combat pathogens, including competitive exclusion for resources, production of cell wall-degrading compounds, and direct parasitization of pathogenic hyphae through invasive hyphal growth, underscoring their potential as effective biocontrol agents against soil-borne diseases in cucurbit crops (González et al. 2020).

The exploration of endophytic fungi has uncovered significant potential for natural antifungal agents, offering innovative solutions for plant disease management. Notably, *Aspergillus terreus*, derived from *Hevea brasiliensis* seeds, exhibited robust inhibitory effects against *Rigidoporus microporus* and *Corynespora cassiicola*, achieving 81% and 64–70% growth suppression in dual culture assays (Chen et al. 2023). Additionally, *A. terreus* completely halted *R. microporus* growth on sterilized rubber tree wood and reduced *C. cassiicola* infection rates by 87–93% on treated leaves. Separate studies on *Diaporthe* sp., a neem (*Azadirachta indica*) stem endophyte, identified antifungal compounds like 8R-acetoxymultiplolide, demonstrating efficacy against plant pathogens such as *Aspergillus niger* and *Fusarium* species, albeit with limited effectiveness against *Candida albicans*. Furthermore, nonpathogenic Penicillium species in neem displayed antifungal properties, pending identification of the responsible compounds. These findings underscore the intricate relationships between endophytic fungi, host plants, and pathogens, indicating host-specific interactions and diverse antifungal mechanisms that warrant further investigation to inform the development of targeted biocontrol strategies and natural antifungal products (Chen et al. 2023).

2.6.2 Fungal Endophytes Against Bacterial Endophytes

Endophytes have been found to produce a diverse array of antibacterial compounds, providing protection to their host plants against bacterial pathogens. These compounds exhibit varying spectra of activity, ranging from broad-spectrum to targeted efficacy against specific microbial groups (Afzal et al. 2019; Yadav and Meena 2021; Seth et al. 2023). For instance, *Javanicin* demonstrates potent activity against a wide range of microbes, with notable effectiveness against *Bacillus* and *Escherichia coli* (Kharwar et al. 2009). Additionally, endophytes produce an array of broadly antimicrobial secondary metabolites such as alkaloids, terpenoids, aliphatic compounds, polyketides, phenylpropanoids, and peptides, as well as acetol, acetic acid, and hexanoic acid. Notably, certain *Microdiplodia* endophytes produce terpenoids like Phomadecalin E and 8α-acetoxyphomadecalin C, which exhibit robust antibacterial properties against *Pseudomonas aeruginosa*, a bacterium capable of causing soft root rot in plants such as *Panax ginseng*, *Arabidopsis*, and *Ocimum basilicum*, and also posing an opportunistic threat to human health.

Fungal endophytes have emerged as valuable sources of broad-spectrum antimicrobial compounds, enhancing plant defense against various pathogens (Islam et al.

2024). *Chaetomium globosum*, for instance, exhibits potent activity against multiple pathogenic microorganisms and disrupts biofilm formation. Similarly, *Penicillium* sp., isolated from *Stephania dielsiana*, demonstrates remarkable antimicrobial efficacy, with MIC values varies from 1.2 to 6 mg/mL against seven animal pathogenic microbes. *Trichoderma harzianum*, a *Rosmarinus officinalis* endophyte, shows significant antimicrobial activity against *Pseudomonas aeruginosa*, *Escherichia coli*, and *Staphylococcus aureus*, highlighting its biocontrol potential against phytopathogenic bacteria (Islam et al. 2024). Other endophytes, such as *Aspergillus fumigatus*, *Diaporthe phaseolorum*, and *A. versicolor*, isolated from tomato plants, produce antibacterial metabolites like acetol, hexanoic acid, and acetic acid, effectively controlling bacterial spot of tomato caused by *Xanthomonas vesicatoria*. Additionally, *Aspergillus* spp. endophytes from *Cupressaceae* plants exhibit different antibacterial action against *Bacillus* species such as *Pseudomonas syringae* and *Erwinia amylovora*. Secondary metabolites like cycloepoxy-lactone, effective against multiple pathogens, hold particular promise in plant defense (Islam et al. 2024). While these antimicrobial substances may be directly produced by endophytes or induced in host plants, further research is needed to elucidate their secretion mechanisms and associated gene expression (Oukala et al. 2021).

2.6.3 Fungal Endophytes Against Plant Parasitic Nematodes

Plant parasitic nematodes (PPNs) pose a significant problem and danger to global agriculture, resulting in high economic losses, particularly in tropical and subtropical regions. These nematodes establish feeding sites on stems and plant roots, facilitating the entry of secondary pathogens and potentially transmitting viruses. Conventional chemical nematicides have limitations due to their nontarget effects on beneficial microbial communities. As a result, researchers are exploring microorganisms that can coexist with crops and inhibit PPN growth (Abd-Elgawad 2024). Fungal endophytes, such as those from the genera *Acremonium*, *Alternaria*, and *Trichoderma*, have shown promise in controlling nematodes through various mechanisms, including nematocidal compound production, parasitization, and trapping. Certain fungal species produce bioactive compounds that impact nematode colonization, while others, like *Chaetomium* and *Clonostachys*, can outcompete pathogens for resources. Root-knot nematodes, specifically *Meloidogyne* species, affect numerous crops worldwide, but various fungal genera have demonstrated inhibitory effects against these nematodes. Furthermore, compounds produced by these fungi can alter plant metabolites, promote growth, or induce resistance to invading nematodes, offering a promising alternative to chemical nematicides (Abd-Elgawad 2024).

Fusarium species are prominent biocontrol agents against nematodes, leveraging bioactive compounds to enhance plant growth, induce systemic resistance, and directly inhibit nematode development. Research has demonstrated that *Fusarium oxysporum* alters plant hormone production and root exudate composition, deterring *Meloidogyne incognita* colonization. This fungus triggers plant resistance by producing unknown compounds, as seen in banana plants resisting *Radopholus similis*. Notably, *F. oxysporum* strain 162 yields 11 nematocidal compounds, including 4-hydroxybenzoic acid, indole-3-acetic acid (IAA), and gibepyrone D, exhibiting lethal effects on nematodes at concentrations as low as 104–134 μg/mL (Sreenayana et al. 2022). IAA's dual function improves plant health and resistance while acting as a toxin. Studies reveal rapid nematode motility decrease and mortality within 24 h of exposure to *F. oxysporum* compounds, primarily targeting sedentary nematodes while sparing nonpathogenic species. This targeted approach preserves beneficial nematodes that prey on pathogenic bacteria, fungi, or crop pests. Other fungal endophytes, such as *Chaetomium*, *Alternaria*, and *Trichoderma*, have also shown promise as seed treatments or inoculants, reducing damage from *Meloidogyne incognita* and root-knot nematodes in cotton and sacha inchi plants, highlighting the potential for integrated nematode management strategies (Sreenayana et al. 2022).

The *Epichloë* genus comprises endophytic fungi renowned for their symbiotic relationships with diverse grass species, characterized by extensive hyphal colonization, predominantly in shoot tissues. These fungi are prolific producers of bioactive compounds, including alkaloids and non-alkaloid secondary metabolites, which confer protective benefits. Notably, *Epichloë coenophialum* has demonstrated efficacy in reducing parasitism by *Meloidogyne marylandi* and *Pratylenchu*s root-symptom nematodes. While the precise mechanism remains unclear, research suggests that alkaloid production plays a crucial role. Interestingly, despite primarily inhabiting shoot tissues, *Epichloë's* protective effects extend to roots, likely facilitated by plant-mediated translocation of bioactive compounds or induced resistance responses, highlighting the complex interactions between endophytic fungi, plants, and nematodes.

Research has identified several fungal strains with potent nematocidal properties. Specifically, *Chaetomium globosum* produces flavipin, a compound exhibiting antioxidant and anti-nematode activity (Bairwa et al. 2023). *Purpureocillium lilacinum*, meanwhile, secretes proteases and chitinases that disrupt nematode egg development, particularly in *Meloidogyne* and *Heterodera* species, and also parasitizes eggs through hyphal penetration. Other fungi, such as *Diaporthe phaseolorum* and *Melanconium betulinum*, yield 3-hydroxypropionic acid, demonstrating selective nematocidal effects against *Meloidogyne incognita* with LD50 concentrations between 12.5 and 15 μg/mL. Additionally, *Geotrichum* sp. derivatives have shown nematocidal activity against *Bursaphelenchus xylophilus* and *Panagrellus redivivus*. Furthermore, fungi from the genera *Epicoccum*, *Dactylonectria*, *Myrothecium*, and *Fusarium* produce bioactive substances with great efficacy against second generation juveniles of *Heterodera glycines*, highlighting the diverse range of fungal metabolites with potential applications in nematode control (Bairwa et al. 2023).

Fungal endophytes, such as *Acremonium implicatum* and *Fusarium solani*, have been identified as a direct pathogens *Meloidogyne* nematodes, utilizing hyphal extension and penetration to target eggs, juveniles, and females (Grabka et al. 2022). Other notable nematodes pathogen, including *Metacordyceps chlamydosporia*, *Arthrobotrys iridis*, and *Hirsutella rhossiliensis*, have been discovered as plant root endophytes, exhibiting a unique lifestyle switch from saprophytic to parasitic upon nematode exposure. These fungi employ various strategies, such as hyphal traps and paralyzing secretions, to capture and infect nematodes. *Metacordyceps chlamydosporia* produces an alkaline serine protease to facilitate egg penetration, while *Hirsutella* spp. produce sticky conidia to attach to and colonize nematodes (Moosavi and Zare 2020). *Esteya vermicola*, a nematophagous endophyte, shows promise in controlling the invasive pinewood nematode. Endophytic fungi have been explored as seed treatments and root inoculants for agricultural nematode control, with further research needed to unlock their full potential. Elucidating bioactive compounds produced by these endophytes will clarify their modes of action and inhibit nematode development. Investigating fungal species' parasitic capabilities against diverse nematode genera will facilitate the development of effective biocontrol methods, offering a promising avenue for sustainable nematode management (Moosavi and Zare 2020).

2.6.4 Fungal Endophytes Role as Plant Viral Diseases

Fungal endophytes offer a multifaceted approach to mitigating viral diseases in plants, primarily by bolstering plant defenses or inhibiting virus transmission through entomopathogenic activities against vector insects. While their anticancer and antiviral properties against human viruses are notable, research has also explored their potential in combating plant viruses. Studies have demonstrated that fungal endophytes, such as *Neotyphodium uncinatum* in meadow ryegrass, can reduce viral infection by deterring aphid vectors with alkaloid production, thereby limiting Barley yellow dwarf virus spread (Bagheri et al. 2021). Similarly, *Beauveria bassiana* strains have protected squash plants against Zucchini yellow mosaic virus. Interestingly, endophyte-induced resistance appears virus-specific, as *Metarhizium anisopliae* and *Trichoderma harzianum* conferred resistance against sugarcane mosaic virus but not Maize chlorotic mottle virus in maize (Jaber and Salem 2014). Environmental factors, such as light intensity, also influence endophyte-mediated resistance, as seen in *Piriformospora indica*-inoculated tomato plants, which repressed *Pepino mosaic* virus under high light conditions. By priming plants with endophytes, researchers propose a promising alternative to conventional insecticides, potentially providing persistent protection against viral infections and reducing the need for harmful chemicals. This eco-friendly approach warrants further investigation to unlock its full potential in plant virus management (Bagheri et al. 2021).

2.6.5 Fungal Endophytes Role Against Mites

Globally, phytophagous mites pose a significant threat to ornamental and agricultural crops, causing substantial harm via feeding and virus transmission, which compromises photosynthesis, plant health, yield, and market value (Liburd et al. 2019). Notably, the two-spotted spider mite (*Tetranychus urticae*) has a remarkably broad host range, affecting over 1000 plant species across 130 families, including commercially important crops like cucurbits, beans, and fruits. However, pesticide resistance in phytophagous mites has become a pressing problem, with *Tetranychus urticae* and European red mite exhibiting alarming levels of resistance to over 90 and 50 compounds, respectively. To mitigate multiple acaricide resistance and protect nontarget beneficial organisms, such as predatory mites, integrated pest management systems require careful strategic application, emphasizing the need for innovative and sustainable solutions to combat these devastating pests (Liburd et al. 2019).

Mycoacaricides, comprising entomopathogenic hypocrealean fungi like *Akanthomyces muscarius*, *Beauveria bassiana*, and *Purpureocillium lilacinum*, offer a promising control tool against pests. These generalist fungi infect insects through conidial germination, appressorium formation, and cuticle penetration, ultimately colonizing internal tissues and producing toxic metabolites (Mascarin et al. 2024). Notably, *Beauveria* and *Metarhizium* have been extensively studied as biocontrol agents against both insects and mites. These fungi can endophytically colonize various host plants through methods like seed soaking, root dip, and foliar spray, facilitating systemic acropetal growth. Research has demonstrated successful foliar endophyte colonization in cotton seeds treated with *Beauveria bassiana* or *Purpureocillium lilacinum*, leading to reduced cotton aphid reproduction. Similarly, composted cabbage waste inoculated with *Clonostachys rosea* achieved 100% endophyte colonization in tomatoes, although it did not significantly impact two-spotted spider mite populations. However, tomato leaf detachment bioassays showed increased two-spotted spider mite mortality following *B. bassiana* inoculation, with spray application yielding the highest mortality and leaf colonization rates (Mascarin et al. 2024). *Cordyceps* and *Trichoderma* are commercially available strains applied as soil drenches and have also effectively reduced TSSM and green peach aphid populations on pepper plants, highlighting the potential of mycoacaricides as a sustainable pest management solution (Grabka et al. 2022).

Research has demonstrated the efficacy of fungal endophytes in controlling phytophagous mites. For instance, tomato seedlings inoculated with *Fusarium solani* exhibited reduced TSSM egg numbers and upregulated defense genes, emitting volatile compounds attractive to the spider mite predator *Macrolophus pygmaeus* (Grabka et al. 2022), suggesting that endophytes employ multiple protective mechanisms, including antibiosis, feeding deterrence, and defense priming. Notably, *Meira argovae* produces argovin, which eliminated citrus rust mite populations at 0.2 mg/mL. *Meira* species have been isolated from diverse plant sources, including Japanese pear, tobacco, Magnolia, and vetiver grass, highlighting their widespread association with plant rhizospheres, phyllospheres, and endospheres. The identification of acaropathogenic *Meira* species underscores their potential as biocontrol agents against phytophagous mites, offering a promising eco-friendly solution for managing these pests.

Hypocrealean fungi, renowned for their entomopathogenic and acaropathogenic properties, hold great promise for biocontrolling insects and mites. Endophytic species, in particular, offer distinct advantages as biocontrol agents due to their ease of application via seed treatment, ability to persist and spread within host plants, and capacity to prime plant defense mechanisms. This confers protection against a wide range of pests, beyond just mites, and enhances resilience against environmental factors like moisture and UV light. Moreover, certain acaropathogens have shown compatibility with predatory mites, even exhibiting synergistic effects in some cases, although potential negative interactions must be considered (Mascarin et al. 2024). As a result, endophytic mycoacaricides are poised to play a vital role in integrated pest management strategies, mitigating phytophagous mite infestations, and combating acaricide resistance, thereby contributing to a more sustainable and holistic approach to pest control (Mascarin et al. 2024).

2.7 Summary and Conclusion

The varied collection of species known as fungal endophytes has a range of ecological and symbiotic roles. Here, we classified endophytes into four functional classes based on host range, colonization and transmission patterns, tissue specificity, and fitness benefits imparted by symbioses. Even though each class's effects on plants might vary greatly, the great majority of endophytes are still poorly understood. In any case, it is evident that these fungi can significantly affect plant survival and fitness in all terrestrial habitats, and as a result, they probably have a big influence on plant biogeography, evolution, and community structure. Endophyte and/or product commercialization can increase agricultural production and plant fitness. We still need to resolve a number of deadlocks pertaining to the fundamental mechanisms of recently introduced fungal endophytes (FE) into plants and eco-agrosystems, despite the market's growing demand for these biological goods. Understanding whether laboratory results are representative of the actual world, especially in harsh climates, will be aided by field applications.

References

Abd-Elgawad MM (2024) Upgrading strategies for managing nematode pests on profitable crops. Plan Theory 13(11):1558

Afzal I, Shinwari ZK, Sikandar S, Shahzad S (2019) Plant beneficial endophytic bacteria: mechanisms, diversity, host range and genetic determinants. Microbiol Res 221:36–49

Alsenani F, Alsufyani M, Taib M, Almalki W, Bdirah F (2023) A comprehensive review on phytoconstituents, bioactivities, and clinical studies on Ficus carica L. (Moraceae) and its role in human health and disease management. NVEO J 10(1):15–43

Anthony JM, Asthana M, Kumar A (2021) Mycotoxins and their consequences in livestock. In: Fungal diseases in animals: from infections to prevention. Springer International Publishing, Cham, pp 15–34

Armengol ES, Harmanci M, Laffleur F (2021) Current strategies to determine antifungal and antimicrobial activity of natural compounds. Microbiol Res 252:126867

Arnold AE, Lutzoni F (2007) Diversity and host range of foliar fungal endophytes: are tropical leaves biodiversity hotspots? Ecology 88(3):541–549

Bacon CW, White JF (2018) Stains, media, and procedures for analyzing endophytes. In Biotechnology of endophytic fungi of grasses, pp 47–56. CRC Press
Bagheri A, Askari Seyahooei M, Fathipour Y (2021) The endophytes. In: Microbial approaches for insect pest management. Springer, pp 151–215
Bairwa A, Dipta B, Mhatre PH, Venkatasalam EP, Sharma S, Tiwari R et al (2023) Chaetomium globosum KPC3: an antagonistic fungus against the potato cyst nematode, *Globodera rostochiensis*. Curr Microbiol 80(4):125
Baldrian P, Bell-Dereske L, Lepinay C, Větrovský T, Kohout P (2022) Fungal communities in soils under global change. Stud Mycol 103(1):1–24
Bard NW, Cronk QC, Davies TJ (2024) Fungal endophytes can modulate plant invasion. Biol Rev 99:1652
Barkodia M, Joshi U, Rami N, Wati L (2018) Endophytes: a hidden treasure inside plant. Int J Chem Stud 6(5):1660–1665
Bills GF, Gloer JB (2016) Biologically active secondary metabolites from the fungi. Microbiol Spectrum 4(6):10–1128
Bvenura C, Kambizi L (2022) Future grain crops. In: Future foods. Academic Press, pp 81–105
Caruso DJ, Palombo EA, Moulton SE, Zaferanloo B (2022) Exploring the promise of endophytic fungi: a review of novel antimicrobial compounds. Microorganisms 10(10):1990
Chandran H, Meena M, Swapnil P (2021) Plant growth-promoting rhizobacteria as a green alternative for sustainable agriculture. Sustainability 13(19):10986. https://doi.org/10.3390/su131910986
Chen L, Xu L, Li X, Wang Y, Feng Y, Huang G (2023) The diseases and pests of rubber tree and their natural control potential: a bibliometric analysis. Agronomy 13(8):1965
Choudhary N, Dhingra N, Gacem A, Yadav VK, Verma RK, Choudhary M et al (2023) Towards further understanding the applications of endophytes: enriched source of bioactive compounds and bio factories for nanoparticles. Front Plant Sci 14:1193573
Clay K (1988) Fungal endophytes of grasses: a defensive mutualism between plants and fungi. Ecology 69(1):10–16
Dhakshinamoorthy M, Packiam KK, Kumar PS, Saravanakumar T (2021) Endophytic fungus *Diaporthe caatingaensis* MT192326 from *Buchanania axillaris*: an indicator to produce biocontrol agents in plant protection. Environ Res 197:111147
Duperron S (2016) Microbial symbioses. Elsevier
El Mansy SM, Nouh FAA, Mousa MK, Abdel-Azeem AM (2020) Endophytic fungi: diversity, abundance, and plant growth-promoting attributes. Agriculturally Important Fungi for Sustainable Agriculture: Volume 1: Perspective for Diversity and Crop Productivity, pp 21–59
Florea S, Schardl CL, Hollin W (2015) Detection and isolation of *Epichloë* species, fungal endophytes of grasses. Curr Protoc Microbiol 38(1):19A–191A
Frank AC, Saldierna Guzmán JP, Shay JE (2017) Transmission of bacterial endophytes. Microorganisms 5(4):70
González V, Armijos E, Garcés-Claver A (2020) Fungal endophytes as biocontrol agents against the main soil-borne diseases of melon and watermelon in Spain. Agronomy 10(6):820
Grabka R, d'Entremont TW, Adams SJ, Walker AK, Tanney JB, Abbasi PA, Ali S (2022) Fungal endophytes and their role in agricultural plant protection against pests and pathogens. Plan Theory 11(3):384
Gray A E (2016) A molecular characterization of the fungal endophytes within the needles of ponderosa pine (Pinus ponderosa)
Grünig CR, Queloz V, Sieber TN, Holdenrieder O (2008) Dark septate endophytes (DSE) of the *Phialocephala fortinii* s.l.–*Acephala applanata* species complex in tree roots: classification, population biology, and ecology. Botany 86(12):1355–1369
Hamzazai A (2020) Factors shaping endophyte communities associated with selected cultivated plants in Arizona. Doctoral dissertation, The University of Arizona
Hardoim PR, Van Overbeek LS, Berg G, Pirttilä AM, Compant S, Campisano A et al (2015) The hidden world within plants: ecological and evolutionary considerations for defining functioning of microbial endophytes. Microbiol Mol Biol Rev 79(3):293–320

Hidayat I (2019) Dark septate endophytes and their role in enhancing plant resistance to abiotic and biotic stresses. Plant Growth Promoting Rhizobacteria for Sustainable Stress Management: Volume 1: Rhizobacteria in Abiotic Stress Management, pp 35–63

Hou QZ, Chen DW, Wang YP, Ehmet N, Ma J, Sun K (2022) Analysis of endophyte diversity of two *Gentiana* plants species and the association with secondary metabolite. BMC Microbiol 22(1):90

Islam T, Haque MA, Barai HR, Istiaq A, Kim JJ (2024) Antibiotic resistance in Plant pathogenic bacteria: recent data and environmental impact of unchecked use and the potential of biocontrol agents as an eco-friendly alternative. Plan Theory 13(8):1135

Jaber LR, Salem NM (2014) Endophytic colonisation of squash by the fungal entomopathogen *Beauveria bassiana* (Ascomycota: Hypocreales) for managing Zucchini yellow mosaic virus in cucurbits. Biocontrol Sci Tech 24(10):1096–1109

Kamran M, Imran QM, Ahmed MB, Falak N, Khatoon A, Yun BW (2022) Endophyte-mediated stress tolerance in plants: a sustainable strategy to enhance resilience and assist crop improvement. Cells 11(20):3292

Kharwar RN, Verma VC, Kumar A, Gond SK, Harper JK, Hess WM et al (2009) Javanicin, an antibacterial naphthaquinone from an endophytic fungus of neem, *Chloridium* sp. Curr Microbiol 58:233–238

Khruengsai S, Pripdeevech P, Tanapichatsakul C, Srisuwannapa C, D'Souza PE, Panuwet P (2021) Antifungal properties of volatile organic compounds produced by *Daldinia eschscholtzii* MFLUCC 19-0493 isolated from *Barleria prionitis* leaves against *Colletotrichum acutatum* and its post-harvest infections on strawberry fruits. PeerJ 9:e11242

Krishnaraj PU, Pasha MA (2017) Metagenome of rhizosphere and endophytic ecosystem. Understanding Host-Microbiome Interactions-An Omics Approach: Omics of Host-Microbiome Association, pp 125–156

Kumar R, Swapnil P, Meena M, Selpair S, Yadav BG (2022) Plant growth-promoting rhizobacteria (PGPR): approaches to alleviate abiotic stresses for enhancement of growth and development of medicinal plants. Sustain For 14(23):15514. https://doi.org/10.3390/su142315514

Kurepin LV, Zaman M, Pharis RP (2014) Phytohormonal basis for the plant growth promoting action of naturally occurring biostimulators. J Sci Food Agric 94(9):1715–1722

Liburd OE, Lopez L, Carrillo D, Revynthi AM, Olaniyi O, Akyazi R (2019) Integrated pest management of mites. In: Integrated management of insect pests. Burleigh Dodds Science Publishing, pp 841–886

Lugtenberg BJ, Caradus JR, Johnson LJ (2016) Fungal endophytes for sustainable crop production. FEMS Microbiol Ecol 92(12):fiw194

Mane RS, Vedamurthy AB (2018) The fungal endophytes: sources and future prospects. J Med Plant Stud 6:121–126

Mascarin GM, Golo PS, de Souza Ribeiro-Silva C, Muniz ER, de Oliveira Franco A, Kobori NN, Fernandes ÉKK (2024) Advances in submerged liquid fermentation and formulation of entomopathogenic fungi. Appl Microbiol Biotechnol 108(1):1–20

Meena M, Yadav G, Sonigra P, Nagda A, Mehta T, Swapnil P, Harish, Marwal A, Kumar S (2023) Multifarious responses of forest soil microbial community toward climate change. Microb Ecol 86:49–74. https://doi.org/10.1007/s00248-022-02051-3

Melnick RL (2010) Endophytic Bacillus spp. of Theobroma cacao: ecology and potential for biological control of cacao diseases

Mengistu AA (2020) Endophytes: colonization, behaviour, and their role in defense mechanism. Int J Microbiol:2020

Moosavi MR, Zare R (2020) Fungi as biological control agents of plant-parasitic nematodes. Plant defence: biological control. Springer, pp 333–384

Naranjo-Ortiz MA, Gabaldón T (2019) Fungal evolution: diversity, taxonomy and phylogeny of the fungi. Biol Rev 94(6):2101–2137

Nelson A, Vandegrift R, Carroll GC, Roy BA (2020) Double lives: transfer of fungal endophytes from leaves to woody substrates. PeerJ 8:e9341

Ngubane NP (2022) Fungal endophyte assemblages associated with twigs of olives in the Core Cape subregion, South Africa. Doctoral dissertation, Stellenbosch: Stellenbosch University

Oita S (2020) Foliar endophyte diversity and community structure: ecological drivers and methodological advances. Doctoral dissertation, The University of Arizona

Oleńska E, Małek W, Wójcik M, Swiecicka I, Thijs S, Vangronsveld J (2020) Beneficial features of plant growth-promoting rhizobacteria for improving plant growth and health in challenging conditions: a methodical review. Sci Total Environ 743:140682

Oukala N, Aissat K, Pastor V (2021) Bacterial endophytes: the hidden actor in plant immune responses against biotic stress. Plan Theory 10(5):1012

Patil RH, Patil MP, Maheshwari VL (2016) Bioactive secondary metabolites from endophytic fungi: a review of biotechnological production and their potential applications. In: Studies in Natural Products Chemistry. Elsevier, Amsterdam, Netherlands, 49:189–205

Popp J, Pető K, Nagy J (2013) Pesticide productivity and food security. a review. Agron Sustain Dev 33:243–255

Rai M, Rathod D, Agarkar G, Dar M, Brestic M, Pastore GM, Junior MRM (2014) Fungal growth promotor endophytes: a pragmatic approach towards sustainable food and agriculture. Symbiosis 62:63–79

Rana KL, Kour D, Kaur T, Devi R, Yadav AN, Yadav N et al (2020) Endophytic microbes: biodiversity, plant growth-promoting mechanisms and potential applications for agricultural sustainability. Antonie Van Leeuwenhoek 113:1075–1107

Ren A, Li C, Gao Y (2011) Endophytic fungus improves growth and metal uptake of *Lolium arundinaceum* Darbyshire ex. Schreb. Int J Phytoremed 13(3):233–243

Riva V, Mapelli F, Bagnasco A, Mengoni A, Borin S (2022) A meta-analysis approach to defining the culturable core of plant endophytic bacterial communities. Appl Environ Microbiol 88(6):e02537–e02521

Rodriguez RJ, White JF Jr, Arnold AE, Redman ARA (2009) Fungal endophytes: diversity and functional roles. New Phytol 182(2):314–330

Ross KM (2016) Resistance to African black beetle (*Heteronychus arator*) in perennial ryegrass (*Lolium perenne* L.) infected with AR1 endophyte. Doctoral dissertation, University of Waikato

Saikkonen K, Young CA, Helander M, Schardl CL (2016) Endophytic Epichloë species and their grass hosts: from evolution to applications. Plant Mol Biol 90:665–675

Sallam N, Ali EF, Seleim MA, Khalil Bagy HM (2021) Endophytic fungi associated with soybean plants and their antagonistic activity against *Rhizoctonia solani*. Egypt J Biol Pest Control 31:1–9

Schouten A (2016) Mechanisms involved in nematode control by endophytic fungi. Annu Rev Phytopathol 54(1):121–142

Segaran G, Sathiavelu M (2019) Fungal endophytes: A potent biocontrol agent and a bioactive metabolites reservoir. Biocatal Agric Biotechnol 21:101284

Sena L, Mica E, Valè G, Vaccino P, Pecchioni N (2024) Exploring the potential of endophyte-plant interactions for improving crop sustainable yields in a changing climate. Front Plant Sci 15:1349401

Seth K, Vyas P, Deora S, Gupta AK, Meena M, Swapnil P, Harish. (2023) Understanding plant-plant growth-promoting rhizobacteria (PGPR) interactions for inducing plant defense. In: Swapnil P, Meena M, Harish, Marwal A, Vijayalakshmi S, Zehra A (eds) Plant-Microbe Interaction—Recent Advances in Molecular and Biochemical Approaches, vol 2. Academic Press, Elsevier, pp 201–226. https://doi.org/10.1016/B978-0-323-91876-3.00010-5

Sieber TN (2002) Fungal root endophytes. In: Plant roots. CRC Press, pp 1369–1418

Sreenayana B, Vinodkumar S, Nakkeeran S, Muthulakshmi P, Poornima K (2022) Multitudinous potential of *Trichoderma* species in imparting resistance against *F. oxysporum* f. sp. *cucumerinum* and *Meloidogyne incognita* disease complex. J Plant Growth Regul 41(3):1187–1206

Suebrasri T, Somteds A, Harada H, Kanokmedhakul S, Jogloy S, Ekprasert J et al (2020) Novel endophytic fungi with fungicidal metabolites suppress sclerotium disease. Rhizosphere 16:100250

Tadych M, Bergen MS, White JF Jr (2014) *Epichloë* spp. associated with grasses: new insights on life cycles, dissemination and evolution. Mycologia 106(2):181–201

Terhonen E, Blumenstein K, Kovalchuk A, Asiegbu FO (2019) Forest tree microbiomes and associated fungal endophytes: functional roles and impact on forest health. Forests 10(1):42

Uzma F, Mohan CD, Siddaiah CN, Chowdappa S (2019) Endophytic fungi: promising source of novel bioactive compounds. In: Advances in Endophytic Fungal Research: Present Status and Future Challenges. Springer Cham, Switzerland, pp 243–265

Yadav G, Meena M (2021) Bioprospecting of endophytes in medicinal plants of Thar Desert: an attractive resource for biopharmaceuticals. Biotechnology Reports 30:e00629. https://doi.org/10.1016/j.btre.2021.e00629

Yadav G, Meena M (2022) Biological control of plant diseases by endophytes. In: Shah MP, Deka D (eds) Endophytic Association: What, Why and How (Volume: Developments in Applied Microbiology and Biotechnology). Academic Press, Elsevier Inc., pp 119–135. https://doi.org/10.1016/B978-0-323-91245-7.00007-9

Yan L, Zhu J, Zhao X, Shi J, Jiang C, Shao D (2019) Beneficial effects of endophytic fungi colonization on plants. Appl Microbiol Biotechnol 103:3327–3340

Yang T, Xiong C, Zhou J, Zhang W, Qian X (2023) Phyllosphere mycobiome: diversity and function. In: Plant Mycobiome: Diversity, Interactions and Uses. Springer International Publishing, Cham, pp 63–120

Endophytic Potassium-Solubilizing Microbes: A Sustainable Approach to Enhancing Crop Nutrition and Productivity

3

Alapati Prema Siva Naga Teja and Baljeet Singh Saharan

Abstract

Potassium (K) is an essential macronutrient for plant growth, yet its limited availability in soils poses a significant challenge to agricultural productivity. Despite its abundance, only 1–2% of soil K is accessible for plant uptake, with the rest locked in mineral forms. While K fertilizers address these deficiencies, their excessive use raises environmental and economic concerns. Sustainable alternatives, such as potassium-solubilizing microbes (KSMs), offer a promising solution by enhancing K bioavailability through mineral dissolution and organic acid production. Microbial species like *Bacillus mucilaginosus* not only release K but also improve soil health by solubilizing additional nutrients. This chapter emphasizes the potential of KSMs to reduce dependence on chemical fertilizers, promote eco-friendly farming, and support sustainable food production systems.

Keywords

Potassium solubilizing microbes · Sustainable agriculture · Nutrient bioavailability · Soil health · Eco-friendly farming

3.1 Introduction

Potassium (K) is a vital macronutrient essential for the growth and development of plants, ranking third in importance after major nutrients like nitrogen (N) and phosphorus (P). As it is a crucial element in the photosynthetic tissues of terrestrial

A. P. S. N. Teja · B. S. Saharan (✉)
Department of Microbiology, Chaudhary Charan Singh Haryana Agricultural University, Hisar, India
e-mail: Baljeetsaharan@hau.ac.in

M. Meena et al. (eds.), *Metabolic Sustainability of Endophytes*, Microorganisms for Sustainability 52, https://doi.org/10.1007/978-981-96-4004-1_3

plants and is particularly important in certain species, such as banana and cotton, which require more K than P or N (Mora et al. 2012). Despite the presence of significant amounts of K in soils, only about 1–2% is readily available for plant uptake, as most is bound to minerals. This limited availability often leads to potassium deficiencies, which pose a major challenge to agricultural production (Meena et al. 2015). Current agricultural practices, including intensive farming and the use of high-yield crop varieties, have significantly depleted soil K reserves. Factors such as leaching, runoff, and soil erosion further exacerbate this issue, making K deficiency a major concern in the coarse-textured soils (Olaniyan et al. 2022). To combat this deficiency, potassium fertilization is recommended, as crops typically respond positively to its application. The highest consumption of potassium fertilizers occurs in countries like the United States, China, Brazil, and India, with common forms including muriate of potash and sulfate of potash. However, the excessive use of these fertilizers raises environmental and economic concerns (Ding et al. 2021). To address these challenges, there is a pressing need for sustainable and cost-effective methods to enhance potassium bioavailability in soils while reducing fertilizer dependence. Recent research highlights the role of potassium-solubilizing microorganisms (KSMs), including arbuscular mycorrhizal fungi and specific bacteria, as effective alternatives for meeting crop K requirements (Boubekri et al. 2021). KSMs have been recognized for their ability to release potassium from soil reserves, thereby reducing the necessity for additional fertilizers (Teotia et al. 2017). Recent findings by (Singh et al. 2023) indicate that KSBs can extract not only potassium but also silicon and aluminum from K insoluble aluminosilicate minerals, which is essential for promoting sustainable agricultural practices. By managing KSB populations, reliance on chemical fertilizers can be reduced, promoting environmental sustainability. These bacteria enhance mineral dissolution rates through microbial respiration and organic matter decomposition, producing carbonic acid that aids in nutrient availability. For instance, *Bacillus* species, such as *B. mucilaginosus*, optimize K and SiO_2 solubilization by releasing organic acids during decomposition. The significance of KSB lies in their ability to increase soluble K concentrations in the soil, which is essential for various physiological processes in plants, including enzyme activation, photosynthesis, and stress tolerance. By leveraging the K-solubilizing capabilities of these microorganisms, it is possible to reduce dependence on chemical K fertilizers and foster sustainable soil management practices, ultimately contributing to eco-friendly agricultural systems. In summary, potassium is a critical nutrient for plant health and productivity, and addressing its deficiency through innovative and sustainable practices, including the use of KSMs, is vital for meeting the increasing global demand for crop production (Fig. 3.1).

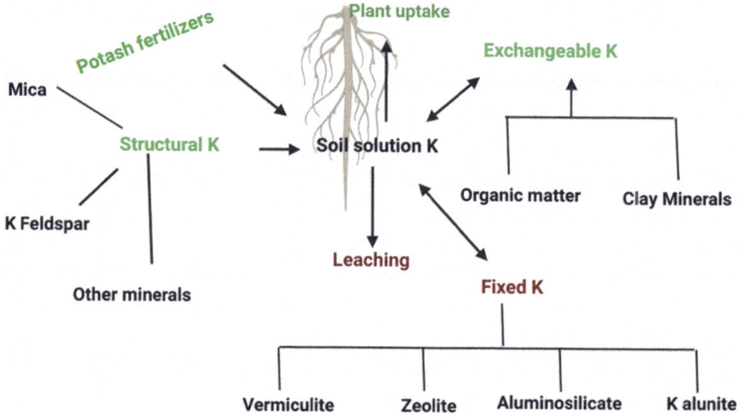

Fig. 3.1 The interplay between various forms of potassium found in the soil

3.2 Soil Plant and Microbes' Interaction

Soil fertility is a measure of the soil's capacity to supply essential nutrients, such as N, P, and K, which are limited in their availability. Beneficial microbes associated with plant roots play a considerable role in enhancing the accessibility of those nutrients to plants (de Zelicourt et al. 2013). The soil is home to a diverse array of microorganisms, including bacteria, protozoa, fungi, algae, and actinomycetes. Bacteria, being the most prevalent, account for approximately 95% of the microbial population. It is estimated that the soil harbors around 60,000 bacterial species, many of which remain unidentified and perform specific functions. The diversity and abundance of these bacteria are influenced by various soil factors, such as soil moisture, organic carbon content, types of plants present and temperature (Glick 2012; Seth et al. 2023). Recent study indicates that many plant species rely on microbial interactions for survival. Plants have the ability to selectively choose their root-associated microorganisms, resulting in distinct microbial communities. Effective plant-microbe interactions need reciprocal recognition and coordinated responses from both plants and microbes (de Zelicourt et al. 2013). Plants communicate with rhizosphere microorganisms by releasing chemical signals, while microbes can trigger plant responses that promote symbiosis (Ma et al. 2016). The interactions between plants and bacteria can be beneficial, harmful, or neutral, with the effects varying depending on soil conditions (Cheng et al. 2010). Beneficial bacteria include nitrogen-fixing nodules, endophytic bacteria that infiltrate plant tissues without harming them, and bacterial species that live freely in the soil (Glick 2012; Chandran et al. 2021; Kumar et al. 2022). Beneficial bacteria in agricultural contexts are those that promote plant development by increasing seed emergence, plant weight, and crop output. Despite our inadequate understanding of soil bacterium-plant interactions, many of these beneficial microorganisms are used commercially in agriculture. Prominent examples include *Burkholderia cepacia* and *Bacillus* spp.

3.3 Mechanism Involved in Potassium (K) Solubilization

Potassium-solubilizing microbes (KSMs) employ a range of strategies to release potassium (K) from silicate minerals within the soil (Table 3.1). One of the primary methods involves generating low-molecular-weight organic acids (OAs) that reduce the pH of the soil (Table 3.2). This acidification, in turn, protonates K-containing minerals, making potassium more soluble (Olaniyan et al. 2022). Additionally, these organic acids can chelate other cations such as Si^{4+}, Mg^{2+}, and Ca^{2+}, indirectly promoting potassium release and increasing its availability to plants (Keshavarz et al. 2013). In soils with low potassium levels, microbial communities often secrete extracellular polymers and macromolecules, including proteins and polysaccharides, which contribute to biofilm formation on mineral surfaces. This biofilm facilitates mineral degradation and bio-weathering, thereby enhancing the availability of bioavailable potassium (Jini et al. 2024). Additionally, potassium solubilization may involve redox reactions, where KSMs engage in electron transfer to metal groups present on mineral surfaces, leading to the breakdown of metal complexes. This electron transfer process is often facilitated by multi-heme cytochromes found in bacterial membranes (Sattar et al. 2019). Various factors, including mineral composition, nutrient availability, and environmental conditions, influence the specific solubilization mechanisms employed by KSMs (Kour et al. 2020) (Fig. 3.2).

3.4 Production of Organic Acids

The solubilization of potassium (K) by microorganisms involves the secretion of organic acids that lower the pH of the rhizosphere, thereby facilitating the release of K from minerals. These organic acids, produced by various bacterial species, promote the dissolution of potassium-bearing minerals such as phlogopite, feldspar, and mica through mechanisms like protonation, chelation, and acidolysis (Kour et al. 2020). For example, potassium-solubilizing rhizobacteria can release

Table 3.1 Mechanisms used different microbes for the solubilization of potassium

Microorganism	Method of solubilization of K	Reference
Sphingomonas, Burkholderia	Acidification, complexation	Uroz et al. (2007)
Bacillus circulans	GY92 lipo-chito oligosaccharids production	Lian et al. (2001)
Paenibacillus glucanolyticus	Organic acids	Sangeeth et al. (2012)
Bacillus mucilaginosus	Acidification, illite solubilization by producing IAA, mica through organic acids	Han and Lee (2005), Basak and Biswas (2009), Abdel-Salam and Shams (2012)
Burkholderia glathei	Siderophores, organic ligands	Calvaruso et al. (2007)
Pseudomonas sp.	Excretion of organic acids	Sheng and Huang (2002)

Table 3.2 Organic acids produced by different type of microbes

Microorganism	Acid produced
Aspergillus terreus and *Aspergillus niger*	Citric, oxalic acids
Bacillus edaphicus	Citric, oxalic, tartaric, succinic, α-ketogluconic acids
B. megaterium, *Pseudomonas* sp.	Lactic, malic, oxalic acids
B. megaterium	Citric acid
B. mucilaginosus	Carboxylic acid, citric, oxalic, tartaric acids
Enterobacter hormaechei	Oxalic, citric acids
Paenibacillus mucilaginosus	Citric, oxalic, tartaric acids
Pseudomonas aeruginosa	Acetic, citric, oxalic acids
Rhizobium tropici strain Q34	Succinic, tartaric, citric acids

Source: Olaniyan et al. (2022)

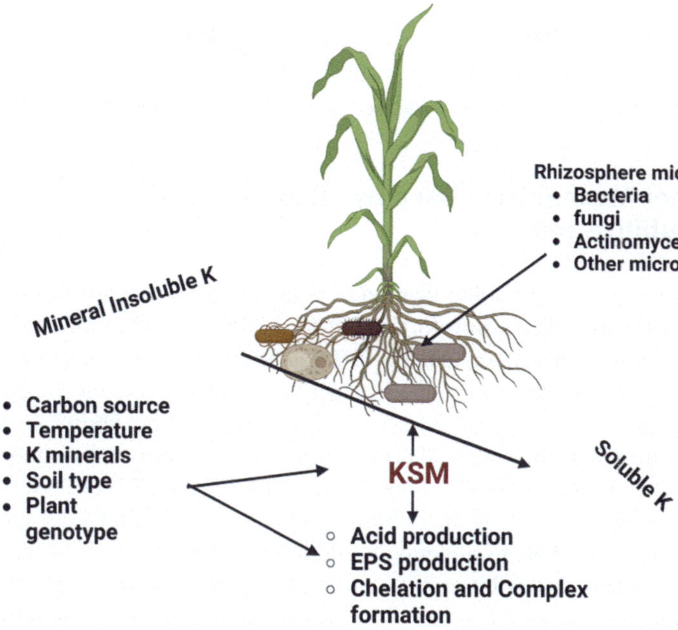

Fig. 3.2 Mechanism involved in K solubilization

potassium ions from phlogopite by chelating aluminum, silicon, and iron, which helps to break down the mineral structure (Etesami et al. 2017). Additionally, *Pseudomonas* and *Bacillus* strains have been shown to produce organic acids such as citric, lactic, and oxalic acids, which effectively solubilize potassium in both liquid cultures and soil environments (Chen et al. 2022) (Fig. 3.3). In alkaline soils, these organic acids also aid in the solubilization of phosphate, further improving nutrient availability (Etesami and Adl 2020). Overall, the production of organic acids by microorganisms plays a significant role in converting insoluble potassium minerals into forms that plants can readily absorb (Yang et al. 2023) (Table 3.2).

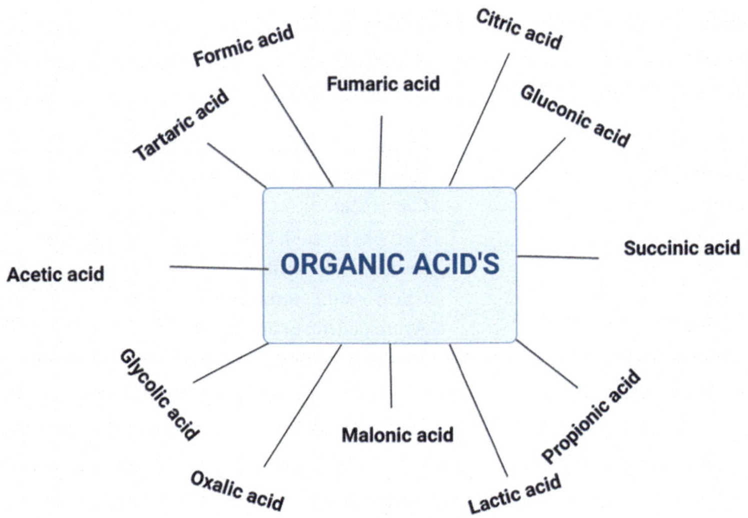

Fig. 3.3 Different types of organic acids produced by microbes in K solubilization

3.5 Other Mechanisms Used by Microbes for K Solubilization

Microorganisms assist in the solubilization of potassium (K) minerals through various indirect pathways. According to Uroz et al. (2009), Sattar et al. (2019), and Basak (2019), these pathways include the direct adhesion to the surface of minerals, chelation of cations that are connected to K silicate, and metal-complexing ligand production, respectively. Efficient KSMs have demonstrated the ability to weather minerals like phlogopite through acid dissolution and aluminum chelation, breaking down the crystal structure (Abou-el-Seoud and Abdel-Megeed 2012). Additionally, microbial metabolic byproducts, including organic ligands, extracellular enzymes, and chelates, also contribute to mineral dissolution by altering the pH of the microenvironment (Sasse et al. 2018; Sindhu et al. 2022). *Priestia megaterium*'s ability to release K has recently been related to extracellular potassium feldspar-binding proteins, including enolase, elongation factor TU which possess potassium feldspar-binding sites (Wu et al. 2023). Another study showed that *Priestia aryabhattai* SK1–7 upregulated antioxidant genes in response to reactive oxygen species (ROS) stress induced by acid production, facilitating K solubilization. The expression of the *maeA-1* gene, encoding malic enzyme, was particularly important as it increased pyruvate production, which scavenged ROS and enhanced potassium feldspar dissolution, simultaneously increasing aluminum and silicon release (Yang et al. 2023).

3.6 Diversification of Microbes Involved in K Solubilization

Soil microorganisms play a vital role in transforming fixed and insoluble potassium into forms that plants can readily absorb (Gundala et al. 2013). Microbial inoculants that solubilize K from minerals and rocks provide significant economic and environmental advantages, promoting plant growth and increasing yields. The first documented instance of microbial involvement in rock potassium solubilization dates to Muentz in 1890 (Muentz 1890). A variety of KSMs, including *Bacillus mucilaginosus*, *Acidithiobacillus ferrooxidans*, *Bacillus circulans*, *Pseudomonas*, and *Burkholderia*, have been identified as effective agents in releasing K from K-bearing minerals (Basak and Biswas 2012). Both fungal and bacterial microorganisms contribute significantly to plant growth by mobilizing insoluble K forms, with their populations varying across different soil types. Rhizosphere microorganisms are particularly important for solubilizing bound soil minerals (Sindhu et al. 2009). Many soil microbes, including those found on mountain rocks, are capable of solubilizing silicate minerals (Gundala et al. 2013). Notably, the subspecies *B. mucilaginosus siliceus* is recognized for its ability to extract K from aluminosilicates and feldspar. This bacterium, along with others like *B. mucilaginosus* strain CS1, has been isolated from both rhizosphere and non-rhizosphere soils, exhibiting inhibitory effects on gram-negative bacteria (Lin et al. 2002). Furthermore, K-solubilizing bacteria (KSB) have been discovered near the roots of several cereal crops which are grown in K and silicate-rich soils (Mikhailouskaya and Tcherhysh 2005). A variety of KSMs, including bacteria, fungi, and actinomycetes, have been documented. Key bacterial genera include *Achromobacter*, *Azospirillum*, *Bacillus*, *Frateuria*, *Klebsiella*, *Pantoea*, *Microbacterium*, *Myroides*, *Paenibacillus*, *Leclercia*, and *Raoultella* (Liu et al. 2020; Baghel et al. 2020) (Table 3.3). These microorganisms produce enzymes that help solubilize K in the soil. Fungal genera involved in this process include *Aspergillus*, *Cladosporium*, *Glomus*, *Penicillium*, and *Rhizophagus* (Lopes-Assad et al. 2010; Islam et al. 2007). Potassium-solubilizing actinomycetes are represented by genera such as *Actinomyces*, *Streptomyces*, and *Streptoverticillium* (Kumar et al. 2018). Furthermore, silicate-solubilizing bacteria have been identified in both rhizosphere and non-rhizosphere soils associated with various crops (Liu et al. 2020).

3.7 Isolation and Identification of KSM's

The Aleksandrov agar medium is widely utilized for isolating and quantifying potassium-solubilizing microorganisms (KSMs). This medium contains essential components such as glucose, magnesium sulfate, ferric chloride, calcium carbonate, calcium phosphate, K-bearing minerals, and agar, following a formulation by Aleksandrov et al. (1967). It was inspired by the Pikovskaya agar plate method for phosphorus solubilizers, a similar plate assay was developed for KSM isolation, where the formation of visible halos around colonies indicates K solubilization. Colonies with clear zones are selected after incubation. To improve the efficiency of

Table 3.3 Microorganisms which are capable of solubilizing K

Microorganism	Niche/Crop
Bacteria	
Azospirillum spp.	Wheat
Achromobacter piechaudii	Maize, sorghum, wheat
Azotobacter chroococcum, *Rhizobium* sp. *Bacillus mucilaginosus*	Maize, wheat
Bacillus sp.	*Triticum*, cotton, *Zea mays*, *Solanum melongena*, and *Brassica napus*
Klebsiella variicola, *Enterobacter asburiae*, *E. cloacae*, *Agrobacterium tumefaciens*, *E. aerogenes*, *Pantoea agglomerans*, *Bacillus circulans*	Tobacco
Bacillus subtilis, *Bacillus megaterium*	Maize
Bacillus methylotrophicus, *Bacillus* spp.	Elephant foot yam
Bacillus mucilaginous	Rice
Burkholderia pyrrocinia, *Burkholderia ubonensis*	Sorghum, tomato, groundnut Bamboo
Fungi	
Rhizophagus irregularis, *Glomus mosseae*	Maize
Aspergillus niger	Sugarcane
Cladosporium herbarum	Wheat
Glomus fasciculatum	Mulberry
A. flavus, *Aspergillus* sp., *A. nidulans*, *A. terreus*, *A. niger*, *Penicillium lilacinum*	Legumes
Actinomycetes	
Streptomyces rochei, *Streptomyces sundarbansensis*	Bitter vine
Streptoverticillium album	Soil

Source: Gautam et al. (2022)

this screening process, researchers recently optimized the method by adding acid-base indicators like phenol red and bromocresol purple to the medium (Rajawat et al. 2016). Some drawbacks are seen in the plate assay method when the halo is faint or absent, necessitating the use of broth assays for direct K solubilization measurement. In these cases, K release is quantified using flame photometry (Boubekri et al. 2021).

Additionally, internal transcribed spacers and 16S rRNA are being utilized to evaluate the genetic diversity of KSMs (Wang et al. 2015). KSMs are typically more concentrated near the roots than in soil far away from the roots soil (Gandhi et al. 2022). These bacteria are isolated using a serial dilution plate technique, as the normal method has some drawbacks, with the usage of modified Aleksandrov medium, as described by Parmar and Sindhu (2013). After incubation, the K solubilization zones are measured, and the morphological characteristics of the isolates are analyzed using optical microscopy after phenol red staining. Physiological characteristics such as organic acid utilization, anaerobic growth, and carbohydrate metabolism are evaluated using methods established by Claus and Berkeley (1986). Furthermore, studies by Subhashini and Kumar (2014) examined the isolates' substrate utilization patterns and their tolerance to variations in salt concentration, temperature, and pH.

3.8 The Role of KSMs in Sustainable Soil Management and Crop Production

Maintaining soil health is essential for sustainable crop production, especially as intensive farming practices often deplete soil nutrients. Implementing an integrated nutrient management system, as suggested by Vlek and Vielhauer (1994), can enhance soil nutrient availability. Biofertilizers play a crucial role in this system by reducing the need for chemical fertilizers, which can be harmful to both plant and soil health (Ghorbani et al. 2008). Economically and environmentally friendly, biofertilizers introduce beneficial microorganisms into the soil, enhancing nutrient content and protecting plants from various abiotic and biotic stresses (Nadeem et al. 2013). Solubility of native nutrients present in the soil can be made available to plants by the application of biofertilizers. They help decrease reliance on chemical fertilizers while maintaining crop yields (Habibi et al. 2011). Although potassium-rich rocks are a cost-effective source of potassium, their slow release limits their agricultural use (Zapata and Roy 2004). However, potassium-solubilizing microorganisms (KSMs), such as *Bacillus mucilaginosus*, can significantly increase soil potassium levels. While simply applying potassium and phosphorus rocks may not significantly enhance soil nutrient content, the use of KSMs has been shown to improve the availability of these elements. For instance, (Supanjani et al. 2006) demonstrated that organic fertilizers containing potassium and phosphorus solubilizers positively impact the growth and yield of crops like peppers. Additionally, combining rock phosphate with biofertilizers increases soil potassium content, as observed by Abou-el-Seoud and Abdel-Megeed (2012). The application of KSMs, such as *Frateuria aurantia* and fungi like *Aspergillus terreus*, has also been shown to improve plant growth by increasing potassium availability in the soil (Prajapati et al. 2013; Patel 2011). Moreover, El-Haddad et al. (2011) found that biofertilizers containing nitrogen-fixing bacteria and phosphorus-solubilizing bacteria significantly reduced nematode infestations, while Wu et al. (2005) reported enhanced corn growth with KSM use. Beyond promoting plant growth, biofertilizers contribute to soil health, as Dasan (2012) demonstrated with *Frateuria aurantia*, which also shows potential as a biological control agent against plant pathogens.

However, the effectiveness of biofertilizers can be influenced by factors such as short shelf life, storage temperature, and packaging quality (Verma et al. 2011). To address these challenges, the support material used in biofertilizer preparation is critical for maintaining viability until field application, with vehicle-based formulations offering longer shelf lives (Brar et al. 2012). Bhattacharyya and Kumar (2000) highlighted that such formulations can retain their quality for up to 6 months. Liquid broth formulations, in particular, have shown high efficacy for potassium-solubilizing biofertilizers (Liu et al. 2012). Utilizing KSMs as biofertilizers not only enhances agricultural productivity but also promotes environmental sustainability, as noted by Sindhu et al. (2010). Bacteria used in biofertilizer production can be easily isolated and cultivated for agricultural applications, offering a sustainable alternative to chemical fertilizers, which can degrade soil health and reduce microbial diversity. For example, Lin et al. (2002) reported a significant increase in K solubility in tomato crops treated with silicate-dissolving bacteria like *Bacillus mucilaginosus*. This treatment led to a 36.6% increase in biomass compared to control groups (Basak and Biswas 2009). Similarly, *Frateuria aurantia* has been shown to significantly boost crop yields (Ramarethinam and Chandra 2005).

3.9 Utilizing Potassium-Solubilizing Microorganisms for Improved Nutrient Management in Agriculture

K is crucial for plant growth, playing a key role in processes such as enzyme activation, turgor pressure maintenance, sugar translocation, and disease resistance (Zhang et al. 2011). Potassium deficiency leads to symptoms like poor root development, slow growth, increased disease susceptibility, delayed maturity, and reduced crop yields. While most soil potassium exists in insoluble forms within minerals like feldspar and mica, biofertilizers provide a sustainable solution to enhance its availability (Zhang and Kong 2014). Biofertilizers, which can be applied to seeds, roots, or soil, improve nutrient mobility, particularly nitrogen, phosphorus, and potassium (NPK), while reducing nutrient leaching and being more cost-effective and environmentally friendly compared to chemical fertilizers (Aseri et al. 2008).

The use of potassium-solubilizing microorganisms (KSMs) as biofertilizers has proven effective in reducing chemical fertilizer use and increasing crop yields. For instance, Meena et al. (2016) showed that KSM inoculation increased potassium availability in soil from 84.8% to 127.9%. Similarly, Singh et al. (2010) found that using waste mica inoculated with *Azotobacter chroococcum* and *Bacillus mucilaginosus* enhanced potassium assimilation in wheat and maize. This bacterial inoculation also boosted nitrogen and phosphorus content in plants (Subhashini 2015). Another study reported that tomato plants inoculated with *Bacillus mucilaginosus* exhibited a 125% increase in biomass and over a 150% increase in potassium and phosphorus uptake (Etesami et al. 2017). Positive results have also been seen in sorghum, Sudan grass, and potatoes (AbdelSalam and Shams 2012). While bacteria have been the primary focus of research, fungi are also being recognized as effective potassium solubilizers. For example, *Penicillium pinophilum* significantly increased pomegranate yield by 116.9% when combined with potassium feldspar (Maity et al. 2019). Additionally, the yeast *Torulaspora globosa* and the ectomycorrhizal fungus *Piloderma* spp. have demonstrated the ability to release potassium from minerals, enhancing plant potassium nutrition (Rosa-Magri et al. 2012; Glowa et al. 2003).

Beyond potassium solubilization, KSMs provide additional benefits by improving the growth of plant by secretion of some phytohormones and additional PGP activities like indole-3-acetic acid, gibberellins and cytokinins (Soumare et al. 2021), solubilization of P (Soumare et al. 2020a), nitrogen fixation (Soumare et al. 2020b), biological control of pathogens (Gopalakrishnan et al. 2020), and ACC production (Boubekri et al. 2021). Acids produced by KSMs can also reduce heavy metal toxicity (Archana et al. 2012) and indirectly enhance potassium solubilization by generating strong acids like H_2SO_4 and HNO_3, which further dissolve potassium-bearing minerals.

Overall, potash biofertilizers are highly effective in promoting nutrient uptake, managing plant stress, extending crop shelf life, maintaining soil pH, improving plant physiological responses, and enhancing soil texture and fertility. The steps involved in the selection and development of K biofertilizer are illustrated in Fig. 3.4.

3.10 Challenges and Future Perspectives

Potash biofertilizers offer a natural and sustainable alternative to chemical fertilizers, helping to meet the growing demand for sustainable agriculture by reducing production costs, simplifying field applications, and minimizing the reliance on chemical inputs. These biofertilizers enhance plant absorption of potassium (K) from the soil, improving overall plant health and productivity through various mechanisms, including soil pH maintenance, nutrient uptake, extending crop shelf life, and stress management.

Research on KSB and actinomycetes has shown their applicability across a range of plants. However, further studies are needed to explore the ecological niches of other plants and their interactions with microbes. It may lead to the development of more efficient K biofertilizers, a better understanding of microbial interactions in agriculture. Currently, only a limited number of KSM-based biofertilizers are available on the market, with most being derived from *Frateuria* species. Examples include Symbion-K, KSoilB®, GreenK, ABTEC Bio-Potash, UPTAKE, and ADVEN, based on *Frateuria aurantia* or related species (Kore et al. 2020). Another product, POTAK, is a liquid formulation based on *Bacillus licheniformis*. Despite the potential of these biofertilizers, they represent a small portion of the global biofertilizer market, where nitrogen (N) and phosphorus (P) biofertilizers dominate, accounting for 79% and 14%, respectively (Soumare et al. 2020b).

The limited use and study of KSMs in agriculture may be due to their slower impact on crop yields and the lack of awareness and knowledge among farmers about the benefits of biological K fertilizers. To enhance the effectiveness and adoption of KSMs, further research is necessary, particularly in areas such as genetic screening of genotypes of plants for their capacity to produce organic acids to utilize soil's insoluble form of K. Recent studies suggest that beneficial microbes can be attracted to hormones or exudates released by plants so we can manipulate plants to produce them (Arif et al. 2020). However, the molecular dialogue between plants and microbes remains poorly understood, highlighting the need for more research in proteomics and metabolomics to optimize plant-microbe interactions (Batista and Singh 2021).

Advancing the engineering of K-solubilizing genes in efficient microbes with strong colonization abilities, as well as developing plants that more effectively

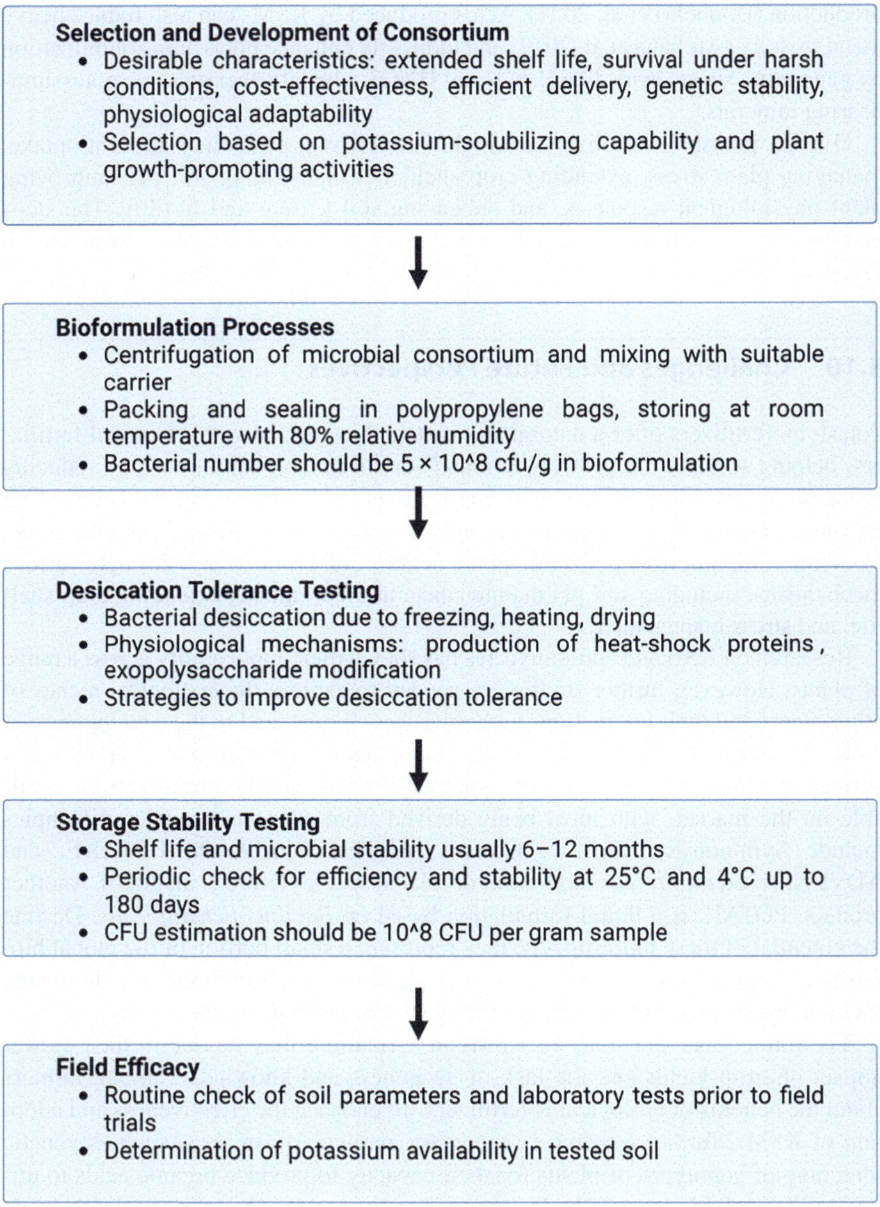

Fig. 3.4 Steps involved in selection and development of potash biofertilizer

utilize K, could be valuable goals for improving agronomic outcomes. Metagenomic approaches can be utilized to screen for those genes which play a key role in acid synthesis as well as release, which could also be promising (Sindhu et al. 2016). Understanding the function of root-associated microbiomes and their practical applications in natural conditions is crucial (Arif et al. 2020). Given the similarity between K and P solubilization processes, developing new slow-release K and P fertilizers using solubilizing microorganisms could offer agronomic and economic benefits (Supanjani et al. 2006; Boubekri et al. 2021).

References

Abdel-Salam MA, Shams AS (2012) Feldspar-K fertilization of potato (*Solanum tuberosum* L.) augmented by biofertilizer. J Agric Environ Sci 12(6):694–699

Abou-el-Seoud II, Abdel-Megeed A (2012) Impact of rock materials and biofertilizations on P and K availability for maize (*Zea mays*) under calcareous soil conditions. Saudi J Biol Sci 19:55–63

Aleksandrov VG, Blagodyr RN, Ilev IP (1967) Liberation of phosphoric acid from apatite by silicate bacteria. Microchem J 29:111–114

Archana DS, Nandish MS, Savalagi VP, Alagawadi AR (2012) Screening of potassium solubilizing bacteria (KSB) for plant growth promotional activity. Bioinfolet 9(4):627–630

Arif I, Batool M, Schenk PM (2020) Plant microbiome engineering: expected benefits for improved crop growth and resilience. Trends Biotechnol 38:1385–1396

Aseri GK, Jain N, Panwar J, Rao AV, Meghwalc PR (2008) Biofertilizers improve plant growth, fruit yield, nutrition, metabolism and rhizosphere enzyme activities of pomegranate (*Punica granatum* L.) in Indian. Thar Desert Sci Hortic 117(2):130–135

Baghel V, Thakur JK, Yadav SS, Manna MC et al (2020) Phosphorus and potassium solubilization from rock minerals by endophytic *Burkholderia* sp. strain FDN2-1 in soil and shift in diversity of bacterial endophytes of corn root tissue with crop growth stage. Geomicrobiol J 37(6):550–563. https://doi.org/10.1080/01490451.2020.1734691

Basak BB (2019) Waste mica as alternative source of plant-available potassium: evaluation of agronomic potential through chemical and biological methods. Nat Resour Res 28:953–965. https://doi.org/10.1007/s11053-018-9430-3

Basak BB, Biswas DR (2012) Modification of waste mica for alternative source of potassium: evaluation of potassium release in soil from waste mica treated with potassium solubilizing bacteria (KSB). LAP LAMBERT Academic Publishing

Basak BB, Biswas DR (2009) Influence of potassium solubilizing microorganism (*Bacillus mucilaginosus*) and waste mica on potassium uptake dynamics by Sudan grass (*Sorghum vulgare* Pers.) grown under two Alfisols. Plant Soil 317:235–255. https://doi.org/10.1007/s11104-008-9805-z

Batista BD, Singh BK (2021) Realities and hopes in the application of microbial tools in agriculture. Microb Biotechnol 14:1258–1268

Bhattacharyya P, Kumar R (2000) Liquid biofertilizer current knowledge and future prospect national seminar on development and use of biofertilizers, biopesticides and organic manures. Bidhan Krishi Viswavidyalaya, Kalyani, pp 10–12

Boubekri K, Soumare A, Mardad I, Lyamlouli K, Hafidi M, Ouhdouch Y, Kouisni L (2021) The screening of potassium- and phosphate-solubilizing actinobacteria and the assessment of their ability to promote wheat growth parameters. Microorganisms 9(3):470. https://doi.org/10.3390/microorganisms9030470

Brar SK, Sarma SJ, Chaabouni E (2012) Shelf-life of biofertilizers: an accord between formulations and genetics. J Biofertil Biopesticide 3:1–2

Calvaruso C, Turpault MP, Leclerc E, Frey-Klett P (2007) Impact of ectomycorrhizosphere on the functional diversity of soil bacterial and fungal communities from a forest stand in relation to nutrient mobilization processes. Microb Ecol 54:567–577

Chandran H, Meena M, Swapnil P (2021) Plant growth-promoting rhizobacteria as a green alternative for sustainable agriculture. Sustain For 13(19):10986. https://doi.org/10.3390/su131910986

Chen Y, Yang H, Shen Z, Ye J (2022) Whole-genome sequencing and potassium-solubilizing mechanism of *Bacillus aryabhattai* SK1-7. Front Microbiol 12:722379. https://doi.org/10.3389/fmicb.2021.722379

Cheng Z, McConkey BJ, Glick BR (2010) Proteomic studies of plant–bacterial interactions. Soil Biol Biochem 42:1673–1684

Claus D, Berkeley CW (1986) The genus *Bacillus*. In: PHA Sneath (ed) Bergey's manual of systematic bacteriology, vol 2. Williams, Wilkins. Balt For 34:1105–1139

Dasan AS (2012) Compatibility of agrochemicals on the growth of phosphorous mobilizing bacteria Bacillus megaterium var. phosphaticum potassium mobilizing bacteria *Frateuria aurantia*. Appl Res Dev Inst J 6:118–134

de Zelicourt A, Al-Yousif M, Hirt H (2013) Rhizosphere microbes as essential partners for plant stress tolerance. Mol Plant 6:242–245

Ding Z, Ali EF, Almaroai YA, Eissa MA, Abeed AHA (2021) Effect of potassium solubilizing bacteria and humic acid on faba bean (*Vicia faba* L.) plants grown on sandy loam soils. J Soil Sci Plant Nutr 21:791–800. https://doi.org/10.1007/s42729-020-00401-z

El-Haddad ME, Mustafa I, Selim SM, El-Tayeb TS, Mahgoob AE, Abdel-Aziz NH (2011) The nematicidal effect of some bacterial biofertilizers on *Meloidogyne incognita* in sandy soil. Braz J Microbiol 42:105–113

Etesami H, Adl SM (2020) Plant Growth-Promoting Rhizobacteria (PGPR) and their action mechanisms in availability of nutrients to plants. In: Kumar M, Kumar V, Prasad R (eds) Phyto-microbiome in stress regulation. Environmental and microbial biotechnology. Springer, Singapore. https://doi.org/10.1007/978-981-15-2576-6_9

Etesami H, Emami S, Alikhani HA (2017) Potassium solubilizing bacteria (KSB): mechanisms, promotion of plant growth, and future prospects—a review. J Soil Sci Plant Nutr 17(4):897–911. https://doi.org/10.4067/S0718-95162017000400005

Gandhi R, Prittesh P, Jinal HN, Chavan SM, Paul D, Amaresan N (2022) Evaluation of the effect of potassium solubilizing bacterial strains on the growth of wheat (*Triticum aestivum* L.). J Plant Nutr 46:1479. https://doi.org/10.1080/01904167.2022.2074859

Gautam SS, Gondwal M, Soni R, Gautam BPS (2022) Potash biofertilizers: current development, formulation, and applications. Trends of Applied Microbiology for Sustainable Economy 481–500. https://doi.org/10.1016/B978-0-323-91595-3.00014-8

Ghorbani R, Koocheki A, Jahan M, Asadi GA (2008) Impact of organic amendments and compost extracts on tomato production and storability in agroecological system. Agron J 28:307–311

Glick BR (2012) Plant growth-promoting bacteria: mechanisms and applications. Scientifica 2012:963401

Glowa KR, Arocena JM, Massicotte HB (2003) Extraction of potassium and/or magnesium from selected soil minerals by piloderma. Geomicrobiol J 20:99–111

Gopalakrishnan S, Srinivas V, Prasanna SL (2020) Chapter 5—streptomyces. In: Amaresan N, Kumar MS, Annapurna K, Kumar K, Sankaranarayanan A (eds) Beneficial microbes in agro-ecology. Academic Press, Amsterdam, pp 55–71

Gundala PB, Chinthala P, Sreenivasulu B (2013) A new facultative alkaliphilic, potassium solubilizing Bacillus sp. SVUNM9 isolated from mica cores of Nellore District, Andhra Pradesh, India: research and reviews. J Microbiol Biotechnol 2:1–7

Habibi A, Heidari G, Sohrabi Y, Badakhshan H, Mohammadi K (2011) Influence of bio, organic and chemical fertilizers on medicinal pumpkin traits. J Med Plants Res 5:55905597

Han HS, Lee KD (2005) Phosphate and potassium solubilizing bacteria effect on mineral uptake, soil availability and growth of eggplant. Res J Agric Biol Sci 1(2):176–180

Islam M, Deora A, Hashidoko Y, Rahman A, Ito T, Tahara S (2007) Isolation and identification of potential phosphate solubilizing bacteria from the rhizoplane of Oryza sativa L. cv. BR29 of Bangladesh. *Zeitschrift für Naturforschung C* 62(1–2):103–110. https://doi.org/10.1515/znc-2007-1-218

Jini D, Ganga VS, Greeshma MB et al (2024) Sustainable agricultural practices using potassium-solubilizing microorganisms (KSMs) in coastal regions: a critical review on the challenges and opportunities. Environ Dev Sustain 26:13641–13664. https://doi.org/10.1007/s10668-023-03199-9

Keshavarz Zarjani J, Aliasgharzad N, Oustan S, Emadi M, Ahmadi A (2013) Isolation and characterization of potassium solubilizing bacteria in some Iranian soils. Arch Agron Soil Sci 59(12):1713–1723. https://doi.org/10.1080/03650340.2012.756977

Kore SA, Tirmali AM, Jadhav AC, Badgujar CC (2020) Formulation of a compatible microbial consortia of nitrogen fixing, phosphate solubilising and potash mobilizing bacteria for optimizing nutrient supplementation to tuberose. Int J Chem Stud 8:3804–3809

Kour D, Rana KL, Kaur T, Yadav N, Halder SK, Yadav AN, Sachan SG, Saxena AK (2020) Potassium solubilizing and mobilizing microbes: biodiversity, mechanisms of solubilization, and biotechnological implication for alleviations of abiotic stress. New and Future Developments in Microbial Biotechnology and Bioengineering, pp 177–202. https://doi.org/10.1016/B978-0-12-820526-6.00012-9

Kumar R, Swapnil P, Meena M, Selpair S, Yadav BG (2022) Plant growth-promoting rhizobacteria (PGPR): approaches to alleviate abiotic stresses for enhancement of growth and development of medicinal plants. Sustain For 14(23):15514. https://doi.org/10.3390/su142315514

Kumar A, Kumar A, Patel H (2018) Role of microbes in phosphorus availability and acquisition by plants. Int J Curr Microbiol App Sci 7(5):1344–1347

Lian B, Prithiviraj B, Souleimanov A, Smith DL (2001) Evidence for the production of chemical compounds analogous to nod factor by the silicate bacterium *Bacillus circulans* GY92. Microbiol Res 156(3):289–292

Lin QM, Rao ZH, Sun YX, Yao J, Xing LJ (2002) Identification and practical application of silicate-dissolving bacteria. Agric Sci China 1(1):81–85

Liu D, Lian B, Dong H (2012) Isolation of *Paenibacillus* sp. and assessment of its potential for enhancing mineral weathering. Geomicrobiol J 29:413–421

Liu Z, Wang H, Xu W, Wang Z (2020) Isolation and evaluation of the plant growth promoting rhizobacterium *Bacillus methylotrophicus* (DD-1) for growth enhancement of rice seedling. Arch Microbiol 202:2169–2179. https://doi.org/10.1007/s00203-020-01934-8

Lopes-Assad ML, Avansini SH, Rosa MM, de Carvalho JRP, Ceccato-Antonini SR (2010) The solubilization of potassium-bearing rock powder by *Aspergillus Niger* in small-scale batch fermentations. Can J Microbiol 56:598–605

Ma Y, Oliveira RS, Freitas H, Zhang C (2016) Biochemical and molecular mechanisms of plantmicrobe-metal interactions: relevance for phytoremediation. Front Plant Sci 7

Maity A, Belgamwar R, Polshettiwar V (2019) Facile synthesis to tune size, textural properties and fiber density of dendritic fibrous nanosilica for applications in catalysis and CO2 capture. Nat Protoc 14:2177–2204

Meena VS, Maurya BR, Verma JP, Meena RS (2016) Potassium solubilizing microorganisms for sustainable agriculture. Springer, New Delhi

Meena VS, Maurya BR, Verma JP, Aeron A, Kumar A, Kim K, Bajpai VK (2015) Potassium solubilizing rhizobacteria (KSR): isolation, identification, and K-release dynamics from waste mica. Ecol Eng 81:340–347. https://doi.org/10.1016/j.ecoleng.2015.04.065

Mikhailouskaya N, Tcherhysh A (2005) K-mobilizing bacteria and their effect on wheat yield. Latvian J Agron 8:154–157

Mora V, Baigorri R, Bacaicoa E, Zamarreño AM, García-Mina JM (2012) The humic acid-induced changes in the root concentration of nitric oxide, IAA and ethylene do not explain the changes in root architecture caused by humic acid in cucumber. Environ Exp Bot 76:24–32

Muentz A (1890) Sur la decomposition des roches et la formation de la terre arable. C R Acad Sci 110:1370–1372

Nadeem SM, Naveed M, Zahir ZA, Asghar HM (2013) Plant-microbe interactions for sustainable agriculture: fundamentals and recent advances. In: Arora NK (ed) Plant-microbe symbiosis: fundamentals and advances. Springer, New Delhi

Olaniyan FT, Alori ET, Adekiya AO, Ayorinde BB, Daramola FY, Osemwegie OO, Babalola OO (2022) The use of soil microbial potassium solubilizers in potassium nutrient availability in soil and its dynamics. Ann Microbiol 72(1):45

Parmar P, Sindhu SS (2013) Potassium solubilization by rhizosphere bacteria: influence of nutritional and environmental conditions. J Microbiol Res 3(1):25–31

Patel BC (2011) Advance method of preparation of bacterial formulation using potash mobilizing bacteria that mobilize potash and make it available to crop plant. WIPO Patent Application WO/2011/154961

Prajapati K, Sharma MC, Modi HA (2013) Growth promoting effect of potassium solubilizing microorganisms on *Abelmoscus esculantus*. Int J Agric Sci 3:181–188

Rajawat MVS, Singh S, Tyagi SP, Saxena AK (2016) Amodifiedplateassay for rapid screening of potassium-solubilizing bacteria. Pedosphere 26:768–773

Ramarethinam S, Chandra K (2005) Studies on the effect of potash solubilizing/mobilizing bacteria *Frateuria aurantia* on Brinjal growth and yield. Pestology 11:35–39

Rosa-Magri MM, Avansini SH, Lopes-Assad ML, Tauk-Tornisielo SM, Ceccato-Antonini SR (2012) Release of potassium from rock powder by the yeast *Torulaspora globosa*. Braz Arch Biol Technol 55:577–582. https://doi.org/10.1590/S1516-89132012000400013

Sangeeth KP, Suseela Bhai R, Srinivasan V (2012) *Paenibacillus glucanolyticus*, a promising potassium solubilizing bacterium isolated from black pepper (*Piper nigrum* L.) rhizosphere

Sasse J, Martinoia E, Northen T (2018) Feed your friends: do plant exudates shape the root microbiome? Trends Plant Sci 23(1):25–41

Sattar A, Naveed M, Ali M, Zahir ZA, Nadeem SM, Yaseen M, Meena VS, Farooq M, Singh R, Rahman M, Meena HN (2019) Perspectives of potassium solubilizing microbes in sustainable food production system: a review. Appl Soil Ecol 133:146–159. https://doi.org/10.1016/j.apsoil.2018.09.012

Seth K, Vyas P, Deora S, Gupta AK, Meena M, Swapnil P, Harish (2023) Understanding plant-plant growth-promoting rhizobacteria (PGPR) interactions for inducing plant defense. In: Swapnil P, Meena M, Harish, Marwal A, Vijayalakshmi S, Zehra A (eds) Plant-microbe interaction—recent advances in molecular and biochemical approaches, vol 2. Academic Press, Elsevier, pp 201–226. https://doi.org/10.1016/B978-0-323-91876-3.00010-5

Sheng XF, Huang WY (2002) Mechanism of potassium release from feldspar affected by the strain NBT of silicate bacterium. Acta Pedol Sin 39:863–871

Sindhu SS, Dua S, Verma MK, Khandelwal A (2010) Growth promotion of legumes by inoculation of rhizosphere bacteria. In: Khan MS, Zaidi A, Musarrat J (eds) Microbes for legume improvement. Springer Wien, New York, pp 195–235

Sindhu SS, Parmar P, Phour M (2013) Nutrient cycling: potassium solubilization by microorganisms and improvement of crop growth. In: Parmar N, Singh A (eds) Geomicrobiology and biogeochemistry: soil biology. Springer-Wien, New York, pp 27–40

Sindhu SS, Parmar P, Phour M, Sehrawat A (2016) Potassium-Solubilizing Microorganisms (KSMs) and Its Effect on Plant Growth Improvement. In: Meena V, Maurya B, Verma J, Meena R. (eds) Potassium Solubilizing Microorganisms for Sustainable Agriculture. Springer, New Delhi. https://doi.org/10.1007/978-81-322-2776-2_13

Sindhu SS, Sehrawat A, Phour M, Kumar R (2022) Nutrient acquisition and soil fertility: Contribution of rhizosphere microbiomes in sustainable agriculture. In Microbial BioTechnology for Sustainable Agriculture, vol 1. Singapore: Springer Nature Singapore, pp 1–41

Sindhu SS, Verma MK, Suman M (2009) Molecular genetics of phosphate solubilization in rhizosphere bacteria and its role in plant growth promotion. In: Khan MS, Zaidi A (eds) Phosphate solubilizing microbes and crop productivity. Nova Science Publishers, New York, pp 199–228

Singh A, Singh B, Chinna GS, Chahal HS, Devi R (2023) Revitalization of potassium solubilizing microbes in food production system: an overview. Agric Rev 44(3):311–319

Singh RG, Behura SK, Kumar R (2010) Litholytic property of Kulattha (Dolichous biflorus) vs potassium citrate in renal calculus disease: a comparative study. The Journal of the Association of Physicians of India, 58:286–289

Soumare A, Boubekri K, Lyamlouli K, Hafidi M, Ouhdouch Y, Kouisni L (2020a) From isolation of phosphate solubilizing microbes to their formulation and use as biofertilizers: status and needs. Front Bioeng Biotechnol 7:425

Soumare A, Diedhiou AG, Arora NK, Al-Ani LKT, Ngom M, Fall S, Hafidi M, Ouhdouch Y, Kouisni L, Sy MO (2021) Potential role and utilization of plant growth promoting microbes in plant tissue culture. Front Microbiol 12:649878

Soumare A, Diedhiou AG, Thuita M, Hafidi M, Ouhdouch Y, Gopalakrishnan S, Kouisni L (2020b) Exploiting biological nitrogen fixation: a route towards a sustainable agriculture. Plan Theory 9:1011

Subhashini DV (2015) Growth promotion and increased potassium uptake of tobacco by potassium-mobilizing bacterium *Frateuria aurantia* grown at different potassium levels in Vertisols. Commun Soil Sci Plant Anal 46:210–220

Subhashini DV, Kumar AV (2014) Phosphate solubilizing *Streptomyces* spp. obtained from the rhizosphere of *Ceriops decandra* of Corangi mangroves. Indian J Agric Sci 84(5):560–564

Supanjani HHS, Jung SJ, Lee KD (2006) Rock phosphate and potassium rock solubilizing bacteria as alternative sustainable fertilizers. Agro Sust Dev 26:233–340

Teotia P, Kumar V, Kumar M, Prasad R, Sharma S (2017) Probiotic microbiome: potassium solubilization and plant productivity. In: Kumar V, Kumar M, Sharma S, Prasad R (eds) Probiotics in agroecosystem. Springer, Singapore, pp 451–467

Uroz S, Calvaruso C, Turpault P, Frey-Klett P (2009) Mineral weathering by bacteria: ecology, actors and mechanisms. Trends Microbiol 17:378–387

Uroz S, Calvaruso C, Turpault MP, Pierrat JC, Mustin C, Frey-Klett P (2007) Effect of the mycorrhizosphere on the genotypic and metabolic diversity of the bacterial communities involved in mineral weathering in a forest soil. Appl Environ Microbiol 73(9):3019–3027

Verma M, Sharma S, Prasad R (2011) Liquid biofertilizers: advantages over carrier-based biofertilizers for sustainable crop production. News L Intl Soc Environ Bot 17:2

Vlek PLG, Vielhauer K (1994) Nutrient management strategies in stressed environments. In: Virmani SM, Katyal JC, Eswaran H, Abrol IP (eds) Stressed ecosystems and sustainable agriculture. Oxford and IBH Publishing Co., New Delhi, pp 203–229

Wang SQ, Zhao X, Xing GX, Yang YC, Zhang M, Chen HK (2015) Improving grain yield and reducing N loss using polymer-coated urea in Southeast China. Agron Sustain Dev 35:1103–1115

Wu SC, Cao ZH, Li ZG, Cheung KC, Wong MH (2005) Effects of biofertilizer containing N-fixer, P and K solubilizers and AM fungi on maize growth: a greenhouse trial. Geoderma 125:155–166

Wu X, Zhao Z, Zhao Z, Zhang Y, Li M, Yu Q (2023) Analysis of the potassium-solubilizing *Priestia megaterium* strain NK851 and its potassium feldspar-binding proteins. Intern J Mol Sci 24:14226. https://doi.org/10.3390/ijms241814226

Yang H, Lu L, Chen Y, Ye J (2023) Transcriptomic analysis reveals the response of the bacterium *Priestia aryabhattai* SK1-7 to interactions and dissolution with potassium feldspar. Appl Environ Microbiol 89(5):e02034–e02022

Zapata F, Roy RN (2004) Use of phosphate rock for sustainable agriculture. FAO and IAEA, Rome

Zhang C, Kong F (2014) Isolation and identification of potassium-solubilizing bacteria from tobacco rhizospheric soil and their effect on tobacco plants. Appl Soil Ecol 82:18–25

Zhang HM, Xue-Yun YANG, Xin-Hua HE, Ming-Gang XU et al (2011) Effect of long-term potassium fertilization on crop yield and potassium efficiency and balance under wheat-maize rotation in China. Pedosphere, 21(2):154–163.

Fungal Endophytes: Secondary Metabolites and Their Utilisation for Sustainable Agriculture

4

Vidhi Jain, Purnamsree Gogoi, Madhu Kamle, Nitish Rai, Pradeep Kumar, and Namita Ashish Singh

Abstract

The use of novel and helpful bioactive compounds is growing in today's ever-changing world to address the many issues that humanity faces, such as environmental stress, biotic and abiotic stresses, water and food scarcity, the emergence of drug-resistant microbes, pollution, and rising rates of infections and diseases. Recent years have seen a rise in the worldwide concern for the development of more sustainable agriculture, and studies have been conducted to understand ecology and investigate the possibility of endophytic interactions in plant growth. Despite their lack of research, fungal endophytes are a powerful source of innovative natural compounds with applications in industry, agriculture, and medi-

V. Jain · N. A. Singh (✉)
Department of Microbiology, Mohanlal Sukhadia University, Udaipur, Rajasthan, India
e-mail: namita.singh@mlsu.ac.in

P. Gogoi
Department of Forestry, North Eastern Regional Institute of Science and Technology, Nirjuli, Arunachal Pradesh, India

M. Kamle
Department of Forestry, North Eastern Regional Institute of Science and Technology, Nirjuli, Arunachal Pradesh, India

Department of Biotechnology, University of Lucknow, Lucknow, Uttar Pradesh, India

N. Rai
Department of Biotechnology, Mohanlal Sukhadia University, Udaipur, Rajasthan, India

Department of Zoology, University of Lucknow, Lucknow, Uttar Pradesh, India

P. Kumar
Department of Botany, University of Lucknow, Lucknow, Uttar Pradesh, India

College of Life Science & Biotechnology, Korea University, Seongbuk-gu, Seoul, Republic of Korea

M. Meena et al. (eds.), *Metabolic Sustainability of Endophytes*, Microorganisms for Sustainability 52, https://doi.org/10.1007/978-981-96-4004-1_4

cine. About 10% of the world's estimated one million plant species have been investigated to date for their remarkably diversified fungal endophytes. The array of bioactive metabolites generated by the fungal endophytes that have been explored so far is addressed in this chapter, with a focus on those that can be used to protect and enhance crops, with the goal to achieve sustainable agriculture.

Keywords

Endophytic fungi · Secondary metabolites · Agriculture · Crop protection

4.1 Introduction

Plants are not isolated living things; rather, they coexist in close proximity with the microorganisms in their environment and particularly living within them. The alliance between plants and fungi is exceptionally widespread. According to fossil records, this union with endophytes and mycorrhizas has persisted for about 400 million years (Hodkinson 2018), beginning when plants began to colonise the soil. This suggests the group's significance in propelling the evolution of terrestrial life (Anjum et al. 2019). The literal meaning of the phrase "endophyte" is "within the plants" (from "endon," which means within, and "phyton," which means plants). Endophytes live solely within plant tissues and can grow within roots, stems, and/or leaves, in contrast to mycorrhizal fungi, which colonise plant roots and spread into the rhizosphere. According to a more recent definition put forth by Le Cocq et al. (2017), endophytes are microorganisms that live inside plant tissues for at least a part of their life cycles and under no circumstances inflict damage against the host plant. Numerous ecological niches, such as the Antarctic and Artic areas, deserts, mangroves, rainforests, and marine and coastal habitats, are home to flourishing endophytes (Meshram et al. 2023).

There are two most important categories of endophytic fungi that reflect variations in evolutionary closeness, classification, plant hosts, and ecological roles: the clavicipitaceous endophytes, which affect several grasses; and the non-clavicipitaceous endophytes, which can be obtained from asymptomatic tissues of nonvascular plants, ferns and allies, conifers, and angiosperms. Rodriguez et al. (2009) categorised clavicipitaceous and non-clavicipitaceous fungi into four distinct classes: class 1 includes all the clavicipitaceous fungi that specifically colonise grasses and they can be found in the aerial part and/or roots of their hosts and are propagated both horizontally and vertically; and classes 2, 3, and 4 comprise non-clavicipitaceous fungi. Endophytes capable of colonising the aerial fraction and roots, and that are transmitted horizontally and vertically, are included in class 2. Class 3, a very diverse group that has only horizontal transmission, consists of endophytes commonly associated with the leaves of tropical tree species. Finally, class 4 comprises the dark septate endophytic (DSE) fungi that have melanin in their septa and occur entirely in the roots of their hosts and display merely horizontal transmission.

Nonetheless, several authors have suggested that these fungal endophytes should be further divided into subclasses that reflect a multitude of aspects, such as the host range, means of reproduction, colonised plant part, mode of transfer, source of nutrition, and capacity to manifest symptoms in the host plant (Purahong and Hyde 2011). The mechanism of reproduction by sexual or asexual spores further categorises endophytic fungi as sexual or asexual. For instance, the endophytes *Epichloë* have been split into the genera *Epichloë* and *Neotyphodium*, which reproduce sexually and asexually, respectively (Schardl and Craven 2003). When vegetative propagules or spores produced by the endophyte are spread to the plant population through air or via some vector, it is called horizontal transmission, while transference of the fungi to the plant progeny via seeds is known as vertical transmission, referred to as seed-transmitted endophytes. Quesada-Moraga et al. (2014) documented the vertical transmission of *B. bassiana* via seeds in experimentally inoculated opium poppy plants using seed soaking. Depending on the area of the plant that is infected, fungal endophytes can be categorised as either foliar or root endophytes and as symptomatic or asymptomatic in terms of how their infection manifests itself. For instance, *Fusarium* spp. was determined to be asymptomatic since it was verified that it did not exhibit any symptoms in the cord roots of banana cultivars (Sikora et al. 2008). In a study conducted by Behie et al. (2015), *Metarhizium* spp. was primarily located in the plant roots, whilst *B. bassiana* and *Pochonia chlamydosporia* were localised within the stems and leaves. Last but not least, based on how they feed, they can be classified as either necrotrophic or biotrophic (Bamisile et al. 2018). Necrotrophic fungus promotes necrosis to grow from dead tissues, whereas biotrophic endophytic fungi obtain nutrients from living tissues.

Remarkably, every one of the roughly 300,000 plant species that inhabit our planet is home to one or more endophytes. Only approximately 10% of the one million diverse fungal species have been reported till date, making endophytic biology the subject of extensive study (Strobel and Daisy 2003). Plant-endophyte interaction is influenced by a number of variables, including the pattern of endophytic infection, the mode of transmission, genetic background, environmental variability, imbalance in nutrient exchange, physiological stress, and senescence (Rodriguez and Redman 2008). Although the benefits of this interaction have long been recognised and debated, comprehensive research assessing the true value offered by these fungi has only lately been carried out (Quesada-Moraga 2020). In terms of microbial diversity, fungal endophytes are a comparatively unexplored source that yields unique natural compounds that can be utilised throughout industry, agriculture, and medicine. This coevolutionary process establishes an advantageous partnership between the host plant and the endophyte. The endophyte receives essential nutrients and shelter from the host plant, and in return, the endophyte produces bioactive compounds that increase the fitness of the host plant. The ability of host plants to tolerate biotic and abiotic stresses, store water, and protect themselves from microbial, pest, and insect attack is greatly enhanced by the bioactive compounds produced by endophytes. Endophytes thus play a crucial part in plant symbiosis by shielding their host plants against pathogenic pressure and challenging environmental circumstances.

A deeper comprehension of this relationship will be crucial to enhancing the ways in which these microorganisms can be used in agriculture to boost crop yields and plant growth, manage pests, reduce pathogen virulence, and/or assist plants in surviving environmental stressors like drought, high pH and temperature swings, heavy metal toxicity, and nutrient shortages (Naik 2019). The production of bioactive compounds by endophytic fungi is also a significant factor in improving crop productivity and food quality, which promotes sustainable agriculture (Lugtenberg et al. 2016).

4.2 Plant and Endophyte Interaction

Endophytic fungi are found in most of the plant species on the planet. The endophytic interaction is characterised as balanced antagonism because the identification of the plant as a host involves induction of virulence mechanisms to facilitate colonisation and the triggering of host defences by these events. The biological communication between the endophyte and the host plant for the establishment of the interaction is intricate and entails the selective expression of fungal genes that produce enzymes and secondary compounds that facilitate colonisation (Yan et al. 2019). When the endophyte's germinative tube gets close to the root, the root loses its apical dominance, and a hyphal penetration apparatus forms. This apparatus introduces infected hyphae into the root cortex, initiating the colonisation process. These actions encourage a balanced stimulation of plant defence genes, and when the fungus reaches the inner cortex, the hyphae then break through the plant cell wall and continue to colonise nearby tissues (endoderm, pericycle, xylem, phloem) of the roots and soil (Khan et al. 2015). In this connection, balance prevails because the fungus receives nutrients from the host plant and providing the host resistance to both biotic and abiotic challenges (Bamisile et al. 2018; Bhupenchandra et al. 2024).

Endophytic colonisation alters the host plant's metabolic pathways leading to increased production of growth hormones, viz. cytokinins, indole-3-acetonitrile, and indole-3-acetic acid, and amplifies nutrient uptake. The capacity of endophytes to sequester, chelate, and degrade metals aids in plant survival in contaminated soils, improving the host plant's fitness and competitiveness. Fungal endophytes contribute significantly to the cycling of nutrients by break down of dead host plants (Sun et al. 2011). Conversely, plant hosts offer nutritive support, spatial structure, and desiccation resistance for endophytes, as well as facilitating their transfer to subsequent host generations. Additionally, plants offer vital nutrients that endophytes need for growth, self-defence, and the completion of their life cycle.

4.2.1 The Interaction's Direct Benefits

Interaction with endophytic fungi enables a plant to utilise more nutrients and produce more phytohormones, which in turn increases biomass production, expands the development of the root system, and increases plant height, volume, reproduction, and productivity. Because of these attributes, they can be termed as biofertilisers. In order to enhance the amount of nutrients that the plant host receives from the soil and organic matter, endophytic fungi can boost the uptake of macronutrients like phosphorus, nitrogen, potassium, and magnesium or micronutrients like zinc, iron, and copper (Rana et al. 2020). Endophytic fungi can produce auxins, gibberellins (GAs), and cytokinins. The primary regulators of plant growth, auxins, have a number of advantageous impacts on the development of shoots and roots, including the commencement of the root formation process, differentiation of vascular tissue, and responses to tropism (Jaroszuk-Ściseł et al. 2014; Chandran et al. 2021; Kumar et al. 2022).

Gibberellins are vital in numerous plant responses that include seed germination, stem extension, reproductive activity, flourishing, fruit development, and ageing (Bomke and Tudzynski 2009). *Trichoderma asperellum* inoculation dramatically decreased the need for phosphorus fertilisation in onion (*Allium cepa*) cultivation (Ortega-Garcia et al. 2015). In a field investigation, Baron et al. (2018) introduced *Aspergillus sydowii* into maize (*Zea mays*) and found that plants which interacted with the fungus accumulated much larger amounts of P in their tissues, even at lower fertilisation dosages.

4.2.2 The Interaction's Indirect Benefits

The synthesis of antibiotics, secondary metabolites, and siderophores, which turn on a plant's defence mechanism and protect the plant from sickness and infections, are among the indirect advantages of interacting with endophytic fungi. Endophytic fungi assist plants in strengthening their self-defence system by promoting the commencement of induced systemic resistance (ISR) pathways, which may overlap with acquired systemic resistance (ASR), because both systems can boost plant growth and guard against pests and diseases (Busby et al. 2016). Microbe-associated molecular patterns (MAMPs), which are identified by plant receptors and cause the so-called MAMP-triggered immunity (MTI), are the first step in the activation of systemic resistance. The fungal cell wall's chitin and β-glucan, as well as secreted enzymes including xylanases, glucanases, and chitinases and their byproducts, are examples of MAMPs that cause MTI (Latz et al. 2018).

Apart from MAMPs, microbes produce effector molecules, such as secondary metabolites, which can promote resistance and activate effector-triggered immunity (ETI) (Poveda et al. 2020). The complex response that results from MAMPs and effectors activating the ASR and ISR pathways includes ion flux, protein phosphorylation and dephosphorylation, production of signalling molecules (e.g. salicylic acid and ethylene) and reactive oxygen species (ROS), and selective expression of genes that support defence responses (e.g. thickening the cell wall of plants, producing phytoalexins and pathogen-related (PR) proteins, and even cell death) (Chadha et al. 2015). According to Kaddes et al. (2019), endophytic fungi are outstanding producers of compounds that exhibit potency against diseases and herbivores. These compounds include alkaloids, steroids, terpenoids, peptides, polyketones, flavonoids, quinols, phenols, chlorinated compounds, and volatile organic compounds.

4.3 Secondary Metabolites Produced by Endophytic Fungi

It has been found that plants colonised by endophytic fungi are highly tolerant to high temperatures and salinities, as well as well adapted to both biotic and abiotic challenges. This is credited to the production of secondary metabolites by fungal cohort. Endophytes have been known to produce a variety of novel and structurally distinct bioactive metabolites such as polyketides, alkaloids, terpenoids, steroids, isocoumarin derivatives, quinines, flavanoids, phenols, phenolic acids, and volatile organic compounds (Rustamova et al. 2019) as depicted in Fig. 4.1. A great place to look for bioactive metabolites is among the well-known fungus genera such as

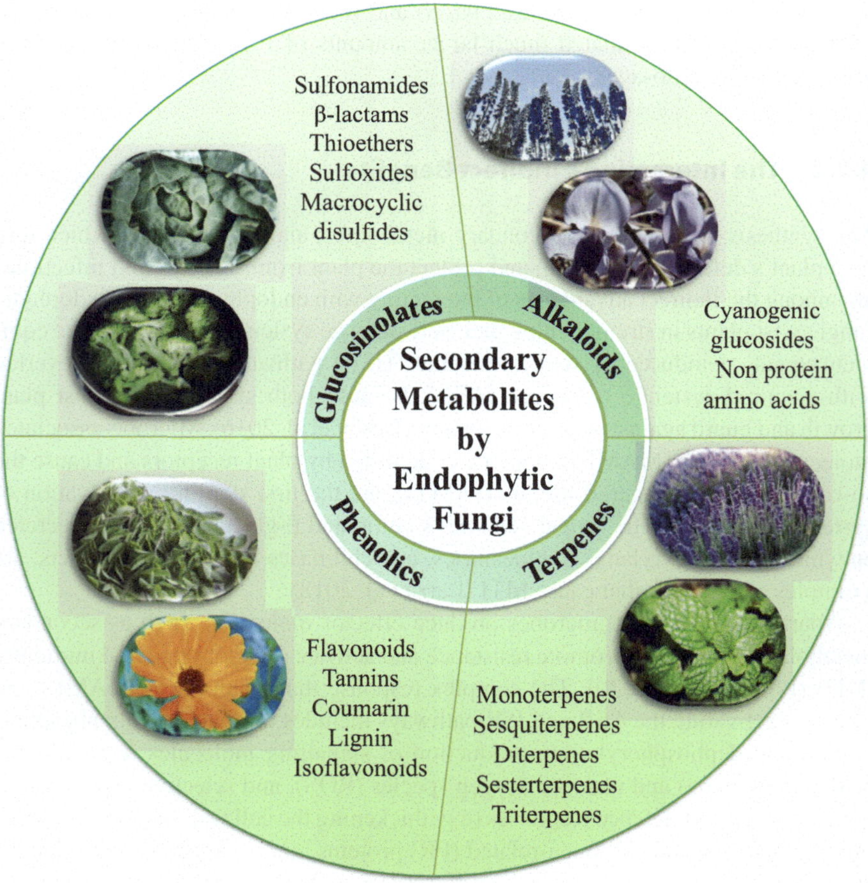

Fig. 4.1 Secondary metabolites synthesised by the endophytic bacteria

Penicillium, *Fusarium*, *Aspergillus*, *Sclerotium*, *Myxormia*, *Alternaria*, *Colletotrichum*, *Cladosporium*, *Diaporthe*, and *Curvularia* (Toghueo 2020). Through the use of techniques such as high-performance liquid chromatography (HPLC), thin-layer chromatography (TLC), nuclear magnetic resonance (NMR), infrared spectroscopy (IR), matrix-assisted laser desorption/ionisation—time of flight (MALDI-TOF), electron spray ionisation (ESI), and fast atom bombardment (FAB), bioactive compounds isolated from the extract of fungal endophytes can be purified and characterised.

4.3.1 Types of Secondary Metabolites

4.3.1.1 Polyketides

As the primary class of secondary metabolites generated by the fungal kingdom, polyketides are diverse in nature and include antibiotics, poisons, antiparasitics, antifungals, and insecticides. Coenzyme A esters, such as acetyl or malonyl CoA, can be converted into polyketides by post-synthetic conversion and decarboxylation condensation using polyketide synthases (Piel 2010). According to Saeed and Qasim (2014), isocoumarins are a class of naturally occurring compounds that are interesting from a biosynthetic, structural, and pharmacological perspective. They are coumarin isomers with a reversed lactone moiety that may have a 6,8-dioxygenated pattern, a 3-(un)substituted phenyl ring, or a 3-alkyl chain (C1–C17). Plants colonised by endophytes produce a number of isocoumarin derivatives that show promise for biocontrol.

4.3.1.2 Alkaloids

Alkaloids are a broad class of nitrogen- and amino-containing molecules that have a variety of natural roles and are found in plants, animals, and microbes. Alkaloids, principally peramine and ergovaline, are produced by grasses infected with *Acremonium* spp. and *E. typhina*. Plants of the families *Asteraceae*, *Boraginaceae*, *Convolvulaceae*, *Fabaceae*, *Lamiaceae*, *Orchidaceae*, and *Poaceae* produce pyrrolizidines, which are alkaloid chemicals, to protect themselves from insect pests. Broad-spectrum insecticides called lolines, which function as both feeding deterrents and metabolic poisons, are secreted by grasses infected with endophytic fungi.

4.3.1.3 Indole Derivatives

Indoles are substances with a five-membered nitrogen-containing pyrrole ring fused to a benzene ring. Recently, 6-isoprenylindole-3-carboxylic acid, a novel indole derivative, was identified from *Colletotrichum* sp. endophytic with *A. annua*. According to Lu et al. (2000), it shows antifungal action against the phytopathogenic fungus *Rhizoctonia cerealis*, *Phytophthora capsici*, and *Gaeumannomyces graminis* var. *tritici*.

4.3.1.4 Terpenoids

Terpenoids are the largest class of natural compounds and are a diversified set of products made up of C5 isoprene units (Christianson 2017). It is known that fungi that colonise plant hosts release terpenoids, primarily sesqui- and diterpenes. The most prevalent subclass of terpenoids among fungal secondary metabolites, sesquiterpenes, are made up of three isoprene units arranged in a molecular formula of $C_{15}H_{24}$. *Talaromyces wortmannii* promotes growth and pathogen resistance of *Brassica campestris* L. by β-caryophyllene production (Yamagiwa et al. 2011).

4.3.1.5 Steroids

Steroids are organic compounds comprised of 17 carbon atoms arranged in four rings. An endophytic fungus called *Chaetomium globosum* ZY-22 that was isolated from *Ginkgo biloba* was used to extract polyhydroxylated steroids (Qin et al. 2009). Phomopsterones A and B, functionalised ergostane-type steroidswere isolated from the plant-derived *Phompsis* sp. (Hu et al. 2017).

4.3.1.6 Quinones

Quinones, such as p- and o-benzoquinones, are a type of chemical compounds with a six-carbon cyclic dione structure and two double bonds. When *Alternaria* spp. and *Phoma* spp. associate with plants, they release Altersolanol A, a highly hydroxylated quinone having antibacterial activity (Yang et al. 1994). Preussomerin N1, palmarumycin CP4a, and palmarumycin CP5, which are Ras farnesyl protein transferase inhibitors, are produced by an endophytic *Coniothyrium* sp. (Krohn et al. 2001).

4.3.1.7 Flavonoids

Plant secondary metabolites, known as flavonoids, have a wide range of important pharmacological properties, including antibacterial, anticancer, antioxidation, and cardiovascular protection. Owing to the chemical makeup of polyhydroxyl compounds, flavonoids possess strong biological antioxidant properties, which confer preventative and therapeutic benefits against a number of common illnesses (Boudaoud et al. 2009). Flavonoids are produced by the endophytic fungi ST22 and SX10 that were isolated from *Ginkgo biloba* L. (Qiu et al. 2010).

4.3.1.8 Phenolic Compounds

Phenolic compounds, which can range from basic phenolic molecules to highly polymerised compounds, are secondary metabolites with benzene rings and one or more hydroxyl substituents. Kaempferol from *Mucor fragilis*, which was separated from *Sinopodophyllum hexandrum*, was identified by Huang et al. (2014). According to Das et al. (2017), endophytic fungi isolated from *Zingiber nimmonii* produced antioxidant compounds such as syringic acid, vanillic acid, para-hydroxybenzoic acid, and caffeic acid.

4.3.1.9 Volatile Organic Compounds (VOCs)

In the midst of the diverse array of secondary metabolites produced by endophytic fungi, over 300 of these molecules are VOCs (Kaddes et al. 2019). According to Korpi et al. (2009), both primary and secondary metabolic pathways produce fungal volatile organic molecules. These are lipophilic low-molecular-weight compounds that vaporise easily at 20 °C and 0.01 KPa. Because volatile organic compounds (VOCs) may travel great distances through air, liquids, and porous soils, they are often referred to as "infochemicals" and function as signalling molecules in both short- and long-distance intercellular contacts. They are categorised into various classes and can be found in a variety of forms, including simple hydrocarbons, alcohols, cyclohexanes, ketones, phenols, thioalcohols, thioesters, and derivatives of benzene (Ortíz-Castro et al. 2009). Fungal species and the ambient circumstances that support their growth determine the qualitative and quantitative composition of volatile organic compounds (VOCs) (Meena et al. 2017). Most fungus groups produce antibiotic-like volatile organic compounds (VOCs), and some have a significant impact on terrestrial carbon cycles, atmospheric chemistry, and subsurface ecology. The endophytic fungus *Muscodor albus*, which was isolated from *Cinnamomum zeylanicum*'s tiny branches, produces a mixture of at least 28 volatile chemicals that efficiently suppress and kill some other fungi and bacteria (Worapong et al. 2001). In comparison to synthetic pesticides and fertilisers, microbial volatile organic compounds are safe, biodegradable, and effective at low concentrations. Therefore, they can be used as an alternative to damaging pesticides, fungicides, and bactericides, and they can play a significant part in sustainable agriculture. These can be prepared for distribution to fields, safe handling, and storage. We are only now starting to see how well-suited VOCs are for biocontrol (Song and Ryu 2013).

4.4 Factors Affecting Metabolite Production

It is necessary to comprehend the possible elements influencing the host plant's growth conditions, as these aspects eventually impact the fungal colonisation and bioactive chemical production. For instance, a certain circumstance may affect the host plants' distribution ranges, which may then affect the species of fungal endophytes, their richness, spore germination, and metabolism. Moreover, a number of variables, such as the host and endophyte genotypes, geographic location, time of year, and environmental conditions, influence the synthesis and extraction of bioactive compounds from endophytic fungi (Morales-Sánchez et al. 2020). The following lists some of the known variables that impact the synthesis of bioactive chemicals.

4.4.1 Atmospheric Moisture and Temperature

Temperature, light, humidity, and other environmental factors can all have an indirect impact on the prevalence of fungal endophytes in plant tissues. According to reports, under conditions of high mean annual atmospheric moisture and low mean annual sunshine hour, fungal endophytes associated with medicinal plants are able to synthesise a large amount of nutrients (Wu et al. 2013). According to recent reports, carrot peels yielded the highest amount of bioactive chemicals with superior antioxidant activity when heated at a high temperature (Nguyen and Le 2018).

4.4.2 The Geographical Location of Host Plants

Geographical location of the host plant and surrounding environment are demonstrated to affect the majority of the fungal endophytic population, as well as their variety and abundance. White snakeroot (*Ageratina altissima*) fungal communities were statistically analysed, and the results showed that they strongly resist both biotic and anthropogenic disturbance (Christian et al. 2016). According to Arnold and Lutzoni (2007), plants growing in the tropical zone yield more bioactive chemicals produced from fungal endophytes than from any other endophyte connected with a zonal plant.

4.4.3 Age of Host Plant Tissue

Numerous fungal endophyte species colonise plant tissues, including dermis, parenchyma, and vascular ducts, across a range of age groups. An investigation on *Calotropis procera* reveals that the highest levels of fungal endophyte colonisation and species richness were seen in older leaves. The investigation additionally demonstrated the existence of *Xylaria* sp. fungal endophytes in immature leaves (Nascimento et al. 2015). Prior research has indicated that the season and kind of tissue (such as a stem, root, leaf, stem bark, or twig) affect the colonisation of fungal endophytes (Mishra et al. 2012).

4.4.4 Genetic Background

The relationship of fungal endophyte with their host plants depends upon the host genotype, genotype of the endophyte, and environment in which they grow (Salam et al. 2017). The host genotype, endophyte genotype, and growth environment all influence the connection between fungal endophytes and their host plants (Salam et al. 2017). The relationship between the fungal endophyte and host plant is determined by the host plant's identification and the fungal reaction, or by small differences in the fungal endophyte's gene expression in response to the host plant's reaction (Sarsaiya et al. 2019). According to a study, strigolactone (SL) secreted by the host plant root may facilitate *Mucor* sp. colonisation in *Arabidopsis thaliana* (Rozpadek et al. 2018). Thus, it can be said that the host plant's genetic makeup affects the fungal endophyte colonisation process, which in turn affects the extraction of bioactive substances derived from fungal endophytes, both quantitatively and qualitatively.

4.5 Applications of Secondary Metabolites Produced by Endophytes in Sustainable Agriculture

Approximately 80% of endophytic fungi produce bioactive substances possessing herbicidal and antibacterial characteristics. Through the production of growth hormones, endophytes can increase biomass and plant growth without stimulating the host's nutritional metabolism or interfering with its ability to absorb nutrients. Endophytes can decrease herbivory through the production of bioactive substances that are poisonous to insects and other animals. They aid host plants in withstanding and surviving environmental stressors like heat, salt, and drought (Gupta et al. 2023). They have a significant impact on the ecological community by reducing the rate of environmental deterioration, biodiversity loss, and contamination of water and land due to excessive use of harmful organic pesticides, industrial sewage, and deadly gases. Endophytic fungi can aid in the development of novel biomolecules for a range of biotechnological applications that will ultimately contribute to the establishment of sustainable agriculture as shown in Fig. 4.2. Various secondary

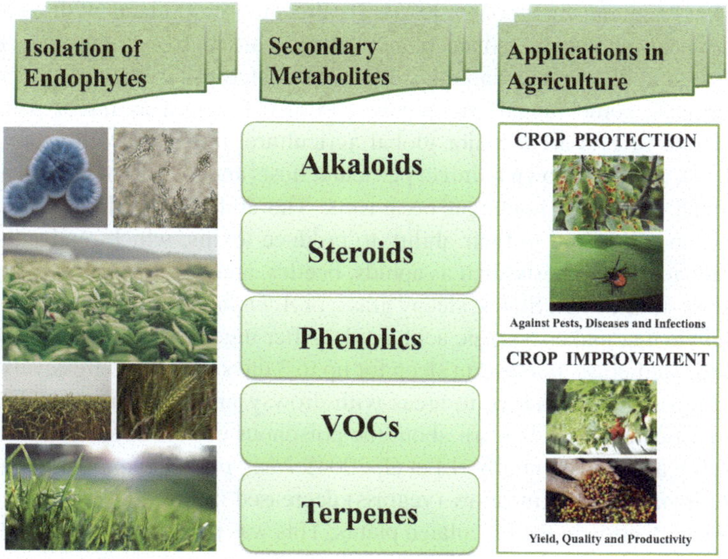

Fig. 4.2 Application of secondary metabolites produced by endophytic fungi in agriculture

metabolites produced by endophytes and its utilisation in different aspects of agriculture has been listed in Table 4.1.

4.5.1 Crop Protection

Plant diseases and pests cause an estimated 30–50% reduction in the world's annual crop production in agriculture; this loss needs to be fought to maintain food security for an ever-increasing human population (Mousa and Raizada 2013). The widespread use of agrochemicals to eradicate plant diseases and control particular plants or animals leads to an increase in agricultural output, quality, and shelf life (Omomowo and Babalola 2019). The microbial community has been observed to be impacted by changes in environmental conditions brought about by the application of these products, and these agents have garnered significant attention in relation to sustainability issues as well as detrimental effects on the environment and human health (Suryanarayanan 2019). The use of beneficial endophytes as biological control agents for crop protection is an underutilised but promising alternative strategy that is gaining traction (De Silva et al. 2019). By employing a variety of strategies, including the synthesis of secondary metabolites, competing with plant pathogens for resources like space, nitrogen, and minerals, inducing systemic resistance in the host plant, and stimulating plant growth, endophytic fungi can function as antagonists of plant pathogens and reduce their activity (Dubey et al. 2019; Vega 2018; Quesada-Moraga 2020).

By reducing the amount of carbon accessible in the rhizosphere, endophytic colonisation may help discourage infections. Additionally, high melanin levels may contribute to the formation of secondary metabolites that are poisonous to herbivores (Mandyam and Jumpponen 2005). Treatment of crop plants with endophytic fungi has been shown to reduce plant damage caused by various insect pests. Through their ability to infect aphids, thrips, lepidopterous larvae, and other cosmopolitan insects, some fungal endophytes can also function as insect pathogenic agents. These insects are a major global agricultural threat. According to Akutse et al. (2014), they are known to infect particular hosts and to present little to no harm to beneficial insects or nontarget organisms. The ability of endophytes to resist insects is closely linked to their ability to produce toxins, which render the plant unpalatable to various pests such as aphids, beetles, and grasshoppers.

In the western United States, sleepy grass, or *Achnatherum robustum*, is home to an endophyte that makes lysergic acid amide. After ingesting only modest amounts of contaminated grass, horses can sleep for up to 3 days before progressively recovering. Animals that are native to areas with drowsy grasses do not eat them, and once domesticated animals learn about the poisonous effects of such grasses, they avoid such plants preferentially (Li et al. 2004). *Neotyphodium uncinatum* inoculation of *Lolium pratense* (meadow ryegrass) decreased the amount of barley yellow dwarf virus infection in the inoculated plants. This was probably because the inoculation produced alkaloids that discouraged viruliferous aphid vectors and hence decreased the spread of the virus infection (Lehtonen et al. 2006). The nematode

Table 4.1 Secondary metabolites produced by endophytic fungi and its applications in agriculture

S.No.	Host plant	Endophyte	Secondary metabolite	Class of secondary metabolite	Application in agriculture	References
1	Rain forests	*Pestalotiopsis microspora*	Ambuic acid	Alkaloids	Active against *Fusarium* species and *Pythium ultimum*	Li et al. (2001)
2	Grasses	*Epichloe* spp.	Alkaloids	Alkaloids	Protection from herbivores	Schardl et al. (2004)
3	*Zea mays*	*Acremonium zeae*	Pyrrocidines A and B	Alkaloids	Antifungal activity against *Aspergillus flavus*	Wicklow et al. (2005)
4	*Picea glauca*	*Phialocephala scopiformis*	Rugulosin Skyrin Emodin	Quinone	Diminish herbivory of eastern spruce budworm *Choristoneura fumiferana*	Miller et al. (2008)
5	Ryegrass	*Epichloe* strain AR37	Epoxy-janthitrems	Alkaloids	Boad pesticidal effects against major pasture pests in New Zealand	Johnson et al. (2013)
6	Lettuce (*Lactuca sativa)* Radish (*Raphanus sativus)*	*Xylaria*	Cytochalasin E	Polyketides	Growth inhibition	Zhang et al. (2014)
7	*Lolium perenne*	*E. Festucae*	Gibberellins	Terpenoids	Plant growth	Dupont et al. (2015)
8	Rice	*Fusarium oxysporum* species complex	4-hydroxybenzoic acid, indole-3-acetic acid (IAA) Gibepyrone D	Indoles Polyketides	Nematocidal activity against Root-knot nematode, *Meloidogyne incognita*	Bogner et al. (2017)
9	Grasses	*Epichloe* spp.	Auxins	Indoles	Plant growth	Schmid et al. (2017)

(continued)

Table 4.1 (continued)

S.No.	Host plant	Endophyte	Secondary metabolite	Class of secondary metabolite	Application in agriculture	References
10	*Picea glauca*	*Phialocephala* sp. *P. scopiformis*	Vermiculine Rugulosin	Quinone	Mitigation of *C. fumiferana* and reduce damage to the foliage	McMullin et al. (2018)
11	*Bethencourtia palmensis*	*Aspergillus* species	Mellein Neoaspergillic acid	Isocoumarin derivative Amines	Inhibited growth of *Alternaria alternata*, *Botrytis cinerea* and *F. Oxysporum*	Morales-Sánchez et al. (2021)
12	*Moringa oleifera*	*Aspergillus terreus*	Ethyl acetate extract	Terpenoids Glycosides Flavonoids	Antifungal activity against fungi causing mucormycosis as *Rhizopus oryzae*, *Mucor racemosus*, *Syncephalastrum racemosum*	Hashem et al. (2022)
13	Kiwi	*Paraphaeosphaeria sporulosa*	Sporulactones A–D Sporulactones E and F	Isobenzofuranones Isocoumarins	Against canker disease caused by *Pseudomonas syringae*	Chen et al. (2022)
14	*Justicia adhatoda*	*Paecilomyces lilacinus*	Phytol Squalene Cyclotrisiloxane Hexamethyl	Phenols Flavonoids	Promotes growth of *Solanum lycopersicum* under heavy metal stress	Musa et al. (2023)

species *Bursaphelenchus xylophilus* and *Panagrellus redivivus* were the targets of nematocidal activity exhibited by three chlorinated, epimeric oxazinane derivatives that were obtained from *Geotrichum* sp. (Li et al. 2007).

Ownley et al. (2008) found that endophytic colonisation of tomato and cotton seedlings, achieved by soaking the seeds in *B. bassiana* conidia, provided protection against the two plant pathogens, *Rhizoctonia solani* and *Pythium myriotylum*. According to Fávaro et al. (2012), the endophyte *Epicoccum nigrum* that was isolated from sugarcane is particularly well-known for its ability to biocontrol diseases like *Pythium* in cotton, *Sclerotinia sclerotiorum* in sunflower, phytoplasma bacteria in apples, and *Monilinia* spp. in peaches and nectarines. Phytophagous mites are globally important pests of ornamental plants and agricultural crops. They inflict harm by eating on plants and spreading viruses, which lowers the plants' ability to photosynthesize, as well as their general health, yield, and market value. Citrus rust mites and other mites are significantly killed by the yeast-like basidiomycete *Meira geulakonigii*, probably as a result of argovin secretion (Paz et al. 2011).

Squash plants injected with *B. bassiana* showed signs of decreased zucchini yellow mosaic virus (ZYMV) disease incidence and severity (Jaber and Salem 2014). After extracting the homodimeric macrolide pyrenophorol from the foliar endophyte *Lophodermium nitens* of white pine, *Pinus strobus* L., Sumarah et al. (2015) found that the extract inhibited the growth of the pine disease *Cronartium ribicola*. Klieber and Reineke (2015) recorded 50% mortality of all larval instars and reduced lifetime of *Tuta absoluta* larvae fed with tomato leaves colonised by *B. bassiana*. Numerous fungal genera, such as *Acremonium*, *Alternaria*, *Arthrobotrys*, *Chaetomium*, *Cladosporium*, *Clonostachys*, *Diaporthe*, *Drechslerella*, *Epichloë*, *Epiccocum*, *Fusarium*, *Gibellulopsis*, *Melanconium*, *Neotyphodium*, *Paecilomyces*, *Phialemonium*, *Phyllosticta*, *Piriformospora*, *Purpureocillium*, *Talaromyces*, and *Trichoderma*, have been reported to have inhibitory effects on *Meloidogyne* species (Ghahremani et al. 2019; Márquez-Dávila et al. 2020).

Phomopsis obscurans and *Phomopsis viticola* were the targets of the actions of cytochalasins H and J produced by the endophytes *Diaphorte miriciae* UFMGCB 7719 and 6350, which are associated with the tropical medicinal plants *Copaifera pubiflora* and *Melocactus ernestii* (Carvalho et al. 2018). Two 10-membered lactones with antifungal activity against a variety of plant pathogenic fungi, such as *Aspergillus niger*, *Botrytis cinerea*, *Cochliobolus heterostrophus*, *Fusarium avenaceum*, *F. moniliforme*, *Ophiostoma minus*, and *Penicillium islandicum*, were produced by *Diaporthe* sp., which was isolated as a stem endophyte of *Azadirachta indica* (Chutulo and Chalannavar 2018). According to Baron et al. (2019), the most well-characterised and often utilised fungi in biological control programmes for the management of arthropods that may behave as pests or disease vectors are *Beauveria bassiana* and *Metarhizium anisopliae*.

In dual culture assays, it was demonstrated that *Trichoderma erinaceum* isolated from ginger and Stemona root inhibited the growth of the southern stem rot disease agent *Sclerotium rolfsii* by 64%, and in pot trials, it reduced infection by 58%. According to extract analysis, *T. erinaceum* produced the polyketide groups glucanase and chitinase, which inhibited *S. rolfsii*'s growth (Suebrasri et al. 2020). At doses of 0.2–2.5 mg/mL, a crude extract of *F. proliferatum*, which was separated from the medicinal plant *Cissus quadrangularis*, inhibited the development of *Rhizoctonia solani* and *F. oxysporum*. Subsequent examination of the crude extract demonstrated the presence of unsaturated alkenes, phenolics, and terpenoids (Singh et al. 2021). It was discovered that *Aspergillus neoniger* extracts separated from the medicinal plant *Ficus carica* may inhibit the growth of *Penicillium avelaneum*, *P. notatum*, and *A. terreus* by at least 80%. The production of aurasperone A and D by *A. neoniger* was determined by analysis using high-performance liquid chromatography and nuclear magnetic resonance spectroscopy (Abdou et al. 2021).

4.5.2 Crop Improvement and Adaptation

Endophytes are regarded as primary sources of bioactive compounds that can be thought of as biofertilisers because they promote plant growth by facilitating nutrient uptake through both the plant root system and the transfer of nitrogen derived from insects. Endophytes also act as a storehouse of unique bioactive secondary metabolites, such as alkaloids, saponins, tannins, phenolic acids, steroids, quinones, and terpenoids (Gouda et al. 2016; Yadav and Meena 2021; Yadav and Meena 2022). In one of the three trials, Maniania et al. (2003) found that spraying *M. anisopliae* at weekly intervals against onion thrips (*Thrips tabaci*) significantly increased onion yields. According to Rodriguez et al. (2008), *Leymus mollis* (dunegrass) plants from various coastal beach habitats in the United States have been found to be symbiotic with *Fusarium culmorum*, an endophyte that, when injected into salt-sensitive non-coastal plants, conferred salt tolerance to those plants. In their study, Fakhro et al. (2010) investigated the effects of inoculating tomato with *P. indica*. They found that *P. indica* colonises tomato roots in soil, increases leaf biomass by as much as 20%, and decreases the severity of Verticillium wilt disease caused by *Verticillium dahliae* by more than 30%.

It has been shown that *Phoma glomerata* (LWL2) and *Penicillium* sp. (LWL3) can form endophytic interaction with cucumber plants, which exhibit markedly higher biomass and improved growth in the presence of salt and water stress. The symbiotic relationship lessened the consequences of sodium toxicity during saline stress and enhanced the absorption of vital minerals like potassium, calcium, and magnesium. Furthermore, it was shown that the fungus modulated the synthesis of salicylic, jasmonic, and abscisic acids, demonstrating that they reprogrammed plant growth in stressful environments (Waqas et al. 2012). Behie and Bidochka (2014) assessed how seven fungal species—*M. robertsii*, *M. guizhouense*, *M. brunneum*, *M. flavoviridae*, *M. acridum*, *B. bassiana*, and *Akanthomyces lecanii*—transferred nitrogen from insects endophytically to four cultures that included two dicots, *Phaseolus vulgaris* (common bean) and *Glycine max* (soybean), and two monocots, *Triticum aestivum* (wheat) and *Panicum virgatum* (switchgrass).

Lopez and Sword (2015) assessed the impact of *Purpureocillium lilacinum* and *B. bassiana* endophytic colonisation on cotton plants and showed how both fungi might increase the dry mass and blossom count of the plants. According to Yamaji et al. (2016), the reason *Clethra barbinervis* can withstand high concentrations of heavy metals is because root fungal endophytes, such as *Phialocephala fortinii*, *Rhizodermea veluwensis*, and *Rhizoscyphus* sp., support the plant's ability to grow, promote K uptake, and reduce the toxicity of heavy metals. Jaber and Enkerli (2017) observed that artificial inoculation of *Beauveria brongniartii*, *B. bassiana*, and *M. brunneum* resulted in an improvement in the height, fresh weight of shoots, and roots of *Vicia faba* plants. The endophytes that live in *Panax ginseng*, including *Aspergillus* sp., *Cladosporium* sp., *Fusarium* sp., *Penicillium* sp., *Verticillium* sp., and *Ascomycete* sp., release triterpenoid compounds, saponins, and ginsenosides that promote root growth and biotic and abiotic stress tolerance (Sahoo et al. 2017). The production of gibberellins and indoleacetic acid by *Aspergillus niger* CSR3 was able to monitor the growth of endogenous rice seedlings. In culture, the endophyte also produced siderophores and solubilised phosphate, demonstrating its potential as a biofertiliser and plant disease suppressor (Lubna et al. 2018).

Jaber (2018) detailed the methodical colonisation of the aerial portion and roots of wheat (*T. aestivum*) by *B. bassiana* and *M. brunneum*, which led to the promotion of plant growth (shoot height, root length, and fresh root and shoot weights). Furthermore, the investigation demonstrated that endophytic colonisation had an adverse effect on *Fusarium culmorum*, a pathogen that is primarily responsible for crown and root rot in wheat. Inoculated with *Glomus intraradices* and *Glomus* sp. strain, lavender plants produced more giberellins and showed excellent drought resistance through increased N and P contents, water contents, and root biomass (Hussain et al. 2018).

Elevated temperatures pose a significant hindrance to crop yielding and microbiological colonisation, leading to significant cellular damage like protein aggregation and disintegration. The fungus *Curvularia protuberata* and its mycovirus, *Curvularia* thermal tolerance virus (CThTV), have been directly connected to the grass *Dichanthelium lanuginosum*'s ability to withstand soil temperatures ranging from 38 to 65 °C in Yellowstone National Park (Kumar et al. 2019). In cacao, barley, and Chinese cabbage plants, *Trichoderma* sp. and *P. indica* demonstrated resistance to salt and drought (Chhipa and Deshmukh 2019). The capacity of *Purpureocillium lilacinum*, *Purpureocillium lavendulum*, and *Metarhizium marquandii* to produce IAA and solubilise P from fluorapatite was also shown by Baron et al. (2020). After testing on plants of soybean, bean, and corn, the chosen strains were found to be able to enhance growth-improving factors such as dry matter (shoot and roots) and the availability of crucial nutrients like P and N. Keswani et al. (2022) examined the biosynthesis of gibberellins, which are advantageous to crop production and are produced by a variety of fungi.

4.6 Conclusion and Future Prospects

The fungal endophytes are an abundant source of novel secondary metabolites. Additionally, fungal endophytes' capacity to enhance plants' resistance to abiotic stressors like heat, salt, drought, and others adds a new dimension to host plants-endophytes interactions. This potential could be extensively investigated or employed in agriculture as a substitute for entomopathogenic fungi's auto-dissemination in flooding situations, as well as to reduce pests and diseases in context of climate change. In any case, it is evident that these fungi can significantly affect plant survival and fitness in all terrestrial habitats, and as a result, they probably have a big influence on the biogeography, evolution, and community structure of plants. The toxic properties of some compounds in plants are of immense value for improving the tolerance to insect pests and offer disease resistance. Due to inadequate isolation and characterisation, the commercial synthesis of bioactive chemicals from endophytes is still in its infancy. The complexities of the endophytic relationship between plants and microbes might be better understood by using contemporary genomic studies such as genome sequencing, comparative genomics, microarray, next-generation sequencing, metagenomics, metatranscriptomics involving meta-omics, and comparative studies. Endophytic fungi are advantageous because they offer numerous direct and indirect benefits to crop plants. Endophytic fungi are a valuable tool for agriculture as evidenced by their usage as an alternative with tremendous potential in agriculture field. Systematic investigation and characterisation of endophytic fungi for their ecological roles are anticipated to reveal new mechanisms controlling a variety of biological processes vital to plant health and provide concrete answers for problems pertaining to agriculture and the environment.

Acknowledgments The authors of the respective universities are grateful to their authorities for supporting this work.

References

Abdou R, Alqahtani AM, Attia HG (2021) Bioactive metabolites of *Aspergillus neoniger*, an endophyte of the medicinal plant *Ficus carica*. Indian J Pharm Sci 83(1):101–109. https://doi.org/10.36468/pharmaceutical-sciences.755

Akutse KS, Fiaboe KK, Van den Berg J, Ekesi S, Maniania NK (2014) Effects of endophyte colonization of *Vicia faba* (Fabaceae) plants on the life-history of leafminer parasitoids *Phaedrotoma scabriventris* (hymenoptera: braconidae) and *Diglyphus isaea* (hymenoptera: eulophidae). PLoS One 22:9(10):e109965. https://doi.org/10.1371/journal.pone.0109965

Anjum R, Afzal M, Baber R, Khan MAJ, Kanwal W, Sajid W, Raheel A (2019) Endophytes: as potential biocontrol agent—review and future prospects. J Agric Sci 11:113. https://doi.org/10.5539/jas.v11n4p113

Arnold AE, Lutzoni F (2007) Diversity and host range of foliar fungal endophytes: are tropical leaves biodiversity hotspots? Ecology 88(3):541–549. https://doi.org/10.1890/05-1459

Bamisile BS, Dash CK, Akutse KS, Keppanan R, Wang L (2018) Fungal endophytes: beyond herbivore management. Front Microbiol 9:544. https://doi.org/10.3389/fmicb.2018.00544

Baron NC, Costa NTA, Mochi DA, Rigobelo EC (2018) First report of *Aspergillus sydowii* and *Aspergillus brasiliensis* as phosphorus solubilizers in maize. Ann Microbiol 68(12):863–870. https://doi.org/10.1007/s13213-018-1392-5

Baron NC, Rigobelo EC, Zied DC (2019) Filamentous fungi in biological control: current status and future perspectives. Chil J Agric Res 79(2):307–315. https://doi.org/10.4067/S0718-58392019000200307

Baron NC, Pollo AD, Rigobelo EC (2020) *Purpureocillium lilacinum* and *Metarhizium marquandii* as plant growth-promoting fungi. Peerj 8:25. https://doi.org/10.7717/peerj.9005

Behie SW, Bidochka MJ (2014) Ubiquity of insect-derived nitrogen transfer to plants by endophytic insect-pathogenic fungi: an additional branch of the soil nitrogen cycle. Appl Environ Microbiol 80(5):1553–1560. https://doi.org/10.1128/AEM.03338-13

Behie SW, Jones SJ, Bidochka MJ (2015) Plant tissue localization of the endophytic insect pathogenic fungi *Metarhizium* and *Beauveria*. Fungal Ecol 13:112–119. https://doi.org/10.1128/AEM.03338-13

Bhupenchandra I, Chongtham SK, Devi AG, Dutta P, Sahoo MR, Mohanty S, Kumar S, Choudhary AK, Devi EL, Sinyorita S, Devi SH, Mahanta M, Kumari A, Devi HL, Josmee RK, Pusparani A, Pathaw N, Gupta S, Meena M, Janmeda P, Seth CS, Arora A, Sahoo A, Swapnil P (2024) Unlocking the potential of arbuscular mycorrhizal fungi: exploring role in plant growth promotion, nutrient uptake mechanisms, biotic stress alleviation, and sustaining agricultural production systems. J Plant Growth Regul 43(9):1–39. https://doi.org/10.1007/s00344-024-11467-9

Bogner CW, Kamdem RS, Sichtermann G, Matthäus C, Hölscher D, Popp J, Proksch P, Grundler FM, Schouten A (2017) Bioactive secondary metabolites with multiple activities from a fungal endophyte. Microb Biotechnol 10(1):175–188. https://doi.org/10.1111/1751-7915.12467

Bomke C, Tudzynski B (2009) Diversity, regulation, and evolution of the gibberellin biosynthetic pathway in fungi compared to plants and bacteria. Phytochemistry 70(15–16):1876–1893. https://doi.org/10.1016/j.phytochem.2009.05.020

Boudaoud H, Debbache N, Berboucha M, Atman D, Chaher N, Atmani D (2009) Flavonoids in human health: from structure to biological activity. Curr Res Nutr Food Sci 5:225–237. https://doi.org/10.2174/157340109790218049

Busby PE, Ridout M, Newcombe G (2016) Fungal endophytes: modifiers of plant disease. Plant Mol Biol 90(6):645–655. https://doi.org/10.1007/s11103-015-0412-0

Carvalho CR, Ferreira-D'silva A, Wedge DE, Cantrell CL, Rosa LH (2018) Antifungal activities of cytochalasins produced by *Diaporthe miriciae*, an endophytic fungus associated with tropical medicinal plants. Can J Microbiol 64(11):835–843. https://doi.org/10.1139/cjm-2018-0131

Chadha N, Mishra M, Rajpal K, Bajaj R, Choudhary DK, Varma A (2015) An ecological role of fungal endophytes to ameliorate plants under biotic stress. Arch Microbiol 197(7):869–881. https://doi.org/10.1007/s00203-015-1130-3

Chandran H, Meena M, Swapnil P (2021) Plant growth-promoting rhizobacteria as a green alternative for sustainable agriculture. Sustain For 13(19):10986. https://doi.org/10.3390/su131910986

Chen Q, Yu JJ, He J, Feng T, Liu JK (2022) Isobenzofuranones and isocoumarins from kiwi endophytic fungus *Paraphaeosphaeria sporulosa* and their antibacterial activity against *Pseudomonas syringae* pv. actinidiae. Phytochemistry 195:113050. https://doi.org/10.1016/j.phytochem.2021.113050

Chhipa H, Deshmukh SK (2019) Fungal endophytes: rising tools in sustainable agriculture production. In: Jha S (ed) Endophytes and secondary metabolites, Reference series in Phytochemistry. Springer, Cham. https://doi.org/10.1007/978-3-319-90484-9-26

Christian N, Sullivan C, Visser ND, Clay K (2016) Plant host and geographic location drive endophyte community composition in the face of perturbation. Microb Ecol 72(3):621–632. https://doi.org/10.1007/s00248-016-0804-y

Christianson DW (2017) Structural and chemical biology of Terpenoid Cyclases. Chem Rev 117(17):11570–11648. https://doi.org/10.1021/acs.chemrev.7b00287

Chutulo EC, Chalannavar RK (2018) Endophytic mycoflora and their bioactive compounds from *Azadirachta indica*: a comprehensive review. J Fungi (Basel) 4(2):42. https://doi.org/10.3390/jof4020042

Das M, Prakash HS, Nalini MS (2017) Antioxidative properties of phenolic compounds isolated from the fungal endophytes of *Zingiber nimmonii* (J. Graham) Dalzell. Front Biol 12:151–162. https://doi.org/10.1007/s11515-016-1441-z

De Silva NI, Brooks S, Lumyong S, Hyde KD (2019) Use of endophytes as biocontrol agents. Fungal Biol Rev 33(2):133–148. https://doi.org/10.1016/j.fbr.2018.10.001

Dubey MK, Meena M, Aamir M, Zehra A, Upadhyay RS (2019) Regulation and role of metal ions in secondary metabolites production by microorganisms. In: Gupta VK, Jogaiah S (eds) Singh HB. Elsevier, New and future developments in microbial biotechnology and bioengineering, pp 559–577. https://doi.org/10.1016/B978-0-444-63504-4.00019-0

Dupont P, Eaton CJ, Wargent JJ, Fechtner S, Solomon P, Schmid J, Day RC, Scott B, Cox MP (2015) Fungal endophyte infection of ryegrass reprograms host metabolism and alters development. New Phytol 208(4):1227–1240. https://doi.org/10.1111/nph.13614

Fakhro A, Andrade-Linares DR, von Bargen S, Bandte M, Büttner C, Grosch R, Schwarz D, Franken P (2010) Impact of *Piriformospora indica* on tomato growth and on interaction with fungal and viral pathogens. Mycorrhiza 20(3):191–200. https://doi.org/10.1007/s00572-009-0279-5

Fávaro LC, Sebastianes FL, Araújo WL (2012) *Epicoccum nigrum* P16, a sugarcane endophyte, produces antifungal compounds and induces root growth. PLoS One 7(6):e36826. https://doi.org/10.1371/journal.pone.0036826

Ghahremani Z, Escudero N, Saus E, Gabaldón T, Sorribas FJ (2019) Pochonia Chlamydosporia induces plant-dependent systemic resistance to *Meloidogyne incognita*. Front Plant Sci 13(10):945. https://doi.org/10.3389/fpls.2019.00945

Gouda S, Das G, Sen SK, Shin HS, Patra JK (2016) Endophytes: a treasure house of bioactive compounds of medicinal importance. Front Microbiol 29(7):1538. https://doi.org/10.3389/fmicb.2016.01538

Gupta A, Meshram V, Gupta M, Goyal S, Qureshi KA, Jaremko M, Shukla KK (2023) Fungal endophytes: microfactories of novel bioactive compounds with therapeutic interventions; a comprehensive review on the biotechnological developments in the field of fungal Endophytic biology over the last decade. Biomolecules 13(7):1038. https://doi.org/10.3390/biom13071038

Hashem AH, Shehabeldine AM, Abdelaziz AM, Amin BH, Sharaf MH (2022) Antifungal activity of Endophytic *Aspergillus terreus* extract against some fungi causing Mucormycosis: ultrastructural study. Appl Biochem Biotechnol 194(8):3468–3482. https://doi.org/10.1007/s12010-022-03876-x

Hodkinson TR (2018) Evolution and taxonomy of the grasses (Poaceae): a model family for the study of species-rich groups. Annu Plant Rev 15:255–294. https://doi.org/10.1002/9781119312994.apr0622

Hu Z, Wu Y, Xie S, Sun W, Guo Y, Li XN, Liu J, Li H, Wang J, Luo Z, Xue Y, Zhang Y (2017) Phomopsterones A and B, two functionalized Ergostane-type steroids from the Endophytic fungus *Phomopsis* sp. TJ507A. Org Lett 19(1):258–261. https://doi.org/10.1021/acs.orglett.6b03557

Huang JX, Zhang J, Zhang XR, Zhang K, Zhang X, He XR (2014) *Mucor fragilis* as a novel source of the key pharmaceutical agents podophyllotoxin and kaempferol. Pharm Biol 52(10):1237–1243. https://doi.org/10.3109/13880209.2014.885061

Hussain SS, Mehnaz S, Siddique KHM (2018) Harnessing the plant microbiome for improved abiotic stress tolerance. In: Egamberdieva D, Ahmad P (eds) Microorganisms for sustainability. Springer Nature, Singapore. https://doi.org/10.1007/978-981-10-5514-0_2

Jaber LR (2018) Seed inoculation with endophytic fungal entomopathogens promotes plant growth and reduces crown and root rot (CRR) caused by *Fusarium culmorum* in wheat. Planta 248(6):1525–1535. https://doi.org/10.1007/s00425-018-2991-x

Jaber LR, Enkerli J (2017) Fungal entomopathogens as endophytes: can they promote plant growth? Biocontrol Sci Technol 27:28–41. https://doi.org/10.1080/09583157.2016.1243227

Jaber LR, Salem NM (2014) Endophytic colonisation of squash by the fungal entomopathogen *Beauveria bassiana* (Ascomycota: Hypocreales) for managing Zucchini yellow mosaic virus in cucurbits. Biocontrol Sci Tech 24:1096–1109. https://doi.org/10.1080/09583157.2014.923379

Jaroszuk-Ściseł J, Kurek E, Trytek M (2014) Efficiency of indoleacetic acid, gibberellic acid and ethylene synthesized in vitro by *Fusarium culmorum* strains with different effects on cereal growth. Biologia 69(3):281–292. https://doi.org/10.2478/s11756-013-0328-6

Johnson LJ, De Bonth ACM, Briggs LR, Caradus JR, Finch SC, Fleetwood DJ, Fletcher LR, Hume DE, Johnson RD, Popay AJ, Tapper BA, Simpson WR, Voisey CR, Card SD (2013) The exploitation of Epichloae endophytes for agricultural benefit. Fungal Divers 60(1):171–188. https://doi.org/10.1007/s13225-013-0239-4

Kaddes A, Fauconnier ML, Sassi K, Nasraoui B, Jijakli MH (2019) Endophytic fungal volatile compounds as solution for sustainable agriculture. Molecules 24(6):1605. https://doi.org/10.3390/molecules24061065

Keswani C, Singh SP, García-Estrada C, Mezaache-Aichour S, Glare TR, Borriss R, Rajput VD, Minkina TM, Ortiz A, Sansinenea E (2022) Biosynthesis and beneficial effects of microbial gibberellins on crops for sustainable agriculture. J Appl Microbiol 132(3):1597–1615. https://doi.org/10.1111/jam.15348

Khan AL, Hussain J, Al-Harrasi A, Al-Rawahi A, Lee IJ (2015) Endophytic fungi: resource for gibberellins and crop abiotic stress resistance. Crit Rev Biotechnol 35(1):62–74. https://doi.org/10.3109/07388551.2013.800018

Klieber J, Reineke A (2015) The entomopathogen *Beauveria bassiana* has epiphytic and endophytic activity against the tomato leaf miner *Tuta absoluta*. J Appl Entomol 140(8):580–589. https://doi.org/10.1111/jen.12287

Korpi A, Jarnberg J, Pasanen AL (2009) Microbial volatile organic compounds. Crit Rev Toxicol 39(2):139–193. https://doi.org/10.1080/10408440802291497

Krohn K, Florke U, John M, Root N, Steingrover K, Aust HJ, Draeger S, Schulz B, Antus S, Simonyi M, Zsila F (2001) Biologically active metabolites from fungi part 16. New preussomerins J, K and L from an endophytic fungus: structure elucidation, crystal structure analysis and determination of absolute configuration by CD calculations. Tetrahedron 57:4343–4348. https://doi.org/10.1016/S0040-4020(01)00320-9

Kumar V, Soni R, Jain L, Dash B, Goel R (2019) Endophytic fungi: recent advances in identification and explorations. In: Singh BP (ed) Advances in endophytic fungal research, fungal biology. Springer International Publishing AG. https://doi.org/10.1007/978-3-030-03589-1_13

Kumar R, Swapnil P, Meena M, Selpair S, Yadav BG (2022) Plant growth-promoting rhizobacteria (PGPR): approaches to alleviate abiotic stresses for enhancement of growth and development of medicinal plants. Sustain For 14(23):15514. https://doi.org/10.3390/su142315514

Latz MAC, Jensen B, Collinge DB, Jorgensen HJL (2018) Endophytic fungi as biocontrol agents: elucidating mechanisms in disease suppression. Plant Ecol Divers 11(5–6):555–567. https://doi.org/10.1080/17550874.2018.1534146

Le Cocq K, Gurr SJ, Hirsch PR, Mauchline TH (2017) Exploitation of endophytes for sustainable agricultural intensification. Mol Plant Pathol 18(3):469–473. https://doi.org/10.1111/mpp.12483

Lehtonen PT, Helander M, Siddiqui SA, Lehto K, Saikkonen K (2006) Endophytic fungus decreases plant virus infections in meadow rye grass (*Lolium pratense*). Biol Lett 2(4):620–623. https://doi.org/10.1098/rsbl.2006.0499

Li JY, Harper JK, Grant DM, Tombe BO, Bashyal B, Hess WM, Strobel GA (2001) Ambuic acid, a highly functionalized Cyclohexenone with antifungal activity from *Pestalotiopsis* spp. and *Monochaetia* sp. Phytochemistry 56(5):463–468. https://doi.org/10.1016/S0031-9422(00)00408-8

Li C, Nan Z, Paul VH, Dapprich P, Liu Y (2004) A new *Neotyphodium* species symbiotic with drunken horse grass (*Achnatherum inebrians*) in China. Mycotaxon 90(1):141–147

Li GH, Yu ZF, Li X, Wang XB, Zheng LJ, Zhang KQ (2007) Nematicidal metabolites produced by the endophytic fungus *Geotrichum* sp. AL4. Chem Biodivers 4(7):1520–1524. https://doi.org/10.1002/cbdv.200790131

Lopez DC, Sword GA (2015) The endophytic fungal entomopathogens *Beauveria bassiana* and *Purpureocillium lilacinum* enhance the growth of cultivated cotton (*Gossypium hirsutum*) and negatively affect survival of the cotton bollworm (*Helicoverpa zea*). Biol Control 89:53–60. https://doi.org/10.1016/j.biocontrol.2015.03.010

Lu H, Zou WX, Meng JC, Hu J, Tan RX (2000) New bioactive metabolites produced by *Colletotrichum* sp., an endophytic fungus in *Artemisia annua*. Plant Sci 151(1):67–73. https://doi.org/10.1016/S0168-9452(99)00199-5

Lubna AS, Hamayun M, Gul H, Lee IJ, Hussain A (2018) *Aspergillus Niger* CSR3 regulates plant endogenous hormones and secondary metabolites by producing gibberellins and indoleacetic acid. J Plant Interact 13(1):100–111. https://doi.org/10.1080/17429145.2018.1436199

Lugtenberg BJJ, Caradus JR, Johnson LJ (2016) Fungal endophytes for sustainable crop production. FEMS Microbiol Ecol 92(12):17. https://doi.org/10.1093/femsec/fiw194

Mandyam K, Jumpponen A (2005) Seeking the elusive function of the root-colonising dark septate endophytic fungi. Stud Mycol 53:173–189. https://doi.org/10.3114/sim.53.1.173

Maniania NK, Sithanantham S, Ekesi S, Ampong-Nyarko K, Löhr B, Matoka CM (2003) A field trial of the entomogenous fungus *Metarhizium anisopliae* for control of onion thrips, *Thripstabaci*. Crop Prot 22(3):553–559. https://doi.org/10.1016/S0261-2194(02)00221-1

Márquez-Dávila K, Arévalo-López L, Gonzáles R, Vega L, Meza M (2020) *Trichoderma* and *Clonostachys* as biocontrol agents against *Meloidogyne incognita* in sacha inchi. Pesqui Agropecu Trop 50:e60890. https://doi.org/10.1590/1983-40632020v5060890

McMullin DR, Nguyen HDT, Daly GJ, Menard BS, Miller JD (2018) Detection of foliar endophytes and their metabolites in *Picea* and *Pinus* seedling needles. Fungal Ecol 31:1–8. https://doi.org/10.1016/j.funeco.2017.09.003

Meena M, Swapnil P, Zehra A, Dubey MK, Upadhyay RS (2017) Antagonistic assessment of *Trichoderma* spp. by producing volatile and non-volatile compounds against different fungal pathogens. Arch Phytopathol Plant Protect 50(13–14):629–648. https://doi.org/10.1080/03235408.2017.1357360

Meshram V, Elazar M, Maymon M, Sharma G, Shawahna R, Charuvi D, Freeman S (2023) Endophytic *Fusarium clavum* confers growth and salt tolerance in *Cucumis melo*. Environ Exp Bot 206:105–153. https://doi.org/10.1016/j.envexpbot.2022.105153

Miller JD, Sumarah MW, Adams GW (2008) Effect of a rugulosin-producing endophyte in *Picea glauca* on *Choristoneura fumiferana*. J Chem Ecol 34(3):362–368. https://doi.org/10.1007/s10886-007-9412-0

Mishra A, Gond SK, Kumar A, Sharma VK, Verma SK, Kharwar RN, Sieber TN (2012) Season and tissue type affect fungal endophyte communities of the Indian medicinal plant *Tinospora cordifolia* more strongly than geographic location. Microb Ecol 64(2):388–398. https://doi.org/10.1007/s00248-012-0029-7

Morales-Sánchez V, Fe Andrés M, Díaz CE, González-Coloma A (2020) Factors affecting the metabolite productions in endophytes: biotechnological approaches for production of metabolites. Curr Med Chem 27(11):1855–1873. https://doi.org/10.2174/0929867326666190626154421

Morales-Sánchez V, Díaz CE, Trujillo E, Olmeda SA, Valcarcel F, Muñoz R, Andrés MF, González-Coloma A (2021) Bioactive metabolites from the endophytic fungus *Aspergillus* sp. SPH2. J Fungi (Basel) 7(2):109. https://doi.org/10.3390/jof7020109

Mousa WK, Raizada MN (2013) The diversity of anti-microbial secondary metabolites produced by fungal endophytes: an interdisciplinary perspective. Front Microbiol 4:65. https://doi.org/10.3389/fmicb.2013.00065

Musa M, Jan FG, Hamayun M, Jan G, Khan SA, Rehman G, Ali S, Lee I-J (2023) An endophytic fungal isolate *Paecilomyces lilacinus* produces bioactive secondary metabolites and promotes growth of *Solanum lycopersicum* under heavy metal stress. Agronomy 13(3):883. https://doi.org/10.3390/agronomy13030883

Naik BS (2019) Functional roles of fungal endophytes in host fitness during stress conditions. Symbiosis 79(2):99–115. https://doi.org/10.1007/s13199-019-00635-1

Nascimento T, Oki Y, Lima D, Almeida-Cortez J, Fernandes GW, Souza-Motta C (2015) Biodiversity of endophytic fungi in different leaf ages of *Calotropis procera* and their antimicrobial activity. Fungal Ecol 14:79–86. https://doi.org/10.1016/j.funeco.2014.10.004

Nguyen VT, Le MD (2018) Influence of various drying conditions on phytochemical compounds and antioxidant activity of carrot peel. Beverages 4(4):80. https://doi.org/10.3390/beverages4040080

Omomowo OI, Babalola OO (2019) Bacterial and fungal endophytes: tiny giants with immense beneficial potential for plant growth and sustainable agricultural productivity. Microorganisms 7(11):481. https://doi.org/10.3390/microorganisms7110481

Ortega-Garcia JG, Montes-Belmont R, Rodriguez-Monroy M, Ramirez-Trujillo JA, Suarez-Rodriguez R, Sepulveda- Jimenez G (2015) Effect of *Trichoderma asperellum* applications and mineral fertilization on growth promotion and the content of phenolic compounds and flavonoids in onions. Sci Hortic 195:8–16. https://doi.org/10.1016/j.scienta.2015.08.027

Ortíz-Castro R, Contreras-Cornejo HA, Macías-Rodríguez L, López-Bucio J (2009) The role of microbial signals in plant growth and development. Plant Signal Behav 4(8):701–712. https://doi.org/10.4161/psb.4.8.9047

Ownley BH, Griffin MR, Klingeman WE, Gwinn KD, Moulton JK, Pereira RM (2008) *Beauveria bassiana*: endophytic colonization and plant disease control. J Invertebr Pathol 98(3):267–270. https://doi.org/10.1016/j.jip.2008.01.010

Paz Z, Bilkis I, Gerson U, Kerem Z, Sztejnberg A (2011) Argovin, a novel natural product secreted by the fungus *Meiraargovae*, is antagonistic to mites. Entomol Exp Appl 140(3):247–253. https://doi.org/10.1111/j.1570-7458.2011.01155.x

Piel J (2010) Biosynthesis of polyketides by trans-AT polyketide synthases. Nat Prod Rep 27:996–1047. https://doi.org/10.1039/B816430B

Poveda J, Abril-Urias P, Escobar C (2020) Biological control of plant-parasitic nematodes by filamentous fungi inducers of resistance: *Trichoderma*, mycorrhizal and endophytic fungi. Front Microbiol 11:992. https://doi.org/10.3389/fmicb.2020.00992

Purahong W, Hyde KD (2011) Effects of fungal endophytes on grass and non-grass litter decomposition rates. Fungal Divers 47(1):1–7. https://doi.org/10.1007/s13225-010-0083-8

Qin JC, Gao JM, Zhang YM, Yang SX, Bai MS, Ma YT, Laatsch H (2009) Polyhydroxylated steroids from an endophytic fungus, *Chaetomium globosum* ZY-22 isolated from *Ginkgo biloba*. Steroids 74(9):786–790. https://doi.org/10.1016/j.steroids.2009.04.011

Qiu M, Xie R, Shi Y, Zhang H, Chen H (2010) Isolation and identification of two flavonoid-producing endophytic fungi from *Ginkgo biloba* L. Ann Microbiol 60:143–150. https://doi.org/10.1007/s13213-010-0016-5

Quesada-Moraga Q (2020) Entomopathogenic fungi as endophytes: their broader contribution to IPM and crop production. Biocontrol Sci Tech 30(9):864–877. https://doi.org/10.1080/09583157.2020.1771279

Quesada-Moraga E, López-Díaz C, Landa BB (2014) The hidden habit of the entomopathogenic fungus *Beauveria bassiana*: first demonstration of vertical plant transmission. PLoS One 9:e89278. https://doi.org/10.1371/journal.pone.0089278

Rana KL, Kour D, Kaur T, Devi R, Yadav AN, Yadav N, Dhaliwal HS, Saxena AK (2020) Endophytic microbes: biodiversity, plant growth-promoting mechanisms and potential applications for agricultural sustainability. Anton Leeuw Int J G Mol Microbiol 113(8):1075–1107. https://doi.org/10.1007/s10482-020-01429-y

Rodriguez R, Redman R (2008) More than 400 million years of evolution and some plants still can't make it on their own: plant stress tolerance via fungal symbiosis. J Exp Bot 59(5):1109–1114. https://doi.org/10.1093/jxb/erm342

Rodriguez RJ, Henson J, Van Volkenburgh E, Hoy M, Wright L, Beckwith F, Kim YO, Redman RS (2008) Stress tolerance in plants via habitat-adapted symbiosis. ISME J 2(4):404–416. https://doi.org/10.1038/ismej.2007.106

Rodriguez RJ, White JF Jr, Arnold AE, Redman RS (2009) Fungal endophytes: diversity and functional roles. New Phytol 182(2):314–330. https://doi.org/10.1111/j.1469-8137.2009.02773.x

Rozpadek P, Domka AM, Nosek M, Ważny R, Jędrzejczyk RJ, Wiciarz M, Turnau K (2018) The role of strigolactone in the cross-talk between *Arabidopsis thaliana* and the endophytic fungus *Mucor* sp. Front Microbiol 9:441. https://doi.org/10.3389/fmicb.2018.00441

Rustamova N, Bobakulov K, Begmatov N, Turak A, Yili A, Aisa HA (2019) Secondary metabolites produced by endophytic *Pantoea ananatis* derived from roots of *Baccharoides anthelmintica* and their effect on melanin synthesis in murine B16 cells. Nat Prod Res 35(5):796–801. https://doi.org/10.1080/14786419.2019.1597354

Saeed A, Qasim M (2014) Total synthesis of cytotoxic metabolite (±)-desmethyldiaportinol from *Ampelomyces* sp. Nat Prod Res 28(3):185–190. https://doi.org/10.1080/14786419.2013.866111

Sahoo S, Sarangi S, Kerry RG (2017) Bioprospecting of endophytes for agricultural and environmental sustainability (Chapter-19). In: Patra JK, Vishnuprasad CN, Das G (eds) Microbial biotechnology. Springer, Singapore, pp 429–458. https://doi.org/10.1007/978-981-10-6847-8_19

Salam N, Khieu T-N, Liu M-J, Vu T-T, Chu-Ky S, Quach N-T, Phi Q-T, Rao MPN, Fontana A, Sarter S, Li W-J (2017) Endophytic actinobacteria associated with *Dracaena cochinchinensis* Lour.: isolation, diversity, and their cytotoxic activities. Biomed Res Int 2017:1–11. https://doi.org/10.1155/2017/1308563

Sarsaiya S, Shi J, Chen J (2019) A comprehensive review on fungal endophytes and its dynamics on Orchidaceae plants: current research, challenges, and future possibilities. Bioengineered 10(1):316–334. https://doi.org/10.1080/21655979.2019.1644854

Schardl C, Craven K (2003) Interspecific hybridization in plant-associated fungi and oomycetes: a review. Mol Ecol 12(11):2861–2873. https://doi.org/10.1046/j.1365294X.2003.01965.x

Schardl CL, Leuchtmann A, Spiering MJ (2004) Symbioses of grasses with seedborne fungal endophytes. Ann Rev Plant Biol 55:315–340. https://doi.org/10.1146/annurev.arplant.55.031903.141735

Schmid J, Day R, Zhang N, Dupont P-Y, Cox MP, Schardl CL, Minards N, Truglio M, Moore N, Harris DR, Zhou Y (2017) Host tissue environment directs activities of an Epichloë endophyte, while it induces systemic hormone and defense responses in its native perennial ryegrass host. Mol Plant-Microbe Interact 30:138–149. https://doi.org/10.1094/MPMI-10-16-0215-R

Sikora RA, Pocasangre L, Zum Felde A, Niere B, Vu TT, Dababat A (2008) Mutualistic endophytic fungi and in-planta suppressiveness to plant parasitic nematodes. Biol Control 46(1):15–23. https://doi.org/10.1016/j.biocontrol.2008.02.011

Singh A, Kumar J, Sharma VK, Singh DK, Kumari P, Nishad JH, Gautam VS, Kharwar RN (2021) Phytochemical analysis and antimicrobial activity of an endophytic *Fusarium proliferatum* (ACQR8), isolated from a folk medicinal plant *Cissus quadrangularis L.* S Afr J Bot 140:87–94. https://doi.org/10.1016/j.sajb.2021.03.004

Song GC, Ryu CM (2013) Two volatile organic compounds trigger plant self-defense against a bacterial pathogen and a sucking insect in cucumber under open field conditions. Int J Mol Sci 14(5):9803–9819. https://doi.org/10.3390/ijms14059803

Strobel G, Daisy B (2003) Bioprospecting for microbial endophytes and their natural products. Microbiol Mol Biol Rev 67(4):491–502. https://doi.org/10.1128/MMBR.67.4.491-502.2003

Suebrasri T, Somteds A, Harada H, Kanokmedhakul S, Jogloy S, Ekprasert J, Lumyong S, Boonlue S (2020) Novel endophytic fungi with fungicidal metabolites suppress sclerotium disease. Rhizosphere 16:100250. https://doi.org/10.1016/j.rhisph.2020.100250

Sumarah MW, Walker A, Seifert KA, Todorov A, Miller JD (2015) Screening of fungal endophytes isolated from eastern white pine needles. In: Jetter R (ed) The formation, structure and activity of phytochemicals. Springer. https://doi.org/10.1007/978-3-319-20397-3_8

Sun X, Guo LD, Hyde KD (2011) Community composition of endophytic fungi in *Acer truncatum* and their role in decomposition. Fungal Divers 47(1):85–95. https://doi.org/10.1007/s13225-010-0086-5

Suryanarayanan TS (2019) Endophytes and weed management: a commentary. Plant Physiol Rep 24(4):576–579. https://doi.org/10.1007/s40502-019-00488-2

Toghueo RMK (2020) Bioprospecting endophytic fungi from *Fusarium* genus as sources of bioactive metabolites. Mycology 11(1):1–21. https://doi.org/10.1080/21501203.2019.1645053

Vega V (2018) The use of fungal entomopathogens as endophytes in biological control: a review. Mycologia 110(1):4–30. https://doi.org/10.1080/00275514.2017.1418578

Waqas M, Khan AL, Kamran M, Hamayun M, Kang SM, Kim YH, Lee IJ (2012) Endophytic fungi produce gibberellins and indoleacetic acid and promotes host-plant growth during stress. Molecules (Basel) 17(9):10754–10773. https://doi.org/10.3390/molecules170910754

Wicklow DT, Roth S, Deyrup ST, Gloer JB (2005) A protective endophyte of maize: *Acremonium zeae* antibiotics inhibitory to *Aspergillus flavus* and *Fusarium verticillioides*. Mycol Res 109(5):610–618

Worapong J, Strobel G, Ford EJ, Li JY, Baird G, Hess WM (2001) *Muscodoralbus* anam. gen. et sp. nov., an endophyte from *Cinnamomum zeylanicum*. Mycotaxon 79:67–79

Wu L, Han T, Li W, Jia M, Xue L, Rahman K, Qin L (2013) Geographic and tissue influences on endophytic fungal communities of *Taxuschinensis* var. mairei in China. Curr Microbiol 66(1):40–48. https://doi.org/10.1007/s00284-012-0235-z

Yadav G, Meena M (2021) Bioprospecting of endophytes in medicinal plants of Thar Desert: an attractive resource for biopharmaceuticals. Biotechnol Rep 30:e00629. https://doi.org/10.1016/j.btre.2021.e00629

Yadav G, Meena M (2022) Biological control of plant diseases by endophytes. In: Shah MP, Deka D (eds) Endophytic association: what, why and how (volume: developments in applied microbiology and biotechnology). Academic Press, Elsevier Inc., pp 119–135. https://doi.org/10.1016/B978-0-323-91245-7.00007-9

Yamagiwa Y, Inagaki Y, Ichinose Y, Toyoda K, Hyakumachi M, Shiraishi T (2011) *Talaromyces wortmannii* FS2 emits β-caryphyllene, which promotes plant growth and induces resistance. J Gen Plant Pathol 77:336–341. https://doi.org/10.1007/s10327-011-0340-z

Yamaji K, Watanabe Y, Masuya H, Shigeto A, Yui H, Haruma T (2016) Root fungal endophytes enhance heavy-metal stress tolerance of *Clethra barbinervis* growing naturally at mining sites via growth enhancement, promotion of nutrient uptake and decrease of heavy-metal concentration. PLoS One 11(12):e0169089. https://doi.org/10.1371/journal.pone.0169089

Yan L, Zhu J, Zhao XX, Shi JL, Jiang CM, Shao DY (2019) Beneficial effects of endophytic fungi colonization on plants. Appl Microbiol Biotechnol 103(8):3327–3340. https://doi.org/10.1007/s00253-019-09713-2

Yang X, Strobel G, Stierle A, Hess WM, Lee J, Clardy J (1994) A fungal endophyte-tree relationship: *Phoma* sp. in *Taxus wallichiana*. Plant Sci 102(1):1–9. https://doi.org/10.1016/0168-9452(94)90017-5

Zhang Q, Xiao J, Sun QQ, Qin JC, Pescitelli G, Gao JM (2014) Characterization of cytochalasins from the endophytic *Xylaria* sp. and their biological functions. J Agric Food Chem 62(45):10962–10969. https://doi.org/10.1021/jf503846z

Role of Fungal Endophytes in Plant Health and Abiotic Stress Management

5

Azha Ufaq Nabi, Faamiya Shajar, Sheikh Tajamul Islam, and Reiaz Ul Rehman

Abstract

The growing global population has escalated the need to enhance food production, highlighting the importance of sustainable agricultural practices. Fungal endophytes, microorganisms residing within plant tissues, offer an eco-friendly replacement to synthetic compounds in agronomy by enhancing crop growth, stress resistance, and perseverance as a reaction to different stressors. These endophytes can boost nutrient absorption, promote growth through hormone production, and provide biocontrol by inhibiting pathogens through direct competition and systemic resistance. Fungal endophytes are classified into distinct groups, each contributing to plant health and growth through mechanisms like producing bioactive compounds, facilitating stress adaptation, and enhancing plant defense systems. Despite extensive studies on their biodiversity and potential, further research is required to fully comprehend their symbiotic functions, evolutionary history, and application in sustainable agriculture. By channelling the benefits of endophytes, future research can offer innovative solutions to alleviate the impact of climate changes on food production and promote sustainable farming practices.

Keywords

Abiotic stress · Biocontrol · Climate change · Fungal endophytes · Stress management · Sustainable farming

A. U. Nabi · F. Shajar · S. T. Islam · R. U. Rehman (✉)
Department of Bioresources, School of Biological Sciences, University of Kashmir, Srinagar, India

M. Meena et al. (eds.), *Metabolic Sustainability of Endophytes*, Microorganisms for Sustainability 52, https://doi.org/10.1007/978-981-96-4004-1_5

5.1 Introduction

The growing global population has significantly increased the requirement for food, driving the need for innovative strategies to sustain diverse life forms. To meet this rising demand, maximizing crop yields and minimizing deficits caused by biotic stresses (like plant pathogens, pests, and nematodes) and abiotic stresses (like salinity, drought, and heavy metals) is important for securing food supplies worldwide (Verma et al. 2022). The excessive usage of synthetic chemicals to protect plants from these stresses has resulted in adverse environmental consequences, including soil and water pollution, harm to non-target creatures, and the development of resistance to chemicals (Pathak et al. 2022). Addressing these plant stressors is critical for safeguarding food security, as well as for economic, environmental, and public health reasons. Therefore, sustainable alternatives are needed to reduce chemical use in agriculture, mitigate their negative impacts, and secure long-term food production (Muhie 2022). Endophytes present a promising and eco-friendly approach that can help increase crop yield while lessening the detrimental impacts of traditional chemical usage used to manage stressors (Chaudhary et al. 2022).

Endophytes are microorganisms that reside within plants without endangering or affecting their hosts (Verma et al. 2021). These organisms are critical in preserving plant health and ecosystem functioning by forming mutually beneficial relationships with their hosts for all or a certain time period of their life span (Jha et al. 2023). The distribution of endophytes relies on two major elements—host plant species and the surrounding environment. They are mostly found in vascular plants, evolving in the phyllosphere and rhizosphere before invading through natural openings or wounds. Endophytes inhabit different plant types like trees and grasses, and they are mostly seen in stems, roots, and leaves. Endophyte transmission can occur either vertically (systemically), from one generation to the other, or horizontally (non-systemically), between unrelated plants (Lugtenberg et al. 2016). Endophytes have garnered interest in agriculture, industry, and biotechnology owing to their potential applications in crop enhancement, biological control, production of enzymes, biological remediation, nutrient cycling, biodegradation, and biotransformation (Fadiji and Babalola 2020). Diverse bioactives, including flavonoids, alkaloids, phenolic acids, quinones, saponins, tannins, steroids, and terpenoids, are also abundant in endophytes. These substances have a number of advantages, such as immunosuppressive, antibacterial, anticancer, antioxidant, anti-malarial, anti-diabetic, anti-inflammatory qualities (Murali et al. 2017). While these bioactive metabolites enhance the plant's resistance to stress, they also hold significant therapeutic prospects (Lugtenberg et al. 2016) (Saad et al. 2018). The usage of fungal endophytes provides a green and low-impact way to manage diverse stresses in plants, helping the plants to acclimatize to stress conditions while minimizing risks to non-target organisms and reducing chemical residues in food.

5.2 Fungal Endophytes

Endophytes are well known for their capability to strengthen stress endurance of plants, nutrient recycling, fitness, and overall growth. Among them, fungal endophytes have garnered the most attention due to their polyphyletic and ecologically significant nature, which includes ascomycetes, basidiomycetes, and anamorphic fungi (Hardoim et al. 2015). The two common routes of transmission—vertical and horizontal—impact the relationship between fungal endophytes and their host plants (Lugtenberg et al. 2016).

In vertical transmission, fungal endophytes are transmitted from parent to the offspring as the fungal hyphae that penetrate the embryo present within the seed. These vertically transmitted endophytes often establish a mutually advantageous association with the host, enhancing disease resistance, nutrient absorption, and environmental stress endurance (Jha et al. 2023). In horizontal transmission, fungal endophytes transmit from plant to plant, mostly via sexual spores produced by the fungi. Unlike vertical transmission, horizontally transmitted endophytes may develop antagonistic or competitive relationships with their hosts (Baron and Rigobelo 2022; Omomowo et al. 2023).

Fungal endophytes also act as biocontrol agents against numerous stresses in host plants by directly inhibiting pathogens by competing for space, food and through antibiosis, and mycoparasitism. They also confer secondary protection by stimulating the plant's defense strategies (Latz et al. 2018). When applied as bioinoculants, fungal endophytes enhance plant performance by increasing photosynthesis, improving growth, facilitating nutrient mineralization, stimulating enzymatic activity, producing phytohormones and bioactive metabolites, and improving defense against environmental stressors (Hassan et al. 2017). Khalil et al. (2021) demonstrated the potential of endophytic fungi (such as *Alternaria tenuissima*, *Aspergillus flavus*, *Penicillium caseifulvum*, *Penicillium commune*, and *Penicillium crustosum*) obtained from the *Ephedra pachyclada* to ameliorate the defense, growth, and soil quality of *Zea mays* by producing bioactive substances, extracellular enzymes (like cellulase, amylase, catalase, and pectinase), and hormones (e.g. indole acetic acid). Because they can produce a broad variety of bioactive chemicals, fungal endophytes are particularly prized for their capability to strengthen plant well-being, soil quality, and resilience to environmental challenges. By cutting down the reliance on toxic fertilizers and pesticides, fungal endophytes offer a sustainable alternative for agricultural practices (Sinno et al. 2020).

5.3 Classification of Fungal Endophytes

Schaechter (2012) explained that endophytic fungi are commonly grouped into two main divisions on the basis of variances in taxonomy, host choices, colonization, transmission, tissue selectivity, etc. The first division comprises of clavicipitaceous endophytes (C-endophytes), which primarily affect certain grasses, and the second division includes non-clavicipitaceous endophytes (NC-endophytes). Rodriguez

Types	Clavicipitaceous Fungi	Non-Clavicipitaceous Fungi		
	Class I	Class II	Class III	Class IV or Dark Septate Endophytes
Roles in Stress management	Increase plant biomass, confer drought tolerance, reduce herbivory	Herbivore resistance, resistant to heat, salinity, drought	Increase host fitness	Trap and eliminate oxygen radicals generated during abiotic stress
Examples	Neotyphodium endophytes	*Piriformospora indica* in Barley	Agaricomycotina, Pucciniomycotina etc.	*Trichoderma harzianum*

Fig. 5.1 Classification of endophytic fungi and their roles in stress management

et al. (2009), however, offered a different perspective on fungal endophyte classification, dividing them into four classes. They recognized two chief classes, clavicipitaceous and non-clavicipitaceous, based on evolutionary evidence and life history characteristics. They further classified non-clavicipitaceous fungi into three classes on the basis of colonization and transmission within the plants and the advantages they offered to the host plant (Fig. 5.1).

5.3.1 Clavicipitaceous Fungi

This class includes both free-living and symbiotic species that associate with insects, fungi, or plants such as grasses, rushes, and sedges (Bacon and White Jr 2000). Different species in this family manufacture alkaloids which are detrimental to both animals and humans. The mycelium of these endophytes grows in the spaces between the cells of culms, rhizomes, and leaf sheaths. It can also be sporadically seen on the leaf blade surfaces (Dugan et al. 2002; Tadych et al. 2007). The consequences of C-endophytes on host plants are summarized below:

- *Insect repellence*: Many clavicipitaceous endophytes improve their host plant's resistance to insect feeding. This benefit is because of the synthesis of alkaloid mycotoxins like loline and peramine that are often linked to insect resistance (Riedell et al. 1991).
- *Mammalian herbivore deterrence*: Some clavicipitaceous endophytes deter feeding by mammalian herbivores through the production of mycotoxins like ergot and lolitrem.
- *Nematode repellence*: Research has shown that these endophytes possess anti-nematode properties. For example, *Acremonium coenophialum*, an endophyte in tall fescue (*F. arundinacea*), reduces nematode populations in field soils (Kimmons et al. 1990).

- *Increased disease resistance*: Certain clavicipitaceous endophytes, such as *Epichloë festucae*, produce compounds like indole derivatives, sesquiterpenes, and diacetamides that restrict the development of disease-causing fungi (Lee 2010).
- *Enhanced ecophysiology of host plants*: These endophytes boost the ecophysiological performance of plants, aiding them in enduring abiotic stresses (Arachevaleta et al. 1989).

5.3.2 Non-Clavicipitaceous Fungi

Most of the studies published on non-clavicipitaceous endophytes generally focus on the localization and concentration of endophytes in healthy leaf tissue, the separation and examination of bioactives, using them as biological control agents, and phylogeny-based recognition and categorization (Higgins et al. 2007). In spite of the paucity of functional research, it has been shown that certain NC-endophytes form mutualistic relationships with plants, providing benefits such as enhanced absorption of nutrients, endurance to stress, and improved growth and yield (Rodriguez et al. 2008b). Previously thought to belong to a single functional group, NC-endophytes are now known to belong to at least three different functional groups. These groups will be discussed in the following sections.

5.3.2.1 Class II Fungal Endophytes

This category of endophytes consists of a diverse range of organisms, all associated to the Dikarya group, with most classified under Ascomycota and a smaller proportion under Basidiomycota. The Ascomycota members are limited to the Pezizomycotina subphylum, representing several different classes, while Class II endophytes in Basidiomycota consist of some species from the Agaricomycotina and Pucciniomycotina subphyla. These endophytes differ from one another in several ways: they typically colonize roots, stems, and leaves; form widespread diseases in plants; are transmitted by seed coats and rhizomes; are scarce in the rhizosphere; provide habitat-specific as well as non-habitat-specific fitness advantages; and generally show elevated rates of infection (90–100%) in plants growing under stressful conditions. Most, if not all, Class II endophytes studied so far promote increased shoot and root biomass in their host plants, possibly due to the induction of phytohormones by the host or the fungi's own hormone production (Tudzynski and Sharon 2002).

Additionally, several Class II endophytes offer some protection to their hosts from fungal infections (Campanile et al. 2007), which could be attributed to their generation of secondary metabolites, their capacity of acting as parasites of other fungi (Samuels et al. 2000), or initiation of systemic resistance (Vu et al. 2006). It may be possible that this disease resistance is due to the pathogen's inability to challenge the endophytes for nutrients or space. However, only some Class II endophytes have been directly studied for their relationships with host defense mechanisms. For example, *Fusarium oxysporum* and a species of *Cryptosporiopsis* provided enhanced biotic stress resistance in *Hordeum vulgare* and *Larix decidua*, mainly because of increased levels of phenolics.

5.3.2.2 Class III Fungal Endophytes

These endophytes are exclusively found in aerial plant tissues, transferred horizontally, and develop infections that are highly localized. They include various fungal species linked with the tropical tree leaves (Gamboa and Bayman 2001) and those found in the aerial tissues of non-vascular plants, seedless vascular plants, conifers, and woody and herbaceous angiosperms in ecosystems extending from tropical to Arctic or Antarctic regions (Stone 1988; Murali et al. 2007). Apart from colonizing photosynthetic and herbaceous tissues, they are prevalent in flowers, fruits, wood, and bark (Tejesvi et al. 2005). Due to their extraordinary diversity within a single host or tissue, understanding the ecological functions of Class III endophytes remains challenging, and these roles are often unexplored. Nonetheless, some studies have introduced Class III endophytes into sterile plants to assess their impacts. Generally, plants inoculated with different Class III endophytes display no significant deviations in growth rate, biomass cumulation, or other quantifiable features (Arnold 2002). Bark endophytes protect trees from Dutch elm disease, and diverse endophyte communities have been shown to reduce injury and leaf mortality caused by *Phytophthora* sp. in *T. cacao*. Although Class III endophytes are sometimes associated with characteristics typical of parasitic or pathogenic organisms, such as high host diversity and horizontal transmission, research shows that they can also function as mutualists. Thus, class III endophytes offer diverse ecological functions as compared to class I and Class II endophytes.

5.3.2.3 Class IV Fungal Endophytes

These fungi are also called as ‘dark septate endophytes’ (DSE). These endophytes are characterized by a dark melanized septa and their limited access to the roots of host plant. Generally, they are ascomycetous fungi, either conidial or sterile, that create melanized structures like inter- and intracellular hyphae and microsclerotia within the roots. DSE exhibit minimal host or habitat specificity and are known to be linked with around 600 plant species, including nonmycorrhizal plants, from different areas across the globe. DSE are mostly found in boreal and temperate forests, particularly in association with the roots of trees and shrubs, particularly conifers. They are not considered pathogenic, as they are generally seen on healthy roots, and thus, are broadly referred to as endophytes because they inhabit living plant tissues while avoiding any noticeable harm. While there is significant knowledge about the ecological roles and symbiotic functions of Class I endophytes, and increasing understanding of Class II and III endophytes, relatively very less is understood regarding the major members of fungi that make up Class IV endophytes. Taxonomic recognition of DSE and their host plants is an important step, but it is now essential to explore their evolutionary history, symbiotic functions, phylogenetic links, and ecological roles. Mandyam and Jumpponen (2005) suggest that DSE interactions are like other mycorrhizal relationships i.e. they are multipurpose, not restricted to nutritional uptake and host development. They suggested that DSE establishment may inhibit pathogens by decreasing the availability of carbon in the rhizosphere and that the raised melanin levels in DSE could be used in producing secondary metabolites that deter herbivory, making plants with Class IV endophytes more competitive.

5.4 Involvement of Endophytes in Plant Health

The variety of endophytes in plants and their roles in ecology, evolutionary biology, and defence strategies against biotic and abiotic stressors through metabolite synthesis have been extensively studied. Endophytic biotechnology can serve as efficient tools to develop plants that are essential for agriculture, industry, and economy. Applying different endophytic organisms associated to plants in the right way can improve agricultural yield, increase endurance to different stress conditions, and enhance metabolite productivity across various plants (Wani et al. 2015).

Important new bioactive substances are also generated by endophytes. Unlike epiphytes or soil microbes, the connection between different endophytes and their host plants is significant in the biosynthesis of different biologically active substances (Strobel 2003). Novel biotechnological uses for endophytic organisms, like phytoremediation and bioremediation, are becoming increasingly popular (Li et al. 2012). Through three different strategies—biofertilization, phytostimulation, and biocontrol—endophytes perform important functions in plants.

5.4.1 Endophytes as a Source of Biologically Active Compounds

Endophytic bacteria release certain physiologically active chemicals or metabolites into their host tissues without leading to any noticeable damage. In the pharmaceutical fields, bioactive chemicals that boost plant resistance against harmful bacteria and are synthesized by various endophytic microbes are also employed as biologically active compounds with anticancer, antibiotic, antiviral, and antidiabetic properties (Guo et al. 2008). Researchers are also looking into the potential medical, agricultural, and industrial uses of different biologically potent substances produced by endophytes, such as terpenoids, steroids, peptides, polyketones, quinols, flavonoids, and insecticides like azadirachtin (Molina et al. 2012).

According to Shankar Naik and Krishnamurthy (2010), several of these bioactive compounds exhibit cytotoxicity, antioxidant, antimicrobial, antineoplastic, antiproliferative, anti-leishmanial, and fuel generation capabilities. Cryptocandin, pestaloside, cryptocin, ecomycins, pestalopyrone, and pseudomycins are a few instances of antifungal compounds generated by endophytic organisms.

Seasons, regions, environmental variables, age, and the plant's tissue all affect the biology of the endophyte and cause significant distinctions in the formation of biologically active compounds. Furthermore, the type and number of metabolites can be influenced by growing conditions and separation techniques (Gunatilaka 2006). According to Aly et al. (2011), the generation of bioactive substances in the laboratory is controlled by different factors, including the type, size, and composition of the plant tissue selected for isolation as well as the pH, incubation period and temperature, agitation, and culture conditions.

5.4.2 Production of Extracellular Enzymes

The synthesis of extracellular enzymes by endophytes is a significant process for endophyte colonization in plants, and this function is crucial to plant health. Extracellular enzymes, also known as exoenzymes, are important industrial constituents in multiple sectors, including food processing, fermentation, and other biotechnological uses. Different forms of extracellular enzymes, like oxidoreductases, hydrolases, lyases, and transferases are synthesized by microbes including bacteria and fungi (Traving et al. 2015). Extracellular enzymes are engaged in the continuous metabolism of different biomolecules, including lignin, proteins, carbohydrates, and organic phosphate, breaking them down into micromolecules that are carried throughout the cells. These enzymes are also involved in the symbiosis process (Strong and Claus 2011). Furthermore, Leo et al. (2016) found that extracellular hydrolyase enzymes boost plant defences against infections. According to Chathurdevi and Gowrie (2016), extracellular enzymes generated by endophytic fungus isolated from medicinal plants can help plants flourish despite unfavourable conditions. Additionally, according to Sunitha et al. (2013), about 50 endophytic fungal strains with the capacity to generate amylase, laccase, cellulase, pectinase, lipase, and protease were isolated and identified. Furthermore, the production of 1-aminocyclopropane-1-carboxylate (ACC) deaminase, amylase, cellulases, esterase, pectinase, protease, lipase, asparaginase, phytase, and xylanase by bacterial endophytes has been studied.

5.4.3 Growth Promoting Activity

The capability of plants to adjust to tough situations and different surroundings, which might restrict their development and growth, is greatly aided by endophytic microorganisms. In order to withstand harsh situations, plants can establish symbiotic relationships with microbes, leading to beneficial outcomes for both partners' evolution and survival (Rodriguez et al. 2009). The capacity of endophytes to synthesize different biologically active substances and enzymes promotes plant development. Numerous plant species are associated to plant growth-promoting microorganisms (PGPM), which can have positive impact on the plants, including augmented growth and reduced vulnerability to damage inflicted by bacteria, viruses, fungi, as well as nematodes (Chandran et al. 2021; Kumar et al. 2022). The primary functions of PGPM are linked with the production of phytohormones like gibberellins, cytokinins, siderophores, phosphate solubilization, nutrient absorption, and inhibitory impact on phytopathogens. Furthermore, PGPM have the ability to cause chemical or physical alterations to plant defence, a phenomenon referred as induced systemic resistance (ISR). A wide range of metabolite compounds generated by endophytic fungi are utilized to enhance plant growth. Endophytic fungi like *Sebacina vermifera, Piriformospora indica*, and various species of *Colletotrichum and Penicillium* are recognized for their superior plant growth-promoting effects, especially during adverse circumstances. Different reactions were seen by PGPM against various plant stresses, conferring resistance against infections, and support for the restoration of injured cells or deteriorated elements. The primary plant growth hormone, indoleacetic acid (IAA), is manufactured by different types of microbes, including plant growth-promoting rhizobacteria (PGPR), and is central to the overall growth and development of plant stem and root cells (Hassan 2017; Seth et al. 2023). According to Fouda et al. (2019), endophytic actinobacteria synthesize plant growth promoters like auxins, cytokinins, gibberellins, and IAA in vitro. Small molecules known as siderophores, strong affinity for iron, are produced by bacteria, fungi, and actinobacteria that thrive in low-iron conditions. It has been investigated that a number of endophytic microbes can manufacture high molecular weight siderophores (Kannahi and Senbagam 2014; Yadav and Meena 2021; Yadav and Meena 2022). According to Rungin et al. (2012), endophytic actinobacteria produce siderophores that regulate plant growth and serve as a defence against pathogens.

5.4.4 Biological Control Organisms

Endophytes are considered as important biocontrol agents, offering an alternative to chemical methods. Endophytic fungi perform a crucial task in managing herbivorous insects in grasses and conifers. *Beauveria bassiana*, an endophytic fungus which is a popular entomopathogen, can be utilized to control borer insects in sorghum. Recombinant endophytes, which can easily inhabit various plants, represent a new approach to restrict various plant pests by expressing diverse anti-pest proteins. According to Hassan et al. (2018), houseflies and mosquitoes can be biocontrolled by using Cu nanoparticles obtained from endophyte, *Streptomyces capillispiralis* Ca-1. Similarly, copper oxide nanoparticles obtained from *Oxalis corniculate* L. plant, which are synthesized by two endophytic actinomycetes, *S. zaomyceticus* Oc-5 and *S. pseudogriseolus* Acv-11, show antimicrobial efficacy against various pathogenic fungi (Hassan et al. 2019).

5.4.5 Bioremediation Activity

Endophytes possess a remarkable capacity to strengthen plant growth in polluted soil by breaking down hazardous compounds. Researchers have noted the bioremediation potential of *Nicotiana tabacum* plants when they inoculate the seeds with endophytes. These plants exhibited enhanced biomass accumulation during heavy metal stress from cadmium (Cd), with higher Cd concentrations in plant tissues as compared to non-inoculated plants. These observations highlight the positive outcomes of endophyte-inoculated seeds on the build-up and absorption of heavy metals. To investigate the function of endophytes in breaking down pollutants like plastics, various endophytic fungal species were tested in agar and broth media consisting of polyester polyurethane (PUR) as the only C-source (Russell et al. 2011). Among them, two isolates of *Pestalotiopsis microspore* were capable to utilize PUR as the C-source under both aerobic and anaerobic conditions, employing the serine hydrolase enzyme to degrade PUR.

5.4.6 Induced Systemic Resistance

Endophytes enhance plant resistance to stressors by triggering defence strategies called as induced systemic resistance (ISR). Different bacterial elements, such as antibiotics, salicylic acid, N-acyl-homoserine lactones, siderophores, jasmonic acid, lipopolysaccharides, and volatile compounds (e.g. acetoin), are responsible for initiating ISR (Bordiec et al. 2011). The endophytic bacterium *Pseudomonas fluorescens* 89B-61 was the first to be observed as inducing ISR to defend cucumber plants from cucumber anthracnose. The endurance of potato plants to the pathogen *Pectobacterium atrosepticum* was enhanced because of the endophyte *Methylobacterium* sp. IMBG290. This enhanced tolerance was associated with alterations in the makeup of the indigenous endophytic population. These modifications were associated to disease tolerance, highlighting the crucial role of the endophytic community in suppressing disease (Pavlo et al. 2011). Endophytic fungi also participate in protective strategies through ISR induction, although less than endophytic bacteria.

5.5 Endophytic Fungi and Host Plant Health

The association of endophytic fungi with healthy host plant is a topic of significant curiosity in ecology and plant biology. Host plant health pertains to the plant's competence to thrive, reproduce, and transmit its hereditary material to future generations (de La Fuente Cantó et al. 2020; Vandenkoornhuyse et al. 2015). Endophytic fungi (EFs) can influence host plant fitness both positively and negatively, on the basis of different factors (Shankar Naik 2019). EFs can improve the intake of essential macronutrients like N and P from the soil, which increases plant growth and as such boost the plant's fitness by increasing its reproductive potential (Bhardwaj et al. 2023). The study of endophytic fungi has illuminated the complex relationship between plants, fungi, and their surroundings. They perform a pivotal function in nutrient cycling, carbon sequestration, and sustaining ecosystem health (Bhardwaj et al. 2023). Often, they establish mutualistic relationship with host plants, offering advantages such as increased growth and increased robustness to stressors like drought or salinity (Kamran et al. 2022). Some endophytic fungi generate specific metabolites that inhibit herbivores and guard plants from pathogens. This can improve plant health by limiting the loss of reserves and vital tissues, like reproductive structures (Akram et al. 2023). Additionally, they can aid plants to compete with neighbouring plants for resources, providing a competitive benefit that enhances plant fitness by enabling better resource access and more vigorous growth (Baron and Rigobelo 2022). However, some EFs can become pathogenic during certain circumstances, harming the host plants. Pathogenic endophytes can reduce plant existence by inducing diseases, decreasing the growth, or even leading to death (Kogel et al. 2006). Moreover, the resources a host plant invests in maintaining its relationship with EFs may be actually derived from other essential functions, potentially lowering overall fitness, especially when the endophytes provide limited benefits (Cheplick 2007). While the benefits of mutualistic plant–EF relationships are well-documented, more research is needed on unprofitable interactions, the equilibrium between endophytic and pathogenic behaviours, and the related costs and benefits of these infections. Comprehending the molecular mechanisms, genetic signals, and physiological characteristics that influence the shift from mutualism to pathogenicity and vice versa could offer valuable insights for biotechnology and agroecology (Morales-Vargas et al. 2024).

Trichoderma species, which are well known for reducing abiotic stress and encouraging plant growth, are among the most researched beneficial EFs. In addition to being widely employed in agriculture as biological controls against diseases, *Trichoderma* spp. also promote root development and nutrient uptake, which as a result boosts plant growth (Fontana et al. 2021). Investigations have revealed that these fungi augment plant resilience against an array of abiotic stressors, including salt, drought, and heavy metal toxicity, by generating enzymes and phytohormones (Fig. 5.2) linked to stress (Zaidi et al. 2014). Besides *Trichoderma*, other EF genera used in agriculture include *Clonostachys rosea* (Supratman et al. 2021), *Piriformospora indica* (Franken 2012), and certain *Fusarium* species. However,

Fig. 5.2 Benefits of fungal endophytic colonization

caution is necessary with some *Fusarium* species, as they can cause plant diseases (Toghueo 2020).

5.6 Colonization Mechanisms of Endophytic Fungi Within the Plants

Since the defining feature of endophytic fungi (EFs) is their capacity to establish within the plants, it is crucial to explore this process in greater depth. EFs employ a variety of complex mechanisms to colonize plant tissues. However, certain processes are common across many interactions. Typically, EFs enter the host through natural apertures like wounds, stomata, or root tips. Additionally, certain fungi might develop specific structures or adhesive molecules to help them to get attached to the surface of the plants (Mengistu 2020).

Although chemotaxis seems to play a role in how EFs identify and interact with their host plants, more insight is needed. Few EFs can rigorously respond to chemical cues produced by the plant, leading them to entry points (Tsai et al. 2020). Once bound to the plant surface, EFs use a variety of ways to penetrate the plant tissues, like secreting enzymes such as cellulases or pectinases to degrade the cell wall (Jha et al. 2023). After entry, EFs can grow either intercellularly or intracellularly, where they may form specific structures like hyphal coils or microsclerotia (Ji et al. 2022; Chambers et al. 2008). Successful colonization by EFs often depends on their ability to intrude or supress the plant defence system. They may produce molecules that restrict the plant's immune reaction, reducing the chances of being recognized as a threat. Additionally, some EFs produce bioactive specialized metabolites that aid in their colonization of plant tissues (Bhardwaj et al. 2023). Once established, EFs can reside within the plant for long span of time, mostly across the entire life of plant. They may grow towards various plant parts like seeds, leaves, stems, and roots (Baron and Rigobelo 2022). This allows EFs to be transferred both vertically (from parent to offspring) and horizontally (between plants) through different routes, such as seeds, root-to-root contact, and aerial spore dispersal (Shahzad et al. 2018).

5.7 The Influence of Endophytic Fungi on Abiotic Stress Adaptations

Drought, salinity, heat, and heavy metals are significant factors that greatly reduce crop yield, leading to global yield decrease exceeding 50% (Gowtham et al. 2020). These stresses can cause irreversible harm, including slowed growth, severe injury, and even lead to death (Murali et al. 2021). They also inhibit plant growth and metabolism, decreasing biomass and the generation of active compounds. Among the many advantages of endophytic fungi, their ability to improve crop adaptation to various abiotic stresses is particularly noteworthy. This potential holds great promise for advancing green farming, ensuring food security, the elimination of starvation and dietary deficiencies, and environmental conservation (Sodhi and Saxena

2023; Jha et al. 2023). Approaches that utilize specific EFs or manipulate their relationships with host plants either via inoculation or genetic manipulation offer assurance for producing stress-resistant crop varieties. For effective agricultural uses, it is crucial to get greater understanding into the processes governing EF–plant interactions, determine the ideal settings for these interactions, and assess their possible impact on the environment and ecosystems. Many of the adverse situations plants endure due to abiotic stress factors entail oxidative stress.

Oxidative stress occurs when damaging oxygen radicals or other compounds that may strip electrons from biomolecules are generated (Schutzendubel and Polle 2002). Biomolecules degrade quickly as a result of this imbalance in their redox state, severely upsetting associated processes. Most organisms possess defence mechanisms against oxidative stress, chiefly through the production of compounds like flavonoids, glutathione, and other antioxidants (Yadav 2010). Endophytic fungi can help safeguard plants in conditions of elevated oxidative stress, thus protecting them from abiotic stress.

5.7.1 Endophytic Fungi and Crop Drought Tolerance

Drought severely limits agricultural crop productivity and presents significant climate-related roadblocks to achieving food security in underdeveloped nations (Al-Turki et al. 2023). Insufficient water availability negatively impacts plants by causing osmotic stress, hindering photosynthesis, restricting nutrient uptake, and the overproduction of ROS. Plants have a broad set of biochemical, physiological, and molecular responses to drought, including closing of stomata, lower rates of photosynthesis, transpiration and release of stress hormones, activation of antioxidant defence systems, etc. (Fig. 5.2) (Gowtham et al. 2022). Research has shown that introducing fungal endophytes into plants substantially alleviates the deteriorating effects of drought stress. This beneficial plant–symbiont relationship enhances drought resistance through improvements in total biomass, leaf morphology and structure, pigments, stomatal conductance, photosynthesis, transpiration, nutrient content, and antioxidant enzyme activity (González-Teuber et al. 2018; Li et al. 2019; Liu and Wei 2021). By enhancing water absorption and retention via a variety of methods, EFs help host plants become more drought-tolerant (Dastogeer et al. 2020). Fungal hyphae can grow deep into the soil, reaching water sources inaccessible to plant roots, thereby maintaining a steady water supply during drought stress (Wahab et al. 2023). Additionally, certain EFs, like *Epichloë* endophytes, may influence the quantities of phytohormones such as abscisic acid (ABA) and cytokinins, which are critical in modulating plant tolerance to drought (Cui et al. 2022). For example, some EFs generate ABA analogues that mimic the action of natural ABA, triggering stomatal closure and lowering transpirational loss of water (Xu et al. 2018). In contrast, other EFs synthesize cytokinins that assist cellular proliferation and growth, strengthening plant tissues against drought-induced damage. Some fungi produce superoxide dismutase (SOD) and catalase (CAT) enzymes that degrade the ROS and thus, safeguarding the cells from oxidative injury (Sadeghi et al. 2020). Moreover, EFs can boost the build-up of osmoprotectants like proline, which aid in maintaining cellular stability during drought (Javed et al. 2022). The mutualistic root endophyte *Piriformospora indica* is known to augment drought resistance in *Hordeum vulgare* by colonizing the plant roots. It improves the electron transport chain and photosystem activity while also boosting protein build-up, which supports primary metabolism, energy regulation, transport, photorespiration, and modifies the metabolism of amino acids within the plant (Ghaffari et al. 2019). Similarly, Miranda et al. (2023) reported that endophytic strains of *Zopfiella erostrata* improved drought resistance through widespread colonizing the roots of *Solanum lycopersicum* and *Triticum aestivum*. These strains formed abundant melanized mycelium in the rhizosphere leading to increased biomass, enhanced nutrient mineralization and water absorption, increased proline build-up, and reduced lipid peroxidation levels in plants exposed to drought.

The capacity of EFs to increase crop resilience to drought holds enormous potential for sustainable agriculture, particularly in areas where water shortage is a concern. Drought-tolerant EFs can be used to inoculate crops, reducing the need for chemical and irrigation inputs and fostering environmental sustainability and climate change resistance. Moreover, EFs can be used to increase agricultural resilience and output by acting as biocontrol agents and biofertilizers. However, challenges remain in fully realizing the potential of EFs for drought tolerance. These include identifying and characterizing drought-tolerant EFs suited to various crop species and environmental conditions, as well as developing effective inoculation methods for large-scale farming (Gowtham et al. 2024).

5.7.2 Protection by Endophytic Fungi Against Toxic Metal-Induced Oxidative Stress

The partnership of host plants with endophytic fungi offers numerous benefits to both, enabling plants to thrive in environments where others cannot survive (El-Shafey et al. 2021). One significant source of oxidative stress affecting plant development is soil contaminated with toxic metals (TMs), an increasing concern due to rising pollution. Industrial activities and mining operations have resulted in cumulation of harmful metal in the atmosphere (Jha et al. 2023; Arcega-Cabrera et al. 2015). TM pollution is a serious issue, particularly due to its adverse effect on crop production, which hinders efforts to eradicate food insecurity and starvation. The toxicity of TMs depends on their chemical reactivity and oxidation state, which are affected by parameters like soil pH, organic matter, and the availability of ions. A feasible approach to alleviating oxidative stress caused by toxic metals is the use of fungal endophytes. Endophytic fungi inhabit plants without inflicting any damage and protecting them from the oxidative damages of TMs (Ignatova et al. 2021). Endophytic fungi offer several mechanisms for protecting plants against TMs.

Making metal flocculants or chelating agents is one technique. Exopolysaccharides, lipids, glycolipids, and proteins that attract and clump metal ions are produced by endophytes. The biosorption and immobilization of metal ions, changes to their redox state, and the synthesis of biopolymers and siderophores that bind to the poisonous metal ions and reduce the oxidative stress are all related to better plant resistance to TMs conferred by the endophytic fungi. Additionally, the endophytes-host plant association promotes better nutrient absorption, improving the plant's resistance to oxidative stress. For example, *Sporobolomyces ruberrimus*, a basidiomycete, shields its host by producing siderophores that chelate excess iron (Fe) without interfering with the uptake of other micronutrients, while also promoting the development of lateral roots. Endophytes can influence the levels of phytohormones in their host. For example, *S. ruberrimus* helps regulate ethylene concentration, enabling the plant to resist oxidative stress caused by excess Fe (Domka et al. 2023). Some endophytes can also synthesize or modulate phytohormones, raising the levels of antioxidants (Verma et al. 2021). *Piriformospora indica* was observed to increase the tolerance of *Nicotiana tabacum* to Cd toxicity by stimulating the expression of photosynthesis-related genes, glutamine synthetase, and peroxidase, thereby reducing the harm caused by Cd exposure. Similarly, Musa et al. (2023) reported that inoculation with the endophyte *Paecilomyces lilacinus* led to increased production of IAA and secondary metabolites, which significantly promoted the growth of *Solanum lycopersicum* and mitigated the harmful impacts of heavy metal stress. Fungal endophytes contribute to the biosorption of heavy metals, thereby alleviating the detrimental effects of heavy metal toxicity. This strategy for bioremediating heavy metals not only reduces their toxicity but also supports the development of a healthier dietary ecosystem (Gowtham et al. 2024).

5.7.3 Endophytic Fungi Enhance Crop Tolerance to Temperature Stress

Thermal stress is generally described as an upsurge in surrounding temperatures by 10–15 °C. This often leads to physiological and cellular damage in plants, affecting protein and cellular stability (Guihur et al. 2022). When exposed to heat, plants trigger complex biochemical and molecular responses, including the activation of heat shock proteins (HSPs), ROS, and enzymatic antioxidants (Songy et al. 2019). These responses also disrupt photosystems, initiate rubisco degradation, and reduce chlorophyll levels (Crafts-Brandner and Law 2000). The transcription factor APETALA 2 (AP2)/ethylene-responsive factor (ERF) is frequently upregulated during heat and drought stress (Rocheta et al. 2014). Heat stress is known to create transcriptional memory, involving genes like ascorbate peroxidase 2 (APX2) and chromatin modifications, such as histone H3K4 hypermethylation (Oberkofler and Bäurle 2022). Plants naturally combat heat stress through signal transduction pathways, enhancing stress tolerance, and acclimatization (Kollist et al. 2019). Moreover, the hyperexpression of rubisco activase (Rca) has been suggested as a way to counter heat-induced rubisco inhibition (Qu et al. 2023).

Scientists have investigated the roles of endophytic fungi in ameliorating heat stress resistance in different crops. For instance, research on soybeans and sunflowers inoculated with *Aspergillus flavus* revealed that the symbiosis reduced heat stress and notably increased proline synthesis, abscisic acid, and phenolic compounds at 40 °C compared to controls (Ismail et al. 2019). Comparable outcomes were noted in soybean and sunflower seedlings administered with *Rhizopus oryzae*, exposed to 25 °C and 40 °C, with both species exhibiting higher quantities of ascorbic acid oxidase, catalase, proline, phenolics, flavonoids, proteins, sugars, and lipids. Additionally, the endophytes increased chlorophyll content, root and shoot lengths, and overall biomass. Endophytes can trigger the build-up of specific metabolites and upregulate enzymes participating in oxidative stress metabolism, as seen in cucumber plants administered with *Thermomyces* sp. and in sunflowers and soybeans with *Aspergillus niger* (Ali et al. 2018). In geothermal soil simulation experiments, tomato plants and panic grass inoculated with *Curvularia protuberata* exhibited improved drought and heat tolerance (Rodriguez et al. 2008a, b). Colonization by fungal endophytes has led to lower the malondialdehyde and hydrogen peroxide levels, thereby minimizing cell damage caused by peroxidation of membrane lipids during cold stress (Poveda et al. 2020). The root endophyte *Piriformospora indica* has also proven its potential to enhance cold stress resistance in host plants by regulating the genes engaged in metabolic pathways, phytohormone signalling, and lipid and inorganic ion transport (Jiang et al. 2021). Additionally, Li et al. (2021) found that *P. indica* improved cold resistance in *Musa acuminata* seedlings by boosting antioxidant capacity, promoting soluble sugar accumulation, and upregulating cold-responsive genes such as CSD1C, Why 1, HOS1, and CBF7-1 in the leaves.

5.7.4 Salt Stress Alleviation Assisted by Endophytic Fungi

Salt stress is among the topmost challenges that adversely impact plant growth and development worldwide, particularly in areas with low precipitation (Hasanuzzaman and Fujita 2022). Prolonged exposure to salinity increases intracellular osmotic pressure, leading to excess ROS production and toxic levels of sodium (Na^+) accumulation in plants. This disrupts the K^+/Na^+ equilibrium, impairs plant metabolism, and damages cellular membranes, chlorophyll, proteins, and nucleic acids (Bomle et al. 2021). Developing efficient management techniques to counter salt stress is essential for endorsing low-impact farming and ensuring ideal plant growth. Fungal endophytes enhance plant's resistance to salt stress by regulating endogenous hormones, regulating defence gene expression, and activating the antioxidant system, which helps lower ROS levels. The mutualistic interaction between fungal endophytes that secrete phytohormones can support plant development and mitigate the adverse effects of salinity stress by amassing antioxidants and proline, regulating phytohormones, forming a physical blockade against salt uptake in roots, and preserving water potential and ionic homeostasis (Fig. 5.2) (Ali et al. 2022). The root mutualistic endophyte *Piriformospora indica* has shown a significant impact on *Hordeum vulgare* biomass and growth by inducing a systemic response that alters the plant's proteome and ion content so as to cope with the salt stress (Ghaffari et al.

2016). Similarly, it has been reported that the endophytic relationship between *Yarrowia lipolytica* and *Zea mays* increased plant growth during salinity stress. This enhancement was attributed to the secretion of exogenous indole acetic acid, regulation of endogenous IAA and abscisic acid, and the production of catalase, peroxidase, and proline. Additionally, plants associated with the endophytic entomopathogen *Beauveria bassiana* experienced improved growth and alleviation of salt stress effects, likely because of proline build-up and the activation of antioxidant enzymes in *Solanum tuberosum* tissues (Tomilova et al. 2023). The fungi like *Ulocladium* sp., *Fusarium avenaceum*, *Chaetomium* sp., and *Fusarium tricinctum* are known to aid plants in mitigating salinity stress. They are proven to boost the capacity of mung bean plants to withstand saline conditions. They were obtained from plants in the Northwestern Himalayas. *F. avenaceum* was found to be the most successful in increasing mung bean output and promoting salt tolerance among them. Through a variety of strategies, such as scavenging ROS, preserving ion balance, and concentrating organic solutes, carbohydrates, proteins, and lipids, these endophytes assist their host plants in withstanding salinity stress (Pandit et al. 2023).

5.8 Future Research Directions and Method Development for Harnessing Beneficial Endophytic Fungi

Although the advantages of utilizing endophytic fungi to enhance crop tolerance to abiotic stress in response to climate shift are evident, significant progress has to be made, particularly in discovering new strains which can be applied to various plants and shifting environmental circumstances. In the coming years, researchers should prioritize investigating the nature of the relationship between EFs and crops under abiotic stress by integrating techniques from microbiology, genetics, molecular biology, metabolomics, and ecology. The aim is to craft procedures for the swift segregation and identification of stress-tolerant strains and to understand the underlying mechanisms through which EFs confer these benefits. Additionally, characterizing the genetic and enzymatic pathways that enable EFs to thrive under severe ecological challenges and form beneficial symbioses with crops is crucial. To accomplish this, microbiological techniques should be employed to isolate, culture, and preserve EF strains, which could help alleviate the impact of climate shifts on agricultural output and combat hunger. Establishing globally accessible EF collections, particularly for vulnerable populations, is essential (Fite et al. 2023). Advanced microscopy techniques should be utilized to better understand the interactions between EFs and host plants at cellular levels (Vaitiekūnaitė et al. 2022; Valdés-Santiago et al. 2021). Different molecular techniques will be vital for characterizing the evolution, phylogeny, genetic potential, and gene expression profiles of advantageous EFs (Subudhi et al. 2018). These tools can also be used to study the genetic, functional, physiological, and systemic responses of plants associated with EFs, focusing on improving plant fitness, crop productivity, and stress tolerance across various environmental settings. Standardized procedures for functional assays, like in vitro analyses, greenhouse trials, and field studies, are needed to explore

EF–plant associations, such as synthesis of growth-promoting factors, pathogen biocontrol, nutrient exchange, and biotic and abiotic stress resistance. These procedures would greatly contribute to creating a global EF strain collection with verified attributes that enhance plant growth and crop yields, alongside data on crop types, environmental conditions, and the specific benefits each EF strain offers. Bioinformatics and genetic engineering represent additional methods for optimizing EF–crop interactions under abiotic stress. Bioinformatic tools can analyse omics data to recognize genes, metabolic pathways, and regulatory networks associated with the fruitful EF–crop interactions, as well as the molecular basis of EF tolerance to harsh environments (Slama et al. 2021). These tools can also predict the potential ecological roles and benefits of specific EF species based on genomic, transcriptomic, proteomic, and metabolomic data. Meanwhile, genetic engineering techniques could be used to manipulate genes or metabolic pathways in both fungi and plants to better understand their roles in symbiosis. The creation of fungal mutants could further clarify gene functions related to colonization, nutrient exchange, and bioactive compound synthesis. In summary, maximizing the benefits of EFs to help crops withstand abiotic stress will require a multidisciplinary approach that addresses the challenge from multiple angles, aiming for short-term, measurable outcomes.

References

Akram S, Ahmed A, He P, He P, Liu Y, Wu Y et al (2023) Uniting the role of endophytic fungi against plant pathogens and their interaction. J Fungi 9(1):72

Ali AH, Abdelrahman M, Radwan U, El-Zayat S, El-Sayed MA (2018) Effect of Thermomyces fungal endophyte isolated from extreme hot desert-adapted plant on heat stress tolerance of cucumber. Appl Soil Ecol 124:155–162

Ali R, Gul H, Rauf M, Arif M, Hamayun M, Husna et al (2022) Growth-promoting endophytic fungus (*Stemphylium lycopersici*) ameliorates salt stress tolerance in maize by balancing ionic and metabolic status. Front Plant Sci 13:890565

Al-Turki A, Murali M, Omar AF, Rehan M, Sayyed RZ (2023) Recent advances in PGPR-mediated resilience toward interactive effects of drought and salt stress in plants. Front Microbiol 14:1214845

Aly AH, Debbab A, Proksch P (2011) Fungal endophytes: unique plant inhabitants with great promises. Appl Microbiol Biotechnol 90:1829–1845

Arachevaleta M, Bacon CW, Hoveland CS, Radcliffe DE (1989) Effect of the tall fescue endophyte on plant response to environmental stress. Agron J 81(1):83–90

Arcega-Cabrera F, Garza-Pérez R, Noreña-Barroso E, Oceguera-Vargas I (2015) Impacts of geochemical and environmental factors on seasonal variation of heavy metals in a coastal lagoon Yucatan, Mexico. Bull Environ Contam Toxicol 94:58–65

Arnold AE (2002) Fungal endophytes in neotropical trees: abundance, diversity, and ecological interactions. In: Ganeshalah R, Shaanker U, Bawa K (eds) Tropical ecosystems: structure, diversity and human welfare, pp 739–743

Bacon CW, White JF Jr (2000) Physiological adaptations in the evolution of endophytism in the Clavicipitaceae. In: Microbial endophytes. CRC Press, pp 251–276

Baron NC, Rigobelo EC (2022) Endophytic fungi: a tool for plant growth promotion and sustainable agriculture. Mycology 13(1):39–55

Bhardwaj M, Kailoo S, Khan RT, Khan SS, Rasool S (2023) Harnessing fungal endophytes for natural management: a biocontrol perspective. Front Microbiol 14:1280258

Bomle DV, Kiran A, Kumar JK, Nagaraj LS, Pradeep CK, Ansari MA et al (2021) Plants saline environment in perception with rhizosphere bacteria containing 1-aminocyclopropane-1-carboxylate deaminase. Int J Mol Sci 22(21):11461

Bordiec S, Paquis S, Lacroix H, Dhondt S, Ait Barka E, Kauffmann S et al (2011) Comparative analysis of defence responses induced by the endophytic plant growth-promoting rhizobacterium *Burkholderia phytofirmans* strain PsJN and the non-host bacterium Pseudomonas syringae pv. pisi in grapevine cell suspensions. J Exp Bot 62(2):595–603

Campanile G, Ruscelli A, Luisi N (2007) Antagonistic activity of endophytic fungi towards *Diplodia corticola* assessed by in vitro and in planta tests. Eur J Plant Pathol 117:237–246

Chambers SM, Curlevski NJ, Cairney JW (2008) Ericoid mycorrhizal fungi are common root inhabitants of non-Ericaceae plants in a south-eastern Australian sclerophyll forest. FEMS Microbiol Ecol 65(2):263–270

Chandran H, Meena M, Swapnil P (2021) Plant growth-promoting rhizobacteria as a green alternative for sustainable agriculture. Sustainability 13(19):10986

Chathurdevi G, Gowrie SU (2016) Endophytic fungi isolated from medicinal plant—a source of potential bioactive metabolites. Int J Curr Pharm Res 8:50–56

Chaudhary P, Agri U, Chaudhary A, Kumar A, Kumar G (2022) Endophytes and their potential in biotic stress management and crop production. Front Microbiol 13:933017

Cheplick GP (2007) Costs of fungal endophyte infection in Lolium perenne genotypes from Eurasia and North Africa under extreme resource limitation. Environ Exp Bot 60(2):202–210

Crafts-Brandner SJ, Law RD (2000) Effect of heat stress on the inhibition and recovery of the ribulose-1, 5-bisphosphate carboxylase/oxygenase activation state. Planta 212:67–74

Cui X, He W, Christensen MJ, Yue J, Zeng F, Zhang X et al (2022) Abscisic acid may play a critical role in the moderating effect of Epichloë endophyte on Achnatherum inebrians under drought stress. J Fungi 8(11):1140

Dastogeer KMG, Chakraborty A, Sarker MSA, Akter MA (2020) Roles of fungal endophytes and viruses in mediating drought stress tolerance in plants. Int J Agric Biol 24:1497–1512

de La Fuente Cantó C, Simonin M, King E, Moulin L, Bennett MJ, Castrillo G, Laplaze L (2020) An extended root phenotype: the rhizosphere, its formation and impacts on plant fitness. Plant J 103(3):951–964

Domka A, Jędrzejczyk R, Ważny R, Gustab M, Kowalski M, Nosek M et al (2023) Endophytic yeast protect plants against metal toxicity by inhibiting plant metal uptake through an ethylene-dependent mechanism. Plant Cell Environ 46(1):268–287

Dugan FM, Sitton JW, Sullivan RF, White JF Jr (2002) The Neotyphodium endophyte of wild barley (*Hordeum brevisubulatum* subsp. *violaceum*) grows and sporulates on leaf surfaces of the host. Symbiosis

El-Shafey NM, Marzouk MA, Yasser MM, Shaban SA, Beemster GT, AbdElgawad H (2021) Harnessing endophytic fungi for enhancing growth, tolerance and quality of rose-scented geranium (*Pelargonium graveolens* (L'Hér) Thunb.) plants under cadmium stress: a biochemical study. J Fungi 7(12):1039

Fadiji AE, Babalola OO (2020) Elucidating mechanisms of endophytes used in plant protection and other bioactivities with multifunctional prospects. Front Bioeng Biotechnol 8:467

Fite T, Kebede E, Tefera T, Bekeko Z (2023) Endophytic fungi: versatile partners for pest biocontrol, growth promotion, and climate change resilience in plants. Front Sustain Food Syst 7:1322861

Fontana DC, de Paula S, Torres AG, de Souza VHM, Pascholati SF, Schmidt D, Dourado Neto D (2021) Endophytic fungi: biological control and induced resistance to phytopathogens and abiotic stresses. Pathogens 10(5):570

Fouda A, Hassan SED, Eid AM, El-Din Ewais E (2019) The interaction between plants and bacterial endophytes under salinity stress. In: Endophytes and secondary metabolites. Springer, Cham, pp 1–18

Franken P (2012) The plant strengthening root endophyte Piriformospora indica: potential application and the biology behind. Appl Microbiol Biotechnol 96:1455–1464

Gamboa MA, Bayman P (2001) Communities of endophytic fungi in leaves of a tropical timber tree (*Guarea guidonia*: Meliaceae) 1. Biotropica 33(2):352–360

Ghaffari MR, Ghabooli M, Khatabi B, Hajirezaei MR, Schweizer P, Salekdeh GH (2016) Metabolic and transcriptional response of central metabolism affected by root endophytic fungus *Piriformospora indica* under salinity in barley. Plant Mol Biol 90:699–717

Ghaffari MR, Mirzaei M, Ghabooli M, Khatabi B, Wu Y, Zabet-Moghaddam M et al (2019) Root endophytic fungus *Piriformospora indica* improves drought stress adaptation in barley by metabolic and proteomic reprogramming. Environ Exp Bot 157:197–210

González-Teuber M, Urzúa A, Plaza P, Bascuñán-Godoy L (2018) Effects of root endophytic fungi on response of *Chenopodium quinoa* to drought stress. Plant Ecol 219:231–240

Gowtham HG, Singh B, Murali M, Shilpa N, Prasad M, Aiyaz M et al (2020) Induction of drought tolerance in tomato upon the application of ACC deaminase producing plant growth promoting rhizobacterium Bacillus subtilis Rhizo SF 48. Microbiol Res 234:126422

Gowtham HG, Singh SB, Shilpa N, Aiyaz M, Nataraj K, Udayashankar AC, Amruthesh KN, Murali M, Poczai P, Gafur A et al (2022) Insight into recent progress and perspectives in improvement of antioxidant machinery upon PGPR augmentation in plants under drought stress: a review. Antioxidants 11:1763. https://doi.org/10.3390/antiox11091763

Gowtham HG, Hema P, Murali M, Shilpa N, Nataraj K, Basavaraj GL et al (2024) Fungal endophytes as mitigators against biotic and abiotic stresses in crop plants. J Fungi 10(2):116

Guihur A, Rebeaud ME, Goloubinoff P (2022) How do plants feel the heat and survive? Trends Biochem Sci 47(10):824–838

Gunatilaka AL (2006) Natural products from plant-associated microorganisms: distribution, structural diversity, bioactivity, and implications of their occurrence. J Nat Prod 69(3):509–526

Guo B, Wang Y, Sun X, Tang K (2008) Bioactive natural products from endophytes: a review. Appl Biochem Microbiol 44:136–142

Hardoim PR, Van Overbeek LS, Berg G, Pirttilä AM, Compant S, Campisano A et al (2015) The hidden world within plants: ecological and evolutionary considerations for defining functioning of microbial endophytes. Microbiol Mol Biol Rev 79(3):293–320

Hasanuzzaman M, Fujita M (2022) Plant responses and tolerance to salt stress: physiological and molecular interventions. Int J Mol Sci 23(9):4810

Hassan SED (2017) Plant growth-promoting activities for bacterial and fungal endophytes isolated from medicinal plant of *Teucrium polium* L. J Adv Res 8(6):687–695

Hassan SED, Fouda A, Radwan AA, Salem SS, Barghoth MG, Awad MA et al (2019) Endophytic actinomycetes Streptomyces spp mediated biosynthesis of copper oxide nanoparticles as a promising tool for biotechnological applications. JBIC J Biol Inorg Chem 24:377–393

Higgins KL, Arnold AE, Miadlikowska J, Sarvate SD, Lutzoni F (2007) Phylogenetic relationships, host affinity, and geographic structure of boreal and arctic endophytes from three major plant lineages. Mol Phylogenet Evol 42(2):543–555

Ignatova L, Kistaubayeva A, Brazhnikova Y, Omirbekova A, Mukasheva T, Savitskaya I et al (2021) Characterization of cadmium-tolerant endophytic fungi isolated from soybean (Glycine max) and barley (Hordeum vulgare). Heliyon 7(11)

Ismail, Hamayun M, Hussain A, Afzal Khan S, Iqbal A, Lee IJ (2019) Aspergillus flavus promoted the growth of soybean and sunflower seedlings at elevated temperature. Biomed Res Int 2019(1):1295457

Javed J, Rauf M, Arif M, Hamayun M, Gul H, Ud-Din A et al (2022) Endophytic fungal consortia enhance basal drought-tolerance in Moringa oleifera by upregulating the antioxidant enzyme (APX) through Heat shock factors. Antioxidants 11(9):1669

Jha P, Kaur T, Chhabra I, Panja A, Paul S, Kumar V, Malik T (2023) Endophytic fungi: hidden treasure chest of antimicrobial metabolites interrelationship of endophytes and metabolites. Front Microbiol 14:1227830

Ji X, Xia Y, Zhang H, Cui JL (2022) The microscopic mechanism between endophytic fungi and host plants: from recognition to building stable mutually beneficial relationships. Microbiol Res 261:127056

Jiang W, Pan R, Buitrago S, Wu C, Abdelaziz ME, Oelmüller R, Zhang W (2021) Transcriptome analysis of Arabidopsis reveals freezing-tolerance related genes induced by root endophytic fungus Piriformospora indica. Physiol Mol Biol Plants 27:189–201

Kamran M, Imran QM, Ahmed MB, Falak N, Khatoon A, Yun BW (2022) Endophyte-mediated stress tolerance in plants: a sustainable strategy to enhance resilience and assist crop improvement. Cells 11(20):3292

Kannahi M, Senbagam N (2014) Studies on siderophore production by microbial isolates obtained from rhizosphere soil and its antibacterial activity

Khalil AMA, Hassan SED, Alsharif SM, Eid AM, Ewais EED, Azab E et al (2021) Isolation and characterization of fungal endophytes isolated from medicinal plant Ephedra pachyclada as plant growth-promoting. Biomol Ther 11(2):140

Kimmons CA, Gwinn KD, Bernard EC (1990) Nematode reproduction on endophyte-infected and endophyte-free tall fescue

Kogel KH, Franken P, Hückelhoven R (2006) Endophyte or parasite–what decides? Curr Opin Plant Biol 9(4):358–363

Kollist H, Zandalinas SI, Sengupta S, Nuhkat M, Kangasjärvi J, Mittler R (2019) Rapid responses to abiotic stress: priming the landscape for the signal transduction network. Trends Plant Sci 24(1):25–37

Kumar R, Swapnil P, Meena M, Selpair S, Yadav BG (2022) Plant growth-promoting rhizobacteria (PGPR): approaches to alleviate abiotic stresses for enhancement of growth and development of medicinal plants. Sustainability 14(23):15514

Latz MA, Jensen B, Collinge DB, Jørgensen HJ (2018) Endophytic fungi as biocontrol agents: elucidating mechanisms in disease suppression. Plant Ecol Divers 11(5–6):555–567

Lee KEUNSUB (2010) The effects of endophytic Fusarium verticillioides on the interactions of maize and its fungal pathogen, Ustilago maydis. PhD, University of Minnesota, Minnesota

Leo VV, Passari AK, Joshi JB, Mishra VK, Uthandi S, Ramesh N et al (2016) A novel triculture system (CC3) for simultaneous enzyme production and hydrolysis of common grasses through submerged fermentation. Front Microbiol 7:447

Li HY, Wei DQ, Shen M, Zhou ZP (2012) Endophytes and their role in phytoremediation. Fungal Divers 54:11–18

Li X, He XL, Zhou Y, Hou YT, Zuo YL (2019) Effects of dark septate endophytes on the performance of *Hedysarum scoparium* under water deficit stress. Front Plant Sci 10:903

Li D, Bodjrenou DM, Zhang S, Wang B, Pan H, Yeh KW et al (2021) The endophytic fungus Piriformospora indica reprograms banana to cold resistance. Int J Mol Sci 22(9):4973

Liu Y, Wei X (2021) Dark septate endophyte improves the drought-stress resistance of *Ormosia hosiei* seedlings by altering leaf morphology and photosynthetic characteristics. Plant Ecol 222(7):761–771

Lugtenberg BJ, Caradus JR, Johnson LJ (2016) Fungal endophytes for sustainable crop production. FEMS Microbiol Ecol 92(12):fiw194

Mandyam K, Jumpponen A (2005) Seeking the elusive function of the root-colonising dark septate endophytic fungi. Stud Mycol 53(1):173–189

Mengistu AA (2020) Endophytes: colonization, behaviour, and their role in defense mechanism. Int J Microbiol 2020(1):6927219

Miranda V, Silva-Castro GA, Ruiz-Lozano JM, Fracchia S, García-Romera I (2023) Fungal endophytes enhance wheat and tomato drought tolerance in terms of plant growth and biochemical parameters. J Fungi 9(3):384

Molina G, Pimentel MR, Bertucci TC, Pastore GM (2012) Application of fungal endophytes in biotechnological processes. Chem Eng Trans 27

Morales-Vargas AT, López-Ramírez V, Álvarez-Mejía C, Vázquez-Martínez J (2024) Endophytic fungi for crops adaptation to abiotic stresses. Microorganisms 12(7):1357

Muhie SH (2022) Novel approaches and practices to sustainable agriculture. J Agric Food Res 10:100446
Murali TS, Suryanarayanan TS, Venkatesan G (2007) Fungal endophyte communities in two tropical forests of southern India: diversity and host affiliation. Mycol Prog 6(3):191–199
Murali M, Mahendra C, Hema P, Rajashekar N, Nataraju A, Sudarshana MS, Amruthesh KN (2017) Molecular profiling and bioactive potential of an endophytic fungus Aspergillus sulphureus isolated from Sida acuta: a medicinal plant. Pharm Biol 55(1):1623–1630
Murali M, Gowtham HG, Singh SB, Shilpa N, Aiyaz M, Niranjana SR, Amruthesh K (2021) Bioprospecting of ACC deaminase producing Rhizobacteria towards sustainable agriculture: a special emphasis on abiotic stress in plants. Appl Soil Ecol 168:104142
Musa M, Jan FG, Hamayun M, Jan G, Khan SA, Rehman G et al (2023) An endophytic fungal isolate Paecilomyces lilacinus produces bioactive secondary metabolites and promotes growth of Solanum lycopersicum under heavy metal stress. Agronomy 13(3):883
Naik BS, Krishnamurthy YL (2010) Endophytes: the real untapped high energy biofuel resource. Curr Sci 98(883):8
Oberkofler V, Bäurle I (2022) Inducible epigenome editing probes for the role of histone H3K4 methylation in Arabidopsis heat stress memory. Plant Physiol 189(2):703–714
Omomowo IO, Amao JA, Abubakar A, Ogundola AF, Ezediuno LO, Bamigboye CO (2023) A review on the trends of endophytic fungi bioactivities. Sci Afr 20:e01594
Pandit AG, Vendan KT, Earanna N (2023) Fungal endophyte mediated salinity stress tolerance in mung bean (Vigna radiata L. Wilczek) Mysore Journal of Agricultural Sciences, (40–47), University of Agricultural Sciences.
Pathak VM, Verma VK, Rawat BS, Kaur B, Babu N, Sharma A et al (2022) Current status of pesticide effects on environment, human health and it's eco-friendly management as bioremediation: a comprehensive review. Front Microbiol 13:962619
Pavlo A, Leonid O, Iryna Z, Natalia K, Maria PA (2011) Endophytic bacteria enhancing growth and disease resistance of potato (Solanum tuberosum L.). Biol Control 56(1):43–49
Poveda J, Zabalgogeazcoa I, Soengas P, Rodríguez VM, Cartea ME, Abilleira R, Velasco P (2020) Brassica oleracea var. acephala (kale) improvement by biological activity of root endophytic fungi. Sci Rep 10(1):20224
Qu L, Gu X, Li J, Guo J, Lu D (2023) Leaf photosynthetic characteristics of waxy maize in response to different degrees of heat stress during grain filling. BMC Plant Biol 23(1):469
Riedell WE, Kieckhefer RE, Petroski RJ, Powell RG (1991) Naturally-occurring and synthetic loline alkaloid derivatives: insect feeding behavior modification and toxicity. J Entomol Sci 26(1):122–129
Rocheta M, Becker JD, Coito JL, Carvalho L, Amâncio S (2014) Heat and water stress induce unique transcriptional signatures of heat-shock proteins and transcription factors in grapevine. Funct Integr Genomics 14:135–148
Rodriguez F, Arsene-Ploetze F, Rist W, Rüdiger S, Schneider-Mergener J, Mayer MP, Bukau B (2008a) Molecular basis for regulation of the heat shock transcription factor σ32 by the DnaK and DnaJ chaperones. Mol Cell 32(3):347–358
Rodriguez RJ, Henson J, Van Volkenburgh E, Hoy M, Wright L, Beckwith F et al (2008b) Stress tolerance in plants via habitat-adapted symbiosis. ISME J 2(4):404–416
Rodriguez RJ, White JF Jr, Arnold AE, Redman ARA (2009) Fungal endophytes: diversity and functional roles. New Phytol 182(2):314–330
Rungin S, Indananda C, Suttiviriya P, Kruasuwan W, Jaemsaeng R, Thamchaipenet A (2012) Plant growth enhancing effects by a siderophore-producing endophytic streptomycete isolated from a Thai jasmine rice plant (Oryza sativa L. cv. KDML105). Antonie Van Leeuwenhoek 102:463–472
Russell JR, Huang J, Anand P, Kucera K, Sandoval AG, Dantzler KW et al (2011) Biodegradation of polyester polyurethane by endophytic fungi. Appl Environ Microbiol 77(17):6076–6084
Saad EL, Salem SS, Fouda A, Awad MA, El-Gamal MS, Abdo AM (2018) New approach for antimicrobial activity and bio-control of various pathogens by biosynthesized copper nanoparticles using endophytic actinomycetes. J Radiat Res Appl Sci 11(3):262–270

Sadeghi F, Samsampour D, Seyahooei MA, Bagheri A, Soltani J (2020) Fungal endophytes alleviate drought-induced oxidative stress in mandarin (Citrus reticulata L.): toward regulating the ascorbate–glutathione cycle. Sci Hortic 261:108991
Samuels GJ, Pardo-Schultheiss R, Hebbar KP, Lumsden RD, Bastos CN, Costa JC, Bezerra JL (2000) Trichoderma stromaticum sp. nov., a parasite of the cacao witches broom pathogen. Mycol Res 104(6):760–764
Schaechter M (ed) (2012) Eukaryotic microbes. Academic Press
Schutzendubel A, Polle A (2002) Plant responses to abiotic stresses: heavy metal-induced oxidative stress and protection by mycorrhization. J Exp Bot 53(372):1351–1365
Seth, K., Vyas, P., Deora, S., Gupta, A.K., Meena, M., Swapnil, P., and Harish. (2023). Understanding plant-plant growth-promoting rhizobacteria (PGPR) interactions for inducing plant defense. In: Swapnil, P., Meena, M., Harish, Marwal, A., Vijayalakshmi, S., and Zehra, A. (eds), Plant-microbe interaction—recent advances in molecular and biochemical approaches. Academic Press, Elsevier, Volume 2, pp. 201–226
Shahzad R, Khan AL, Bilal S, Asaf S, Lee IJ (2018) What is there in seeds? Vertically transmitted endophytic resources for sustainable improvement in plant growth. Front Plant Sci 9:24
Shankar Naik B (2019) Functional roles of fungal endophytes in host fitness during stress conditions. Symbiosis 79(2):99–115
Sinno M, Ranesi M, Gioia L, d'Errico G, Woo SL (2020) Endophytic fungi of tomato and their potential applications for crop improvement. Agriculture 10(12):587
Slama HB, Cherif-Silini H, Bouket AC, Silini A, Alenezi FN, Luptakova L et al (2021) Biotechnology and bioinformatics of endophytes in biocontrol, bioremediation, and plant growth promotion. In: Endophytes: mineral nutrient management, Volume 3, pp 181–205
Sodhi GK, Saxena S (2023) Plant growth promotion and abiotic stress mitigation in rice using endophytic fungi: advances made in the last decade. Environ Exp Bot 209:105312
Songy A, Fernandez O, Clément C, Larignon P, Fontaine F (2019) Grapevine trunk diseases under thermal and water stresses. Planta 249:1655–1679
Stone JK (1988) Fine structure of latent infections by Rhabdocline parkeri on Douglas-fir, with observations on uninfected epidermal cells. Can J Bot 66(1):45–54
Strobel GA (2003) Endophytes as sources of bioactive products. Microbes Infect 5(6):535–544
Strong PJ, Claus H (2011) Laccase: a review of its past and its future in bioremediation. Crit Rev Environ Sci Technol 41(4):373–434
Subudhi E, Sahoo RK, Dey S, Das A, Sahoo K (2018) Unraveling plant-endophyte interactions: an omics insight. In: Endophytes and secondary metabolites, pp 1–19
Sunitha VH, Devi DN, Srinivas C (2013) Extracellular enzymatic activity of endophytic fungal strains isolated from medicinal plants. World J Agric Sci 9(1):01–09
Supratman U, Suzuki T, Nakamura T, Yokoyama Y, Harneti D, Maharani R et al (2021) New metabolites produced by endophyte Clonostachys rosea B5–2. Nat Prod Res 35(9):1525–1531
Tadych M, Bergen M, Dugan FM, White JF Jr (2007) Evaluation of the potential role of water in spread of conidia of the Neotyphodium endophyte of Poa ampla. Mycol Res 111(4):466–472
Tejesvi MV, Mahesh B, Nalini MS, Prakash HS, Kini KR, Subbiah V, Shetty HS (2005) Endophytic fungal assemblages from inner bark and twig of Terminalia arjuna W. & A. (Combretaceae). World J Microbiol Biotechnol 21:1535–1540
Toghueo RMK (2020) Bioprospecting endophytic fungi from Fusarium genus as sources of bioactive metabolites. Mycology 11(1):1–21
Tomilova OG, Kryukova NA, Efimova MV, Kolomeichuk LV, Kovtun IS, Glupov VV (2023) The endophytic entomopathogenic fungus Beauveria bassiana alleviates adverse effects of salt stress in potato plants. Horticulturae 9(10):1140
Traving SJ, Thygesen UH, Riemann L, Stedmon CA (2015) A model of extracellular enzymes in free-living microbes: which strategy pays off? Appl Environ Microbiol 81(21):7385–7393
Tsai AYL, Oota M, Sawa S (2020) Chemotactic host-finding strategies of plant endoparasites and endophytes. Front Plant Sci 11:1167
Tudzynski B, Sharon A (2002) Biosynthesis, biological role and application of fungal phytohormones. In: Industrial applications. Springer Berlin Heidelberg, Berlin, Heidelberg, pp 183–211

Vaitiekūnaitė D, Bružaitė I, Snitka V (2022) Endophytes from blueberry (Vaccinium sp.) fruit: characterization of yeast and bacteria via label-free surface-enhanced Raman spectroscopy (SERS). Spectrochim Acta A Mol Biomol Spectrosc 275:121158

Valdés-Santiago L, Vargas-Bernal R, Herrera-Pérez G, Colli-Mull JG, Ordaz-Arias A (2021) Application of two-photon microscopy to study Sclerotium cepivorum Berk Sclerotia isolated from naturally infested soil and produced in vitro. Curr Microbiol 78:749–755

Vandenkoornhuyse P, Quaiser A, Duhamel M, Le Van A, Dufresne A (2015) The importance of the microbiome of the plant holobiont. New Phytol 206(4):1196–1206

Verma H, Kumar D, Kumar V, Kumari M, Singh SK, Sharma VK et al (2021) The potential application of endophytes in management of stress from drought and salinity in crop plants. Microorganisms 9(8):1729

Verma A, Shameem N, Jatav HS, Sathyanarayana E, Parray JA, Poczai P, Sayyed RZ (2022) Fungal endophytes to combat biotic and abiotic stresses for climate-smart and sustainable agriculture. Front Plant Sci 13:953836

Vu T, Hauschild R, Sikora RA (2006) Fusarium oxysporum endophytes induced systemic resistance against Radopholus similis on banana. Nematology 8(6):847–852

Wahab A, Muhammad M, Munir A, Abdi G, Zaman W, Ayaz A et al (2023) Role of arbuscular mycorrhizal fungi in regulating growth, enhancing productivity, and potentially influencing ecosystems under abiotic and biotic stresses. Plan Theory 12(17):3102

Wani ZA, Ashraf N, Mohiuddin T, Riyaz-Ul-Hassan S (2015) Plant-endophyte symbiosis, an ecological perspective. Appl Microbiol Biotechnol 99:2955–2965

Xu L, Wu C, Oelmüller R, Zhang W (2018) Role of phytohormones in Piriformospora indica-induced growth promotion and stress tolerance in plants: more questions than answers. Front Microbiol 9:1646

Yadav SK (2010) Heavy metals toxicity in plants: an overview on the role of glutathione and phytochelatins in heavy metal stress tolerance of plants. S Afr J Bot 76(2):167–179

Yadav G, Meena M (2021) Bioprospecting of endophytes in medicinal plants of Thar Desert: an attractive resource for biopharmaceuticals. Biotechnol Rep 30:e00629

Yadav G, Meena M (2022) Biological control of plant diseases by endophytes. In: Shah MP, Deka D (eds) Endophytic association: what, why and how (volume: developments in applied microbiology and biotechnology). Academic Press, Elsevier Inc, pp 119–135

Zaidi NW, Dar MH, Singh S, Singh US (2014) Trichoderma species as abiotic stress relievers in plants. In: Biotechnology and biology of Trichoderma. Elsevier, pp 515–525

Fungal Endophytes: A Resource for Producing Extracellular Enzymes

6

Abhishek Sahoo, Indurekha Meher, Sushree Pralipta Chhotaray, Mukesh Meena, Prashant Swapnil, Jaya Arora, and Andleeb Zehra

Abstract

Fungal endophytes are endosymbionts that coexist asymptomatically with plants during various phases of growth and development. They attracted a lot of interest as a valuable and mostly unexplored resource for the manufacture of extracellular enzymes. These symbiotic fungi increase the resilience and health of plants by producing a wide variety of enzymes that are vital for different biochemical processes. Extracellular enzymes are produced by the proliferation of microorganisms and are involved in a variety of biological and environmental activities that take place outside of cells. This chapter illustrates that some key enzymes produced by the endophytic fungi are cellulases, ligninase, pectinases, amylases, xylanases, proteases, lipases, phytases, chitinases, polyphenol oxidases, peroxidase, laccase, etc. These enzymes have critical roles in agriculture like in bioremediation, waste degradation and detoxification, and industries such as biomass conversion, biofuel production, and pharmaceuticals. Various factors like host plant species, signaling chemical, age of plant, pH, temperature, moisture con-

A. Sahoo · M. Meena (✉)
Laboratory of Phytopathology and Microbial Biotechnology, Department of Botany, Mohanlal Sukhadia University, Udaipur, Rajasthan, India
e-mail: drmukeshmeena321@mlsu.ac.in

I. Meher · S. P. Chhotaray · P. Swapnil (✉)
School of Basic Science, Department of Botany, Central University of Punjab, Bhatinda, Punjab, India
e-mail: prashant.swapnil@cup.edu.in

J. Arora
Laboratory of Biomolecular Technology, Department of Botany, Mohanlal Sukhadia University, Udaipur, Rajasthan, India

A. Zehra
MI College, Male, Maldives

M. Meena et al. (eds.), *Metabolic Sustainability of Endophytes*, Microorganisms for Sustainability 52, https://doi.org/10.1007/978-981-96-4004-1_6

tent, nutrient availability, and fungal strain influence the tendency of enzyme production. Additionally, this chapter discusses the adaptability of fungal endophytes to various environmental conditions and methods of enhancing production of enzymes by optimization of fermentation condition, advanced genetic engineering methods like CRISPR/cas9 and CRISPRi technology, and co-cultivation. Significant improvements in environmentally friendly technologies, industrial processes, and sustainable technologies can result from using the enzymatic capability of fungal endophytes. This study emphasizes the need of continuing research to identify new enzymes from fungal endophytes with enhanced functions and a wide range of applications.

Keywords

Fungal endophytes · Extracellular enzymes · Cellulase · Laccase · Genetic engineering · CRISPR/cas9 · CRISPRi

Abbreviations

AAO	Aryl-alcohol oxidases
AChE	Acetylcholinesterase
BGCs	Biosynthetic gene clusters
BGL	β-glucosidase
C	Carbon
CMCase	Carboxymethyl cellulose
CRISPR/cas9	Clustered regularly interspaced short palindromic repeats associated protein 9
CRISPRi	Clustered regularly interspaced short palindromic repeats interference
DNA	Deoxyribonucleic acid
E^0	Redox potential
FPase	Filter paper cellulase
HIV	Human immunodeficiency virus
HPLC	High performance liquid chromatography
IU/gds	International units of enzyme activity per gram of dry solid
kDa	Kilo dalton
L-ASNase	L-asparaginase
MnP	Manganese peroxidases
MS	Mass spectrometry
N	Nitrogen
NMR	Nuclear magnetic resonance
P	Phosphorus
PAHs	Polycyclic aromatic hydrocarbons

PPOs	Polyphenol oxidases
RNA	Ribonucleic acid
ROS	Reactive oxygen species
SbmF	Submerged fermentation
SDS-PAGE	Sodium dodecyl sulfate polyacrylamide gel electrophoresis
SSF	Solid-state fermentation
SSSF	Semi-solid-state fermentation
TAC	Total anthocyanin content
TPC	Total phenolic content
UFPLC	Ultra-fast preparative liquid chromatography
VP	Versatile peroxidases

6.1 Introduction

Endophytic fungi inhabit internally in plant tissues either intercellularly or intracellularly, without causing any apparent harm to their hosts (Yan et al. 2015). They possess symbiotic relationship with their host. Fungal endophytes coevolved and adapted with their host throughout their life cycle, which is why they are ubiquitous in plants like bryophytes, pteridophytes, gymnosperms, and angiosperms. Fungal endophytes are found in almost every tissues of host, e.g., leaves, stems, roots, buds, bark, fruits, ovules, seeds, tubers, xylem, and phloem (Breen 1994). Endophytic fungi have been the subject of extensive research due to their potential benefits to plants, including increased resistance to pathogens, improved tolerance to abiotic stress, and enhanced nutrient uptake. They have the ability to produce extracellular enzymes. These enzymes have various applications in the food industry, textile industry, leather industry, agriculture, beverage, and human health sectors (Mishra et al. 2017). Endophytic fungi secrete a wide variety of extracellular enzymes and bioactive secondary metabolites of great biotechnological interests (Khan et al. 2016). Extracellular enzymes are responsible for breaking down the simple as well as complex organic matter and release phosphorus (P), nitrogen (N), and carbon (C) into the atmosphere and soil (Sunitha et al. 2012). Enzymes are highly effective biocatalysts for a wide array of reactions. Microorganisms are a viable alternative source of enzymes due to their ability to be cultured in large quantities rapidly through fermentation. They exhibit extensive biochemical diversity and can be genetically manipulated with relative ease (Raveendran et al. 2018). Endophytic fungi, in particular, secrete numerous extracellular enzymes, including cellulases, lipases, amylases, proteinases, pectinases, and laccases. Fungal enzymes play pivotal roles in biodegradation and hydrolysis, processes essential for protecting against pathogen invasion and extracting nutrients from the host plant (Sunitha et al. 2012; Yadav and Meena 2021, 2022). Their primary functions involve the hydrolysis of food substances and pathogen defense. Fungal enzymes are also extensively utilized in the beverage, food, confectionery, textile, and leather industries to streamline the processing of raw materials. Moreover, these enzymes typically exhibit greater stability compared to those sourced from other origins (Yadav 2019).

6.2 Extracellular Enzymes: Key Players in Physiological Functioning

Fungal endophytes are fungi that live inside the tissues of plants without causing any apparent harm to the host (Sunitha et al. 2012). Extracellular enzymes are secreted by microbes, including fungi, bacteria, and actinomycetes, into their surrounding environment. These enzymes play a crucial role in the breakdown of complex organic molecules into simpler forms, which can then be absorbed and utilized by the fungus. In context of fungi, extracellular enzymes play crucial role in growth and development along with pathogenicity. Extracellular enzymes degrade variety of complex molecules, solubilize phosphorus, degrade xenobiotics, and provide antimicrobial potential to their host (Yuan et al. 2017).

6.3 Types of Extracellular Enzyme

Fungal endophytes produce a variety of extracellular enzymes, including phytase, lipase, cellulase, amylase, laccase, protease, pectinase, and ligninase. A few extracellular enzymes are discussed here.

6.3.1 Cellulase

Endophytic fungi reside within plant tissues without causing visible harm to their host (Kogel et al. 2006). Some of these fungi can produce cellulase and other enzymes, such as xylanase, β-xylosidase, endo-β-1,4-glucosidase, β-1,4-glucosidase, and endoglucanase, which are essential for the degradation of cellulose, a complex carbohydrate found in plant cell walls. In recent years, microbial enzymes have garnered increasing attention due to their remarkable efficiency and wide range of applications in diverse physical and chemical environments (Singh et al. 2016). Endophytic fungi capable of producing cellulase have attracted interest for their potential uses in fields like agriculture, biofuel production, and bioremediation. Here is some information about cellulase producing endophytic fungi. Numerous fungi are known to produce cellulase enzymes. Solid state cellulase production by *Melanoporia* sp. CCT 7736 was reported from coconut (*Cocos nucifera* L.) shell powder (Oliveira et al. 2016). Without any kind of pretreatment, fermentation was carried out for 24 hours, and good results were obtained. Maximal activity was observed at pH 6.5 and 60 °C. Majority of cellulases of industrial importance show maximum activity at acidic pH, but in this case, maximum activity after optimization was observed at 7.5 IU/gds (international units of enzyme activity per gram of dry solid). Isolated cellulase has good stability without any cryopreservation (Oliveira et al. 2016). Eight fungal endophytes were isolated from the different plant parts of *Michelia champaca* (family Magnoliaceae). Screening and evaluation of their antifungal, anticancer, and acetylcholinesterase (AChE) inhibitory activities show promising results. Antifungal potential of fungal extract was checked against two phytopathogenic fungi, while investigation of the EtOAc extract of endophytic *Colletotrichum gloeosporioides* results in finding of a new compound (2-phenylethyl 1*H*-indol-3-yl-acetate) and seven already known compounds: 4-hydroxy-benzamide, 2(2-aminophenyl) acetic acid, 2(4-hydroxyphenyl) acetic acid, 2(2-hydroxyphenyl) acetic acid, cyclo-(*S**-Pro-*S**-Val), cyclo-(*S**-Pro-*S**-Tyr), and uracil. All chemical structures were determined by using 1D and 2D NMR, as well as MS analysis. Compounds 1–8 were tested in vitro for their antifungal and AChE inhibitory properties. Comparable to that of the positive control (nystatin), compound 1 showed encouraging action against *Cladosporium cladosporioides* and *C. sphaerospermum*. Utilizing apple pomace as a substrate, the impact of several inducers on *Aspergillus niger* NRRL-567's bioproduction of cellulase and hemicellulase was studied (Dhillon et al. 2012). The fermentation process resulted in the rapid production of various cellulase enzymes, including xylanase, β-glucosidase (BGL), carboxymethyl cellulose (CMCase), and FPase (filter paper cellulase), with peak activity occurring between 48 and 72 hours. $CuSO_4$ and veratryl alcohol were used, and after 48 hours of incubation, the increased FPase and BGL activity of 133.68 ± 5.44 IU/gram dry substrate (gds) and 60.09 ± 3.43 IU/gds, respectively, were detected. Lactose produced the greater CMCase activity of 172.31 ± 14.21 IU/gds following a 48-hour incubation period. Similarly, after 72 hours of fermentation, greater xylanase activity of 1412.58 ± 27.9 IU/gds was found with veratryl alcohol. This work provides insight into the quick bioproduction of hemicellulase and fungal cellulase from low-cost waste, apple pomace, when combined with various inducers. Extracellular β-glucosidase was produced using four strains of *Aspergillus*: *A. niger* CDBB-H-176, *A. niger* CDBB-H-175, *A. niger* ATCC 9642, and *Aspergillus terreus* CDBB-H-194 (Alarid-García and Escamilla-Silva 2017). Maximizing the activity of β-glucosidase was achieved by optimizing the growth medium parameters through the use of an orthogonal experimental design (L9). The ideal circumstances, which applied to all four strains, were as follows: 30 °C, pH 6.0, 200 rpm orbital agitation, and 0.5% (w/v) sucrose concentration were the measured values. *A. niger* CDBB-H-175 was the most prolific strain, yielding 701.2 U/mL. In a subsequent phase, we adjusted (L18) the nutrient concentration in the culture medium to see if this alteration would boost β-glucosidase production. For *A. niger* CDBB-H-175, the ideal conditions were as follows (%, w/v): NH_4NO_3, 0.1; $NH_4H_2PO_4$, 0.1; $NaNO_3$, 0.3; KCl, 0.3; KH_2PO_4, 0.15; $MgSO_4 \cdot 7H_2O$, 0.05; yeast extract, 0.1. In these settings, β-glucosidase production was 1207.9 U/mL. Using enzymatic assays, the enzyme was characterized. It was discovered that the four microorganisms used for the experiment generated β-glucosidase at an optimal pH of 5.0 and temperature of 65 °C. Only for *A. niger* CDBB-H-175 and CDBB-H-176 did we find the Michaelis–Menten constants (Km); they were 2.7 and 2.2 mM, respectively. Studies suggested that when utilizing agroindustrial wastes like wheat bran, as a carbon source for submerged (SbmF) or solid-state fermentation (SSF), the filamentous fungus *Aspergillus caespitosus* was an efficient generator of both intracellular and extracellular invertases (Alegre et al. 2009). Encouraging the formation of extracellular enzyme under SSF at 30 °C for 72 hours involved humidifying the substrate with SR salt solution (1:1, w/v). When employing wheat

Table 6.1 List of some cellulase-producing endophytic fungi

Name of endophyte	Host	Reported by
Colletotrichum gloeosporioides	*Michelia champaca*	Chapla et al. (2014)
Melanoporia sp.	*Cocos nucifera*	Oliveira et al. (2016)
Penicillium oxalicum	*Taxus cuspidate*	Li et al. (2021)
Aspergillus sp.	*Sponge*	El-Gendy and El-Bondkly (2010)
Pestalotiopsis microspora	*Forest trees*	Goukanapalle et al. (2020)
Acremonium sp.	*Sugarcane*	de Almeida et al. (2011)
Aspergillus flavus		Onofre et al. (2013)
Myceliophthora thermophile		de Cassia Pereira et al. (2016)
Aspergillus sp.	*Spartina* sp.	Liang et al. (2012)
Alternaria alternate	*Croton oblongifolius*	Wipusaree et al. (2011)
Alternaria alternate	*Asclepias sinaica*	Fouda et al. (2015)
Cladosporium cladosporioides	*Lawsonia inermis*	Amirita et al. (2012)
Bjerkandera sp.	*Drimys winteri*	Oses et al. (2006)
Acremonium sp.	*Acrostichum aureum*	Maria et al. (2005)
Alternaria sp.		
Aspergillus sp.	*Acanthus ilicifolius*	
Pestalotiopsis sp.		
Penicillium sp.	*Centella asiatica*	Devi et al. (2012)
Acremonium terricola	*Opuntia ficus-indica*	Bezerra et al. (2012)

bran under SSF, the extracellular activity was approximately 5.5 times higher than when using the same carbon source in SbmF (Khanna medium). On the other hand, compared to wheat bran alone, the amount of enzyme produced by wheat bran plus oatmeal was 2.2 times greater. The addition of phosphate and nitrogen sources had an impact on the enzymatic production. The extracellular activity in SbmF and SSF decreased when glucose was added, but the intracellular form that was formed in SbmF increased three to five times. The extracellular and intracellular enzymes synthesized in SbmF displayed maximum activity at 60 °C, while the invertase produced in SSF revealed its optimal temperature at 50 °C. At pH 4.0–6.0, all enzymatic forms demonstrated their maximum activity, and they remained stable for an hour at 50 °C. A few examples of very well-known cellulase-producing fungi are cited here (Table 6.1).

6.3.2 Ligninase

Lignin is a major component of plant cell walls, second only to cellulose in abundance. It accounts for about 20–30% of the dry weight of wood (Abdel-Raheem and Shearer 2002). Lignin is a three-dimensional hydrophobic polymer derived from the random coupling of phenylpropanoid precursors, including coniferyl, sinapyl, and p-coumaryl alcohols (Eriksson et al. 2012). Ligninases, such as manganese peroxidase and lignin peroxidase, facilitate the breakdown of lignin, making cellulose and other nutrients more accessible. Fungal endophytes that produce ligninase are capable of growing within plant tissues and releasing enzymes that degrade this complex polymer found in plant cell walls. These fungi are useful in several areas, such as biotechnology and bioremediation, and have the potential to contribute significantly to the breakdown of lignin. Fungal endophytes that produce ligninase are fungi that can grow inside plant tissues and release enzymes that can break down the complex organic polymer called lignin that is present in plant cell walls. It is believed that white-rot fungi are the only known organisms that can totally break down lignin into carbon dioxide and water, even if the mechanism of lignin degradation is still not entirely understood (Risna and Suhirman 2002). These fungi may play a significant role in the degradation of lignin and have applications in biotechnology and bioremediation, among other fields.

6.3.3 Pectinases

Plant cell walls contain pectin, a complex polysaccharide that is broken down by pectinases. These enzymes can lead to the degradation of plant tissues. Fungal endophytes, microscopic organisms that reside within plant tissues without harming their host, can produce pectinase, which breaks down the pectin in plant cell walls. Pectinases are a class of enzymes responsible for degrading pectin, a chain molecule with a rhamnogalacturonan backbone found in the primary cell walls of terrestrial plants (Anisa and Girish 2014). These pectinases are widely used in the wine and fruit juice industries. They are widely used in the degumming of plant rough fibers, the treatment of pectic wastewaters, and textile processing. Pectinases make almost 25% of the food enzymes manufactured worldwide, and demand for them is rising daily. *Rhizopus* sp. and *Trichoderma viride* were isolated from farmyard manure and agricultural waste in the current study using pectin agar medium. Using pectin agar medium and pectin broth medium, respectively, pure cultures of these fungi were examined for pectinolytic activity. These fungi were producing pectinolytic enzymes, as evidenced by the zone of clearance in the plate test method and the clearing of turbidity in the pectin broth medium. The turbidometric and polygalacturonic assay results demonstrated these fungi's capacity to manufacture pectinolytic enzymes.

Endophytic fungi from *Aloe vera* and *Ocimum sanctum* were isolated and evaluated for their antifungal activity against *Helminthosporium maydis*, *Fusarium oxysporum*, *Colletotrichum falcatum*, and *Rhizoctonia solani*. The synthesis of extracellular enzymes, including amylase, cellulase, chitinase, pectinase, laccase, lipase, and urease, was further assessed in the isolated fungi (Yadav et al. 2015). Eighteen endophytic fungi, which differ morphologically from one another, were isolated from the leaves, stem, and roots of *Ocimum sanctum* and *Aloe vera* throughout the investigation. *Fusarium oxysporum* is the target of antagonistic action displayed by AVR1 and AVR3. Only AVR3 was discovered to be positive for chitinase production, while T2S1 and AVL1 were found to be positive for amylase, TL1, AVR1, and AV2L1 for cellulase production. Only AVL1 was found to be positive for lipase production, and TL2, T2R1, AVL1, AVL4, T2L1, and TL3 were positive for urease production. No isolates were found to be positive for laccase production. AVR3, TL2, TL3, TS1, AV2R1, T2S1, and T2L1 were positive for pectinase production. The study's findings imply that endophytic fungus connected to *O. sanctum* and *Aloe vera* has abundant sources of enzymes and possible carriers of antibacterial action.

Some fungal strains were obtained from various materials, such as fruits and soil (Abdullah et al. 2018). The pectin hydrolysis zone served as the basis for qualitative screening. Meanwhile, submerged fermentation was used to perform quantitative screening. The strains with the highest pectinolytic potential were chosen from among all of the strains, identified, and given the code *Aspergillus niger* ABT-5. It was assessed how various fermentation media affected the synthesis of pectinase. The maximum amount of pectinase was produced in the M5 medium with 10 g wheat bran, nutritional medium comprising (g/l) of $(NH_4)_2SO_4$ 6.0, K_2HPO_4 6.0, KH_2PO_4 6.0, and $MgSO_4.7H_2O$ 0.1. Other crucial physicochemical factors, such as the length of the incubation period, the temperature, the volume of the medium, the size of the inoculum, and the sources of carbon and nitrogen, were also tuned for the synthesis of pectinase. The optimal conditions for pectinase production (15.5 U/mL) were determined to be pH 6, a temperature of 30 °C, and a medium volume of 50 mL, following 72 hours of incubation. The greatest sources of carbon and nitrogen were later determined to be fructose and urea. The main component of agro-industrial leftovers is complex polysaccharides, which promote microbial development and the generation of enzymes of industrial significance. One of the most extensively distributed enzymes in bacteria, fungus, and plant is pectinase. In the submerged fermentation process of *Aspergillus niger*, Czapek media mixed with orange waste peel as a carbon source presented the best enzymatic production (Ahmed et al. 2016). After 5 days of fermentation, a partial optimization culture revealed the highest enzyme output (117.1 ± 3.4 μM/mL/min) at 30 °C in an orange waste peel medium with a pH of 5.5 and a 4% substrate concentration. Ion exchange chromatography and ammonium sulfate precipitation were used to purify the generated enzyme. Using a gel filtration chromatographic approach, a purification fold of 5.59 with specific activity and recovery percentages of 97.2 U/mg and 12.96% were attained, respectively. SDS-PAGE revealed that the isolated pectinase from *A. niger* has a molecular weight of 30 kDa. According to the pectinase activity profile, the isolated enzyme functioned best at pH 7 and 55 °C temperature. The enzyme is valuable in many industrial sectors, including the juice and textile industries, because it can produce maximal amounts of pectinase when there is a less expensive substrate present at low concentrations.

6.3.4 Proteases

Proteases are enzymes that catalyze the hydrolysis of proteins. Fungal endophytes may produce proteases to acquire nitrogen and other essential nutrients from plant proteins. These protease-producing fungal endophytes can colonize plant tissues

Table 6.2 Some examples of fungal endophytes known to produce proteases

Name of endophyte	Host	Reported by
Mortierella hyalina	*Osbeckia stellata*	Bhagobaty and Joshi (2012)
Paecilomyces variabilis	*Osbeckia chinensis*	
Penicillium sp.	*Schima khasiana*	
Penicillium sp.	*Camellia caduca*	

and secrete proteases to break down proteins. Below are examples of fungal endophytes known for their ability to produce proteases (see Table 6.2).

6.3.5 Lipases

Lipases are enzymes that catalyze the breakdown of lipids (insoluble substrates) into fatty acids and glycerol. Fungal endophytes secrete lipases to break down lipids, enabling them to access energy stored in plant lipid reserves. The hydrolysis of triglycerides into glycerol and free fatty acids is catalyzed by lipases (triacylglycerol hydrolases, E.C. 3.1.1.3), a type of hydrolase (Gupta et al. 2007). These extracellular, water-soluble enzymes act on insoluble lipid substrates and specifically catalyze the hydrolysis of long-chain triglycerides (Singh and Mukhopadhyay 2012). Lipases can act on a variety of substrates including natural oils, synthetic triglycerides, and esters of fatty acids (Aravindan et al. 2007). Several authors have documented the existence of *Candida rugosa* lipase isoforms (Benjamin and Pandey 1998; Jaeger et al. 1994). Extracellular lipase that is thermostable is produced by the thermophilic fungi *Aspergillus terreus*, *Rhizopus homothallicus*, and *Mucor pusillus*. *Mucor* sp. generates an extracellular, inducible, alkalophilic, and thermostable lipase, which is unusual in lipases that remain stable at high temperatures and alkaline environments. Few studies using molds containing thermostable and alkalophilic lipase have been published to date (Ginalska et al. 2004). A variety of industrial applications show that lipases have emerged as one of the top biocatalysts/bio-accelerators, with a proven capacity to support the multibillion dollar underutilized bioindustry (Dhillon et al. 2012; Joseph et al. 2008). A variety of environments, such as oil-contaminated soils, vegetable oil waste sites, dairy product factories, seeds, and decaying food, are home to fungi that can manufacture lipases (Ko et al. 2005; Sharma et al. 2001). Molecular biology techniques have advanced rapidly, and the advent of more dependable high-throughput screening methods has further enhanced the applicability of lipase for chemical synthesis. Nowadays, a wild-type enzyme is modified to fit a specific purpose in order to create second-generation biocatalysts (Goddard and Reymond 2004; Otten and Quax 2005; Reetz 2001; Schmidt and Bornscheuer 2005). The value of lipase for organic synthesis has increased due to the quick advancement of molecular biology techniques and the availability of more dependable high-throughput screening methods. Some techniques have been developed to obtain higher conversions for highly specific enzymes for each application, improving the possibility of industrial applications for lipase (Franken et al. 2010; Soccol and Vandenberghe 2003). *Candida rugosa* lipases have great significance for their diverse biotechnological potential (Dhillon et al. 2012). *Rhizopus oryzae*, *Rhizopus microsporus*, and *Rhizopus stolonifer* comprise the three main groups of *Rhizopus* species. A His134 and a Leu234 have been substituted by an Asn and a Leu, respectively, in *R. oryzae* lipase, *Rhizopus delemar* lipase, and *Rhizopus javanicus* lipase (Minning et al. 1998). It has been revealed that the lipase gene from *R. stolonifer* shares 84% of its amino acid sequence with that of *R. oryzae* lipase. The molecular characterization of *R. microspores* lipase, however, has not been reported (Yu et al. 2009). The distinct pharmacological characteristics of *Yarrowia lipolytica* (YLLIP2) lipase make it a prime option for enzyme replacement therapy. It is not inhibited by bile salts and exhibits maximum activity at low pH levels. YLLIP2 shares a gene family with *Thermomyces lanuginosus* lipase, a widely recognized lipase with numerous uses in detergent production and biotechnological procedures (Aloulou et al. 2007).

Lipase has a wide range of industrial applications, such as the synthesis of biopolymers and biodiesel, as well as the manufacturing of medications, agrochemicals, flavors, and cosmetics (Haki and Rakshit 2003). The stability, selectivity, and broad substrate specificity of fungal lipases have drawn particular attention from the industrial sector (Dutra et al. 2008; Griebeler et al. 2011; Ray 2012). The food, detergent, chemical, and pharmaceutical sectors have come to rely more and more on lipases due to their unique capacity to accomplish certain chemical transformations (Franken et al. 2010; Gupta et al. 2007). Fungal extracellular enzymes can be extracted easily, which significantly reduces the cost and makes this source preferable over bacteria. Lipase gained a great deal of attention as biocatalysts in several industrial processes such as detergents, baking, cheese making, oils, and fats cleaning from hard surface, leather, and paper processing applications in food industry (Reetz 2002). Here are some examples of lipase-producing endophytic fungi (Table 6.3).

Table 6.3 Lipase producing endophytic fungi

Endophyte	Host plant	Reported by
Phyllosticta sp.	*Adhatoda vasica*	Amirita et al. (2012)
Curvularia brachyspora		
Drechslera hawaiiensis		
Curvularia vermiformis	*Coleus aromaticus*	Amirita et al. (2012)
Mortierella hyalina	*Osbeckia stellata*	Bhagobaty and Joshi (2012)
Paecilomyces variabilis	*Osbeckia chinensis*	
Penicillium sp.	*Camellia caduca*	
Penicillium sp.	*Schima khasiana*	
Alternaria sp.	*Acrostichum aureum*	
Alternaria chlamydosporus	*Acanthus ilicifolius*	Maria et al. (2005)
Pestalotiopsis		
Aspergillus sp.		
Fusarium sp.	*Acrostichum aureum*	Maria et al. (2005)

6.3.6 Amylases

Amylases hydrolyze starch, a polysaccharide, into simpler sugars, allowing fungal endophytes to utilize plant carbohydrate reserves. Amylase-producing fungal endophytes colonize plant tissues and produce enzymes, such as amylases, that break down starches and carbohydrates. Amylases, including α-amylases, β-amylases, and glucoamylases, are among the most important enzyme groups in biotechnology. Microbial amylolytic enzymes are widely used in industries like food, chemicals, detergents, and textiles to hydrolyze starch into various sugar solutions. The breakdown of starch primarily involves three key enzymes: α-amylase (1,4 α-glucan glucanohydrolase, EC 3.2.1.1), β-amylase (1,4 α-glucan maltohydrolase, EC 3.2.1.2), and glucoamylase (1,4 α-glucan glucohydrolase, EC 3.2.1.3) (Pandey et al. 2000). Several fungal species are known to produce glucoamylases, and industrial amylases are frequently produced from *Aspergillus* and *Rhizopus* species (Pandey et al. 2000). Among various endophytic fungi, *Fusicoccum* sp. BCC4124 exhibited significant amylolytic activity when grown on media enriched for multienzyme induction and agro-industrial substrates (Champreda et al. 2007). The activities of both α-amylase and α-glucosidase were markedly enhanced in the presence of maltose and starch. α-amylase was purified to apparent electrophoretic homogeneity and demonstrated robust hydrolytic activity on soluble starch. Notably, this enzyme possesses desirable characteristics for biotechnological applications, such as high activity at 70 °C and resistance to glucose inhibition up to 1 mol/L.

The endophyte *Cylindrocephalum* sp. isolated from the medicinal plant *Alpinia calcarata* (Haw.) Roscoe was cultured in liquid medium to evaluate amylase production under various conditions. Optimal amylase production was achieved at 30 °C and pH 7.0. Of the different carbon and nitrogen sources tested, the combination of 1.5% maltose and 0.3% sodium nitrate yielded the highest amylase production (Sunitha et al. 2012). The investigation of De Oliveira et al. (2016) compared the production and the catalytic properties of amylolytic enzymes obtained from the fungi *Lichtheimia ramosa* (mesophilic) and *Thermoascus aurantiacus* (thermophilic). After 96 hours of cultivation, wheat bran supplemented with nutritional solution (pH 4.0) showed the maximum amylase production in both fungi, reaching 417.2 U/g of dry substrate (or 41.72 U/mL) for *L. ramosa* and 144.5 U/g of dry substrate (or 14.45 U/mL) for *T. aurantiacus*. At pH 6.0 and 60 °C, the enzymes exhibited increased catalytic activity. Between pH 3.5–10.5 and pH 4.5–9.5, respectively, the amylases produced by *L. ramosa* and *T. aurantiacus* were stable. After an hour of incubation, the amylase of *L. ramosa* remained steady at 55 °C, but the amylase of *T. aurantiacus* retained 60% of its initial activity in the same settings. When ethanol was present, both enzymes were active. Different types of starch were hydrolyzed by the enzymes; with corn starch produced the best results. There was potential for saccharification and dextrinizing in the enzyme complex that *L. ramosa* generated. The primary hydrolysis product released by the enzymatic extract made by the fungus *T. aurantiacus* was glucose monomers, which only showed saccharifying capability. Here are some examples of fungal endophytes known to produce amylases (Table 6.4).

Table 6.4 Here are some examples of fungal endophytes known to produce amylases

Endophyte name	Host plant	Reference
Preussia minima	*Eremophilia longifolia*	Zhang et al. (2010)
Alternaria sp.		
Cladosporium cladosporioides	*Costus igneus*	Amirita et al. (2012)
Curvularia brachyspora	*Lawsonia inermis*	
Drechslera hawaiiensis	*Adhatoda vasica*	
Nigrospora sphaerica	*Costus igneus*	
Colletotrichum crassipes	*Lawsonia inermis*	
Pestalotiopsis sp.	*Acanthus ilicifolius*	Maria et al. (2005)
Fusarium sp.	*Acrostichum aureum*	
Alternaria sp.		
Acremonium sp.		
Cylindrocephalum sp.	*Alpina calcarata*	Sunitha et al. (2012)
Discosia sp.	*Calophyllum inophyllum*	Hegde et al. (2011)
Paecilomyces variabilis	*Osbeckia chinensis*	Bhagobaty and Joshi (2012)
Colletotrichum gloeosporioides	*Costus igneus*	Amirita et al. (2012)
Phyllosticta sp.	*Lawsonia inermis*	
Xylaria sp.	*Coleus aromaticus*	

6.3.7 Phytases

Phytases (myo-inositol hexakisphosphate phosphohydrolases) catalyze the hydrolysis of phytic acid (myo-inositol hexakisphosphate), the predominant form of organic phosphorus in plant seeds and grains (Wodzinski and Ullah 1996). Since phytic acid is indigestible by monogastric animals, its excretion contributes to phosphorus pollution (Wang et al. 2020). As a result, phytases play a crucial role in animal nutrition and environmental management. Fungal endophytes, owing to their symbiotic association with plants, represent a promising source of extracellular phytases. They offer advantages such as stability, specificity, and production efficiency. Fungal endophytes encompass a wide range of taxa, primarily Ascomycota and Basidiomycota. These fungi inhabit various plant species, including crops and trees, suggesting a vast untapped resource for novel phytases. Studies have isolated phytase-producing endophytes from diverse environments, indicating their adaptive mechanisms and potential applications in different agro-ecological zones. Endophytic fungal phytases exhibit a variety of biochemical properties, including optimal pH and temperature ranges, substrate specificity, and resistance to proteolysis. These enzymes generally show optimal activity at acidic to neutral pH and moderate temperatures, aligning with the conditions in the gastrointestinal tracts of animals (Corrêa and de Araújo 2020). Molecular studies have revealed the presence of several isoforms of phytases within a single fungal species, contributing to their functional diversity.

Incorporation of endophytic fungal phytases in animal feed improves phosphorus availability, enhancing growth and reducing feed costs. This application is particularly significant for poultry and swine industries (El Enshasy et al. 2018). Phytase-producing endophytes can be utilized to enhance plant growth and phosphorus uptake, promoting sustainable agriculture. These endophytes could be integrated into biofertilizers to improve soil health and crop productivity. By reducing phytic acid content in animal manure, endophytic fungal phytases help mitigate phosphorus pollution, contributing to environmental conservation efforts (Balaban et al. 2016; Kour et al. 2020). Phytases can be used to reduce phytic acid content in plant-based foods, improving their nutritional value (Chen and Xu 2023). This is especially beneficial in the processing of cereals and legumes.

6.3.8 Chitinases

Chitinases (EC 3.2.1.14) are glycosyl hydrolases that break down chitin, a β-1,4-linked polymer of N-acetylglucosamine (Churklam and Aunpad 2020). These enzymes are widespread in nature, produced by bacteria, fungi, plants, and animals. Among fungi, endophytic species—those that colonize plant tissues without causing harm—are particularly noteworthy for their production of extracellular chitinases. These enzymes not only aid in the degradation of fungal pathogens but also enhance the endophyte's survival and competitiveness within the plant host. Fungal endophytes show considerable diversity in both the types and activities of chitinases they produce (Muthu Narayanan et al. 2022). These enzymes can be categorized based on their mode of action into endochitinases, which cleave internal bonds of chitin chains, and exochitinases, which cleave from the nonreducing ends (Rathore and Gupta 2015). Molecular studies have revealed a wide range of chitinase genes in fungal endophytes, with significant variation in their sequences, indicating adaptation to different ecological niches and functions (Wang et al. 2021). The action of chitinases involves the hydrolysis of β-1,4-glycosidic bonds in chitin, resulting in the production of chitooligosaccharides and monomeric N-acetylglucosamine. The catalytic mechanism typically involves a conserved active site with key amino acid residues facilitating the cleavage process (Aam et al. 2010; Churklam and Aunpad 2020). Structural studies have shown that chitinases possess a catalytic domain and, in some cases, one or more chitin-binding domains that enhance substrate affinity and enzyme efficiency (Oyeleye and Normi 2018).

Chitinases produced by fungal endophytes contribute to plant defense by degrading the cell walls of pathogenic fungi, thereby inhibiting their growth and infection. These enzymes enable endophytes to utilize chitin as a carbon and nitrogen source, enhancing their survival and growth within plant tissues (Govinda Rajulu et al. 2011; Hamid et al. 2013). Chitinases may play a role in maintaining the balance between endophytes and their host plants, modulating immune responses and promoting mutualistic interactions (Kumar et al. 2018). Endophytic chitinases can influence the microbial community within the plant, suppressing pathogenic species while promoting beneficial microbes. Chitinases from fungal endophytes are explored as biocontrol agents against fungal pathogens, reducing the reliance on chemical pesticides and enhancing crop protection (De Silva et al. 2019). These enzymes have applications in the bioconversion of chitinous waste into valuable products, including biofuels, fertilizers, and pharmaceuticals (Taokaew and Kriangkrai 2023). Chitinases hold potential in the development of antifungal therapies, given their ability to target fungal cell walls selectively.

6.3.9 Polyphenol Oxidases

One of the key contributions of fungal endophytes is the production of extracellular enzymes, including polyphenol oxidases (PPOs). PPOs catalyze the oxidation of phenolic compounds to quinones, which subsequently polymerize to form brown or black pigments, playing a crucial role in plant defense mechanisms (Thipyapong et al. 2007). Polyphenol oxidases are part of a family of enzymes that include tyrosinases, catechol oxidases, and laccases, all of which contain copper at their active sites, essential for their catalytic function (Janusz et al. 2020; Marusek et al. 2006; Mayer 2006). The primary function of PPOs is to oxidize monophenols to o-diphenols and further to o-quinones. These quinones are highly reactive and can undergo nonenzymatic reactions to form complex polymers involved in plant defense against pathogens and pests (Taranto et al. 2017).

Fungal endophytes produce a variety of extracellular enzymes, including PPOs, which are pivotal for the symbiotic relationship with their host plants. These enzymes play a significant role in modulating the plant's defense mechanisms and enhancing its ability to cope with environmental stresses. Several studies have reported the isolation and characterization of PPOs from fungal endophytes, demonstrating their potential in various biotechnological applications. The catalytic mechanism of PPOs involves the binding of phenolic substrates to the active site, followed by the oxidation of these substrates to quinones (Taranto et al. 2017). This process involves a series of electron transfer reactions mediated by the copper ions present at the active site. The quinones produced can participate in cross-linking reactions, leading to the formation of protective barriers against invading pathogens.

The production of quinones by PPOs can inhibit the growth of pathogenic microorganisms, making fungal endophyte-derived PPOs potential candidates for biocontrol agents in agriculture. PPOs can oxidize a wide range of phenolic pollutants, suggesting their application in the bioremediation of contaminated environments (Latz et al. 2018). The ability of PPOs to catalyze oxidation reactions can be harnessed in various industrial processes, including the synthesis of pharmaceuticals and the production of bio-based materials.

6.3.10 Oxidases and Peroxidases

Endophytic fungi produce a diverse array of extracellular enzymes, including oxidases and peroxidases, which play significant ecological and industrial roles. Oxidases catalyze the transfer of electrons from a substrate to oxygen, resulting in the formation of hydrogen peroxide or water. Peroxidases, in contrast, use hydrogen peroxide to oxidize various substrates (Kyomuhimbo et al. 2023). Among the oxidases and peroxidases produced by fungal endophytes are: (1) Laccases: Laccases (EC 1.10.3.2) are multicopper oxidases that can oxidize phenolic compounds. They play a crucial role in lignin degradation and are involved in the detoxification of lignin-derived phenolics in plant tissues (Rivera-Hoyos et al. 2013). (2) Manganese peroxidases (MnP): Manganese peroxidases (EC 1.11.1.13) catalyze the oxidation of Mn(II) to Mn(III), which subsequently oxidizes a wide range of organic substrates. These enzymes are important in the degradation of lignin and other recalcitrant organic materials (Gold et al. 2000). (3) Versatile peroxidases (VP): Versatile peroxidases (EC 1.11.1.16) exhibit the combined substrate specificities of lignin peroxidases and manganese peroxidases, making them highly effective in lignocellulosic biomass degradation (Andlar et al. 2018). (4) Aryl-alcohol oxidases (AAO): Aryl-alcohol oxidases (EC 1.1.3.7) produce hydrogen peroxide by oxidizing aromatic alcohols. This hydrogen peroxide is utilized by peroxidases for the degradation of complex organic molecules (Cleveland et al. 2021).

Fungal endophytes secrete oxidases and peroxidases in response to environmental signals such as the presence of lignocellulosic substrates. The production of these enzymes is often regulated by a combination of transcriptional and post-transcriptional mechanisms, involving complex signaling pathways that respond to oxidative stress and substrate availability (Bhadra et al. 2022; Jha et al. 2023). Oxidases and peroxidases produced by fungal endophytes are employed in the bioremediation of polluted environments. These enzymes can degrade a variety of environmental pollutants, including polycyclic aromatic hydrocarbons (PAHs), synthetic dyes, and industrial effluents (Baker et al. 2019). In the pulp and paper industry, these enzymes are used for the biobleaching of paper pulp, reducing the need for harsh chemical treatments and thus minimizing environmental pollution (Singh et al. 2019). The degradation of lignocellulosic biomass by fungal oxidases and peroxidases is crucial for the production of biofuels. These enzymes enhance the saccharification process, improving the yield of fermentable sugars for bioethanol production (Fernández-Fueyo et al. 2014). Fungal endophytes and their extracellular enzymes are sources of novel bioactive compounds. Laccases and peroxidases are utilized in the synthesis of fine chemicals and pharmaceutical intermediates due to their high specificity and eco-friendly nature.

6.3.11 Xylanase

Enzymes are gaining popularity due to their industrial applications; however, their high production costs limit widespread use. One promising approach to address this issue is utilizing lignocellulosic residues from agro-industrial settings as a component in the production medium. This strategy encourages research into alternative sources for enzyme production. Thus, as carbon sources for *Aspergillus* sp. fermentation, natural sunflower stalks (*Helianthus annuus* L.) and their related alkaline and autohydrolysis pretreatment liquors were used (de Souza et al. 2020). The results using the alkaline and autohydrolysis liquors showed xylanase activity of 2.97 and

4.91 U/mL, respectively, whereas the natural residue had the highest value (17.48 U/mL). Yet, the natural sunflower stalk + liquors (alkaline and autohydrolysis, separately) blend significantly enhanced enzyme production, 6.34 and 11.98 U/mL, in contrast to employing each liquor as the sole source encouragement. Sunflower stalks and their liquors are potentially stimulating supplies for xylanase development, in contrast to commercially pure xylan. Ironwood trees were shown to be the source of xylanase by HL.47F.216, endophytic fungi isolated from traditional medicinal plants (Tasia and Melliawati 2017). Cellulase is produced by HL.51F.235 from pin-flower trees, whereas CBN.6F.29, which is derived from *Madagascar periwinkle*, generates both xylanase and cellulase. The clear zone diameter of HL.47F.216 was 2.5 cm, and its xylanase activity was 0.262 U/mL at 50 °C and pH 7. At pH 5 and 70 °C, HL.51F.235 displayed a clear zone diameter of 2.4 cm and cellulase activity of 0.239 U/mL. Cellulase activity measured by CBN.6F.29 was 2.8 cm and 0.394 U/mL (pH 5, 40 °C), while xylanase activity measured 2.3 cm and 0.439 U/mL (pH 8, 70 °C). Low molecular weights of 20 kDa and 37–50 kDa were observed in xylanase from HL.47F.216 and CBN.6F.29, respectively. The cellulases from HL.51F.235 and CBN.6F.29 have respective molecular weights of 25 and 50 kDa, and 100 kDa. Fungal isolate CBN.6F.29 belongs to the class Coelomycetes, according to macroscopic and microscopic identification; *Acremonium* sp. was identified as HL.47F.216, and *Aspergillus nigri* as HL.51F.235. Worldwide, oil palm plants yield thousands of tons of empty fruit bunches each year. The field of biotechnology may find use for these lignocellulosic feedstocks. Using filamentous fungi that generate xylanases and ligninases during the biopulping process to make cellulose is an alternate method of enhancing pulp. Studies suggest *Aspergillus* sp. LPB-5 produced an enzymatic cocktail with low cellulase activity, 13.41 U/g lignin peroxidase, and 54.32 U/g xylanase activity in solid-state biopulping (Colonia et al. 2019). Enzymatic biobleaching was applied to pulps that had received biological, thermal, and chemical pretreatments in order to compare them. By combining biopulping with biobleaching, the lignin content dropped by 36.80%, the hemicellulose content was decreased by 26.27%, and the pulp output was 74.36%, with 36.56% digestibility. Alkaline and biobleaching together removed 81.97% hemicellulose and 93.89% lignin, with a 73.59% digestibility. Enzymatic biobleaching increased the pulp digestibility of all pretreatments. In the end, mild pretreatments would reduce the need for harsh chemicals, and the development of a biopretreatment to remove hemicellulose and alter the interface between lignin and carbohydrate complexes offered a benign solution with considerable environmental benefits. Here are some examples of fungal endophytes known to produce xylanases (Table 6.5).

6.3.12 Laccase

The groundbreaking work of Yoshida in 1883 laid the foundation for understanding the enzyme laccase, which was initially discovered in the latex of *Rhus vernicifera* (Bertrand et al. 2002; Yoshida 1883). Laccase is classified as an oxidoreductase,

Table 6.5 Here are some examples of fungal endophytes known to produce xylanases

Endophyte name	Host plant	Reference
Mortierella hyalina	*Osbeckia stellate*	Bhagobaty and Joshi (2012)
Paecilomyces variabilis	*Osbeckia chinensis*	
Penicillium sp.	*Camellia caduca*	
Acremonium zeae	*Zea mays*	Bischoff et al. (2009)
Acremonium terrícola *Aspergillus japonicus* *Pestalotiopsis guepinii* *Cladosporium sphaerospermum* *cladosporioides* *Xylaria* sp. 2 *Xylaria* sp. 1 *Tetraploa aristata* *Phomopsis archeri* *Penicillium aurantiogriseum* *Nigrospora sphaerica* *Paecilomyces variabilis* *Monodictys castaneae* *Fusarium lateritium*	*Opuntia ficus-indica* Mill.	Bezerra et al. (2012)

specifically one that acts on diphenols and related substances, using oxygen as the electron acceptor (Espina et al. 2021). This multi-copper enzyme catalyzes the oxidation of a wide range of organic and inorganic substrates, including phenolic compounds, aromatic amines, and dyes, utilizing molecular oxygen as the electron acceptor and producing water as a byproduct (Rivera-Hoyos et al. 2013). Laccases are produced by various organisms, including fungi, plants, and bacteria, and they play significant roles in several biological processes such as lignin degradation, pigment biosynthesis, stress defense, and pathogenesis in plants and fungi. In fungi, for example, laccases are involved in the breakdown of lignin, a complex aromatic polymer in wood, thus playing a crucial role in the carbon cycle (Arregui et al. 2019; Rivera-Hoyos et al. 2013). Laccase synthesis like the synthesis of many enzymes in eukaryotic organisms, is a multistep process involving transcription, translation, and several post-translational modifications (Fig. 6.1). The gene family of laccases in fungi exhibits a high degree of multiplicity, with various species encoding multiple laccase genes that give rise to numerous isoenzymes (Rodríguez et al. 2008). This is evident in the characterization studies of fungal laccases, which have identified multiple laccase genes in different species (Table 6.6). The versatility of laccases lies in their ability to modulate their redox potential (E^0), making them suitable for bioremediation applications, including delignification, dye decolorization, and degradation of pollutants like polycyclic aromatic hydrocarbons (PAHs) (Wong 2009; Xu et al. 1998). The addition of mediators further expands their utility in industrial processes. Elucidating the genetic basis of laccase production reveals complex regulatory mechanisms influenced by environmental factors, such as nutrient availability and inducing agents. The presence of multiple laccase gene families in fungi suggests functional diversity during their life cycle. Structurally, laccases exhibit a conserved folding pattern with cupredoxin-like

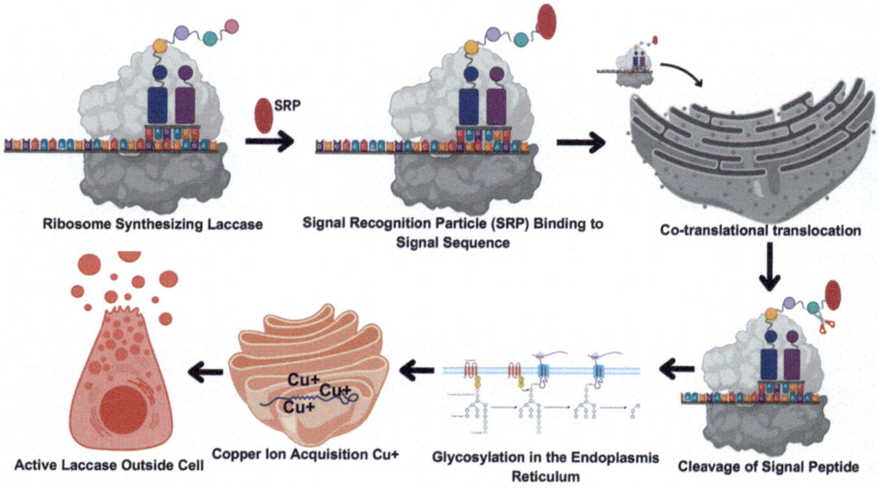

Fig. 6.1 Synthesis, processing, and secretion of laccase

Table 6.6 Multiple laccase genes in different fungal species

Fungus producing laccase	Gene families	Reference
Trametes villosa	*lcc1*, *lcc2*	Yaver et al. (1996)
Trametes sanguinea	*lcc1*, *lcc2*, *lcc3*, *lcc4*, *lcc5*	Hoshida et al. (2001)
Rhizoctonia solani	*lcc1*, *lcc2*, *lcc3*, *lcc4*	Wahleithner et al. (1996)
Trametes sp. I62	*lcc1*, *lcc2*, *lcc3*	Mansur et al. (1997)
Trametes sp. AH28-2	*lccA*, *lccB*, *lccC*	Xiao et al. (2006)
Gaeumannomyces graminis	*LAC1*, *LAC2*	Litvintseva and Henson (2002)
Lentinus sajor-caju	*lac1*, *lac2*, *lac3*, *lac4*	Soden et al. (2002)
Pleurotus eryngii	*pel3*, *pel4*	Rodríguez et al. (2008)
Pleurotus ostreatus	*poxa1a*, *poxa1b*, *poxC*, *poxa2*, *poxa3a*, *poxa3b*	Palmieri et al. (2003)

domains and copper coordination sites (Arregui et al. 2019). Crystallography studies have provided insights into their tertiary structure, although challenges remain in resolving the structures of various laccase isoforms (Rivera-Hoyos et al. 2013). Mechanistically, laccases catalyze substrate oxidation coupled with molecular oxygen reduction, resulting in water generation.

Laccases contribute to the degradation of lignin, a complex aromatic polymer in plant cell walls that provides structural integrity and resistance to microbial attack. By oxidizing the phenolic compounds in lignin, laccases initiate a breakdown process that is crucial for carbon cycling in ecosystems. This activity is particularly prominent in white-rot fungi, which are among the few organisms capable of substantial lignin decomposition (Abdel-Hamid et al. 2013; Agustin et al. 2021). The ability to degrade lignin allows fungi to access cellulose and hemicellulose in plant materials, facilitating nutrient cycling and promoting decay processes in forest ecosystems. In fungi, laccases are involved in the synthesis of melanins and other pigments. These pigments are produced through the oxidative polymerization of phenolic compounds, a reaction catalyzed by laccases (Janusz et al. 2020). Pigments like melanins confer protection against ultraviolet radiation, extreme temperatures, and oxidative stress. They also play roles in fungal virulence and the formation of reproductive structures, influencing fungal reproduction and dispersal.

Laccases enhance fungal resistance to various environmental stresses, including microbial competition, by catalyzing the production of defensive compounds. For instance, they are involved in the detoxification of phenolic toxins produced by plants or competing microbes. This enzymatic activity supports the colonization and survival of fungi under adverse conditions, such as in the presence of antimicrobial agents or heavy metals, thus enhancing their ecological resilience and adaptability (Janusz et al. 2020; Viswanath et al. 2014). Laccases contribute to the virulence of pathogenic fungi by facilitating the invasion of host tissues. This is achieved through the oxidation of host phenolic compounds, which can alter the host's cell structure and immune responses. In plant pathogens, laccases might participate in the breakdown of lignin-like barriers, enabling fungal penetration into plant tissues. In human pathogens, such as *Cryptococcus neoformans*, laccases are essential for the synthesis of melanin, which is involved in protecting the fungus from the host's immune system (Camacho et al. 2019; Voelz and May 2010).

6.4 Applications of Extracellular Enzymes from Fungal Endophytes

The remarkable world of fungi, particularly fungal endophytes, is revolutionizing the sphere of industrial applications, from detergent formulations to breakthroughs in the pharmaceutical industry and innovative food production techniques. By producing bioactive compounds, they confer plant protection against pathogens and insects, thereby functioning as antibiotics, biocontrol agents, and biofertilizers, without eliciting immediate adverse effects on their host plants (Hawar 2022). This symbiotic relationship highlights the potential of fungal endophytes in enhancing plant stress tolerance, insect resistance, and defense mechanisms, underscoring their significance in promoting agricultural sustainability and ecological balance. Research on fungal endophytes has far-reaching applications across various fields, including agriculture, where they contribute to pest control and antimicrobial activities, and biotechnology, where they enable the production of enzyme inhibitors and secondary metabolites (Akram et al. 2023; El-Bondkly et al. 2021). Moreover, their involvement in phytoremediation, biological control, gene expression regulation, and mycorrhizal associations underscores their versatility. This chapter will explore the integral role of fungal endophytes and their extracellular enzymes across multiple industries. It begins with an understanding of these enzymes and their producers, delving into their application within the detergent industry to improve enzyme-based detergents, agricultural advancements through enhanced root growth and antimicrobial activity, and their significant contributions to the food and pharmaceutical industries by optimizing enzyme activity for better product formulations. Further discussions will extend to their value in the textile industry and biofuel production. Harnessing the potential of these enzymes, we can revolutionize our approach to addressing modern challenges and developing innovative solutions.

6.4.1 Industrial Applications

Endophytic fungi like *Pestalotiopsis microspora*, *Aspergillus oryzae*, and *Penicillium chrysogenum* are prolific producers of amylases, which are crucial in starch breakdown processes (Bhadra et al. 2022). A wide array of enzymes is produced by different strains of endophytic fungi, which include cellulases, β-glucosidase, and endoglucanase, vital for cellulose degradation (El-Gendi et al. 2021). This capability is not limited to a few fungi but spans various genera, demonstrating the extensive potential of these organisms in industrial applications. For instance, *Periconia* sp. and *Aspergillus versicolor* are known for their β-glucosidase production, while *Alternaria alternata* has been reported to produce gelatinase (Bhadra et al. 2022). The industrial relevance of these enzymes extends beyond simple degradation activities. Laccases, produced by fungi like *Hormonema* sp. and *Neofusicoccum* sp., are utilized in bioremediation and detoxification processes due to their ability to oxidize phenolic compounds (Wang et al. 2019). Moreover, lipases from *Rhizopus oryzae* and *Lasiodiplodia theobromae* (Riyadi et al. 2024) find applications in the food and pharmaceutical industries, aiding in diverse biochemical reactions. Microbial enzymes are favored over plant and animal enzymes because of their stability, catalytic activity, and ease of production and optimization (Singh et al. 2016). They are highly specific, accelerating only particular reactions by lowering the activation energy without undergoing any permanent change, making them indispensable in industrial applications. Endophytic fungi like *Fusarium*, *Trichoderma*, *Aspergillus*, and *Alternaria* are among the most common producers of these industrially important enzymes (Bhadra et al. 2022). Their capabilities extend to the production of ligninolytic enzyme with potential biotechnological applications in lignocellulosic biorefineries, highlighting their role in advancing sustainable industrial practices (Bhadra et al. 2022; Choudhary et al. 2023).

Fungal endophytes, known for their robust enzyme-producing capabilities, contribute significantly to the detergent industry. The production of lipases, a type of enzyme extracted from endophytic *Bacillus pumilus* WSS5, has been specifically highlighted for its application in detergent formulations (Khan et al. 2023). These enzymes are crucial for breaking down fats and greases, which are common components of household and industrial stains. Proteases also play a pivotal role in the detergent sector. They are responsible for breaking down protein-based stains, making them an essential ingredient in many laundry detergents. The use of proteases in the detergent industry accounts for a substantial portion of the market for commercial enzymes, underscoring their importance in enhancing the cleaning efficacy of detergents (Bezerra et al. 2021; Sudheep et al. 2017; Uzma et al. 2016).

6.4.1.1 Food Industry Applications

Endophytic fungi are prolific producers of various enzymes that have broad applications in the food industry. Amylases, for instance, are extensively utilized in the processing of starch into sugars, playing a crucial role in industries such as brewing and baking. These enzymes are notably produced by endophytic fungi like *Pestalotiopsis microspora*, *Aspergillus oryzae*, and *Penicillium chrysogenum* (Fouda et al. 2015; Joel and Bhimba 2012). Similarly, pectinases, which are vital for the clarification of fruit juices and wines, are produced by strains such as *Pestalotiopsis* sp., *Alternaria alternata*, and *Penicillium glandicola* (Bezerra et al. 2012; Fouda et al. 2015; Maria et al. 2005). Tannases, another significant enzyme obtained from *Aspergillus niger* and *Aspergillus fumigatus*, are employed in the production of tea and coffee, enhancing flavors and reducing bitterness (Cavalcanti et al. 2017).

Enhancing Food Quality and Safety

The impact of endophytic fungi extends beyond mere processing to significantly enhancing the nutritional and safety aspects of food. Inoculation of upland rice with *Trichoderma zelobreve* PBMP16, for example, has shown remarkable improvements in the rice's nutritional quality. This includes increases in phenolic compounds, anthocyanin, antioxidants, and overall nutrient uptake, surpassing even those levels found in chemically fertilized counterparts (Gateta et al. 2023). These enhancements contribute to the overall health benefits of the food while also potentially reducing the need for synthetic fertilizers and chemicals in agriculture. Moreover, the general trend indicated by research shows that plants treated with endophytic fungi consistently exhibit higher total phenolic content (TPC), total anthocyanin content (TAC), and antioxidant activity compared to non-inoculated controls. This not only improves the plant's resistance to diseases but also elevates the nutritional profile of the food products derived from these plants (Gateta et al. 2023).

Stability and Industrial Viability

The industrial application of enzymes requires that they maintain stability under various conditions. Endophytic fungal enzymes often exhibit superior stability compared to those derived from other sources, making them particularly valuable in industrial processes where conditions such as temperature and pH can vary widely (Sunitha et al. 2013). For instance, thermostable enzymes from endophytic fungi like glucoamylase from *Streptosporangium* sp. and tannase from *Pestalotiopsis guepini* demonstrate enhanced performance in industrial settings due to their high thermostability (de Sena et al. 2014; Stamford et al. 2002).

6.4.1.2 Pharmaceutical Applications

Enzymatic Contributions to Therapeutics

Fungal endophytes are renowned for their ability to produce a diverse array of enzymes and secondary metabolites that have significant therapeutic applications. These include enzymes such as amylases, lipases, and cellulases, which not only act directly as antimicrobials but also facilitate the creation of intermediate molecules with antimicrobial activity (Adrio and Demain 2014). The broad spectrum of bioactive compounds produced by these fungi underscores their potential in developing new pharmaceuticals.

Key Pharmaceutical Compounds from Fungal Endophytes

The endophyte *Fusarium proliferatum* TQN5T, isolated from *Dysosma versipellis*, is a notable producer of podophyllotoxin (PTOX) (Nguyen et al. 2023). This compound is highly valued in pharmaceuticals for its potential applications, particularly in cancer treatment (Dame et al. 2016). Another significant contribution is from the endophytic fungus *Curvularia* sp. LCJ413, which produces L-asparaginase (L-ASNase). This enzyme is crucial in the treatment of acute lymphocytic leukemia, highlighting the medical importance of fungal endophytes (Kathiravan et al. 2023). Similarly, L-glutaminase extracted from *Aspergillus oryzae* (PLNSN et al. 2014) exhibits promising anti-retroviral activities, making it a potent agent against HIV by reducing glutamine levels which are crucial for the virus's replication processes (Roberts and McGregor 1991).

Therapeutic Enzymes in Clinical Microbiology

The production of extracellular enzymes by fungal endophytes holds great promise in clinical microbiology. These enzymes have been identified to have various therapeutic applications, including antimicrobial, antitumor, and anti-hypertensive activities. Their role extends beyond simple biological activities, offering potential enhancements in the efficacy of existing drugs and treatments (Bhadra et al. 2022; Kachroo 2022). Chitinase, another fungal enzyme, has shown potential in antibacterial and antifungal applications, and its derivatives have been recognized for their antitumor activities and wound healing properties (Hamid et al. 2013). Furthermore, enzymes like β-glucuronidase have found their niche in the medical industry due to their stability and effectiveness in various therapeutic applications (Zhang et al. 2020). Proteases and lipases from endophytic fungi such as *Myrcia guianensis* have been effective in hydrolyzing biofilms formed by *Staphylococcus aureus*, demonstrating their potential in combating bacterial infections that are resistant to conventional treatments (Matias et al. 2021). Additionally, lipases have been used to target and eliminate protozoan parasites such as *Leishmania amazonensis*, offering a new approach to treating complex infections (Alves et al. 2018).

6.4.2 Applications in Agriculture

6.4.2.1 Bioactive Metabolites and Their Roles

Fungal endophytes produce a variety of bioactive metabolites such as enzymes, alkaloids, polypeptides, polyketides, and terpenoids, which have extensive applications in agriculture. These compounds can enhance plant resistance against a wide range of pathogens and pests, thereby reducing the need for chemical pesticides (Xia et al. 2022). The extracellular enzymes secreted by fungal endophytes, including chitinases, cellulases, β-1,3-glucanases, pectinases, glucanases, and proteases, play a crucial role in agricultural plant health. These enzymes can degrade the cell walls of pathogens or inhibit spore germination, effectively protecting plants from various biotic stresses (Bhadra et al. 2022; Muthu Narayanan et al. 2022). Fungal endophytes enhance plant growth by improving nutrient absorption and utilization. They produce plant hormones such as auxin, ethylene, gibberellin, and cytokinin, which contribute to the overall vigor and productivity of agricultural crops (Bhardwaj et al. 2023). Fungal endophytes protect their host plants by competing with pathogenic microorganisms for habitat and food sources. This competition involves rapid colonization of the plant tissues, which prevents pathogen invasion and nutrient depletion necessary for their growth. The ability of endophytic fungi to produce toxic compounds plays a significant role in defending host plants against insects and animals. These compounds make the plants less palatable or even toxic to potential pests, providing an ecological advantage to the crops (Hawar 2022; Chandran et al. 2021).

6.4.2.2 Bioremediation and Waste Degradation

Fungal endophytes are also utilized in the detoxification and bioremediation of industrial and agricultural wastes. Their capability to degrade polluting compounds without harming the environment is particularly valuable in sustainable agriculture practices (Sharma and Kumar 2021). Microorganisms, including fungal endophytes, are instrumental in the degradation of agricultural waste, transforming it into less harmful substances. Lignocellulolytic enzymes from these fungi offer an alternative method for organic matter degradation, turning plant biomass into useful byproducts such as biofuels (Bhadra et al. 2022; Dashtban et al. 2010).

6.4.3 Environmental Implications

6.4.3.1 Bioremediation and Detoxification Capabilities

Fungal endophytes exhibit substantial potential in environmental management through their production of extracellular enzymes. These enzymes are instrumental in bioremediation processes, crucial for detoxifying or removing harmful industrial and agricultural waste and other polluting compounds (Hawar 2022). The ability of these fungi to produce enzymes such as lignocelluloses, amylase, cellulase, ligninase, pectinase, and xylanase enhances the decomposition process of plant materials, which can significantly impact the carbon and nitrogen cycles (Sunitha et al. 2013).

6.4.3.2 Degradation of Polyaromatic Hydrocarbons

Endophytic fungi are capable of degrading complex organic pollutants, including polyaromatic hydrocarbons (PAHs), using a suite of extracellular enzymes like hydrolytic enzymes (xylanases and cellulases) and ligninolytic enzymes (laccases, ligninases, and peroxidases). This ability positions them as valuable agents in the remediation of environments contaminated with these hazardous substances (Bhadra et al. 2022; Mtui 2012).

6.4.3.3 Sustainable Industrial Applications

The use of lignocellulolytic enzymes from fungal endophytes in various industries, including biofuel production and food technology, presents a more sustainable alternative to traditional methods that often involve harsh chemicals and generate significant waste (Sperandio and Ferreira Filho 2019). These enzymes facilitate the breakdown of lignocellulose, a recalcitrant solid complex formed from the combination of lignin and hemicellulose, thus aiding in the conversion of plant biomass into valuable products (Benatti and Polizeli 2023; Bhadra et al. 2022).

6.5 Factors Influencing Enzyme Production by Fungal Endophytes

Fungal endophytes are harmless plant-dwelling microorganisms that generate a variety of extracellular enzymes with both practical and ecological uses. First and foremost, they aid in the digestion of complex organic substances, which in turn helps ecosystems cycle nutrients. By aiding in the breakdown of plant cell wall constituents and liberating vital nutrients for the host plant, extracellular enzymes also significantly contribute to the development of plants and improve their ability to withstand stress. Numerous variables affect the synthesis of enzymes by fungal endophytes (Fontana et al. 2021; Kumar et al. 2022). Firstly, the precise host plant species is important because various plants have diverse endophytic communities, resulting in variances in enzyme activity suited to break down certain substances found in the host. Second, the internal environment of the plant, including temperature, pH, moisture content, and nutrient availability, can have a significant influence on the synthesis of enzymes. The ideal circumstances for each type of fungus to express its enzymes might differ. Furthermore, due to genetic and evolutionary considerations, various strains within a fungal endophyte species may demonstrate variances in enzyme synthesis. The intricate interaction of variables defining the enzymatic activity of fungal endophytes in the host plant is highlighted by the critical roles that stress responses, nutrient availability, and the age and growth phase of the endophyte play in modifying enzyme synthesis.

6.5.1 Host Plant Factors

Microorganisms known as fungal endophytes invade the interior plant tissues of their hosts despite seeming to harm them. The selectivity of fungal endophytes toward certain host plant species is one of their notable characteristics. Mutualistic or symbiotic links between the endophytes along with their host plants result in this specialization through a process of co-evolution. The relationships between fungal endophytes with their hosts are significantly shaped by the biochemical makeup of the host plant tissues (Saikkonen et al. 1998; Seth et al. 2023). Many different types of organic chemicals, including as carbohydrates, lipids, proteins, secondary metabolites, and signaling molecules, are produced by plants. These substances function inside plant tissues as structural elements, signaling molecules, and sources of nutrients. Additionally, the signaling chemicals that the plant releases can influence the fungal endophyte's ability to produce enzymes either directly or indirectly. Endophyte physiology and behavior can be influenced by molecular signals from plants, including defense-related chemicals, reactive oxygen species (ROS), and phytohormones (Ji et al. 2022; Bhupenchandra et al. 2024). By activating or repressing certain genes associated with enzyme production, these signaling molecules allow endophytes to precisely regulate their metabolic activity in response to signals from plants. The complex relationships between fungal endophytes with their host plants are shaped by the health and age of the plants, which in turn influences the dynamics of enzyme synthesis. Because healthy plants have ideal food availability and metabolic conditions, they create an environment that is favorable to endophyte colonization and enzyme production. Stressed plants, however, can launch defense mechanisms, such the synthesis of reactive oxygen species or antimicrobial chemicals, which can affect the behavior of endophytes and the patterns of enzyme synthesis. Additionally, the accessibility of substrates and signaling molecules for endophytes is altered as plants age due to modifications in tissue structure, root exudates, and senescence (Geisen et al. 2021; Chauhan et al. 2023). Throughout a plant's life cycle, for example, the composition of root exudates might change, which can affect endophyte recruitment and activity in the rhizosphere and the internal tissues. Determining the functions of endophytes in plant health, resistance to stress, and functioning of ecosystems requires an understanding of these complex interactions.

6.5.2 Environmental Factor

The synthesis of enzymes by fungal endophytes is significantly impacted by environmental conditions, which are indicative of the complex interactions that occur within fungal symbionts, the plants they inhabit, and the ecosystem that surrounds it. One important environmental factor that has a direct impact on the kinetics of processes catalyzed by enzymes is temperature. Enzymes have optimal temperatures at which they function at their highest levels. Elevations or decreases in temperature can have an effect on metabolic processes by reducing substrate binding or denaturing enzyme activity. Three thermostable enzymes were synthesized by two different endophytic fungal strains, *Botryosphaeria* sp. AM01 and *Saccharicola* sp. EJC04: endoglucanase, β-glucosidase, and xylanase. Among these enzymes, xylanase exhibited best activity from 60 °C to 75 °C, whereas β-glucosidase showed the maximum activity throughout a wide pH range of 3.5 to 8.5 (Marques et al. 2018). Another important environmental component that affects the stability and function of enzymes is pH. Each enzyme has an ideal pH range for action, and they all tolerate different amounts of acidity or alkalinity. The three-dimensional structure of the enzyme can be damaged by pH variations, which can also reduce the enzyme's catalytic effectiveness. In order to adjust to the pH ranges that are present in the tissues of the host plant or in the surrounding soil, fungi endophytes may develop enzymes with various pH optima (Suz et al. 2014). Fungal species such as *Aspergillus chlamydosporus*, *Fusarium* sp., *Pestalotiopsis* sp., *Acremonium* sp., and *Aspergillus* sp. produce extracellular enzymes like lipase, chitinase, protease, cellulase, laccase, amylase, and tyrosinase, which were all used to measure the pH stability of the samples. The isolates exhibited maximum activity of xylanase and protease at pH 9, and cellulase at pH 7 (Maria et al. 2005). The tannase enzyme that *A. fumigatus* and *A. niger* generated was suited for a variety of industrial applications since it remained stable at pH 4.0 and 30 °C (Cavalcanti et al. 2017). The metabolism and physiology of the host plant are influenced by light, and this can have an impact on the nutritional requirements and biochemical makeup of plant tissues. In order to satisfy the metabolic demands imposed by the modified host plant circumstances, fungal endophytes may adjust their enzyme synthesis in response to variations in the availability of light.

6.5.3 Genetic Factors

The capacity of fungal endophytes to produce enzymes is largely determined by their genetic composition, which also affects how well they can grow inside the tissues of host plants and interact with their surroundings. Variations in the genetic diversity of fungal populations can result in unique sets of genes encoding enzymes with varied substrate specificities and catalytic efficiencies, which can cause variations in the genetic makeup of enzyme production (El-Gendy and El-Bondkly 2010). Moreover, transcription factors, regulatory elements, and signaling pathways, all of which are impacted by a combination of environmental and genetic variables, tightly control the expression of these genes. Genetic variations in endophyte strains can lead to varied production of enzymes in response to different environmental factors or host plant relationships. Certain endophyte strains, for instance, can include gene duplications or deletions that increase or decrease the production of particular enzymes involved in metabolism, food acquisition, or stress response. Furthermore, new enzyme-encoding genes can be introduced by horizontal gene transfer events between endophyte strains or other microbes in the plant microbiome, which increases the variety of enzyme manufacturing capacities (Raghav et al. 2022). Apart from genetic alterations, other changes that can affect gene expression patterns without changing the underlying DNA sequence include histone modifications, DNA methylation, and noncoding RNA-mediated regulation. Temperature, pH, nutrition availability, and signals from the host plant are examples of environmental cues that can cause epigenetic modifications in endophytes that affect the expression patterns of certain enzymes (Verma et al. 2023). Stress exposure, for example, may cause epigenetic changes that trigger stress-responsive genes and pathways, leading to the overexpression of enzymes implicated in defense and stress tolerance. In summary, the complex interaction between genetic and epigenetic variables, in addition to environmental stimuli, determines how fungal endophytes produce enzymes.

6.6 Methods for Enhancing Enzyme Production by Fungal Endophytes

6.6.1 Optimization of Fermentation Conditions

Recent studies have confirmed ability of various endophytic fungi to produce broad spectrum of secondary metabolites, including enzymes, antibiotics, anticancer, and vital medicinal compounds like taxol, vinblastine, immune suppressants, camptothecin, vincristine, etc., identical to their symbiotic plant partner, i.e., hosts (Madhusudhan et al. 2015). Although utilizing these filamentous fungi seems economically feasible and a sustainable alternative for production of these important bioactive compounds, these fungal members do not satiate commercially desired enzyme production naturally. To achieve this large-scale biosynthesis, extraction and purification of these compounds, fermentation serves as an efficient technique (Asomadu et al. 2024). Fermentation is a biological process that enables organisms to extract energy via anaerobic breakdown of glucose (Ross et al. 2002). Fungal members are often associated with alcoholic fermentation (Sharma et al. 2024). Even though fermentation is an oldest biotechnological tool employed by fungi, exploiting their extracellular enzyme production capability demands optimization of traditional fermentation technique. Fermentation optimization is, in a way, a customized alteration of bioreactor environment, i.e., temperature, pH, inoculant concentration, etc., in order to attain maximum production (Rusu et al. 2023).

Recent modifications, innovations, and biotechnological advancements toward fermentation optimization, targeting maximum productivity, have propelled the onset of submerged fermentation (SmF) and solid-state fermentation (SSF) (Madhusudhan et al. 2015). SmF is carried out in a closed liquid medium, such as a shake flask comprising sterilized carbon, nitrogen, and other nutrient sources, monitored at specifically optimized parameters involving pH, temperature, pO_2, etc. (Mokrani and Nabti 2024). SmF is frequently associated with airlift bioreactor and possesses shorter fermentation period (Zhang et al. 2024). Alternatively, fungi respond superfluously toward SSF, yielding enhanced productivity accompanying lesser energy input and negligible foam emission (Mokrani and Nabti 2024). Further, it abruptly reduces wastewater management procedures since SSF tends to cultivate fungi on solid substrate devoid of any extra moisture after the requirement of growth and development of target organism is satisfied (Zhang et al. 2024). Moreover, SSF ensures relatively easy and low-cost retrieval of products than SmF, since the former employs solvent extraction methods using methanol, ethanol, chloroform, water, dimethyl-sulfoxide, etc., while the latter utilizes advanced chromatographic techniques like HPLC, UFPLC, column chromatography, etc. (Mokrani and Nabti 2024). Apart from the advantages, SSF put forth certain challenges, i.e., long fermentation period, risk of contamination, and being a labor-intensive process, thereby seeking more advancements in bioreactor technologies for better customization of optimized fermentation.

Such optimization conditions involving alkaline pH, pO_2, 25 °C temperature, L-asparagine as N_2 source, and starch as carbon source have encouraged accelerated production of alpha-amylase enzyme by *Preussia minima*, a fungal endophyte of Australian *Eremophilia longifolia* (Zaferanloo et al. 2014). Recent discoveries have established that semi-solid-state fermentation (SSSF) can enhance extracellular enzyme production multifold (Fu et al. 2024). They employed homogenous culture of an endophytic fungus, *Penicillium rubens*, into SSSF of *Scutellaria baicalensis* Georgi roots at optimized temperature (30 °C) and inoculation dose to achieve 15–16 fold increase in baicalein and wogonin yield, which can foster food and nutraceutical industries. Hence, every fungal member needs different requirements of conditions for production of different concerned biomolecule, thus demanding more comprehensive and intense research into fermentation optimization.

6.6.2 Genetic Engineering and Mutagenesis

Besides fermentation optimization, white biotechnology is currently becoming a more promising strategy, thrusting genetic manipulation of living organisms for increased enzyme production (Yadav 2019). This industrial biotechnological intervention comprehensively deals with commercially desirable fungal strain development in vitro via onset of mutagenesis, gene deletion, gene editing, gene knockout, genetic modifications, transgenic development, genome shuffling, over expression of transcription factors, protein and metabolite engineering, etc. (Fig. 6.2).

Some findings have confirmed encouraging potential of mutagenicity toward scalable production in industrial fermentations (Adrio and Demain 2006). A recent discovery has displayed that Co^{60} gamma-radiation-induced mutant *Trichoderma afroharzianum* efficiently produces multitude of enzymes, i.e., highest cellulase and xylanase, relatively than wild types (Askari et al. 2024). Moreover, nuclease-based CRISPR/cas9 and CRISPRi technologies are becoming vital breakthroughs in the field of genetic engineering lately (Asogwa et al. 2024; Zhou et al. 2024). For instance, genetic manipulation via Cre/loxP followed by gene editing via CRISPR/cas9 on *Saccharomyces cerevisiae* has proved to be an economical and proficient alternative for large-scale L-lactic acid production (Long et al. 2024). Not only yeast, CRISPR/cas9 gene editing has so far achieved momentous advancements in commercial production of extracellular enzyme or secondary metabolites in *Trichoderma reesei*, *Aspergillus nidulans*, *A. niger*, *A. oryzae*, and many others (Oliveira et al. 2024). Likewise, employing genetic engineering and mutagenesis in filamentous endophyte cultures is emerging as promising strategies to enhance extracellular enzyme production.

Furthermore, In vivo engineering of *E. coli* via deletion of *yjgB* gene and overexpression of *nadK* gene, along with manipulation of several metabolic pathways like pentose phosphate pathway, succeeded in the development of new strains with marked

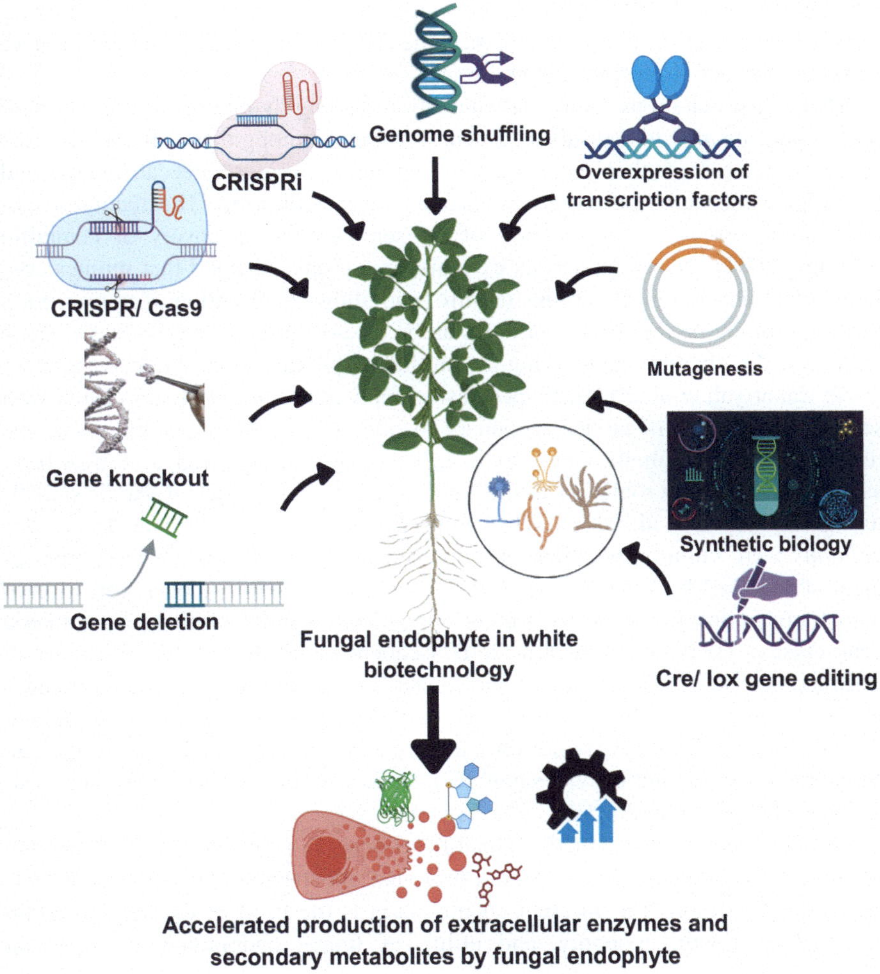

Fig. 6.2 Enhancing fungal endophyte capabilities in white biotechnology

β-carotene synthesizing ability (Wu et al. 2020). Thus, the integration of biotechnology with systems biology, specifically focusing on metabolomic study, has emerged as a novice discipline named synthetic biology, which exhibits promising marketable yields (Rusu et al. 2023). As a catalyzing technique for a better sustainable future, more intense research should be targeted toward fungal white biotechnology.

6.6.3 Co-cultivation with Other Microorganisms

A key challenge in artificially producing extracellular fungal enzymes is the prominent decline in production with continuous subculturing of axenic monocultures

(Madhusudhan et al. 2015). Most probably due to the accumulation of certain inhibitory substances in the culture medium, some differential methylation inhibition, or certain mutations (Isah et al. 2018).

There exist numerous ecological interactions among living organisms, i.e., symbiosis, parasitism, commensalism, amensalism, etc. Among these interactions, the synergistic effects of other living organisms on fungi in a co-culture, can be exploited to mitigate the persistent risks associated with subculturing. The theory of co-cultivation arose as a consequence of the serendipitous discovery of Penicillin (Fleming 2001). Presently, it is an extensively exploited method that mimics ecological and physiological settings in vitro, and allows distribution of transcription factors, promotors, regulators, and insulators, etc., of different origins in the same culture media, thereby intensifying the productivity (Rusu et al. 2023; Wang et al. 2022; Zakariyah et al. 2024). Its persuasive applications are often associated with scalable production of novice secondary metabolites, further extending to occurrence of cryptic metabolites, and can be achieved using fungal-fungal, fungal-host, as well as fungal-microbial cocultures, nullifying the limitations of axenic subculturing (Zakariyah et al. 2024). When fungi *Epicoccum dendrobii* and *Aspergillus nidulans* were cultured together, an unprecedented loss-of-function and replacement of gene *VeA* by *VeA1* gene of *E. dendrobii* origin resulted in overexpression of *SclB* transcription factor, which further upregulated a series of silent biosynthetic gene clusters (BGCs) and altered the consequent metabolite profile (Wang et al. 2022). Likewise, fungal-microbial and fungal-host cocultures have yield 49 and 7 new secondary metabolites, respectively, until 2022 (Xu et al. 2023). The applications of fungal cocultures are not only limited to secondary metabolites; rather, an enumerated list of vital enzymes and acids have also attained discovery and scalability so far (Xu et al. 2023).

Still, the scanty research on the fungal-host coculture represents a desired enigma awaiting to be addressed. Moreover, a profound understanding of appropriate fermentation conditions for different cocultivation is the need of the hour to attain sustainable growth, economy, and utility of fungal endophytes in industrial advancements.

6.7 Challenges and Limitations

The successful cultivation and isolation of fungal endophytes encounter significant challenges due to their intricate and specific growth requirements, impeding the scalability of enzyme production for industrial purposes. Despite their potential in enzyme synthesis, fungal endophytes confront various hurdles, including low inherent enzyme production and secretion rates, necessitating the optimization of growth conditions to enhance yields and diminish production expenses. Additionally, the repetitive subculturing of endophytic strains may result in diminished secondary metabolite production, leading to unstable yields in pure cultures, thereby complicating their utilization in consistent enzyme production and other biotechnological endeavors (Bhadra et al. 2022). Moreover, ongoing debates persist regarding the genetic origin of metabolites synthesized by endophytes. Some researchers posit that endophytes mimic the metabolic pathways of their host plants, while others propose the existence of distinct and autonomous biosynthetic routes for these metabolites within endophytic organisms. Additionally, the molecular mechanisms underlying plant-endophyte interactions remain poorly understood, posing a significant barrier to harnessing these organisms for efficient enzyme production (Tiwari and Bae 2022). Furthermore, while endophytic fungi have demonstrated potential in producing lytic enzymes that increase host resistance against pathogens, it is unclear whether these enzymes can consistently act as elicitors of systemic resistance in plants (Xia et al. 2022). This ambiguity impedes the comprehension and manipulation of metabolite production for specific applications, underscoring the necessity for further investigation to unravel the biosynthetic mechanisms underlying fungal endophytes.

6.8 Conclusion and Future Prospects

The exploration of fungal endophytes throughout this study has illuminated their critical role in revolutionizing industries, showcasing their profound impact on detergents, food, pharmaceuticals, sustainable agriculture, and environmental management. Leveraging the bioactive compounds and thermostable enzymes produced by these microorganisms, industries are moving toward more efficient, eco-friendly processes. From enhancing enzyme-based detergents to contributing valuable bioactive compounds for pharmaceutical applications, the versatility of fungal endophytes underscores their potential as indispensable biotechnological tools. Addressing challenges in maximizing enzyme production and harnessing the full spectrum of benefits offered by fungal endophytes is essential for future advancements. Strategies such as enhancing biosynthetic and secretory capabilities, optimizing culture conditions, exploring biodiversity, and developing new biopesticides from endophyte metabolites are proposed. Despite challenges in cultivation, isolation, and consistent application, advanced techniques like multi-omics approaches and genetic engineering hold promise for overcoming these hurdles. The integration of these approaches stands as a testament to the promising future of fungal endophytes in addressing contemporary challenges across diverse sectors, highlighting their significance in natural ecosystems and innovative applications.

Acknowledgments The authors would like to extend their sincere appreciation to the funding agency Science and Engineering Research Board (SERB), State University Research Excellence (SURE), New Delhi, India. All the authors acknowledge their host institute for infrastructure support.

Conflict of Interest The authors declare no conflict of interest.

Funding This research was supported by the funding agency Science and Engineering Research Board (SERB), State University Research Excellence (SURE) (File Number: SUR/2022/005216).

The research was also supported by Ministry of Education and the State Project Directorate-Rashtriya Uchchatar Shiksha Abhiyaan (SPD-RUSA), Rajasthan, through the RUSA 2.0 project.

Data Availability Not applicable.

References

Aam BB, Heggset EB, Norberg AL, Sørlie M, Vårum KM, Eijsink VG (2010) Production of chitooligosaccharides and their potential applications in medicine. Mar Drugs 8(5):1482–1517

Abdel-Hamid AM, Solbiati JO, Cann IK (2013) Insights into lignin degradation and its potential industrial applications. Adv Appl Microbiol 82:1–28

Abdel-Raheem A, Shearer C (2002) Extracellular enzyme production by freshwater ascomycetes. Fungal Divers 11:1–19

Abdullah R, Jafer A, Nisar K, Kaleem A, Iqtedar M, Iftikhar T, Saleem F, Naz S (2018) Process optimization for pectinase production by locally isolated fungal strain using submerged fermentation. Biosci J 34(4):1025–1032

Adrio JL, Demain AL (2006) Genetic improvement of processes yielding microbial products. FEMS Microbiol Rev 30(2):187–214

Adrio JL, Demain AL (2014) Microbial enzymes: tools for biotechnological processes. Biomol Ther 4(1):117–139

Agustin MB, de Carvalho DM, Lahtinen MH, Hilden K, Lundell T, Mikkonen KS (2021) Laccase as a tool in building advanced lignin-based materials. ChemSusChem 14(21):4615–4635

Ahmed I, Zia MA, Hussain MA, Akram Z, Naveed MT, Nowrouzi A (2016) Bioprocessing of citrus waste peel for induced pectinase production by *Aspergillus niger*; its purification and characterization. J Radiat Res Appl Sci 9(2):148–154

Akram S, Ahmed A, He P, He P, Liu Y, Wu Y, Munir S, He Y (2023) Uniting the role of endophytic fungi against plant pathogens and their interaction. J Fungi 9(1):72

Alarid-García C, Escamilla-Silva EM (2017) Comparative study of the production of extracellular β-glucosidase by four different strains of *Aspergillus* using submerged fermentation. Prep Biochem Biotechnol 47(6):597–610

Alegre ACP, Polizeli MLTM, Terenzi HF, Jorge JA, Guimarães LHS (2009) Production of thermostable invertases by *Aspergillus caespitosus* under submerged or solid state fermentation using agroindustrial residues as carbon source. Braz J Microbiol 40:612–622

Aloulou A, Rodriguez JA, Puccinelli D, Mouz N, Leclaire J, Leblond Y, Carrière F (2007) Purification and biochemical characterization of the LIP2 lipase from *Yarrowia lipolytica*. Biochim Biophys Acta (BBA)-Mol Cell Biol Lipids 1771(2):228–237

Alves DR, Morais SM d, Tomiotto-Pellissier F, Vasconcelos FR, Freire F d CO, Silva ING d, Cataneo AHD, Miranda-Sapla MM, Pinto GAS, Conchon-Costa I (2018) Leishmanicidal and fungicidal activity of lipases obtained from endophytic fungi extracts. PLoS One 13(6):e0196796

Amirita A, Sindhu P, Swetha J, Vasanthi NS, Kannan KP (2012) Enumeration of endophytic fungi from medicinal plants and screening of extracellular enzymes. World J Sci Technol 2(2):13–19

Andlar M, Rezić T, Marđetko N, Kracher D, Ludwig R, Šantek B (2018) Lignocellulose degradation: an overview of fungi and fungal enzymes involved in lignocellulose degradation. Eng Life Sci 18(11):768–778

Anisa SK, Girish K (2014) Pectinolytic activity of *Rhizopus* sp. and *Trichoderma viride*. Int J Res Pure Appl Microbiol 4(2):28–31

Aravindan R, Anbumathi P, Viruthagiri T (2007) Lipase applications in food industry. Indian J Biotechnol 6(2):141–158

Arregui L, Ayala M, Gómez-Gil X, Gutiérrez-Soto G, Hernández-Luna CE, Herrera De Los Santos M, Levin L, Rojo-Domínguez A, Romero-Martínez D, Saparrat MC (2019) Laccases: structure, function, and potential application in water bioremediation. Microb Cell Factories 18:1–33

Askari H, Soleimanian-Zad S, Kadivar M, Shahbazi S (2024) Creating a novel genetic diversity of *Trichoderma afroharzianum* by γ-radiation for xylanase-cellulase production. Heliyon 10(7):e28349

Asogwa IS, Eze CO, Onah AI, Agbo CA, Evurani SA, Attama AA (2024) New options in drug discovery: endophyte technology. Int J Sci Res Arch 11(2):302–321

Asomadu RO, Ezeorba TPC, Ezike TC, Uzoechina JO (2024) Exploring the antioxidant potential of endophytic fungi: a review on methods for extraction and quantification of total antioxidant capacity (TAC). 3 Biotech 14(5):1–18

Baker P, Tiroumalechetty A, Mohan R (2019) Fungal enzymes for bioremediation of xenobiotic compounds. In: Recent advancement in white biotechnology through fungi: volume 3: perspective for sustainable environments. Springer, Switzerland, pp 463–489

Balaban NP, Suleimanova AD, Valeeva LR, Chastukhina IB, Rudakova NL, Sharipova MR, Shakirov EV (2016) Microbial phytases and phytate: exploring opportunities for sustainable phosphorus management in agriculture. Am J Mol Biol 7(1):11–29

Benatti ALT, Polizeli MDLTDM (2023) Lignocellulolytic biocatalysts: the main players involved in multiple biotechnological processes for biomass valorization. Microorganisms 11(1):162

Benjamin S, Pandey A (1998) Candida rugosa lipases: molecular biology and versatility in biotechnology. Yeast 14(12):1069–1087

Bertrand T, Jolivalt C, Briozzo P, Caminade E, Joly N, Madzak C, Mougin C (2002) Crystal structure of a four-copper laccase complexed with an arylamine: insights into substrate recognition and correlation with kinetics. Biochemistry 41(23):7325–7333

Bezerra JDP, Santos MGS, Svedese VM, Lima DMM, Fernandes MJS, Paiva LM, Souza-Motta CM (2012) Richness of endophytic fungi isolated from *Opuntia ficus-indica* Mill.(Cactaceae) and preliminary screening for enzyme production. World J Microbiol Biotechnol 28:1989–1995

Bezerra VHS, Cardoso SL, Fonseca-Bazzo Y, Silveira D, Magalhaes PO, Souza PM (2021) Protease produced by endophytic fungi: a systematic review. Molecules 26(22):7062

Bhadra F, Gupta A, Vasundhara M, Reddy MS (2022) Endophytic fungi: a potential source of industrial enzyme producers. 3 Biotech 12(4):86

Bhagobaty RK, Joshi SR (2012) Enzymatic activity of fungi endophytic on five medicinal plant species of the pristine sacred forests of Meghalaya, India. Biotechnol Bioprocess Eng 17:33–40

Bhardwaj M, Kailoo S, Khan RT, Khan SS, Rasool S (2023) Harnessing fungal endophytes for natural management: a biocontrol perspective. Front Microbiol 14:1280258

Bhupenchandra I, Chongtham SK, Devi AG, Dutta P, Sahoo MR, Mohanty S, Kumar S, Choudhary AK, Devi EL, Sinyorita S, Devi SH, Mahanta M, Kumari A, Devi HL, Josmee RK, Pusparani A, Pathaw N, Gupta S, Meena M, Janmeda P, Seth CS, Arora A, Sahoo A, Swapnil P (2024) Unlocking the potential of arbuscular mycorrhizal fungi: exploring role in plant growth promotion, nutrient uptake mechanisms, biotic stress alleviation, and sustaining agricultural production systems. J Plant Growth Regul 43(9):1–39. https://doi.org/10.1007/s00344-024-11467-9

Bischoff KM, Wicklow DT, Jordan DB, de Rezende ST, Liu S, Hughes SR, Rich JO (2009) Extracellular hemicellulolytic enzymes from the maize endophyte *Acremonium zeae*. Curr Microbiol 58:499–503

Breen JP (1994) *Acremonium* endophyte interactions with enhanced plant resistance to insects. Annu Rev Entomol 39(1):401–423

Camacho E, Vij R, Chrissian C, Prados-Rosales R, Gil D, O'Meally RN, Cordero RJ, Cole RN, McCaffery JM, Stark RE (2019) The structural unit of melanin in the cell wall of the fungal pathogen *Cryptococcus neoformans*. J Biol Chem 294(27):10471–10489

Cavalcanti RMF, de Oliveira Ornela PH, Jorge JA, Guimarães LHS (2017) Screening, selection and optimization of the culture conditions for tannase production by endophytic fungi isolated from Caatinga. J Appl Biol Biotechnol 5(1):001–009

Champreda V, Kanokratana P, Sriprang R, Tanapongpipat S, Eurwilaichitr L (2007) Purification, biochemical characterization, and gene cloning of a new extracellular thermotolerant and glucose tolerant maltooligosaccharide-forming α-amylase from an endophytic ascomycete *Fusicoccum* sp. BCC4124. Biosci Biotechnol Biochem 71(8):2010–2020

Chandran H, Meena M, Swapnil P (2021) Plant growth-promoting rhizobacteria as a green alternative for sustainable agriculture. Sustainability 13(19):10986. https://doi.org/10.3390/su131910986

Chapla VM, Zeraik ML, Leptokarydis IH, Silva GH, Bolzani VS, Young MCM, Pfenning LH, Araújo AR (2014) Antifungal compounds produced by *Colletotrichum gloeosporioides*, an endophytic fungus from *Michelia champaca*. Molecules 19(11):19243–19252

Chauhan P, Sharma N, Tapwal A, Kumar A, Verma GS, Meena M, Seth CS, Swapnil P (2023) Soil microbiome: diversity, benefits and interactions with plants. Sustainability 15(19):14643. https://doi.org/10.3390/su151914643

Chen W, Xu D (2023) Phytic acid and its interactions in food components, health benefits, and applications: a comprehensive review. Trends Food Sci Technol 104201:104201

Choudhary N, Dhingra N, Gacem A, Yadav VK, Verma RK, Choudhary M, Bhardwaj U, Chundawat RS, Alqahtani MS, Gaur RK (2023) Towards further understanding the applications of endophytes: enriched source of bioactive compounds and bio factories for nanoparticles. Front Plant Sci 14:1193573

Churklam W, Aunpad R (2020) Enzymatic characterization and structure-function relationship of two chitinases, LmChiA and LmChiB, from *Listeria monocytogenes*. Heliyon 6(7):e04252

Cleveland M, Lafond M, Xia FR, Chung R, Mulyk P, Hein JE, Brumer H (2021) Two *Fusarium* copper radical oxidases with high activity on aryl alcohols. Biotechnol Biofuels 14(1):138

Colonia BSO, Woiciechowski AL, Malanski R, Letti LAJ, Soccol CR (2019) Pulp improvement of oil palm empty fruit bunches associated to solid-state biopulping and biobleaching with xylanase and lignin peroxidase cocktail produced by *Aspergillus* sp. LPB-5. Bioresour Technol 285:121361

Corrêa TLR, de Araújo EF (2020) Fungal phytases: from genes to applications. Braz J Microbiol 51(3):1009–1020

Dame ZT, Silima B, Gryzenhout M, van Ree T (2016) Bioactive compounds from the endophytic fungus *Fusarium proliferatum*. Nat Prod Res 30(11):1301–1304

Dashtban M, Schraft H, Syed TA, Qin W (2010) Fungal biodegradation and enzymatic modification of lignin. Int J Biochem Mol Biol 1(1):36

de Almeida MN, Guimarães VM, Bischoff KM, Falkoski DL, Pereira OL, Gonçalves DS, de Rezende ST (2011) Cellulases and hemicellulases from endophytic *Acremonium* species and its application on sugarcane bagasse hydrolysis. Appl Biochem Biotechnol 165:594–610

de Cassia Pereira J, Travaini R, Marques NP, Bolado-Rodríguez S, Martins DAB (2016) Saccharification of ozonated sugarcane bagasse using enzymes from *Myceliophthora thermophila* JCP 1-4 for sugars release and ethanol production. Bioresour Technol 204:122–129

De Oliveira APA, Silvestre MA, Garcia NFL, Alves-Prado HF, Rodrigues A, da Paz MF, Fonseca GG, Leite RSR (2016) Production and catalytic properties of amylases from *Lichtheimia ramosa* and *Thermoascus aurantiacus* by solid-state fermentation. Sci World J 2016:7323875

de Sena AR, Dos Santos AC d B, Gouveia MJ, de Mello MRF, Leite TCC, Moreira KA, de Assis SA (2014) Production, characterization and application of a thermostable tannase from *Pestalotiopsis guepinii* URM 7114. Food Technol Biotechnol 52(4):459–467

De Silva NI, Brooks S, Lumyong S, Hyde KD (2019) Use of endophytes as biocontrol agents. Fungal Biol Rev 33(2):133–148

de Souza JB, Michelin M, Amâncio FL, Brazil OAV, Maria de Lourdes TM, Ruzene DS, Silva DP, Mendonça M d C, López JA (2020) Sunflower stalk as a carbon source inductive for fungal xylanase production. Ind Crop Prod 153:112368

Devi NN, Prabakaran JJ, Wahab F (2012) Phytochemical analysis and enzyme analysis of endophytic fungi from *Centella asiatica*. Asian Pac J Trop Biomed 2(3):S1280–S1284

Dhillon GS, Kaur S, Brar SK, Verma M (2012) Potential of apple pomace as a solid substrate for fungal cellulase and hemicellulase bioproduction through solid-state fermentation. Ind Crop Prod 38:6–13

Dutra JC, da Terzi SC, Bevilaqua JV, Damaso MC, Couri S, Langone MA, Senna LF (2008) Lipase production in solid-state fermentation monitoring biomass growth of *Aspergillus niger* using digital image processing. Appl Biochem Biotechnol 147:63–75

El Enshasy H, Dailin DJ, Abd Manas NH, Azlee NİW, Eyahmalay J, Yahaya SA, Abd Malek R, Siwapiragam V, Sukmawati D (2018) Current and future applications of phytases in poultry industry: a critical review. J Adv VetBio Sci Tech 3(3):65–74

El-Bondkly EAM, El-Bondkly AAM, El-Bondkly AAM (2021) Marine endophytic fungal metabolites: a whole new world of pharmaceutical therapy exploration. Heliyon 7(3):e06362

El-Gendi H, Saleh AK, Badierah R, Redwan EM, El-Maradny YA, El-Fakharany EM (2021) A comprehensive insight into fungal enzymes: structure, classification, and their role in mankind's challenges. J Fungi 8(1):23

El-Gendy MMA, El-Bondkly AMA (2010) Production and genetic improvement of a novel antimycotic agent, saadamycin, against dermatophytes and other clinical fungi from endophytic *Streptomyces* sp. Hedaya48. J Ind Microbiol Biotechnol 37(8):831–841

Eriksson KEL, Blanchette RA, Ander P (2012) Microbial and enzymatic degradation of wood and wood components. Springer Science & Business Media, Springer Berlin, Heidelberg, p 407

Espina G, Atalah J, Blamey JM (2021) Extremophilic oxidoreductases for the industry: five successful examples with promising projections. Front Bioeng Biotechnol 9:710035

Fernández-Fueyo E, Ruiz-Dueñas FJ, Martínez AT (2014) Engineering a fungal peroxidase that degrades lignin at very acidic pH. Biotechnol Biofuels 7:1–12

Fleming A (2001) On the antibacterial action of cultures of a *Penicillium*, with special reference to their use in the isolation of B. *influenzae*. Bull World Health Organ 79:780–790

Fontana DC, de Paula S, Torres AG, de Souza VHM, Pascholati SF, Schmidt D, Dourado Neto D (2021) Endophytic fungi: biological control and induced resistance to phytopathogens and abiotic stresses. Pathogens 10(5):570

Fouda AH, Hassan SE-D, Eid AM, Ewais EE-D (2015) Biotechnological applications of fungal endophytes associated with medicinal plant *Asclepias sinaica* (Bioss.). Ann Agric Sci 60(1):95–104

Franken LPG, Marcon NS, Treichel H, Oliveira D, Freire DM, Dariva C, Destain J, Oliveira JV (2010) Effect of treatment with compressed propane on lipases hydrolytic activity. Food Bioprocess Technol 3:511–520

Fu J-X, Jiao J, Gai Q-Y, He X-J, Fu Y-J, Feng X, Gao J (2024) Improvement of Baicalein and Wogonin yields from *Scutellaria baicalensis* Georgi roots fermented by a novel endophytic fungus capable of producing cellulase and β-glucosidase. Food Bioprocess Technol 17:1–15

Gateta T, Nacoon S, Seemakram W, Ekprasert J, Theerakulpisut P, Sanitchon J, Suwannarach N, Boonlue S (2023) The potential of endophytic fungi for enhancing the growth and accumulation of phenolic compounds and anthocyanin in Maled Phai rice (*Oryza sativa* L.). J Fungi 9(9):937

Geisen S, ten Hooven FC, Kostenko O, Snoek LB, van Der Putten WH (2021) Fungal root endophytes influence plants in a species-specific manner that depends on plant's growth stage. J Ecol 109(4):1618–1632

Ginalska G, Bancerz R, Korniłłowicz-Kowalska T (2004) A thermostable lipase produced by a newly isolated *Geotrichum*-like strain, R59. J Ind Microbiol Biotechnol 31(4):177–182

Goddard J-P, Reymond J-L (2004) Enzyme assays for high-throughput screening. Curr Opin Biotechnol 15(4):314–322

Gold MH, Youngs HL, Gelpke MDS (2000) Manganese peroxidase. In: Metal ions in biological systems. CRC Press, Boca Raton, Taylor & Francis, pp 607–634

Goukanapalle PKR, Kanderi DK, Rajoji G, Shanthi Kumari BS, Bontha RR (2020) Optimization of cellulase production by a novel endophytic fungus *Pestalotiopsis microspora* TKBRR isolated from Thalakona forest. Cellulose 27:6299–6316

Govinda Rajulu MB, Thirunavukkarasu N, Suryanarayanan TS, Ravishankar JP, El Gueddari NE, Moerschbacher BM (2011) Chitinolytic enzymes from endophytic fungi. Fungal Divers 47:43–53

Griebeler N, Polloni AE, Remonatto D, Arbter F, Vardanega R, Cechet JL, Di Luccio M, de Oliveira D, Treichel H, Cansian RL (2011) Isolation and screening of lipase-producing fungi with hydrolytic activity. Food Bioprocess Technol 4:578–586

Gupta N, Sahai V, Gupta R (2007) Alkaline lipase from a novel strain *Burkholderia multivorans*: statistical medium optimization and production in a bioreactor. Process Biochem 42(4):518–526

Haki GD, Rakshit SK (2003) Developments in industrially important thermostable enzymes: a review. Bioresour Technol 89(1):17–34

Hamid R, Khan MA, Ahmad M, Ahmad MM, Abdin MZ, Musarrat J, Javed S (2013) Chitinases: an update. J Pharm Bioallied Sci 5(1):21–29

Hawar SN (2022) Extracellular enzyme of endophytic fungi isolated from *Ziziphus spina* leaves as medicinal plant. Int J Biomater 2022:2135927

Hegde SV, Ramesha A, Srinvas C (2011) Optimization of amylase production from an endophytic fungi *Discosia* sp. Isolated from *Calophyllum inophyllum*. J Agric Technol 7(3):805–813

Hoshida H, Nakao M, Kanazawa H, Kubo K, Hakukawa T, Morimasa K, Akada R, Nishizawa Y (2001) Isolation of five laccase gene sequences from the white-rot fungus *Trametes sanguinea* by PCR, and cloning, characterization and expression of the laccase cDNA in yeasts. J Biosci Bioeng 92(4):372–380

Isah T, Umar S, Mujib A, Sharma MP, Rajasekharan PE, Zafar N, Frukh A (2018) Secondary metabolism of pharmaceuticals in the plant in vitro cultures: strategies, approaches, and limitations to achieving higher yield. Plant Cell Tissue Organ Cult (PCTOC) 132:239–265

Jaeger K-E, Ransac S, Dijkstra BW, Colson C, van Heuvel M, Misset O (1994) Bacterial lipases. FEMS Microbiol Rev 15(1):29–63

Janusz G, Pawlik A, Świderska-Burek U, Polak J, Sulej J, Jarosz-Wilkołazka A, Paszczyński A (2020) Laccase properties, physiological functions, and evolution. Int J Mol Sci 21(3):966

Jha P, Kaur T, Chhabra I, Panja A, Paul S, Kumar V, Malik T (2023) Endophytic fungi: hidden treasure chest of antimicrobial metabolites interrelationship of endophytes and metabolites. Front Microbiol 14:1227830

Ji X, Xia Y, Zhang H, Cui J-L (2022) The microscopic mechanism between endophytic fungi and host plants: from recognition to building stable mutually beneficial relationships. Microbiol Res 261:127056

Joel EL, Bhimba BV (2012) Production of alpha amylase by mangrove associated fungi *Pestalotiopsis microspora* strain VB5 and *Aspergillus oryzae* strain VB6. Indian J Geo-Mar Sci 41(3):279–283

Joseph B, Ramteke PW, Thomas G (2008) Cold active microbial lipases: some hot issues and recent developments. Biotechnol Adv 26(5):457–470

Kachroo H (2022) Endophytes: an asset for extracellular hydrolases production. In: Endophyte biology. Apple Academic Press, Taylor & Francis, New York, pp 149–183

Kathiravan A, Udayan E, Gnanadoss JJ (2023) Optimizing the culture conditions for L-asparaginase production from endophytic fungus *Curvularia* sp. LCJ413 through conventional and statistical approach. J Exp Biol Agric Sci 11(1):62–74

Khan AL, Al-Harrasi A, Al-Rawahi A, Al-Farsi Z, Al-Mamari A, Waqas M, Asaf S, Elyassi A, Mabood F, Shin J-H (2016) Endophytic fungi from frankincense tree improves host growth and produces extracellular enzymes and indole acetic acid. PLoS One 11(6):e0158207

Khan SS, Verma V, Rasool S (2023) Purification and characterization of lipase enzyme from endophytic *Bacillus pumilus* WSS5 for application in detergent industry. Biocatal Agric Biotechnol 50:102681

Ko WH, Wang IT, Ann PJ (2005) A simple method for detection of lipolytic microorganisms in soils. Soil Biol Biochem 37(3):597–599

Kogel K-H, Franken P, Hückelhoven R (2006) Endophyte or parasite–what decides? Curr Opin Plant Biol 9(4):358–363

Kour D, Kaur T, Yadav N, Rastegari AA, Singh B, Kumar V, Yadav AN (2020) Phytases from microbes in phosphorus acquisition for plant growth promotion and soil health. In: New and future developments in microbial biotechnology and bioengineering. Elsevier, Amsterdam, Netherlands, pp 157–176

Kumar M, Brar A, Yadav M, Chawade A, Vivekanand V, Pareek N (2018) Chitinases—potential candidates for enhanced plant resistance towards fungal pathogens. Agriculture 8(7):88

Kumar R, Swapnil P, Meena M, Selpair S, Yadav BG (2022) Plant growth-promoting rhizobacteria (PGPR): approaches to alleviate abiotic stresses for enhancement of growth and development of medicinal plants. Sustainability 14(23):15514. https://doi.org/10.3390/su142315514

Kyomuhimbo HD, Feleni U, Haneklaus NH, Brink H (2023) Recent advances in applications of oxidases and peroxidases polymer-based enzyme biocatalysts in sensing and wastewater treatment: a review. Polymers 15(16):3492

Latz MA, Jensen B, Collinge DB, Jørgensen HJ (2018) Endophytic fungi as biocontrol agents: elucidating mechanisms in disease suppression. Plant Ecol Divers 11(5–6):555–567

Li H, Dou M, Wang X, Guo N, Kou P, Jiao J, Fu Y (2021) Optimization of cellulase production by a novel endophytic fungus *Penicillium oxalicum* R4 isolated from *Taxus cuspidata*. Sustainability 13(11):6006

Liang X, Huang Y, Hua D, Zhang J, Xu H, Li Y, Zhang X (2012) Cellulase production by *Aspergillus* sp. on rice grass (*Spartina* spp.) under solid-state fermentation. Afr J Microbiol Res 6(39):6785–6792

Litvintseva AP, Henson JM (2002) Characterization of two laccase genes of *Gaeumannomyces graminis var. graminis* and their differential transcription in melanin mutants and wild type. Mycol Res 106(7):808–814

Long Y, Han X, Meng X, Xu P, Tao F (2024) A robust yeast chassis: comprehensive characterization of a fast-growing *Saccharomyces cerevisiae*. MBio 15:e03196

Madhusudhan MC, Bharathi TR, Prakash HS (2015) Isolation and purification of bioactive metabolites from fungal endophytes–a review. Curr Biochem Eng 2(2):111–117

Mansur M, Suárez T, Fernández-Larrea JB, Brizuela MA, Gonzalez AE (1997) Identification of a laccase gene family in the new lignin-degrading basidiomycete CECT 20197. Appl Environ Microbiol 63(7):2637–2646

Maria GL, Sridhar KR, Raviraja NS (2005) Antimicrobial and enzyme activity of mangrove endophytic fungi of southwest coast of India. J Agric Technol 1(1):67–80

Marques NP, de Cassia Pereira J, Gomes E, da Silva R, Araújo AR, Ferreira H, Rodrigues A, Dussán KJ, Bocchini DA (2018) Cellulases and xylanases production by endophytic fungi by solid state fermentation using lignocellulosic substrates and enzymatic saccharification of pretreated sugarcane bagasse. Ind Crop Prod 122:66–75

Marusek CM, Trobaugh NM, Flurkey WH, Inlow JK (2006) Comparative analysis of polyphenol oxidase from plant and fungal species. J Inorg Biochem 100(1):108–123

Matias RR, Sepúlveda AMG, Batista BN, de Lucena JMVM, Albuquerque PM (2021) Degradation of *Staphylococcus aureus* biofilm using hydrolytic enzymes produced by amazonian endophytic fungi. Appl Biochem Biotechnol 193:2145–2161

Mayer AM (2006) Polyphenol oxidases in plants and fungi: going places? A review. Phytochemistry 67(21):2318–2331

Minning S, Schmidt-Dannert C, Schmid RD (1998) Functional expression of *Rhizopus oryzae* lipase in *Pichia pastoris*: high-level production and some properties. J Biotechnol 66(2–3):147–156

Mishra VK, Passari AK, Chandra P, Leo VV, Kumar B, Uthandi S, Thankappan S, Gupta VK, Singh BP (2017) Determination and production of antimicrobial compounds by *Aspergillus clavatonanicus* strain MJ31, an endophytic fungus from *Mirabilis jalapa* L. using UPLC-ESI-MS/MS and TD-GC-MS analysis. PLoS One 12(10):e0186234

Mokrani S, Nabti E-H (2024) Recent status in production, biotechnological applications, commercial aspects, and future prospects of microbial enzymes: a comprehensive review. Int J Agric Sci Food Technol 10(1):006–020

Mtui GY (2012) Lignocellulolytic enzymes from tropical fungi: types, substrates and applications. Sci Res Essays 7(15):1544–1555

Muthu Narayanan M, Ahmad N, Shivanand P, Metali F (2022) The role of endophytes in combating fungal-and bacterial-induced stress in plants. Molecules 27(19):6549

Nguyen GT, Nguyen HTH, Tran HT, Tran HT, Ho AN, Tran QH, Pham NB (2023) Enhanced podophyllotoxin production of endophyte *Fusarium proliferatum* TQN5T by host extract and phenylalanine. Appl Microbiol Biotechnol 107(17):5367–5378

Oliveira S d R, Maciel TC, de Oliveira Sancho S, Rodrigues S (2016) Solid-state production of cellulase by *Melanoporia* sp. CCT 7736: a new strain isolated from coconut shell (*Cocos nucifera* L.). Bioresour Bioprocess 3:9

Oliveira LR, Gonçalves AR, Quintela ED, Colognese L, Cortes MV d CB, de Filippi MCC (2024) Genetic engineering of filamentous fungi: prospects for obtaining fourth-generation biological products. Appl Microbiol 4(2):794–810

Onofre SB, Silva GC, Mattiello SP, Groth D, Malagi I (2013) Production of cellulolytic enzymes by *Aspergillus flavus* using solid state fermentation based on sugarcane bagasse. Am J Biochem 3(2):25–28

Oses R, Valenzuela S, Freer J, Baeza J, Rodríguez J (2006) Evaluation of fungal endophytes for lignocellulolytic enzyme production and wood biodegradation. Int Biodeterior Biodegrad 57(2):129–135

Otten LG, Quax WJ (2005) Directed evolution: selecting today's biocatalysts. Biomol Eng 22(1–3):1–9

Oyeleye A, Normi YM (2018) Chitinase: diversity, limitations, and trends in engineering for suitable applications. Biosci Rep 38(4):BSR2018032300

Palmieri G, Cennamo G, Faraco V, Amoresano A, Sannia G, Giardina P (2003) Atypical laccase isoenzymes from copper supplemented *Pleurotus ostreatus* cultures. Enzym Microb Technol 33(2–3):220–230

Pandey A, Nigam P, Soccol CR, Soccol VT, Singh D, Mohan R (2000) Advances in microbial amylases. Biotechnol Appl Biochem 31(2):135–152

PLNSN SD, Naveen M, Saroj Y, Siddalingeshwara KG (2014) Role of carbon source for the production of L-glutaminase from *Aspergillus oryzae*. J Drug Deliv Ther 4(5):193–196

Raghav D, Jyoti A, Siddiqui AJ, Saxena J (2022) Plant-associated endophytic fungi as potential bio-factories for extracellular enzymes: progress, challenges and strain improvement with precision approaches. J Appl Microbiol 133(2):287–310

Rathore AS, Gupta RD (2015) Chitinases from bacteria to human: properties, applications, and future perspectives. Enzyme Res 2015:791907

Raveendran S, Parameswaran B, Ummalyma SB, Abraham A, Mathew AK, Madhavan A, Rebello S, Pandey A (2018) Applications of microbial enzymes in food industry. Food Technol Biotechnol 56(1):16

Ray A (2012) Application of lipase in industry. Asian J Pharm Technol 2(2):33–37

Reetz MT (2001) Combinatorial and evolution-based methods in the creation of enantioselective catalysts. Angew Chem Int Ed 40(2):284–310

Reetz MT (2002) Lipases as practical biocatalysts. Curr Opin Chem Biol 6(2):145–150

Risna RA, Suhirman. (2002) Ligninolytic enzyme production by Polyporaceae from Lombok, Indonesia. Fungal Divers 9:123–134

Rivera-Hoyos CM, Morales-Álvarez ED, Poutou-Piñales RA, Pedroza-Rodríguez AM, RodrÍguez-Vázquez R, Delgado-Boada JM (2013) Fungal laccases. Fungal Biol Rev 27(3–4):67–82

Riyadi FA, Alam MZ, Salleh MN, Salleh HM, Hidayatullah IM, Hara H (2024) Characterization of a thermostable-organic solvent-tolerant lipase from thermotolerant *Rhizopus* sp. strain PKC12B2 isolated from palm kernel cake. Case Stud Chem Environ Eng 9:100721

Roberts J, McGregor WG (1991) Inhibition of mouse retroviral disease by bioactive glutaminase-asparaginase. J Gen Virol 72(2):299–305

Rodríguez E, Ruiz-Dueñas FJ, Kooistra R, Ram A, Martínez ÁT, Martínez MJ (2008) Isolation of two laccase genes from the white-rot fungus *Pleurotus eryngii* and heterologous expression of the pel3 encoded protein. J Biotechnol 134(1–2):9–19

Ross RP, Morgan S, Hill C (2002) Preservation and fermentation: past, present and future. Int J Food Microbiol 79(1–2):3–16

Rusu AV, Trif M, Rocha JM (2023) Microbial secondary metabolites via fermentation approaches for dietary supplementation formulations. Molecules 28(16):6020

Saikkonen K, Faeth SH, Helander M, Sullivan TJ (1998) Fungal endophytes: a continuum of interactions with host plants. Annu Rev Ecol Syst 29(1):319–343

Schmidt M, Bornscheuer UT (2005) High-throughput assays for lipases and esterases. Biomol Eng 22(1–3):51–56

Seth K, Vyas P, Deora S, Gupta AK, Meena M, Swapnil P, Harish (2023) Understanding plant-plant growth-promoting rhizobacteria (PGPR) interactions for inducing plant defense. In: Swapnil P, Meena M, Harish, Marwal A, Vijayalakshmi S, Zehra A (eds) Plant-microbe interaction—recent advances in molecular and biochemical approaches, vol 2. Academic Press, Elsevier, United Kingdom, pp 201–226. https://doi.org/10.1016/B978-0-323-91876-3.00010-5

Sharma P, Kumar S (2021) Bioremediation of heavy metals from industrial effluents by endophytes and their metabolic activity: recent advances. Bioresour Technol 339:125589

Sharma R, Chisti Y, Banerjee UC (2001) Production, purification, characterization, and applications of lipases. Biotechnol Adv 19(8):627–662

Sharma A, Sharma P, Deepak B, Mona S (2024) Microbes as environmental allies: from wastewater treatment to sustainable biofuel production. In: Harnessing microbial potential for multifarious applications. Springer, Singapore, pp 359–376

Singh AK, Mukhopadhyay M (2012) Overview of fungal lipase: a review. Appl Biochem Biotechnol 166:486–520

Singh R, Kumar M, Mittal A, Mehta PK (2016) Microbial enzymes: industrial progress in 21st century. 3 Biotech 6:1–15

Singh G, Kaur S, Khatri M, Arya SK (2019) Biobleaching for pulp and paper industry in India: emerging enzyme technology. Biocatal Agric Biotechnol 17:558–565

Soccol CR, Vandenberghe LP (2003) Overview of applied solid-state fermentation in Brazil. Biochem Eng J 13(2–3):205–218

Soden DM, O'callaghan J, Dobson ADW (2002) Molecular cloning of a laccase isozyme gene from *Pleurotus sajor-caju* and expression in the heterologous *Pichia pastoris* host. Microbiology 148(12):4003–4014

Sperandio GB, Ferreira Filho EX (2019) Fungal co-cultures in the lignocellulosic biorefinery context: a review. Int Biodeterior Biodegrad 142:109–123

Stamford TLM, Stamford NP, Coelho L, Araujo JM (2002) Production and characterization of a thermostable glucoamylase from *Streptosporangium* sp. endophyte of maize leaves. Bioresour Technol 83(2):105–109

Sudheep NM, Marwal A, Lakra N, Anwar K, Mahmood S (2017) Fascinating fungal endophytes role and possible beneficial applications: an overview. In: Plant-microbe interactions in agro-ecological perspectives: volume 1: fundamental mechanisms, methods and functions. Springer, Singapore, pp 255–273

Sunitha VH, Ramesha A, Savitha J, Srinivas C (2012) Amylase production by endophytic fungi *Cylindrocephalum* sp. isolated from medicinal plant *Alpinia calcarata* (Haw.) Roscoe. Braz J Microbiol 43(3):1213

Sunitha VH, Nirmala Devi D, Srinivas C (2013) Extracellular enzymatic activity of endophytic fungal strains isolated from medicinal plants. World J Agric Sci 9(1):1–9

Suz LM, Barsoum N, Benham S, Dietrich H-P, Fetzer KD, Fischer R, García P, Gehrman J, Kristöfel F, Manninger M (2014) Environmental drivers of ectomycorrhizal communities in Europe's temperate oak forests. Mol Ecol 23(22):5628–5644

Taokaew S, Kriangkrai W (2023) Chitinase-assisted bioconversion of chitinous waste for development of value-added chito-oligosaccharides products. Biology 12(1):87

Taranto F, Pasqualone A, Mangini G, Tripodi P, Miazzi MM, Pavan S, Montemurro C (2017) Polyphenol oxidases in crops: biochemical, physiological and genetic aspects. Int J Mol Sci 18(2):377

Tasia W, Melliawati R (2017) Cellulase and xylanase production from three isolates of indigenous endophytic fungi. IOP Conf Ser: Earth Environ Sci 101(1):012035

Thipyapong P, Stout MJ, Attajarusit J (2007) Functional analysis of polyphenol oxidases by antisense/sense technology. Molecules 12(8):1569–1595

Tiwari P, Bae H (2022) Endophytic fungi: key insights, emerging prospects, and challenges in natural product drug discovery. Microorganisms 10(2):360

Uzma F, Konappa NM, Chowdappa S (2016) Diversity and extracellular enzyme activities of fungal endophytes isolated from medicinal plants of Western Ghats, Karnataka. Egypt J Basic Appl Sci 3(4):335–342

Verma A, Tiwari H, Singh S, Gupta P, Rai N, Kumar Singh S, Singh BP, Rao S, Gautam V (2023) Epigenetic manipulation for secondary metabolite activation in endophytic fungi: current progress and future directions. Mycology 14(4):275–291

Viswanath B, Rajesh B, Janardhan A, Kumar AP, Narasimha G (2014) Fungal laccases and their applications in bioremediation. Enzyme Res 2014:163242

Voelz K, May RC (2010) Cryptococcal interactions with the host immune system. Eukaryot Cell 9(6):835–846

Wahleithner JA, Xu F, Brown KM, Brown SH, Golightly EJ, Halkier T, Kauppinen S, Pederson A, Schneider P (1996) The identification and characterization of four laccases from the plant pathogenic fungus *Rhizoctonia solani*. Curr Genet 29:395–403

Wang F, Xu L, Zhao L, Ding Z, Ma H, Terry N (2019) Fungal laccase production from lignocellulosic agricultural wastes by solid-state fermentation: a review. Microorganisms 7(12):665

Wang L, Shah AM, Liu Y, Jin L, Wang Z, Xue B, Peng Q (2020) Relationship between true digestibility of dietary phosphorus and gastrointestinal bacteria of goats. PLoS One 15(5):e0225018

Wang C, Zeng Z-Q, Zhuang W-Y (2021) Comparative molecular evolution of chitinases in ascomycota with emphasis on mycoparasitism lifestyle. Microb Genom 7(9):000646

Wang G, Ran H, Fan J, Keller NP, Liu Z, Wu F, Yin W-B (2022) Fungal-fungal cocultivation leads to widespread secondary metabolite alteration requiring the partial loss-of-function VeA1 protein. Sci Adv 8(17):eabo6094

Wipusaree N, Sihanonth P, Piapukiew J, Sangvanich P, Karnchanatat A (2011) Purification and characterization of a xylanase from the endophytic fungus *Alternaria alternata* isolated from the Thai medicinal plant, *Croton oblongifolius* Roxb. Afr J Microbiol Res 5(31):5697–5712

Wodzinski RJ, Ullah AHJ (1996) Phytase. In: Advances in applied microbiology, vol 42. Elsevier, pp 263–302

Wong DW (2009) Structure and action mechanism of ligninolytic enzymes. Appl Biochem Biotechnol 157:174–209

Wu Y, Yan P, Li Y, Liu X, Wang Z, Chen T, Zhao X (2020) Enhancing β-carotene production in *Escherichia coli* by perturbing central carbon metabolism and improving the NADPH supply. Front Bioeng Biotechnol 8:585

Xia Y, Liu J, Chen C, Mo X, Tan Q, He Y, Wang Z, Yin J, Zhou G (2022) The multifunctions and future prospects of endophytes and their metabolites in plant disease management. Microorganisms 10(5):1072

Xiao YZ, Hong YZ, Li JF, Hang J, Tong PG, Fang W, Zhou CZ (2006) Cloning of novel laccase isozyme genes from *Trametes* sp. AH28-2 and analyses of their differential expression. Appl Microbiol Biotechnol 71:493–501

Xu F, Berka RM, Wahleithner JA, Nelson BA, Shuster JR, Brown SH, Palmer AE, Solomon EI (1998) Site-directed mutations in fungal laccase: effect on redox potential, activity and pH profile. Biochem J 334(1):63–70

Xu S, Li M, Hu Z, Shao Y, Ying J, Zhang H (2023) The potential use of fungal co-culture strategy for discovery of new secondary metabolites. Microorganisms 11(2):464

Yadav AN (2019) Fungal white biotechnology: conclusion and future prospects. In: Recent advancement in white biotechnology through fungi: volume 3: perspective for sustainable environments. Springer Nature, Switzerland, pp 491–498

Yadav G, Meena M (2021) Bioprospecting of endophytes in medicinal plants of Thar Desert: an attractive resource for biopharmaceuticals. Biotechnol Rep 30:e00629. https://doi.org/10.1016/j.btre.2021.e00629

Yadav G, Meena M (2022) Biological control of plant diseases by endophytes. In: Shah MP, Deka D (eds) Endophytic association: what, why and how (volume: developments in applied microbiology and biotechnology). Academic Press, Elsevier Inc., pp 119–135. https://doi.org/10.1016/B978-0-323-91245-7.00007-9

Yadav R, Singh AV, Joshi S, Kumar M (2015) Antifungal and enzyme activity of endophytic fungi isolated from *Ocimum sanctum* and *Aloe vera*. Afr J Microbiol Res 9(29):1783–1788

Yan JF, Broughton SJ, Yang SL, Gange AC (2015) Do endophytic fungi grow through their hosts systemically? Fungal Ecol 13:53–59

Yaver DS, Xu F, Golightly EJ, Brown KM, Brown SH, Rey MW, Schneider P, Halkier T, Mondorf K, Dalboge H (1996) Purification, characterization, molecular cloning, and expression of two laccase genes from the white rot basidiomycete *Trametes villosa*. Appl Environ Microbiol 62(3):834–841

Yoshida H (1883) LXIII.—chemistry of lacquer (Urushi). Part I. Communication from the chemical society of Tokio. J Chem Soc Trans 43:472–486

Yu X-W, Wang L-L, Xu Y (2009) *Rhizopus chinensis* lipase: gene cloning, expression in *Pichia pastoris* and properties. J Mol Catal B Enzym 57(1–4):304–311

Yuan Y, Feng H, Wang L, Li Z, Shi Y, Zhao L, Feng Z, Zhu H (2017) Potential of endophytic fungi isolated from cotton roots for biological control against verticillium wilt disease. PLoS One 12(1):e0170557

Zaferanloo B, Bhattacharjee S, Ghorbani MM, Mahon PJ, Palombo EA (2014) Amylase production by *Preussia minima*, a fungus of endophytic origin: optimization of fermentation conditions and analysis of fungal secretome by LC-MS. BMC Microbiol 14:1–12

Zakariyah RF, Ajijolakewu KA, Ayodele AJ, Folami-A BI, Samuel EP, Otuoze SO, Abdulrauf LB, Ahmed RN (2024) Progress in endophytic fungi secondary metabolites: biosynthetic gene cluster reactivation and advances in metabolomics. Bull Natl Res Cent 48(1):44

Zhang JY, Tao LY, Liang YJ, Chen LM, Mi YJ, Zheng LS, Wang F, She ZG, Lin YC, To KK, Fu LW (2010) Anthracenedione derivatives as anticancer agents isolated from secondary metabolites of the mangrove endophytic fungi. Mar Drugs 8(4):1469–1481

Zhang Q, Gao B, Xiao Y, Yang H, Wang Y, Du L, Zhu D (2020) Purification and characterization of a novel β-glucuronidase precisely converts glycyrrhizin to glycyrrhetinic acid 3-O-mono-β-D-glucuronide from plant endophytic *Chaetomium globosum* DX-THS3. Int J Biol Macromol 159:782–792

Zhang Z, Chen X, Gao L (2024) New strategy for the biosynthesis of alternative feed protein: single-cell protein production from straw-based biomass. GCB Bioenergy 16(2):e13120

Zhou L, Höfte M, Hennessy RC (2024) Does regulation hold the key to optimizing lipopeptide production in *Pseudomonas* for biotechnology? Front Bioeng Biotechnol 12:1363183

Harnessing Endophytic Fungi for Effective Phytoremediation: Mechanisms and Applications

7

Andleeb Zehra and Mukesh Meena

Abstract

Endophytic fungi, residing within plant tissues without causing harm, play a crucial role in enhancing phytoremediation—a sustainable method for removing pollutants from the environment. These fungi form symbiotic relationships with their host plants, promoting growth, stress tolerance, and the degradation of harmful substances. This study explores the multifaceted interactions between plants and endophytic fungi in the context of phytoremediation, particularly in heavy metal and organic pollutant environments. Key findings highlight the ability of endophytic fungi to enhance plant resilience, nutrient uptake, and pollutant degradation through enzymatic activity and metal sequestration. The fungi's role in improving phytoremediation outcomes includes increased metal tolerance and absorption, facilitating the transformation of toxic substances into less harmful forms, and bolstering overall plant health. By leveraging these natural interactions, phytoremediation offers a promising approach to restoring contaminated ecosystems, reducing pollution, and supporting environmental sustainability.

Keywords

Endophytic fungi · Nutrient uptake · Organic pollutants · Phytoremediation · Phytotoxicity · Plant-endophytic fungi interactions

A. Zehra (✉)
MI College, Malé, Maldives
e-mail: andleeb.zehra@micollege.edu.mv

M. Meena (✉)
Laboratory of Phytopathology and Microbial Biotechnology, Department of Botany, Mohanlal Sukhadia University, Udaipur, Rajasthan, India
e-mail: drmukeshmeena321@mlsu.ac.in

M. Meena et al. (eds.), *Metabolic Sustainability of Endophytes*, Microorganisms for Sustainability 52, https://doi.org/10.1007/978-981-96-4004-1_7

7.1 Introduction

Endophytic fungi, microorganisms that colonize the internal tissues of plants without causing harm, establish a symbiotic relationship with their host plants. These fungi play a crucial role in promoting plant health and growth by enhancing nutrient uptake, stress tolerance, and the production of secondary metabolites (Alam et al. 2021). They inhabit a wide range of plant hosts, including herbs, shrubs, and trees, and are categorized as either obligate or facultative endophytes based on their interaction with the host plant (Mushtaq et al. 2023). The rapid rise in industrial activities over recent decades has led to significant environmental challenges, including the accumulation of heavy metals in soil and water due to waste discharge from various sources (Bhat et al. 2022). Phytoremediation has emerged as a promising method to address these environmental issues by utilizing plants to remove or neutralize pollutants through mechanisms such as phytoextraction, phytostabilization, and rhizofiltration (Oubohssaine and Dahmani 2024). These processes help detoxify heavy metals and organic pollutants, thereby improving soil and water quality. Endophytic fungi, in particular, contribute to this process by enhancing the plant's ability to tolerate and absorb pollutants, and by producing enzymes that degrade harmful substances (Kafle et al. 2022; Morales-Vargas et al. 2024).

Pollutants in the environment, including heavy metals, organic compounds, and microplastics, present significant risks to ecosystems and human health (Shetty et al. 2023). Heavy metals, such as lead, mercury, and cadmium, can accumulate in living organisms and disrupt biological processes, leading to serious health issues (Mitra et al. 2022). Organic pollutants like pesticides and industrial chemicals can persist in the environment and bioaccumulate in the food chain, posing long-term ecological and health risks (Encarnação et al. 2019). Additionally, nutrient pollution and microplastics further exacerbate environmental problems, affecting water quality and marine life (Ziani et al. 2023; Garai et al. 2024).

Given these challenges, understanding and optimizing phytoremediation techniques that involve endophytic fungi is critical for effective pollution management. This study aims to explore the role of endophytic fungi in enhancing phytoremediation, focusing on their interactions with plants in environments contaminated with heavy metals and organic pollutants. By investigating how these fungi contribute to pollutant degradation, metal tolerance, and overall plant health, the study seeks to identify potential strategies for improving phytoremediation outcomes and supporting sustainable environmental management.

7.2 Pollutant Types, Their Impacts, and Sequestration Methods

Pollutants in the environment arise from various sources and significantly impact ecosystems and human health. They can be categorized into several types, each with distinct characteristics and challenges (Manisalidis et al. 2020). Heavy metals, such as lead, mercury, cadmium, and arsenic, often enter the environment through industrial processes, mining, and agriculture (Tchounwou et al. 2012). These metals can accumulate in organisms, leading to toxic effects that can disrupt biological processes and cause serious health issues (Mitra et al. 2022). Organic pollutants, including pesticides, herbicides, and industrial chemicals like polychlorinated biphenyl (PCBs) and dioxins, persist in the environment and pose risks to wildlife and humans. They can bioaccumulate in the food chain, resulting in long-term ecological damage and health problems such as cancer and endocrine disruption (Encarnação et al. 2019). Nutrient pollution, primarily from nitrogen and phosphorus, results from agricultural runoff and wastewater discharge. Excessive nutrients can cause eutrophication in aquatic ecosystems, leading to algal blooms and oxygen depletion, which harm aquatic life (Yang et al. 2008). Microplastics, small plastic particles from the degradation of larger plastic debris or microbeads, are pervasive in marine environments. They are ingested by aquatic organisms, causing physical harm and potentially introducing toxic chemicals into the food web (Osman et al. 2023).

To address these issues, various sequestration methods are employed. Phytoremediation uses plants to absorb, concentrate, and degrade pollutants from soil and water, while bioremediation leverages microorganisms to transform pollutants into less harmful substances. Chemical methods involve precipitating, adsorbing, or neutralizing pollutants, often in wastewater treatment. Physical methods, such as filtration and encapsulation, physically remove or contain pollutants, preventing further environmental contamination (Khan et al. 2022). Understanding and implementing these methods are crucial for reducing pollution's adverse effects and restoring environmental balance.

7.3 Plant-Fungal Endophyte Interactions in Phytoremediation: Processes and Outcomes

7.3.1 Role of Endophytic Fungi in Phytoremediation

Endophytic fungi reside within the tissues of plants without causing harm, forming a mutualistic relationship with their hosts. These fungi enhance the phytoremediation process through several mechanisms (Nandy et al. 2020; Fig. 7.1). Endophytic fungi help plants cope with environmental stresses, such as high levels of pollutants. By improving the plant's resilience, fungi enable plants to survive and continue functioning in contaminated environments (Ameen et al. 2024). Fungi can enhance the plant's ability to absorb essential nutrients from the soil, which is crucial in polluted areas where nutrient availability might be limited. This improved nutrient uptake supports healthier plant growth and better overall performance in remediation tasks (Rashid et al. 2016). Many endophytic fungi produce enzymes capable of breaking down complex organic pollutants, such as hydrocarbons and pesticides. These enzymes transform harmful substances into simpler, less toxic forms that are more readily absorbed and processed by the plants (Dinakarkumar et al. 2024).

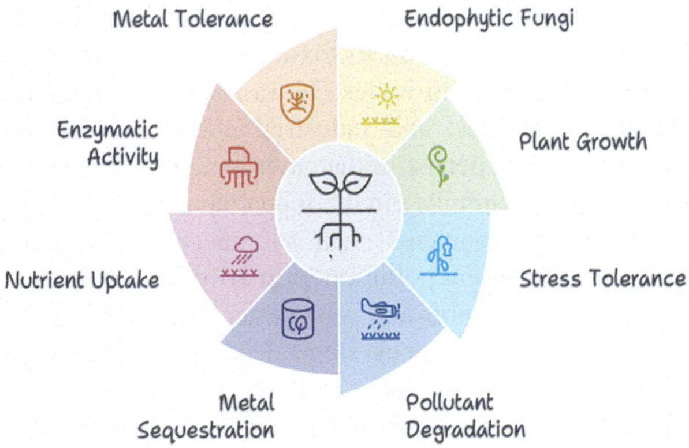

Fig. 7.1 Enhancing phytoremediation with endophytic fungi

7.3.2 Outcomes of Plant-Endophytic Fungi Interactions

The interaction between plants and endophytic fungi plays a crucial role in enhancing phytoremediation, offering an effective, natural solution for reducing pollution and promoting environmental sustainability. Endophytic fungi reside within plant tissues, forming mutualistic relationships and aiding in pollutant removal through various mechanisms (Table 7.1). The enzymatic activity of endophytic fungi accelerates the breakdown of organic pollutants. This process transforms these substances into less harmful by-products, which can then be taken up by the plant and further degraded or neutralized (Singh and Kumar 2023). Endophytic fungi can enhance the plant's tolerance to heavy metals, such as lead, cadmium, and mercury. They facilitate the uptake and sequestration of these metals within plant tissues, reducing their bioavailability in the soil and mitigating their toxic effects (Babu et al. 2014a; Sharma and Kumar 2021). The symbiotic relationship between plants and endophytic fungi boosts overall plant health by increasing resilience to environmental stressors and optimizing nutrient availability. This leads to enhanced growth and productivity, which in turn improves the efficiency of the phytoremediation process (Verma et al. 2022).

The interaction between plants and endophytic fungi in phytoremediation offers a promising solution for cleaning up contaminated environments. By harnessing these natural processes, phytoremediation can effectively reduce pollution, restore ecological balance, and contribute to sustainable environmental management. This approach not only addresses contamination but also supports ecosystem health and plant biodiversity.

Table 7.1 Plant species and their associated endophytic fungi, along with their roles in phytoremediation of heavy metals

Plant species	Endophytic fungi	Role in phytoremediation	References
Lolium perenne L.	*Neotyphodium lolii*	Enhances tolerance to excess zinc (Zn)	Monnet et al. (2005)
Zea mays	*Peyronellaea*	Facilitates absorption and accumulation of heavy metals	Shen et al. (2013)
Cajanus cajan (L) Huth	*Scleroderma citrinum*, *Amanita muscaria*, and *Lactarius rufus*	Engages in metal interaction, aids in phytochelation and glutathione production	Garg and Aggarwal (2011)
Dysphania ambrosioides L.	FT2G59 (*Penicillium*) and FT2G7 (*P. columnaris*)	Provides protection against heavy metal toxicity	Li et al. (2016)
Pinus sabiniana L.	*Scleroderma citrinum*, *Amanita muscaria*, and *Lactarius Rufus*	Offers protection against heavy metals	Krupa and Kozdrój (2007)
Festuca arundinacea Schreb.	*Neotyphodium coenophialum*	Evaluates efficiency of phytoextraction	Yin et al. (2014)
Lolium arundinaceum (Schreb.) Darbysh.	*Neotyphodium coenophialum*	Assesses phytoextraction efficiency in hydroponic systems (accumulation of Na^+)	Yin et al. (2014)
Salix variegata Franch	*Chromosporium* sp., *Fusarium* sp., and *Gonatobotrys* sp.	Exhibits cadmium (Cd) resistance	An et al. (2015)
Solanum nigrum L.	*Colletotrichum* sp., *Alternaria* sp., and *Fusarium* sp.	Shows tolerance to cadmium (Cd)	Khan et al. (2017a)
S. nigrum Korean ecotype	*Fusarium tricinctum* RSF-4L and *Alternaria alternata* RSF-6L	Facilitates phytostabilization of heavy metals	Khan et al. (2017b)
Dysphania ambrosioides L. (Mosyakin & Clemants)	*Plectosphaerella* sp., *Cladosporium* sp., and *Verticillium* sp.	Demonstrates heavy metal tolerance	Li et al. (2016)

7.4 Endophytic Fungi: Crucial in Mitigating Metal Toxicity

Industrial activities, wastewater discharge, sewage systems, and other anthropogenic processes introduce significant amounts of heavy metals into the environment, which pose serious risks to both human health and ecological stability (Singh et al. 2023; Adnan et al. 2022). The remediation of these contaminated sites is often costly and labor-intensive. Recently, there has been growing interest in the interactions between plants, microorganisms, and metals due to three primary reasons: certain microbes can tolerate high levels of metal toxicity, they can remediate contaminated soils and water, and they promote plant growth in polluted environments

(Tang et al. 2024). While much of the research has centered on endophytic bacteria, endophytic fungi, despite being less studied, offer significant promise. These fungi can immobilize toxic metals and metalloids by forming metal oxalates or by binding with melanin-like polymers (Dell'Anno et al. 2022). Fungi, compared to bacterial augmentation—which involves introducing lab-grown microbial cultures to enhance contaminant degradation—have an advantage in competing more effectively with native bacterial microflora (Li et al. 2020). Metal accumulation and bioabsorption can occur via both nonliving biomass (derived from microbes or plants) and living cells, making the study of metal-resistant fungal endophytes critical for phytoremediation (Priya et al. 2022; Ojuederie and Babalola 2017; Dubey et al. 2019). Several endophytes have been identified in wastelands polluted by heavy metals, particularly in soils contaminated with lead (Pb) and zinc (Zn). These microbes have shown potential for use in phytoremediation due to their ability to thrive in such harsh conditions (Li et al. 2012). Furthermore, the roots of plants growing in these contaminated areas have a high prevalence of arbuscular mycorrhizal fungi (AMF), particularly Glomus sp. and *Acaulospora* sp. AMF enhances arsenic (As) uptake in the hyperaccumulating fern *Pteris vittata* L. (Trotta et al. 2006) by increasing the expression of As translocation factors. Because arsenate and phosphate are chemically similar, *P. vittata* L. absorbs arsenic through its phosphate uptake mechanism (Wang et al. 2002), and mycorrhizae are more efficient at absorbing these elements than the plant's root system alone. Pathogenic fungi have been identified in association with the plant *Acacia auriculiformis* (Jiang et al. 2008), while various fungal species have also been isolated from its rhizosphere soil (del Pilar Ortega-Larrocea et al. 2010). Additionally, endophytic yeasts have been found in *Brassica chinensis*, where they enhance the bioaccumulation of heavy metals such as lead (Pb), zinc (Zn), copper (Cu), and cadmium (Cd) (Wang et al. 2013; Deng et al. 2011, 2012). Research indicates that the metal extraction capacity of certain yeasts is higher than that of other microbial species (Wang et al. 2013; Chandran et al. 2021).

Several endophytic fungi, such as *Trichoderma* sp., *Cladosporium* sp., *Lasiodiplodia* sp., *Fusarium* sp., and *Paecilomyces* sp., have been isolated from *Portulaca* sp. (Babu et al. 2014b; Deng et al. 2014; Khan et al. 2014). Additionally, *Microsphaeropsis* sp. *LSE10*, isolated from *Solanum nigrum* L., has demonstrated the ability to absorb up to 247.5 mg/g of cadmium (Cd) (Xiao et al. 2010). The fungus *Exophiala pisciphila* is capable of absorbing Pb and Cd, accounting for up to 20% and 5% of its dry weight, respectively (Zhang et al. 2008). Introducing *Cryptococcus* sp. *CBSB78* has been shown to enhance the phytoextraction of Cd, Pb, and Zn by *Brassica alboglabra* L. in multi-metal-contaminated soils (Deng et al. 2012). Research by Tong et al. (2017) revealed that the frequency of endophytic infection may increase over the course of phytoremediation. However, species like *Bothriochloa ischaemum* and *Festuca rubra* have shown a positive correlation between endophyte infection rates and the reversal of Cd levels, while *Imperata cylindrica* and *Elymus dahuricus* exhibited increased Pb tolerance post-infection (Tong et al. 2017).

7.5 Endophytic Fungi in the Remediation of Organic Pollutants

The endophytic fungus *Pestalotiopsis palmarum* BM-04 has shown resilience in extreme conditions of salinity and crude oil contamination (Naranjo-Briceño et al. 2013). Similarly, *Pestalotiopsis microspora* has demonstrated the ability to degrade plastic polyester polyurethane (PUR) in *Dendrobium* species, while *Phomopsis liquidambari* has been identified for its effectiveness in breaking down polycyclic aromatic hydrocarbons (PAHs) in *Bischofia polycarpa* (Chen et al. 2013; Dai et al. 2010). Additionally, *Ceratobasidium stevensii* has been observed to degrade phenolic acids in *Citrullus lanatus* (Xiao et al. 2014). Endophyte-infected plants have shown enhanced capabilities in degrading petroleum hydrocarbons compared to their noninfected counterparts (Xiao et al. 2014). For example, *Festuca arundinacea* and *Festuca pratensis* in the Poaceae family exhibit increased biomass accumulation and more efficient degradation of total petroleum hydrocarbons (TPH) when infected with endophytic fungi (Soleimani et al. 2010b). In the rhizosphere of these plants, the removal rates for PAHs and TPH are 80–84% and 64–72%, respectively, with higher TPH removal observed in endophyte-infected plants. Endophyte-assisted removal of TPH from oil-contaminated soils is emerging as a promising green technology. This method is associated with increased root biomass and the release of water-soluble phenols into the soil, which aids the enzymatic breakdown of n-alkanes, specifically those with carbon chain lengths of C10–C25 (Soleimani et al. 2010a). Research indicates that endophytic fungi enhance rhizodeposition in plants, influencing microbial mineralization in the soil, without altering the diversity of the microbial community (Van Hecke et al. 2005). To commercialize this innovative approach, it is essential to understand the potential of the mineralization or biotransformation process. The "recovery" phase in any complex biochemical process is crucial for its cost-effectiveness. Chemoheterotrophic metabolism and the filamentous branching hyphal network have been found to be superior for the recovery of metals and metalloids (Liang and Gadd 2017). The reduction of metalloid oxyanions to elemental metalloids is key to the recovery of arsenic (As), selenium (Se), tellurium (Te), and antimony (Sb) (Pierart et al. 2015; Wilson et al. 2010; Gadd 1993).

7.6 Reversing Phytotoxicity Through Endophytic Fungi

Phytotoxicity is a significant challenge in phytoremediation, as it can disrupt the metabolic interactions between plants and their endophytes. Nevertheless, endophytes play a vital role in alleviating phytotoxicity by producing organic acids, iron chelators, siderophores, and various enzymes that degrade pollutants (Soleimani et al. 2010a). While bacterial endophytes are commonly associated with reducing phytotoxicity, fungal endophytes also contribute to the removal of organic pollutants and metal extraction. One such endophyte, *Piriformospora indica*, supports its host plants under severe physiological and nutritional stress by solubilizing soil phosphorus, facilitated through the action of endogenous acid phosphatase. This fungus also enhances nitrate uptake and modulates carbon and sulfur metabolism in plants (Oelmüller et al. 2009). In a study by Li et al. (2012), an isolate of *Phoma* sp. showed higher sensitivity to zinc (Zn) toxicity compared to *Alternaria* sp. and *Peyronellaea* sp., based on their relative metal tolerance and abundance in mine-contaminated wasteland soils. This indicates that the degree of metal tolerance may vary significantly across different fungal species.

7.7 Biotechnological Strategies and Impacts of Endophytic Fungi in Phytoremediation

For effective and eco-friendly phytoremediation using endophytes, several factors are critical. These include selecting and genetically engineering potent endophytic microorganisms, developing transgenic plants, ensuring niche differentiation, maintaining host specificity, leveraging fungal metabolite-driven reactions, enhancing pollutant absorption and tolerance, and promoting root and shoot biomass growth. Additional factors are preserving root structure integrity and regulating pH, salinity, and pollutant-induced toxicity within the rhizosphere. Managing microbial communities and sustaining balanced symbiotic relationships are also crucial (Kidd et al. 2009; Yadav and Meena 2021, 2022). Transgenic technology is becoming a valuable approach to augment the inherent phytoremediation abilities of plants (Chatterjee et al. 2013). Through genetic engineering, it is possible to enhance the natural capacity for pollutant degradation and detoxification by over-expressing specific genes or traits linked to the sequestration of particular contaminants (Schäffner et al. 2002). Nonetheless, it is imperative that such biotechnological methods comply with safety standards and ensure reproducibility (Schäffner et al. 2002). Protoplast fusion is an advanced biotechnological technique utilized in phytoremediation research. For example, the self-fusing CBRF59T3 strain from *Mucor* sp. CBRF59 has been employed to boost metal resistance, significantly increasing the bioavailability of cadmium (Cd) and lead (Pb) in contaminated soils (Deng et al. 2013).

7.8 Leveraging Endophyte-Plant Relationships to Alleviate Xenobiotic Toxicity

Interactions between endophytes and plants present several benefits for mitigating xenobiotic toxicity, although they also come with certain challenges. Endophytic microorganisms exhibit advantageous characteristics such as the absence of tissue-specific competition between species, long-term viability, the ability to tap into the host plant's metabolite resources, and the ease with which they can be genetically modified (Wu et al. 2021). However, these interactions initiate with the recognition of an "aggregation signal," and successful colonization is dependent on achieving a delicate equilibrium between the plant's immune defenses and the virulence factors of the endophyte. The final phase of plant-microbe interaction in phytoremediation involves eliminating contaminants from the environment through a series of intricate biochemical and physiological processes (Kafle et al. 2022). Evaluating these phytoremediation processes requires a comprehensive three-step analysis involving "metagenomic, metatranscriptomic, and metaproteomic" studies to better understand plant-microbe interactions and their role in enhancing natural remediation strategies (Kafle et al. 2022). Techniques such as tagging with fluorescent proteins, overexpressing microbial genes responsible for pollutant degradation, enzyme-linked immunosorbent assay (ELISA), and chromatin immunoprecipitation sequencing (ChIP-Seq) are valuable for conducting in-depth proteomic and genomic studies. These methods are especially useful for verifying the vast amounts of data collected from field experiments (Gade and Kalvakolanu 2012). To fully comprehend the mechanisms underlying endophyte-assisted phytoremediation, it is essential to evaluate a wide range of parameters and environmental factors under robust monitoring systems (Kafle et al. 2022). The intricate network of microbial communities distributed throughout different plant parts, along with their interactions within the plant-microbiome-soil-pollutant matrix, can only be thoroughly investigated through next-generation sequencing and advanced techniques such as real-time, high-throughput functional phenotyping of whole plants (Matchado et al. 2021).

7.9 Conclusion

This study underscores the significant potential of endophytic fungi in enhancing phytoremediation, a vital process for mitigating environmental pollution and promoting sustainability. Endophytic fungi, by forming symbiotic relationships with host plants, contribute substantially to the resilience and efficiency of phytoremediation processes. Their role in promoting plant growth, enhancing nutrient uptake, and degrading organic pollutants, as well as their ability to increase metal tolerance and uptake, demonstrates their multifaceted impact on environmental cleanup efforts. Key findings highlight that endophytic fungi improve plant health and productivity under pollutant stress, enhance the degradation of harmful substances through enzymatic activity, and facilitate the sequestration of heavy metals, thus reducing their bioavailability and toxicity. These interactions not only boost the overall efficiency of phytoremediation but also support ecosystem recovery and environmental sustainability. The study also emphasizes the importance of optimizing biotechnological strategies to harness the full potential of endophytic fungi. Future research should focus on advancing genetic engineering techniques, understanding the complex dynamics of plant-fungal interactions, and developing more effective methods for pollutant removal and detoxification. By integrating these strategies, we can enhance the effectiveness of phytoremediation, contribute to restoring contaminated environments, and address the growing challenges of pollution and environmental degradation. In conclusion, leveraging the natural abilities of endophytic fungi offers a promising and eco-friendly approach to environmental remediation. Continued exploration and innovation in this field are essential for developing sustainable solutions to combat pollution and promote a healthier planet.

7.10 Future Prospects

Future prospects for research into endophytic fungi and their role in phytoremediation hold exciting potential for advancing environmental sustainability and pollution management. As the field evolves, several key directions offer promising opportunities to enhance the efficacy and application of phytoremediation techniques. One major area of interest is the development and optimization of biotechnological strategies. Advances in genetic engineering could play a crucial role in enhancing the capabilities of endophytic fungi. By identifying and manipulating specific genes associated with pollutant degradation and metal sequestration, researchers can potentially create more efficient fungal strains that offer improved performance in diverse environmental conditions. This genetic customization could lead to the creation of transgenic plants that are better equipped to handle heavy metal and organic pollutant contamination, thus enhancing the overall effectiveness of phytoremediation efforts.

Another promising direction involves a deeper understanding of the complex interactions between endophytic fungi, plants, and pollutants. Advanced techniques such as metagenomics, metatranscriptomics, and metaproteomics are crucial for unraveling the intricate dynamics of these interactions. These methods can provide insights into the microbial communities within plant tissues, their functional roles in pollutant degradation, and the mechanisms through which they influence plant health and growth. Such comprehensive analyses can lead to the identification of new fungal strains with superior bioremediation capabilities and the development of tailored strategies for different types of pollutants. The integration of next-generation sequencing and high-throughput functional phenotyping is also expected to drive significant progress. These technologies allow for the detailed exploration of fungal and plant responses to environmental stressors and pollutants, enabling researchers to track and analyze the real-time impact of endophytic fungi on phytoremediation processes. This knowledge can help in refining existing approaches and developing new methodologies for more effective environmental cleanup.

Additionally, future research should focus on the scalability and practical application of endophyte-assisted phytoremediation technologies. This includes exploring cost-effective methods for cultivating and deploying endophytic fungi in large-scale remediation projects, as well as assessing the long-term sustainability and ecological impact of these approaches. Ensuring that these technologies can be implemented effectively in various environmental contexts, from industrial sites to agricultural lands, will be crucial for their widespread adoption. Furthermore, addressing potential challenges related to the stability and efficacy of endophytic fungi in different environmental conditions will be important. Studies investigating the resilience of these fungi to fluctuating environmental factors, such as changes in soil pH, moisture, and pollutant concentrations, will contribute to the development of more robust and adaptable phytoremediation solutions.

In conclusion, the future of endophyte-assisted phytoremediation is poised for significant advancements driven by biotechnological innovations, detailed understanding of plant-microbe interactions, and practical applications of emerging technologies. By leveraging these developments, researchers and practitioners can enhance the effectiveness of phytoremediation, contribute to the restoration of contaminated environments, and support global efforts toward environmental sustainability.

Acknowledgments The authors would like to extend their sincere appreciation to the funding agency Science and Engineering Research Board (SERB), State University Research Excellence (SURE), New Delhi, India. All the authors acknowledge their host institute for infrastructure support.

Conflict of Interest The authors declare no conflict of interest.

Funding This research was supported by the funding agency Science and Engineering Research Board (SERB), State University Research Excellence (SURE) (File Number: SUR/2022/005216). The research was also supported by the Ministry of Education and the State Project Directorate-Rashtriya Uchchatar Shiksha Abhiyaan (SPD-RUSA), Rajasthan, through the RUSA 2.0 project.

Data Availability Not applicable.

References

Adnan M, Xiao B, Xiao P, Zhao P, Li R, Bibi S (2022) Research progress on heavy metals pollution in the soil of smelting sites in China. Toxics 10(5):231

Alam B, Lǐ J, Gě Q, Khan MA, Gōng J, Mehmood S, Yuán Y, Gǒng W (2021) Endophytic fungi: from symbiosis to secondary metabolite communications or vice versa? Front Plant Sci 12:791033

Ameen M, Mahmood A, Sahkoor A, Zia MA, Ullah MS (2024) The role of endophytes to combat abiotic stress in plants. Plant Stress 12:100435

An H, Liu Y, Zhao X, Huang Q, Yuan S, Yang X, Dong J (2015) Characterization of cadmium-resistant endophytic fungi from *Salix variegata* Franch. In three gorges reservoir region, China. Microbiol Res 176:29–37

Babu AG, Shim J, Bang KS, Shea PJ, Oh BT (2014a) *Trichoderma virens* PDR-28: a heavy metal-tolerant and plant growth-promoting fungus for remediation and bioenergy crop production on mine tailing soil. J Environ Manag 132:129–134

Babu AG, Shim J, Shea PJ, Oh BT (2014b) *Penicillium aculeatum* PDR-4 and *Trichoderma* sp. PDR-16 promote phytoremediation of mine tailing soil and bioenergy production with sorghum-sudangrass. Ecol Eng 69:186–191

Bhat SA, Bashir O, Haq SAU, Amin T, Rafiq A, Ali M, Américo-Pinheiro JHP, Sher F (2022) Phytoremediation of heavy metals in soil and water: an eco-friendly, sustainable and multidisciplinary approach. Chemosphere 303:134788

Chandran H, Meena M, Swapnil P (2021) Plant growth-promoting rhizobacteria as a green alternative for sustainable agriculture. Sustainability 13(19):10986. https://doi.org/10.3390/su131910986

Chatterjee S, Mitra A, Datta S, Veer V (2013) Phytoremediation protocols: an overview. In: Plant-based remediation processes. Springer, Heidelberg, pp 1–18

Chen Y, Wang HW, Li L, Dai CC (2013) The potential application of the endophyte *Phomopsis liquidambari* to the ecological remediation of long-term cropping soil. Appl Soil Ecol 67:20–26

Dai CC, Tian LS, Zhao YT, Chen Y, Xie H (2010) Degradation of phenanthrene by the endophytic fungus *Ceratobasidium stevensii* found in *Bischofia polycarpa*. Biodegradation 21:245–255

del Pilar Ortega-Larrocea M, Xoconostle-Cazares B, Maldonado-Mendoza IE, Carrillo-Gonzalez R, Hernández-Hernández J, Garduño MD, Lopez-Meyer M, Gomez-Flores L, González-Chávez MDCA (2010) Plant and fungal biodiversity from metal mine wastes under remediation at Zimapan, Hidalgo, Mexico. Environ Pollut 158(5):1922–1931

Dell'Anno F, Rastelli E, Buschi E, Barone G, Beolchini F, Dell'Anno A (2022) Fungi can be more effective than bacteria for the bioremediation of marine sediments highly contaminated with heavy metals. Microorganisms 10(5):993

Deng Z, Cao L, Huang H, Jiang X, Wang W, Shi Y, Zhang R (2011) Characterization of Cd-and Pb-resistant fungal endophyte *Mucor* sp. CBRF59 isolated from rapes (*Brassica chinensis*) in a metal-contaminated soil. J Hazard Mater 185(2–3):717–724

Deng Z, Wang W, Tan H, Cao L (2012) Characterization of heavy metal-resistant endophytic yeast *Cryptococcus* sp. CBSB78 from rapes (*Brassica chinensis*) and its potential in promoting the growth of *Brassica* spp. in metal-contaminated soils. Water Air Soil Pollut 223:5321–5329

Deng Z, Zhang R, Shi Y, Hu L, Tan H, Cao L (2013) Enhancement of phytoremediation of Cd- and Pb-contaminated soils by self-fusion of protoplasts from endophytic fungus Mucor sp. CBRF59. Chemosphere 91(1):41–47. https://doi.org/10.1016/j.chemosphere.2012.11.065

Deng Z, Zhang R, Shi Y, Hu LA, Tan H, Cao L (2014) Characterization of Cd-, Pb-, Zn-resistant endophytic *Lasiodiplodia* sp. MXSF31 from metal accumulating *Portulaca oleracea* and its potential in promoting the growth of rape in metal-contaminated soils. Environ Sci Pollut Res 21:2346–2357

Dinakarkumar Y, Gnanasekaran R, Reddy GK, Vasu V, Balamurugan P, Murali G (2024) Fungal bioremediation: an overview of the mechanisms, applications and future perspectives. Environ Chem Ecotoxicol 6:293

Dubey MK, Meena M, Aamir M, Zehra A, Upadhyay RS (2019) Regulation and role of metal ions in secondary metabolites production by microorganisms. In: Singh HB, Gupta VK, Jogaiah S (eds) New and future developments in microbial biotechnology and bioengineering. Elsevier, pp 559–577. https://doi.org/10.1016/B978-0-444-63504-4.00019-0

Encarnação T, Pais AA, Campos MG, Burrows HD (2019) Endocrine disrupting chemicals: impact on human health, wildlife and the environment. Sci Prog 102(1):3–42

Gadd GM (1993) Microbial formation and transformation of organometallic and organometalloid compounds. FEMS Microbiol Rev 11(4):297–316

Gade P, Kalvakolanu DV (2012) Chromatin immunoprecipitation assay as a tool for analyzing transcription factor activity. Transcr Regul Methods Protoc 809:85–104

Garai S, Bhattacharjee C, Sarkar S, Moulick D, Dey S, Jana S, Dhar A, Roy A, Mondal K, Mondal M, Mukherjee S (2024) Microplastics in the soil–water–food nexus: inclusive insight into global research findings. Sci Total Environ 173891:173891

Garg N, Aggarwal N (2011) Effects of interactions between cadmium and lead on growth, nitrogen fixation, phytochelatin, and glutathione production in mycorrhizal *Cajanus cajan* (L.). Millsp. J Plant Growth Regul 30:286–300

Jiang M, Cao L, Zhang RJLIAM (2008) Effects of acacia (*Acacia auriculaeformis* A. Cunn)-associated fungi on mustard (*Brassica juncea* (L.) Coss. var. foliosa Bailey) growth in Cd-and Ni-contaminated soils. Lett Appl Microbiol 47(6):561–565

Kafle A, Timilsina A, Gautam A, Adhikari K, Bhattarai A, Aryal N (2022) Phytoremediation: mechanisms, plant selection and enhancement by natural and synthetic agents. Environ Adv 8:100203

Khan AL, Waqas M, Hussain J, Al-Harrasi A, Lee IJ (2014) Fungal endophyte *Penicillium janthinellum* LK5 can reduce cadmium toxicity in *Solanum lycopersicum* (*Sitiens* and *Rhe*). Biol Fertil Soils 50:75–85

Khan AR, Waqas M, Ullah I, Khan AL, Khan MA, Lee IJ, Shin JH (2017a) Culturable endophytic fungal diversity in the cadmium hyperaccumulator *Solanum nigrum* L. and their role in enhancing phytoremediation. Environ Exp Bot 135:126–135

Khan AR, Ullah I, Waqas M, Park GS, Khan AL, Hong SJ, Ullah R, Jung BK, Park CE, Ur-Rehman S, Lee IJ (2017b) Host plant growth promotion and cadmium detoxification in *Solanum nigrum*, mediated by endophytic fungi. Ecotoxicol Environ Saf 136:180–188

Khan AU, Khan AN, Waris A, Ilyas M, Zamel D (2022) Phytoremediation of pollutants from wastewater: a concise review. Open Life Sci 17(1):488–496

Kidd P, Barceló J, Bernal MP, Navari-Izzo F, Poschenrieder C, Shilev S, Clemente R, Monterroso C (2009) Trace element behaviour at the root–soil interface: implications in phytoremediation. Environ Exp Bot 67(1):243–259

Krupa P, Kozdrój J (2007) Ectomycorrhizal fungi and associated bacteria provide protection against heavy metals in inoculated pine (*Pinus sylvestris* L.) seedlings. Water Air Soil Pollut 182:83–90

Li HY, Li DW, He CM, Zhou ZP, Mei T, Xu HM (2012) Diversity and heavy metal tolerance of endophytic fungi from six dominant plant species in a Pb–Zn mine wasteland in China. Fungal Ecol 5(3):309–315

Li X, Li W, Chu L, White JF Jr, Xiong Z, Li H (2016) Diversity and heavy metal tolerance of endophytic fungi from *Dysphania ambrosioides*, a hyperaccumulator from Pb–Zn contaminated soils. J Plant Interact 11(1):186–192

Li Q, Liu J, Gadd GM (2020) Fungal bioremediation of soil co-contaminated with petroleum hydrocarbons and toxic metals. Appl Microbiol Biotechnol 104(21):8999–9008

Liang X, Gadd GM (2017) Metal and metalloid biorecovery using fungi. Microb Biotechnol 10(5):1199–1205

Manisalidis I, Stavropoulou E, Stavropoulos A, Bezirtzoglou E (2020) Environmental and health impacts of air pollution: a review. Front Public Health 8:14

Matchado MS, Lauber M, Reitmeier S, Kacprowski T, Baumbach J, Haller D, List M (2021) Network analysis methods for studying microbial communities: a mini review. Comput Struct Biotechnol J 19:2687–2698

Mitra S, Chakraborty AJ, Tareq AM, Emran TB, Nainu F, Khusro A, Idris AM, Khandaker MU, Osman H, Alhumaydhi FA, Simal-Gandara J (2022) Impact of heavy metals on the environment and human health: novel therapeutic insights to counter the toxicity. J King Saud Univ Sci 34(3):101865

Monnet F, Vaillant N, Hitmi A, Sallanon H (2005) Photosynthetic activity of *Lolium perenne* as a function of endophyte status and zinc nutrition. Funct Plant Biol 32(2):131–139

Morales-Vargas AT, López-Ramírez V, Álvarez-Mejía C, Vázquez-Martínez J (2024) Endophytic fungi for crops adaptation to abiotic stresses. Microorganisms 12(7):1357

Mushtaq S, Shafiq M, Tariq MR, Sami A, Nawaz-ul-Rehman MS, Bhatti MHT, Haider MS, Sadiq S, Abbas MT, Hussain M, Shahid MA (2023) Interaction between bacterial endophytes and host plants. Front Plant Sci 13:1092105

Nandy S, Das T, Tudu CK, Pandey DK, Dey A, Ray P (2020) Fungal endophytes: futuristic tool in recent research area of phytoremediation. S Afr J Bot 134:285–295

Naranjo-Briceño L, Pernía B, Guerra M, Demey JR, De Sisto Á, Inojosa Y, González M, Fusella E, Freites M, Yegres F (2013) Potential role of oxidative exoenzymes of the extremophilic fungus *Pestalotiopsis palmarum* BM-04 in biotransformation of extra-heavy crude oil. Microb Biotechnol 6(6):720–730

Oelmüller R, Sherameti I, Tripathi S, Varma A (2009) *Piriformospora indica*, a cultivable root endophyte with multiple biotechnological applications. Symbiosis 49:1–17

Ojuederie OB, Babalola OO (2017) Microbial and plant-assisted bioremediation of heavy metal polluted environments: a review. Int J Environ Res Public Health 14(12):1504

Osman AI, Hosny M, Eltaweil AS, Omar S, Elgarahy AM, Farghali M, Yap PS, Wu YS, Nagandran S, Batumalaie K, Gopinath SC (2023) Microplastic sources, formation, toxicity and remediation: a review. Environ Chem Lett 21(4):2129–2169

Oubohssaine M, Dahmani I (2024) Phytoremediation: harnessing plant power and innovative technologies for effective soil remediation. Plant Stress 14:100578

Pierart A, Shahid M, Séjalon-Delmas N, Dumat C (2015) Antimony bioavailability: knowledge and research perspectives for sustainable agricultures. J Hazard Mater 289:219–234

Priya AK, Gnanasekaran L, Dutta K, Rajendran S, Balakrishnan D, Soto-Moscoso M (2022) Biosorption of heavy metals by microorganisms: evaluation of different underlying mechanisms. Chemosphere 307:135957

Rashid MI, Mujawar LH, Shahzad T, Almeelbi T, Ismail IM, Oves M (2016) Bacteria and fungi can contribute to nutrients bioavailability and aggregate formation in degraded soils. Microbiol Res 183:26–41

Schäffner A, Messner B, Langebartels C, Sandermann H (2002) Genes and enzymes for in-planta phytoremediation of air, water and soil. Acta Biotechnol 22(1–2):141–151

Sharma P, Kumar S (2021) Bioremediation of heavy metals from industrial effluents by endophytes and their metabolic activity: recent advances. Bioresour Technol 339:125589

Shen M, Liu L, Li DW, Zhou WN, Zhou ZP, Zhang CF, Luo YY, Wang HB, Li HY (2013) The effect of endophytic *Peyronellaea* from heavy metal-contaminated and uncontaminated sites on maize growth, heavy metal absorption and accumulation. Fungal Ecol 6(6):539–545

Shetty SS, Deepthi D, Harshitha S, Sonkusare S, Naik PB, Madhyastha H (2023) Environmental pollutants and their effects on human health. Heliyon 9(9):e19496

Singh VK, Kumar A (2023) Secondary metabolites from endophytic fungi: production, methods of analysis, and diverse pharmaceutical potential. Symbiosis 90(2):111–125

Singh BJ, Chakraborty A, Sehgal R (2023) A systematic review of industrial wastewater management: evaluating challenges and enablers. J Environ Manag 348:119230

Soleimani M, Afyuni M, Hajabbasi MA, Nourbakhsh F, Sabzalian MR, Christensen JH (2010a) Phytoremediation of an aged petroleum contaminated soil using endophyte infected and non-infected grasses. Chemosphere 81(9):1084–1090

Soleimani M, Hajabbasi MA, Afyuni M, Mirlohi A, Borggaard OK, Holm PE (2010b) Effect of endophytic fungi on cadmium tolerance and bioaccumulation by *Festuca arundinacea* and *Festuca pratensis*. Int J Phytoremediation 12(6):535–549

Tang H, Xiang G, Xiao W, Yang Z, Zhao B (2024) Microbial mediated remediation of heavy metals toxicity: mechanisms and future prospects. Front Plant Sci 15:1420408

Tchounwou PB, Yedjou CG, Patlolla AK, Sutton DJ (2012) Heavy metal toxicity and the environment. In: Molecular, clinical and environmental toxicology: volume 3: environmental toxicology. Springer, pp 133–164

Tong J, Miaowen C, Juhui J, Jinxian L, Baofeng C (2017) Endophytic fungi and soil microbial community characteristics over different years of phytoremediation in a copper tailings dam of Shanxi. China Sci Total Environ 574:881–888

Trotta A, Falaschi P, Cornara L, Minganti V, Fusconi A, Drava G, Berta G (2006) Arbuscular mycorrhizae increase the arsenic translocation factor in the as hyperaccumulating fern *Pteris vittata* L. Chemosphere 65(1):74–81

Van Hecke MM, Treonis AM, Kaufman JR (2005) How does the fungal endophyte Neotyphodium coenophialum affect tall fescue (*Festuca arundinacea*) rhizodeposition and soil microorganisms? Plant Soil 275:101–109

Verma A, Shameem N, Jatav HS, Sathyanarayana E, Parray JA, Poczai P, Sayyed RZ (2022) Fungal endophytes to combat biotic and abiotic stresses for climate-smart and sustainable agriculture. Front Plant Sci 13:953836

Wang J, Zhao FJ, Meharg AA, Raab A, Feldmann J, McGrath SP (2002) Mechanisms of arsenic hyperaccumulation in *Pteris vittata*. Uptake kinetics, interactions with phosphate, and arsenic speciation. Plant Physiol 130(3):1552–1561

Wang W, Deng Z, Tan H, Cao L (2013) Effects of Cd, Pb, Zn, Cu-resistant endophytic *Enterobacter* sp. CBSB1 and *Rhodotorula* sp. CBSB79 on the growth and phytoextraction of *Brassica* plants in multimetal contaminated soils. Int J Phytoremediation 15(5):488–497

Wilson SC, Lockwood PV, Ashley PM, Tighe M (2010) The chemistry and behaviour of antimony in the soil environment with comparisons to arsenic: a critical review. Environ Pollut 158(5):1169–1181

Wu W, Chen W, Liu S, Wu J, Zhu Y, Qin L, Zhu B (2021) Beneficial relationships between endophytic bacteria and medicinal plants. Front Plant Sci 12:646146

Xiao X, Luo S, Zeng G, Wei W, Wan Y, Chen L, Guo H, Cao Z, Yang L, Chen J, Xi Q (2010) Biosorption of cadmium by endophytic fungus (EF) *Microsphaeropsis* sp. LSE10 isolated from cadmium hyperaccumulator *Solanum nigrum* L. Bioresour Technol 101(6):1668–1674

Xiao Y, Dai CC, Wang XX, Liu FY, Wang HW, Li XG (2014) Effect of the endophyte *Ceratobasidium stevensii* on 4-HBA degradation and watermelon seed germination. Afr J Microbiol Res 8:1535–1543

Yadav G, Meena M (2021) Bioprospecting of endophytes in medicinal plants of Thar Desert: an attractive resource for biopharmaceuticals. Biotechnol Rep 30:e00629. https://doi.org/10.1016/j.btre.2021.e00629

Yadav G, Meena M (2022) Biological control of plant diseases by endophytes. In: Shah MP, Deka D (eds) Endophytic association: what, why and how (volume: developments in applied microbiology and biotechnology). Academic Press, Elsevier Inc., pp 119–135. https://doi.org/10.1016/B978-0-323-91245-7.00007-9

Yang XE, Wu X, Hao HL, He ZL (2008) Mechanisms and assessment of water eutrophication. J Zhejiang Univ Sci B 9:197–209

Yin L, Ren A, Wei M, Wu L, Zhou Y, Li X, Gao Y (2014) Neotyphodium coenophialum-infected tall fescue and its potential application in the phytoremediation of saline soils. Int J Phytoremediation 16(3):235–246

Zhang Y, Zhang Y, Liu M, Shi X, Zhao Z (2008) Dark septate endophyte (DSE) fungi isolated from metal polluted soils: their taxonomic position, tolerance, and accumulation of heavy metals in vitro. J Microbiol 46:624–632

Ziani K, Ioniță-Mîndrican CB, Mititelu M, Neacșu SM, Negrei C, Moroșan E, Drăgănescu D, Preda OT (2023) Microplastics: a real global threat for environment and food safety: a state of the art review. Nutrients 15(3):617

Endophytic Fungi: Enhancing Sustainable Agriculture Through Biostimulation

8

Priyanshu Sao, Poonam Verma, and Amadou Hamadoun Dicko

Abstract

Endophytic fungi are present in most, if not all, of the plant species on Earth. By means of this relationship, they settle inside plant tissues without exhibiting any symptoms, which benefits the host plant as well as themselves. In recent years, the issue of creating more sustainable agriculture has drawn attention from all around the world. To better understand ecology and investigate the possible function of entophytic interactions in plant development, research has been conducted. The benefits of endophytic colonisation have been extensively studied up to now. This review gathers these studies based on the direct (acquired nutrients and phytohormone production) and indirect (induced resistance, antibiotic and secondary metabolite production, siderophore production, and defence against biotic and abiotic stresses) benefits of entophytic colonisation.

Keywords

Endophytic fungi · Biocontrol · Biofertilisation · Plant growth · Sustainable agriculture · Endophytic colonisation · Plant growth · Worldwide

P. Sao
Department of Biotechnology and Microbiology, Faculty of Sciences, ISBM University, Gariyaband, Chhattisgarh, India

P. Verma (✉)
Department of Biotechnology and Microbiology, Faculty of Sciences, ISBM University, Gariyaband, Chhattisgarh, India

A. H. Dicko
Faculty of Agronomy and Animal Health, University of Segou, Campus of Sebougou, Segou, Mali

M. Meena et al. (eds.), *Metabolic Sustainability of Endophytes*, Microorganisms for Sustainability 52, https://doi.org/10.1007/978-981-96-4004-1_8

8.1 Introduction

There are several connections between fungi and plants. According to fossil evidence, endophytes and mycorrhizas have been a part of this collaboration ever since plants first appeared on Earth more than 400 million years ago. This shows the significance of the group in the evolution of this process (Rodriguez et al. 2009; Rai et al. 2014; Anjum et al. 2019). Although the positive features of this relationship have long been discussed and acknowledged, in-depth research has only just assessed the real advantage offered by these fungi (Busby et al. 2016; Card et al. 2016; Vega 2018; Quesada-Moraga 2020). Though several aspects of the fungal endophytic connection still need further research, fortunately, this topic is advancing (Aly et al. 2011; Chadha et al. 2015; Khan et al. 2015). Modern scientific study has been mostly motivated by the growing realisation of the necessity for more sustainable agriculture. Enhancing sustainable agriculture should contribute to the prevention and reduction of future environmental damage (Jaber and Enkerli 2017; Baron et al. 2020). Examples of these, cited by Van Lexmond et al. (2015), include fish, crustaceans, zooplankton in the aquatic environment, cattle, birds, and beneficial microbes in the soil. Moreover, overuse of fertiliser leads to greenhouse gas emissions, soil acidification, aquifer and water reservoir contamination, heavy metal accumulation, and eutrophication of rivers and lakes (Savci 2012; Kulkarni and Goswami 2019; Yadav and Meena 2022).

Endophytes are kept apart from other plant tissues including leaves, stems, roots, and seeds. They synthesise a wide variety of enzymes, such as chitinases, cellulases, β-1,3-glucanases, pectinases, glucanases, and proteases (Rajulu et al. 2010; Ben Abdallah et al. 2019; Gao et al. 2010). Enzymes of this kind can destroy the cell membrane of infections or halt the germination of spores. This approach protects the host from biotic stress and effectively stops phytopathogen development. Of the 45 endophytic bacteria Zhu and She isolated from the *Ammodendron bifolium* plant, 40% showed significant activity for the synthesis of amylase and cellulose, while 13.3% and 53.3% of the bacteria, respectively, displayed lipase and protease activity (Ying Zhu and Associates, November 2017). An endophytic strain of *Actinoplanes missouriensis* that secretes chitinase inhibited the destruction of cell walls, plasmolysis, and hyphae of the pathogen *Plectosporium tabacinum* (El-Tarabily 2003). Streptomyces produced several lytic enzymes in cocoa witches' broom disease that acted as antagonists to *M. perniciosa* (Macagnan et al. 2008).

Learning about the symbiotic relationship between plants and soil microbiota, as well as the synergistic mode of action that represents a positive interaction, has greatly aided the search for alternative processes that could be used to lessen or even replace the application of pesticides and fertilizers in order to develop more sustainable agriculture (Carneiro et al. 2015; Ahmad et al. 2018). Beyond the potential use of inundate inoculation of microorganisms in crops as conventional biological control agents (BCAs), research has been conducted in this sector. *Beauveria bassiana* and *Metarhizium anisopliae* are the most well-known and frequently used entomopathogenic fungi in biological control programs to manage arthropods that may behave as pests or disease vectors, for example (Baron et al. 2019; Quesada-Moraga 2020).

Still, as other research has demonstrated, this fungus and other recognised entomopathogens may collaborate directly with plants to provide protection. In this case, fungi can compete with the pathogens for resources (space, carbon, nitrogen, and minerals), produce metabolites (antibiotics, volatile compounds, and enzymes), induce systemic resistance in the plant, and promote plant growth as antagonists of plant pathogens. This reduces the pathogen activity, as claimed by Vega et al. (2009), Vidal and Jaber (2015), Vega (2018), Jaber (2018), and Quesada-Moraga (2020). This research primarily aims to investigate the endophytic fungal communities of various plant species, particularly those with agronomic importance.

Endophytes research not only reveals fascinating aspects of the ecology and the interactions between these microbes and plants, but it also facilitates understanding of the potential advantages of these interactions and the elements that need more research to create sustainable human practices, particularly in agriculture. Well-characterised endophytic fungi are more likely to be used as biostimulants, biofertilisers, and biocontrol agents. This paper will investigate features of the endophytic lifestyle and the state of scientific investigation on the possible application of endophytic organisms for crop development and commercialisation as agricultural bioproducts.

8.2 Fungal Endophytes: Definition, Classification, Distribution, and Biodiversity

Many definitions of the word "endophytic" have been proposed, and disagreements over them have been intense. Once introduced by De Bary in 1886, it was widely used to characterise any organism that resides inside (éndon) of a plant (phyton). As reported by Hyde and Soytong (2008), Rodriguez et al. (2009), Hardoim et al. (2015), and other sources, the term "endophytic" now refers to microorganisms that reside inside plant tissues and do not cause disease symptoms in their host. These microorganisms include bacteria, fungi, viruses, protozoa, and even microalgae. Given that some organisations feel the term "endophyte" should only be used to describe habitat rather than function, scientists are currently arguing it in considerable depth (Hardoim et al. 2015). Claiming that endophytic bacteria are those that offer some benefit to their host and that saprophytes or latent or dormant pathogens should not be categorised as "endophytic," some groups even propose renaming the organisms as "mutualistic endophytes" (Card et al. 2016). According to a more recent definition put out by Le Cocq et al. (2017), endophytes are microorganisms that spend at least part of their life cycles inside plant tissues and never cause damage to the host plant. This suggests that fungi currently classified as endophytes, but that may pose a threat to a plant host in the future, should have their classification changed. Balanced host resistance and the activation of virulence mechanisms for colonisation are necessary to identify the plant as a host, so endophytic partnerships are typified by this kind of conflict (Schulz et al. 2015). As long as this relationship is balanced, the fungus helps the host plant in return for nutrients, such as resistance to biotic and abiotic challenges (Bamisile et al. 2018). Hyde and Soytong (2008) classify endophytic fungi as either nonclavicipitaceous fungi, which are not related to grasses, or clavicipitaceous fungi, which are often connected to grasses. In 2009, Rodriguez et al. divided fungus, both clavicipitaceous and nonclavicipitaceous, into four groups (Table 8.1). All clavicipitaceous fungi found in the roots or aerial portion of their hosts, and that specifically colonise grasses, are included in Class 1. Nonclavicipitaceous fungi are limited to Classes 2, 3, and 4, yet they spread in all directions.

Class 2 endophytes can colonise the roots as well as the aerial portion and can proliferate both vertically and horizontally. Class 3 includes endophytes, a very varied group that is only spread horizontally and is often observed on the leaves of tropical tree species. Not insignificantly, Class 4 consists of the dark septate endophytes of fungi. Melanin in their septa allows these endophytes to exclusively propagate horizontally through their hosts' roots. In light of their capacity to colonise and infect insects and to develop into symptomless endophytes on a broad range of plant species, endophytic entomopathogenic fungi should be classified separately, according to recent research by Lugtenberg et al. (2016).

Table 8.1 Summary of characteristics of endophytic fungi classes according to Rodriguez et al. (2009)

Class of endophytes	Main fungal genera	Common host/ colonised tissues	Transmission	Reference
Class 1	Epichloe, Metarhizium, Claviceps and others	Grasses/ shoot and roots	Horizontally and vertically	Faeth and Saari (2012) and Behie and Bidochka (2014)
Class 2	Phylum Ascomycota Penicillium, Aspergillus, Fusarium, Colletotrichum, Trichoderma, Beauveria, Purpureocillium, and others, Phylum Basidiomycota Xylaria spp.	Great host range/roots, stem, and leaves	Horizontally and vertically	Rodriguez et al. (2009), Hiruma et al. (2018), Waqas et al. (2012), Khan et al. (2008), Dash et al. (2018), and Lopez and Sword (2015)
Class 3	Sobreposition with Class 2 endophytes in many cases. It depends on the host, location of infection in the plant, and mode of transmission	Mainly tropical trees/leaves	Horizontally only	Rodriguez et al. (2009)
Class 4	Curvularia, Alternaria, Phialocephala, Deschlera, Ophiosphaerella, Cladosporium, and others	Great host range/roots	Horizontally only	Rodriguez et al. (2009), Hamayun et al. (2010), and Spagnoletti et al. (2017)

Endophytic fungi should be classified based on their function, variety, and ecology, according to Bamisile et al. (2018), who gathered information from multiple investigations. The reproductive strategy utilised by spores determines whether they are sexual or asexual. Moreover, according to the transmission method, they are transmitted either vertically or horizontally. When an endophyte produces vegetative propagules or spores, and these are spread to the plant population by air or another vector, horizontal transmission takes place. By means of seeds, the fungus is transmitted vertically to the progeny of plants (Gimenez et al. 2007; Aly et al. 2011; Lugtenberg et al. 2016). Folar or root endophytes are the two types of fungal endophytes; symptomatic or asymptomatic infections are determined by the manner the infection manifests itself. Necrotrophic and biotrophic are the two types into which they can be separated (Kemen and Jones 2012). While biotrophic endophytic fungi feed on living tissues, necrotrophic fungi encourage the emergence of necrosis from dead tissues. The connection between grasses and clavicipitaceous fungus (Class 1) has been studied in more detail, particularly with regard to the *Epichlöe* genus (Hyde and Soytong 2008; Rodriguez et al. 2009; Card et al. 2014; Zhang et al. 2015; Lugtenberg et al. 2016; Chitnis et al. 2020). According to Card et al. (2016), the primary advantage these fungi provide to plants is the generation of secondary metabolites, primarily alkaloids, which build up in plant tissues and exhibit bioactivity against a variety of vertebrates, invertebrates, and other diseases, including fungi. Furthermore, resistance to abiotic stresses can be offered by these metabolites. According to Johnson et al. (2013) and Finch et al. (2016), growing these grasses is the current global commercialisation of *Epichloe* species.

More study has been done on Class 2 nonclavicipitaceous fungus, which include several taxonomically distinct species that are members of the subkingdom Dikarya, which comprises the phyla Ascomycota and Basidiomycota. These representatives have an endophytic connection with plants and possible advantages that might be looked at in an agricultural context (Rodriguez et al. 2009). Regarding their diversity and prevalence, a lot is yet unknown. Endophytic fungi have been found in several environments, including scorching deserts, temperate, tropical, and polar conditions (Arnold and Lutzoni 2007; Arnold 2008). Regarding the description of the diversity and distribution of fungal endophytes over large geographic regions, only a few generalisations, such as the one that states endophytic fungal diversity is higher in the tropics than in higher latitudes, can be verified, as reviewed by Chadha et al. (2015) and Lugtenberg et al. (2016). Moreover, tropical environments have higher percentages of endophytic species, which are categorised into fewer classifications.

8.3 Unique Aspects of Endophytic Colonisation and Endophytic Biodiversity Description

The range of endophytic fungi associated with plants can vary greatly depending on the surroundings (Vega et al. 2010). This means that even within the same species of plants, the assemblage of fungal endophytes within their tissues may fluctuate if the physiological condition of each plant individual varies (Aly et al. 2011). Moreover, the features of the fungal endophytic population may be influenced by the age of the plant (Sieber 2007). Though certain endophytes are quite specialised and only found inside a few plants, fungal endophytes can colonise a wide variety of plant species. Furthermore, there can be a specific colonised area of the plant (Aly et al. 2011; Bamisile et al. 2018). According to Aly et al. (2011), compared to more probable antagonistic horizontally transferred fungi, vertically transmitted fungi appear to have more mutualistic plant interactions. It is a complex and little-understood process how the endophyte and the host plant establish a genetic link. Selectively expressed fungal genes produce enzymes and secondary metabolites that support colonisation (Yan et al. 2019). With the endophyte's germinative tube approaching the root, the root's apical dominance is lost, and a hyphal penetration apparatus (aspersorium) is produced, which brings infected hyphae into the root cortex and initiates the colonisation process (Khan et al. 2015). These processes, as claimed by Yan et al. (2019) and Khan et al. (2015), support a balanced activation of plant defence genes. The endoderm, pericycle, xylem, and phloem are among the nearby tissues of the roots and soil that the fungus colonises after penetrating the plant cell wall and reaching the inner cortex.

Many signs show how closely the plant and its fungal endophytes are related. First of all, the plant has no defence mechanism against these endophytes. Second, some endophytes can produce metabolites that the plant needs (Germaine et al. 2004; Kusari et al. 2012). Finally, occasionally vertical transmission via seeds allows the reproductive processes of the fungal and plant hosts to occur concurrently (Aly et al. 2011). The fungal endophytic population of plants is unquestionably a difficult task, one that the scientific community is increasingly favouring. According to Rashmi et al. (2019) it is hard to detect these bacteria in plant tissues, and it is rare to see the hyphae. In the past, the endophytic population has been evaluated by separating fungal species that are found only in sterile plant tissues. As such, standard culture media are employed, and adjustments are made as needed, such as increasing the amount of water in the medium to lessen osmotic stress and encourage the development of exploratory hyphae. Murphy et al. (2018) further suggest including plant extracts into the culture media. Several biotic and abiotic factors can affect the diversity of endophytic fungi in a host plant when using cultivation-dependent techniques. These elements are thought to hinder the description of this diversity (Ribeiro et al. 2018; Rashmi et al. 2019; Chen et al. 2020). Studies have demonstrated the great efficacy of genetic methods in identifying the endophytic population of several plant species. The need for culture has been eliminated from the study of microbiomes by bulk DNA sequencing of plant material. Several endophytic species that were unknown before have been found, thanks to this method (Brader et al. 2017; Deyett and Rolshausen 2020).

8.4 What Functions May Endophytic Fungus Fulfil in Terms of Biocontrol, Biofertilisation, and Stress Resilience?

Moreover, the process of desertification, soil salinity, and water supply is all declining in some parts of the world. The problems are related to deforestation, excessive soil use, and ineffective irrigation methods (Chadha et al. 2015; Lugtenberg et al. 2016). Selected crossbreeding of native plants to produce new plant species is one way to address these problems. Stress tolerance is much increased by the mutualistic relationship between bacteria and plants. Still little is known about the underlying genetic processes controlling this connection (Chadha et al. 2015). found that during the process of domestication, the fungal endophyte population seen in wild plants may undergo significant changes and loss of individuals. Fungus thus suffers from the loss of their safe home, while plants lose out on the chance to have a potential companion that could help them better withstand environmental challenges. See Lugtenberg et al. (2016) for a more thorough examination of the factors contributing to the endophyte reduction in plant breeding. Endophytes are hidden within the plant, where they have access to the nutrients required for fungal survival and development, with less competition from other microorganisms (Fig. 8.1).

In return for a protected habitat, fungi therefore enhance the general health and reproductive success of plants (Khan et al. 2015; Lugtenberg et al. 2016). Direct and indirect effects of endophytic fungi colonising plants differ in a number of ways (Berg 2009). Plant growth promotion is mostly facilitated by the uptake of nutrients and the synthesis of phytohormones. Conversely, growth is seen to be indirectly facilitated by the capacity to resist both biotic and abiotic stresses, such as infection resistance (Hardoim et al. 2015; Souza and dos Santos 2017). With the challenges in plant breeding for different crops and the restricted use of genetic modification in several countries because of the lack of understanding of its effects on human, animal, and environmental well-being, the development of natural alternatives shows great potential for achieving more sustainable agriculture (Chadha et al. 2015). Use of endophytic fungus is a relatively new concept. The several ways that these microorganisms may both directly and indirectly support plant health will be covered in the following parts. Furthermore, its prospective application in agricultural activities will be discussed shortly. A summary of the techniques described below is given in Fig. 8.2. Direct advantages of working with endophytic fungus include increased phytohormone levels in the plant and improved nutrient absorption. These factors are immediately related to higher biomass production, faster root system development, taller plants, heavier reproduction, and yield. These advantages support their classification as biofertilisers (Bamisile et al. 2018).

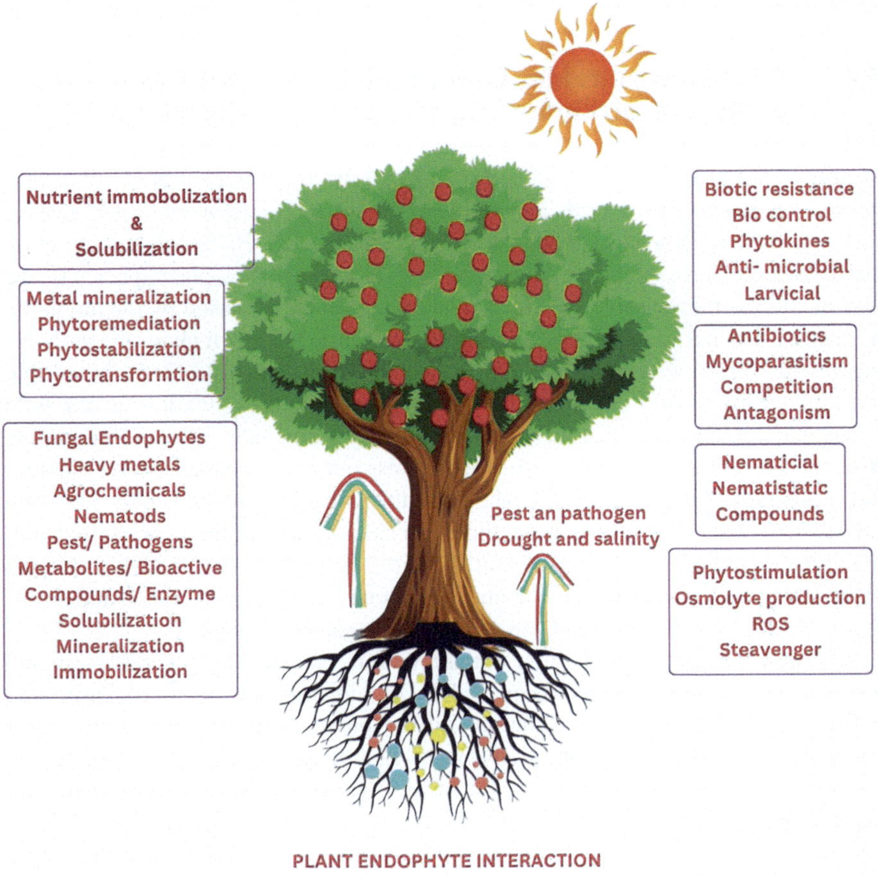

Fig. 8.1 The roles that endophytic fungi play in biocontrol, biofertilisation, and stress tolerance

Fig. 8.2 Benefits of endophytic colonisation by fungi

8.5 Important Considerations Surrounding the Use of Endophytic Fungus in Agriculture

It is common knowledge that some endophytes that promote plant growth may be harmful to people and other animals. Although bacterial strains, such as those from several genera like *Burkholderia*, *Enterobacter*, *Herbaspirillum*, *Ochrobactrum*, *Pseudomonas*, *Ralstonia*, *Staphylococcus*, and *Stenotrophomonas*, have been the focus of most research on the detrimental effects of endophytic microorganisms on humans, the potential of endophytic fungi in this regard is more encouraging (Berg et al. 2005; Mendes et al. 2013). Because the pathogenic bacteria may live in the soil and infect plants of several kingdoms, they can enter plant tissues through tainted manure, irrigation water, seeds, or animals (Mendes et al. 2013).

Studies on the effects of fungi—specifically, *Epichloë* (=*Neotyphodium*) species—that live inside grasses have demonstrated negative effects on vertebrates' health. In plant tissues, these fungi produce and store alkaloids. Many invertebrates and some vertebrates—especially cattle—are poisoned by these compounds (Faeth 2002; Gimenez et al. 2007; Faeth and Saari 2012). Application of these endophytic fungi in agricultural fields has been the subject of much study. Analysing the alkaloid profile generated by every species of hazardous endophytes was the main goal of this work. Finding those species that, in the field, can defend against herbivores—more especially, invertebrates—without endangering vertebrates was the aim. Nevertheless, little study has been done to record how these chemicals affect vertebrates; hence, it is believed that their toxicity to these animals is low (Faeth 2002). Two different strains of *Epichloë festucae* (=*Neotyphodium lolii*) colonised ryegrass that the researchers in Finch et al. (2013) fed to rye cows. Still, the scientists think the amounts of the fungi's alkaloids found in the milk fall into the range that is deemed safe for human consumption. Besides, the animals showed no signs of toxicity. Mycotoxins produced by grass endophytes and other endophytic fungi may be harmful to human health (Chitnis et al. 2020). There is strong motivation to carefully evaluate the use of some bacteria as biological products since they may enter plants endophytically and become dangerous to people and other animals. Whether a disease affects plants or people, Murphy et al. (2018) contend that study is necessary to prevent the selection of any recognised disease.

8.6 The Rapid, Easy, and Potentially Fruitful Use of Endophytes as Bioproducts

The benefits of employing fungal (and other biological) treatments in agriculture are already well known: improved biosafety; reduced risks to human and environmental health; pest specificity (i.e. not affecting beneficial microbes or insects); high efficiency even at low concentrations; controlled multiplication of plants and microbiota; absence of chemical resistance; prevention of pest resistance selection; and application in integrated or conventional pest management systems (Berg 2009). As biological agents, fungi have lately become more and more popular as a long-term substitute for chemical control, particularly in the battle against entomopathogens.

The primary reasons restricting their utilisation are still issues with their field application as well as the low humidity and UV radiation typical of agricultural settings (Vega 2018). Because they live in plants, fungi, in their position as endophytes, are able to go beyond the issues that have historically reduced their worth; furthermore, those fungi that can spread vertically—a feature that has not been much studied—show even more benefits. According to Card et al. (2016), this result implies that, from a commercial standpoint, complicated formulations and field application methods will not be required.

The capacity of endophytes to alter illness depends on a number of variables, including host plants, pathogens, and biotic and abiotic environmental conditions, which raises another modern worry. Endophytes' level of hazard is influenced by pH, temperature, and humidity among other factors. The concurrent interaction of the endophyte and the plant's microbiome may also help to explain why plants react differently to infections (Busby et al. 2016). Plants may react differently to the same illness within a species, and even within a species, a plant's response to the same pathogen may shift based on the host plant and the particular infection.

The second scenario explains how the host plant's defences are intimately linked to the response variability, which influences endophytic invasion in turn. As such, the endophyte used in the vaccination could affect how severe the illness is. However, not much research directly shows context dependence by changing certain variables while keeping others unchanged (Busby et al. 2016). The effective development of the method to use endophytic fungi in agricultural operations depends on a thorough analysis of the effects of these elements. Considering the commercial setting of biological agents, this kind of research is particularly crucial. Growing worldwide knowledge of biological products' use and the fact that more and more people are learning about their composition, benefits, and economic viability have led to an increase in their use. In 2004, the worldwide biocontrol market was predicted to be valued at $588 million (Berg 2009). As per DunhamTrimmer (2020), Latin America is expected to contribute $800 million of the $5 billion in growth expected this year. With sales of more than 120 million USD in 2019 and an even more optimistic market projection for 2020, Brazil has the greatest rate of adoption of organic products worldwide. Products on Brazil's market have quadrupled in number since 2017.

8.7 Conclusions and Outlooks for the Future

While using fungi as biological control agents has been done for a while, its application has recently expanded dramatically. Scientific investigation towards the use of these organisms for this purpose has been driven by worries about the environmental effects of agricultural activities and the demand for improved food options free of harmful chemical substances. A particularly positive step towards attaining the required sustainability in agriculture is the potential of some fungal species, including those frequently used for pest management, to endophytically colonise crops. Probably no artificial substance can offer as many advantageous interactions as these microorganisms can. The numerous direct and indirect advantages that endophytic fungi offer to agricultural plants are confirmed by this study. Endophytic fungi have enormous promise and are valuable as a research and commercial tool when used in place of biocontrol, biostimulation, and biofertilisation.

References

Ahmad M, Pataczek L, Hilger TH, Zahir ZA, Hussain A, Rasche F, Schafleitner R, Solberg SO (2018) Perspectives of microbial inoculation for sustainable development and environmental management. Front Microbiol 9:26. https://doi.org/10.3389/fmicb.2018.02992

Aly AH, Debbab A, Proksch P (2011) Fungal endophytes: unique plant inhabitants with great promises. Appl Microbiol Biotechnol 90(6):1829–1845. https://doi.org/10.1007/s00253-011-3270-y

Anjum R, Afzal M, Baber R, Khan MAJ, Kanwal W, Sajid W, Raheel A (2019) Endophytes: as potential biocontrol agent—review and future prospects. J Agric Sci 11:113

Arnold A (2008) Hidden within our botanical richness, a treasure trove of fungal endophytes. Plant Press 32:13–15

Arnold AE, Lutzoni F (2007) Diversity and host range of foliar fungal endophytes: are tropical leaves biodiversity hotspots? Ecology 88(3):541–549. https://doi.org/10.1890/05-1459

Bamisile BS, Dash CK, Akutse KS, Keppanan R, Wang LD (2018) Fungal endophytes: beyond herbivore management. Front Microbiol 9:11. https://doi.org/10.3389/fmicb.2018.00544

Baron NC, Rigobelo EC, Zied DC (2019) Filamentous fungi in biological control: current status and future perspectives. Chil J Agri Res 79(2):307–315. https://doi.org/10.4067/S0718-58392019000200307

Baron NC, Pollo AD, Rigobelo EC (2020) Purpureocillium lilacinum and metarhizium marquandii as plant growth-promoting fungi. Peerj 8:25. https://doi.org/10.7717/peerj.9005

Behie SW, Bidochka MJ (2014) Ubiquity of insect-derived nitrogen transfer to plants by endophytic insect-pathogenic fungi: an additional branch of the soil Nitrogen cycle. Appl Environ Microbiol 80(5):1553–1560. https://doi.org/10.1128/AEM.03338-13

Ben Abdallah RA, Jabnoun-Khiareddine H, Daami-Remadi M (2019) Secondary metabolites of plant growth promoting Rhizomicroorganisms. In: Exploring the beneficial endophytic microorganisms for plant growth promotion and crop protection: elucidation of some bioactive secondary metabolites involved in both effects. Springer, Singapore, pp 319–352

Berg G (2009) Plant-microbe interactions promoting plant growth and health: perspectives for controlled use of microorganisms in agriculture. Appl Microbiol Biotechnol 84(1):11–18. https://doi.org/10.1007/s00253-009-2092-7

Berg G, Eberl L, Hartmann A (2005) The rhizosphere as a reservoir for opportunistic human pathogenic bacteria. Environ Microbiol 7(11):1673–1685. https://doi.org/10.1111/j.1462-2920.2005.00891.x

Brader G, Compant S, Vescio K, Mitter B, Trognitz F, Ma LJ, Sessitsch A (2017) Ecology and genomic insights into plant-pathogenic and plant-nonpathogenic endophytes. Annu Rev Phytopathol 55(1):61–83. https://doi.org/10.1146/annurevphyto-080516-035641

Busby PE, Ridout M, Newcombe G (2016) Fungal endophytes: modifiers of plant disease. Plant Mol Biol 90(6):645–655. https://doi.org/10.1007/s11103-015-0412-0

Card SD, Faville MJ, Simpson WR, Johnson RD, Voisey CR, Acm DB, Hume DE (2014) Mutualistic fungal endophytes in the Triticeae-survey and description. FEMS Microbiol Ecol 88(1):94–106. https://doi.org/10.1111/1574-6941.12273

Card S, Johnson L, Teasdale S, Caradus J (2016) Deciphering endophyte behaviour: the link between endophyte biology and efficacious biological control agents. FEMS Microbiol Ecol 92(8):19. https://doi.org/10.1093/femsec/fiw114

Carneiro FF, Pignati WA, Rigotto RM, Silva-Augusto LG, Pinheiro ARO, Faria NMX, Friedrich K, Mello MSC (2015) Segurança alimentar e nutricional e saúde. In: Carneiro FF, Augusto LGS, Rigotto RM, Friedrich K, Campos A (eds) Dossiê Abrasco: um alerta sobre os impactos dos agrotóxicos na saúde. Expressão Popular, Rio de Janeiro/São Paulo, pp 46–89

Chadha N, Mishra M, Rajpal K, Bajaj R, Choudhary DK, Varma A (2015) An ecological role of fungal endophytes to ameliorate plants under biotic stress. Arch Microbiol 197(7):869–881. https://doi.org/10.1007/s00203-015-1130-3

Chen WH, Wu SJ, Sun XL, Feng KM, Rahman K, Tan HY et al (2020) High-throughput sequencing analysis of endophytic fungal diversity in Cynanchum sp. S Afr J Bot 134:349–358. https://doi.org/10.1016/j.sajb.2020.04.010

Chitnis VR, Suryanarayanan TS, Nataraja KN, Prasad SR, Oelmuller R, Shaanker RU (2020) Fungal endophyte-mediated crop improvement: the way ahead. Front Plant Sci 11:10. https://doi.org/10.3389/fpls.2020.561007

Dash CK, Bamisile BS, Keppanan R, Qasim M, Lin YW, Ul Islam S, Hussain M, Wang LD (2018) Endophytic entomopathogenic fungi enhance the growth of *Phaseolus vulgaris* L. (*Fabaceae*) and negatively affect the development and reproduction of *Tetranychus urticae* Koch (Acari: Tetranychidae). Microb Pathog 125:385–392. https://doi.org/10.1016/j.micpath.2018.09.044

De Bary A (1886) Ueber einige Sclerotinien und Sclero. Botanische Zeitung 44:377–474

Deyett E, Rolshausen PE (2020) Endophytic microbial assemblage in grapevine. FEMS Microbiol Ecol 96(5):fiaa053. https://doi.org/10.1093/femsec/fiaa053

DunhamTrimmer (2020) Global biocontrol market overview: trends, drivers & insights. http://dunhamtrimmer.com/?page_id=58

El-Tarabily KA (2003) An endophytic chitinase-producing isolate of *Actinoplanes missouriensis*, with potential for biological control of root rot of lupin caused by *Plectosporium tabacinum*. Aust J Bot 51:257–266. https://doi.org/10.1071/BT02107

Faeth SH (2002) Are endophytic fungi defensive plant mutualists? Oikos 98(1):25–36. https://doi.org/10.1034/j.1600-0706.2002.980103.x

Faeth SH, Saari S (2012) Fungal grass endophytes and arthropod communities: lessons from plant defence theory and multitrophic interactions. Fungal Ecol 5(3):364–371. https://doi.org/10.1016/j.funeco.2011.09.003

Finch SC, Thom ER, Babu JV, Hawkes AD, Waugh CD (2013) The evaluation of fungal endophyte toxin residues in milk. N Z Vet J 61(1):11–17. https://doi.org/10.1080/00480169.2012.704626

Finch SC, Pennell CGL, Kerby JWF, Cave VM (2016) Mice find endophyte-infected seed of tall fescue unpalatable – implications for the aviation industry. Grass Forage Sci 71(4):659–666. https://doi.org/10.1111/gfs.12203

Gao F, Dai C, Liu X (2010) Mechanisms of fungal endophytes in plant protection against pathogens. Afr J Microbiol Res 4:1346–1351

Germaine K, Keogh E, Garcia-Cabellos G, Borremans B, van der Lelie D, Barac T, Dowling DN (2004) Colonisation of poplar trees by gfp expressing bacterial endophytes. FEMS Microbiol Ecol 48(1):109–118. https://doi.org/10.1016/j.femsec.2003.12.009

Gimenez C, Cabrera R, Reina M, Gonzalez-Coloma A (2007) Fungal endophytes and their role in plant protection. Curr Org Chem 11(8):707–720. https://doi.org/10.2174/138527207780598765

Hamayun M, Khan SA, Khan AL, Rehman G, Kim YH, Iqbal I et al (2010) Gibberellin production and plant growth promotion from pure cultures of Cladosporium sp MH-6 isolated from cucumber (*Cucumis sativus* L.). Mycologia 102(5):989–995. https://doi.org/10.3852/09-261

Hardoim PR, van Overbeek LS, Berg G, Pirttila AM, Compant S, Campisano A, Doring M, Sessitsch A (2015) The hidden world within plants: ecological and evolutionary considerations for defining functioning of microbial endophytes. Microbiol Mol Biol Rev 79(3):293–320. https://doi.org/10.1128/MMBR.00050-14

Hiruma K, Kobae Y, Toju H (2018) Beneficial associations between Brassicaceae plants and fungal endophytes under nutrient-limiting conditions: evolutionary origins and host-symbiont molecular mechanisms. Curr Opin Plant Biol 44:145–154. https://doi.org/10.1016/j.pbi.2018.04.009

Hyde KD, Soytong K (2008) The fungal endophyte dilemma. Fungal Divers 33:163–173

Jaber LR (2018) Seed inoculation with endophytic fungal entomopathogens promotes plant growth and reduces crown and root rot (CRR) caused by Fusarium culmorum in wheat. Planta 248(6):1525–1535. https://doi.org/10.1007/s00425-018-2991-x

Jaber LR, Enkerli J (2017) Fungal entomopathogens as endophytes: can they promote plant growth? Biocontrol Sci Tech 27(1):28–41. https://doi.org/10.1080/09583157.2016.1243227

Johnson LJ, De Bonth AC, Briggs LR, Caradus JR, Finch SC, Fleetwood DJ, Card SD (2013) The exploitation of epichloae endophytes for agricultural benefit. Fungal Divers 60(1):171–188. https://doi.org/10.1007/s13225-013-0239-4

Kemen E, Jones JD (2012) Obligate biotroph parasitism: can we link genomes to lifestyles? Trends Plant Sci 17(8):448–457. https://doi.org/10.1016/j.tplants.2012.04.005

Khan SA, Hamayun M, Yoon H, Kim HY, Suh SJ, Hwang SK, Kim JM, Lee IJ, Choo YS, Yoon UH (2008) Plant growth promotion and penicillium citrinum. BMC Microbiol 8(1):10. https://doi.org/10.1186/1471-2180-8-231

Khan AL, Hussain J, Al-Harrasi A, Al-Rawahi A, Lee IJ (2015) Endophytic fungi: resource for gibberellins and crop abiotic stress resistance. Crit Rev Biotechnol 35(1):62–74. https://doi.org/10.3109/07388551.2013.800018

Kulkarni S, Goswami A (2019, February) Effect of excess fertilizers and nutrients: a review on impact on plants and human population. In: Proceedings of International Conference on Sustainable Computing in Science, Technology and Management (SUSCOM), Amity University Rajasthan, Jaipur-India

Kusari S, Hertweck C, Spiteller M (2012) Chemical ecology of endophytic fungi: origins of secondary metabolites. Chem Biol 19(7):792–798. https://doi.org/10.1016/j.chembiol.2012.06.004

Le Cocq K, Gurr SJ, Hirsch PR, Mauchline TH (2017) Exploitation of endophytes for sustainable agricultural intensification. Mol Plant Pathol 18(3):469–473

Lopez DC, Sword GA (2015) The endophytic fungal entomopathogens *Beauveria bassiana* and *Purpureocillium lilacinum* enhance the growth of cultivated cotton (*Gossypium hirsutum*) and negatively affect survival of the cotton bollworm (*Helicoverpa zea*). Biol Control 89:53–60. https://doi.org/10.1016/j.biocontrol.2015.03.010

Lugtenberg BJJ, Caradus JR, Johnson LJ (2016) Fungal endophytes for sustainable crop production. FEMS Microbiol Ecol 92(12):17. https://doi.org/10.1093/femsec/fiw194

Macagnan D, Romeiro RDS, Pomella AW, Desouza JT (2008) Production of lytic enzymes and siderophores, and inhibition of germination of basidiospores of *Moniliophthora (ex Crinipellis) perniciosa* by phylloplane actinomycetes. Biol Control 47:309–314. https://doi.org/10.1016/j.biocontrol.2008.08.016

Mendes R, Garbeva P, Raaijmakers JM (2013) The rhizosphere microbiome: significance of plant beneficial, plant pathogenic, and human pathogenic microorganisms. FEMS Microbiol Rev 37(5):634–663. https://doi.org/10.1111/1574-6976.12028

Murphy BR, Doohan FM, Hodkinson TR (2018) From concept to commerce: developing a successful fungal endophyte inoculant for agricultural crops. J Fungi 4(1):11. https://doi.org/10.3390/jof4010024

Quesada-Moraga E (2020) Entomopathogenic fungi as endophytes: their broader contribution to IPM and crop production. Biocontrol Sci Technol 30(9):864–877. https://doi.org/10.1080/09583157.2020.1771279

Rai M, Rathod D, Agarkar G, Dar M, Brestic M, Pastore GM, Marostica MR (2014) Fungal growth promotor endophytes: a pragmatic approach towards sustainable food and agriculture. Symbiosis 62(2):63–79. https://doi.org/10.1007/s13199-014-0273-3

Rajulu MBG, Thirunavukkarasu N, Suryanarayanan TS, Ravishankar JP, El Gueddari NE, Moerschbacher BM (2010) Chitinolytic enzymes from endophytic fungi. Fungal Divers 47:43–53. https://doi.org/10.1007/s13225-010-0071-z

Rashmi M, Kushveer JS, Sarma VV (2019) A worldwide list of endophytic fungi with notes on ecology and diversity. Mycosphere 10(1):798–1079. https://doi.org/10.5943/mycosphere/10/1/19

Ribeiro ADS, Polonio JC, Costa AT, Dos Santos CM, Rhoden SA, Azevedo JL, Pamphile JA (2018) Bioprospection of culturable endophytic fungi associated with the ornamental plant pachystachys lutea. Curr Microbiol 75(5):588–596. https://doi.org/10.1007/s00284-017-1421-9

Rodriguez RJ, White JF Jr, Arnold AE, Redman RS (2009) Fungal endophytes: diversity and functional roles. New Phytol 182(2):314–330. https://doi.org/10.1111/j.1469-8137.2009.02773.x

Savci S (2012) An agricultural pollutant: chemical fertilizer. Int J Environ Sci Dev 3(1):73. https://doi.org/10.7763/IJESD.2012.V3.191

Schulz B, Haas S, Junker C, Andrée N, Schobert M (2015) Fungal endophytes are involved in multiple balanced antagonisms. Curr Sci 109(1):39–45

Sieber TN (2007) Endophytic fungi in forest trees: are they mutualists? Fungal Biol Rev 21(2–3):75–89

Souza BD, dos Santos TT (2017) Endophytic fungi in economically important plants: ecological aspects, diversity and potential biotechnological applications. J Bioenerg Food Sci 4(2):113–126. https://doi.org/10.18067/jbfs.v4i2.121

Spagnoletti FN, Tobar NE, Di Pardo AF, Chiocchio VM, Lavado RS (2017) Dark septate endophytes present different potential to solubilize calcium, iron and aluminum phosphates. Appl Soil Ecol 111:25–32. https://doi.org/10.1016/j.apsoil.2016.11.010

Van Lexmond MB, Bonmatin JM, Goulson D, Noome DA (2015) Worldwide integrated assessment on systemic pesticides. Environ Sci Pollut Res 22(1):1–4. https://doi.org/10.1007/s11356-014-3220-1

Vega V (2018) The use of fungal entomopathogens as endophytes in biological control: a review. Mycologia 110(1):4–30. https://doi.org/10.1080/00275514.2017.1418578

Vega FE, Goettel MS, Blackwell M, Chandler D, Jackson MA, Keller S, Koike M, Maniania NK, Monzon A, Ownley BH (2009) Fungal entomopathogens: new insights on their ecology. Fungal Ecol 2(4):149–159. https://doi.org/10.1016/j.funeco.2009.05.001

Vega FE, Simpkins A, Aime MC, Posada F, Peterson SW, Rehner SA, Arnold AE (2010) Fungal endophyte diversity in coffee plants from Colombia, Hawaii, Mexico and Puerto Rico. Fungal Ecol 3(3):122–138. https://doi.org/10.1016/j.funeco.2009.07.002

Vidal S, Jaber LR (2015) Entomopathogenic fungi as endophytes: plant–endophyte–herbivore interactions and prospects for use in biological control. Curr Sci 109(1):46–54

Waqas M, Khan AL, Kamran M, Hamayun M, Kang SM, Kim YH, Lee IJ (2012) Endophytic fungi produce gibberellins and Indoleacetic Acid and promotes host-plant growth during stress. Molecules 17(9):10754–10773. https://doi.org/10.3390/molecules170910754

Yadav G, Meena M (2022) Biological control of plant diseases by endophytes. In: Shah MP, Deka D (eds) Endophytic association: what, why and how (volume: developments in applied microbiology and biotechnology). Academic Press, Elsevier Inc., pp 119–135. https://doi.org/10.1016/B978-0-323-91245-7.00007-9

Yan L, Zhu J, Zhao XX, Shi JL, Jiang CM, Shao DY (2019) Beneficial effects of endophytic fungi colonization on plants. Appl Microbiol Biotechnol 103(8):3327–3340. https://doi.org/10.1007/s00253-019-09713-2

Zhang W, Card S, McGill C, Mace W, Christensen M, Matthew C (2015) The seed of life–investigating epichloë embryo colonisation. In: Proceedings of the 9th international symposium on fungal endophytes of grasses. AgriBio, Centre for AgriBioscience, Melbourne, p 129

Medicinal Plants and Their Silent Endophytic Fungal Allies

9

Nisha Parashar, Muskan Punjabi, Charu Sharma, Payal Chaturvedi, and Mohamed Ahmed Mosa

Abstract

People worldwide encounter various diseases like cancer, diabetes, tuberculosis, and many more. In order to provide assistance and alleviation from these diseases, it is critical to extract new and beneficial compounds. Traditionally, medicinal plants have been used to extract bioactive compounds to cure a wide variety of diseases. The microbial communities of these plants are also capable of producing bioactive secondary metabolites, which are useful for medicinal purposes and also promote growth of the host plant. Due to the significant, economical, and environmentally beneficial contributions of endophytic fungi, researchers are actively working on isolation of endophytic fungi from unexplored plants and elucidating the biological potential of their wide variety of secondary metabolites. Some such fungal endophytes and their bioactive compounds are *Aspergillus awamori* from *Acacia nilotica* with potent antidiabetic property, *Taxomyces andreanae* from *Taxus brevifolia* showing anticancer activity, and *Phomopsis* sp. from *Garcinia* sp. with antituberculosis potential. This chapter provides an overview of the secondary metabolites produced by endophytic fungi and their use in the treatment of human illnesses.

Keywords

Anticancerous · Antidiabetic · Anti-HIV · Endophytes · Medicinal plants · Secondary metabolites

N. Parashar · M. Punjabi · C. Sharma · P. Chaturvedi (✉)
Department of Microbiology and Biotechnology, IIS (Deemed to be University), Jaipur, Rajasthan, India

M. A. Mosa
Nanotechnology & Advanced Nano-Materials Laboratory (NANML), Plant Pathology Research Institute, Agricultural Research Center, Giza, Egypt

M. Meena et al. (eds.), *Metabolic Sustainability of Endophytes*, Microorganisms for Sustainability 52, https://doi.org/10.1007/978-981-96-4004-1_9

9.1 Introduction

People are being threatened by an increasing number of emerging diseases such as cancers, AIDS, diabetes, neurodegenerative diseases, and so on. Therefore, it is essential to search for new and valuable pharmacologically important compounds to provide support and relief in many human ailments (Guo et al. 2008). As a source of medicinal bioactive substances, plants have been used to treat a variety of diseases. It is widely believed that microorganisms naturally interact with plants. Although the presence of living microorganisms within plants has been recognized for quite some time, it was not until the ecological significance of these organisms and their ability to produce metabolites that can alter the host plant's physiology or hold pharmacological value was unveiled that this topic began to garner substantial attention (Mattoo and Nonzom 2021). The majority of microorganisms that inhabit plants play a crucial role in supporting their growth; however, some remain neutral in their effects (Mendes et al. 2013; Philippot et al. 2013). Highlighting the intricate relationship between flora and the microbial world, this synergy not only underscores plants' therapeutic potential but also opens new avenues for exploring how microorganisms contribute to and enhance the medicinal qualities of plants.

A single plant species is believed to host thousands of microbes. Microbes found around the plant or on plant tissue, known as the rhizosphere and phyllosphere microbial residents, are classified as epiphytes. Those inhabiting within plant tissues, such as leaves, roots, or stems, are referred to as endophytes based on their specific location within the plant species (Turner et al. 2013). Many of them possibly produce bioactive molecules that plants can leverage to fight infections, and some of these compounds have been found to be valuable in the development of novel drugs. Endophytes are known to produce hundreds of natural products, such as alkaloids, terpenoids, flavonoids, steroids, and other compounds. Many of these products serve diverse functional roles, including acting as antibiotics, anticancer agents, biological control agents, and other bioactive chemicals (Guo et al. 2008). In addition to their pharmacological potential, endophytes play a significant role in agriculture. They enhance plant health and productivity by producing compounds that protect against pathogens and pests, stimulate systemic defense mechanisms, and promote growth and development. These interactions improve crop resilience, reduce reliance on chemical fertilizers and pesticides, and support sustainable agricultural practices. Studies have demonstrated that endophytes can produce bioactive compounds that enhance stress tolerance in plants, promote nutrient uptake, and increase resistance to environmental stresses such as drought and salinity (Watts et al. 2023).

Given the multifunctional roles of these microbes in producing valuable compounds for both plant defense and potential drug development, the present chapter highlights the latest advancements in research on endophytes. It explores their application in developing new medical compounds and sustainable agricultural practices. Emphasizing the need for further research and clinical trials, this chapter underscores the importance of natural drug discovery and agricultural biotechnology to address global health and agricultural challenges.

9.1.1 Endophytes

The earliest definition of an endophyte was given by De Bary (1866), who stated that "any organism that grows within plant tissues is referred to as an endophyte". The term "endophytic" refers to microorganisms, including bacteria, fungi, viruses, protozoa, and even microalgae, that live inside plant tissues and do not induce signs of illness in their host (Petrini 1991). Endophytes, typically bacteria or fungi, are symbiotic residents that play essential roles in plant health, growth, stress tolerance, and disease resistance.

9.1.2 Endophytic Fungi

Fungi are a mixotrophic category of organisms with a wide range of life phases and symbiotic relationships with a broad range of autotrophic organisms. Endophytic fungi can produce a wide range of bioactive chemicals, including molecules with pharmacological properties that are equal to or similar to those found in plants. They generate a variety of metabolites belonging to many chemical classes, including phenolic chemicals, alkaloids, flavonoids, steroids, and terpenoids. Fungal endophytes reside entirely within plant tissues and may grow within roots, stems, and/or leaves, emerging to sporulate at plant or host-tissue senescence and play various roles such as to produce secondary metabolites, pest control, resistance to biotic and abiotic stresses.

9.1.3 Types of Endophytic Fungi

Based on phylogeny and life cycle characteristics, endophytic fungi are split into two groups and four classes, Clavicipitaceous (Class 1) endophytes and non-Clavicipitaceous (Classes 2, 3, and 4) endophytes (Rodriguez et al. 2009) (Fig. 9.1).

9.1.3.1 Clavicipitaceous

Clavicipitaceous endophytes or Class 1 endophytes are a tiny group of Clavicipitaceous taxa that are restricted to certain cool- and warm-season grasses and are phylogenetically linked (Jalgaonwala et al. 2011). These endophytes often enhance plant biomass, confer drought tolerance, and produce chemicals that are toxic to animals, thereby reducing herbivory (Clay 1988). They colonize into the shoot and rhizome tissue. Transmission of these endophytes is primarily vertical with maternal plants passing fungi on to offspring via seed infections (Saikkonen et al. 2002). Three categories of Clavicipitaceous endophytes were identified by Clay and Schardl (2002). These categories include mixed interaction and asymptomatic endophytes (Types II and III), as well as symptomatic and pathogenic species (Type I) (Rodriguez et al. 2009).

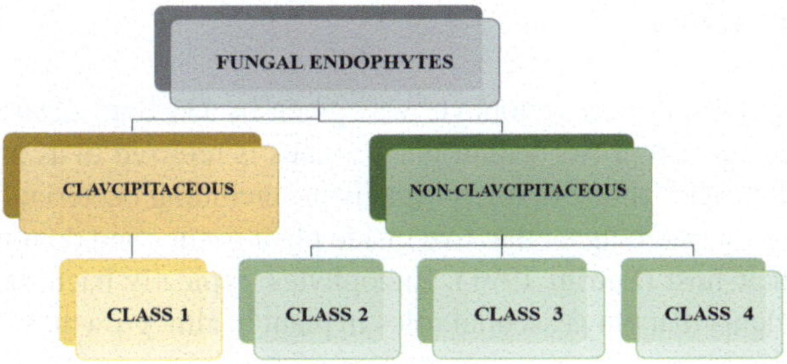

Fig 9.1 Classification of fungal endophytes

9.1.3.2 Non-Clavicipitaceous

Non-Clavicipitaceous is located in asymptomatic tissues of nonvascular plants, such as ferns, allies, conifers, and angiosperms, and is restricted to the Ascomycota or Basidiomycota group. These endophytes are categorized into three functional classes—Classes 2, 3, and 4—based on their host colonization patterns, mechanisms of transmission between host generations, levels of plant biodiversity, and ecological functions. Class 2 endophytes can grow in both aboveground and belowground tissues. They retain broad host range and colonize into the root, shoot, and rhizome tissues of the host plants. Transmission in host is both vertical and horizontal and the diversity of this class is extremely limited. In Class 3, endophytes colonize internal tissue of shoot and have a broad host range. They are transmitted horizontally in host in an individual host plant. The diversity of this class is quite high. Class 4 endophytes are known for their ability to extensively colonize tissues in host plants, primarily in the roots. The host range is broad like the other two non-Clavicipitaceous classes. Transmission of both Class 3 and Class 4 is horizontal in the host tissue (Rodriguez et al. 2009).

9.1.4 Interaction of Fungal Endophytes with Their Host Plants

Endophytes have been identified in a wide variety of plant species across diverse environments, from the Arctic to tropical rainforests (Tan and Zou 2001). The fossil record suggests that the mutualistic relationship between host plants and endophytes has existed since the emergence of higher plants millions of years ago (Gutierrez et al. 2012). These microorganisms have been found in various parts of plants, including leaves, stems, roots, fruits, seeds, and bark (Zhang et al. 2006). Although endophytic fungi are more prevalent than endophytic bacteria, both are common plant-associated microorganisms (Spiteller 2015; Yadav and Meena 2021). With over 300,000 species of higher plants on Earth, it is suggested that each plant species may host a substantial variety of endophytic bacterial and fungal strains, contributing significantly to microbial diversity in plant communities.

The interaction between endophytes and their host plants is believed to be mutually beneficial. Endophytes can enhance their hosts' tolerance to abiotic stressors such as low pH, heavy metals, or temperature variations (Tan and Zou 2001; Jia et al. 2016). Research by Zhang et al. (2006) showed that plants infected with endophytes exhibit improved resilience to environmental stresses compared to those without endophytes. Endophytes promote plant growth and development by producing phytohormones and growth-promoting substances and by improving the absorption of essential nutrients like nitrogen and phosphorus (Tan and Zou 2001; Zhang et al. 2006; Jia et al. 2016; Ludwig-Müller 2015). The colonization patterns of endophytic fungi can vary, with more systematic and extensive colonization occurring in roots rather than aerial parts of plants. The colonization of aerial structures often relies on the host's apoplastic fluid to facilitate the spread of endophytic fungi (Schulz et al. 2002). Endophytic fungal species can exhibit three types of relationships with their host plants depending on the physiological state or specific conditions: beneficial symbiotic or mutualistic endophytes, neutral commensalistic or virulent endophytes, and harmful or virulent pathogens (Kogel et al. 2006).

9.2 Application of Natural Products from Endophytic Fungi

Recent findings have shown that endophytes have the ability to imitate the metabolic processes of plants and generate secondary metabolites or their derivatives. Endophytes have been discovered to have the capacity to excrete a diverse range of bioactive compounds, including alkaloids, flavonoids, steroids, terpenoids, peptides, and polyketides. These secondary metabolites exhibit a wide spectrum of biological activities, making them valuable for various applications in medicine, agriculture, and industry (Fig. 9.2). Certain metabolites have intriguing and pleiotropic pharmacological activity, such as antioxidant, anticancer, antibacterial, and antidiabetic effects (Wen et al. 2022). The capacity to act as antioxidants is one of their key characteristics. A diverse range of chemicals produced by medicinal plants has the ability to scavenge free radicals. These molecules include phenolic acids, flavonoids, quinones, coumarins, lignans, tannins, alkaloids, amines, vitamins, terpenoids, and other naturally occurring compounds. Antioxidant compounds are effective in combating the harmful effects of reactive oxygen species (ROS) and oxygen-derived free radicals and prevent cellular degradation, DNA damage, and cancer development. They also show promise in preventing and treating diseases associated with ROS, such as cancer, hypertension, atherosclerosis, cardiovascular diseases, diabetes mellitus, neurodegenerative diseases, and aging. Pestacin and isopestacin were obtained from the endophytic *Pestalotiopsis microspora* originating from Papua New Guinea. According to Harper et al. (2003), pestacin exhibited greater antioxidant activity than Trolox, which is a derivative of vitamin E. Antioxidant activities have been reported in endophytic fungi isolated from *Eugenia jambolana* (Yadav et al. 2014), *Rhodiola rosea* (Cui et al. 2015), *Dillenia indica* L. (Kumar and Prasher 2024), among others. Although there have been many studies on the antioxidant activities and phenolic contents in plants, no comparative examinations have been conducted on their endophytic communities. Therefore,

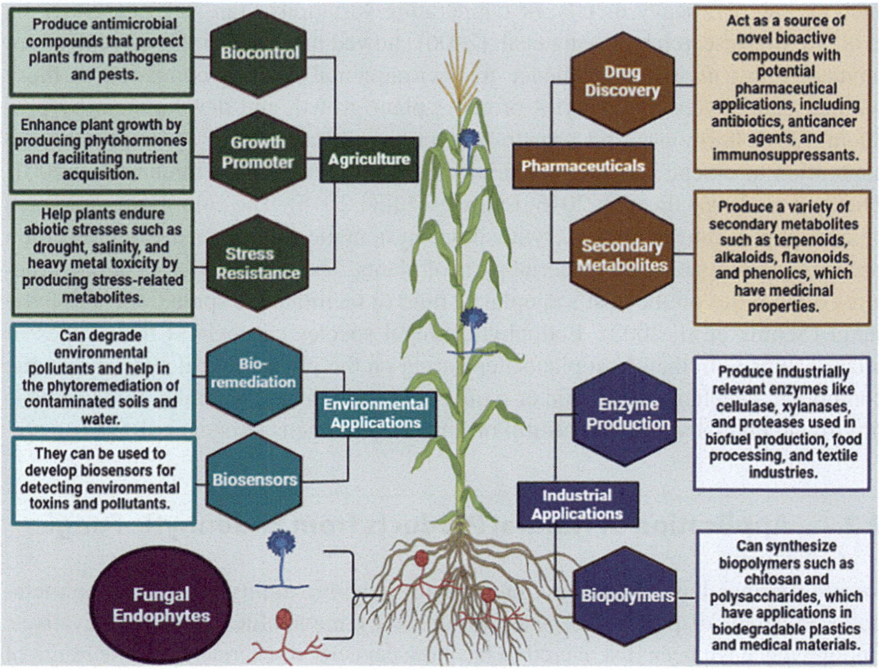

Fig. 9.2 Various applications of fungal endophytes

further research can be conducted to investigate the endophytic antioxidant component.

The role of endophytic metabolites in sustainable agriculture is also particularly noteworthy, as they contribute to plant growth enhancement, disease resistance, and phytoremediation. Consequently, the exploration and utilization of endophytic fungi and their bioactive metabolites offer promising solutions to global health and agricultural challenges. The coming sections of this chapter will deal with the details of pharmacological and agricultural applications, underscoring the importance of continued research and development in this field.

9.2.1 Fungal Endophytes as a Drug Source for Diseases

Increased global health concerns include the advent of antibiotic resistance as well as the emergence of new communicable diseases and serious noncommunicable ailments. Literature suggests that bioactive compounds produced by fungal endophytes have varied functional roles, such as antibiotics, anticancer drugs, biological control agents, and other that can be a solution to these diseases (WHO 2009) (Table 9.1). The role of these endophytes in managing diseases such as diabetes, cancer, viral, bacterial, and fungal diseases caused by pathogenic microorganisms are described in the following sections.

9.2.1.1 Diabetes

Diabetes mellitus is among the most prevalent systemic diseases globally, resulting from the body's resistance to insulin or insufficient insulin production and is prevalent globally. It is regarded as a frequent endocrine dysfunction condition that, by upsetting glucose homeostasis, results in serious consequences such as retinopathy and nephropathy and can be a leading cause of death, particularly when one considers the potentially lethal consequences (Banday et al. 2020).

Through bioactivity-targeted screenings, we can search a vast and highly diversified chemical bank for possible therapeutic compounds. Several plants have been identified as having antidiabetic qualities, and their effectiveness is dependent on their ability to block the enzymes α-amylase and α-glucosidase (Nair et al. 2013). In addition to plants, an endophytic fungus *Aspergillus awamori* isolated from stem and leaves of *Acacia nilotica* has been reported as a rich source of an antidiabetic peptide. This peptide exhibited dual inhibition of α-amylase and α-glucosidase activity (Singh and Kaur 2016). Khan et al. (2019) isolated eight endophytic fungi from leaves of Mango (*Mangifera indica*), Neem (*Azadirachta indica*), and Jambolan (*Syzygium cumini* L.). Aqueous extract of all eight endophytic fungi exhibited inhibition of α-amylase activity between the range of 29.62% and 62%. Yan et al. (2020) discovered that four isobenzofuranone derivatives, derived from fermenting the endophytic fungus *Epicoccum nigrum* SCNU-F0002 isolated from the fresh fruit of the mangrove plant *Acanthus ilicifolius* L., exhibit α-glucosidase inhibitory and antioxidant activities. These fungal endophytes can be used as a feasible and cost-effective source of natural bioactive compounds with potent antidiabetic action.

9.2.1.2 Cancer

Cancer ranks as the second leading cause of death globally and has become a major health concern in the twenty-first century due to its high incidence rate. According to GLOBOCAN 2022, there were 20 million new cancer cases and nearly 10 million cancer-related deaths. Demographic predictions suggest that by 2050, the annual number of new cancer cases will rise to 35 million, marking a 77% increase from the 2022 level (Bray et al. 2024). The rising incidence of cancer can be addressed with promising new natural compounds derived from fungal endophytic flora. These compounds are safe for health, are biocompatible, have fewer toxicity issues, and are less likely to develop resistance compared to conventional anticancer agents. They offer alternative options to chemotherapeutic drugs and potential solutions for cancer treatment (Kousar et al. 2022). Endophytic fungi isolated from *Taxus chinensis* var. *mairei*, *Podophyllum hexandrum*, *Camptotheca acuminate*, and *Catharanthus roseus* produce anticancer drugs like taxol, podophyllotoxin, camptothecin, vinblastine, and vincristine, respectively (Nadeem et al. 2012; Palem et al. 2015; Qiao et al. 2017; Ran et al. 2017). Taxol, the most successful and effective antitumor drug ever discovered, was first reported to be produced from endophytic fungus *Taxomyces andreanae*, isolated from the *Taxus brevifolia* plant (Stierle et al. 1993). In 2017, Kasaei et al. reported a taxol-producing *Cladosporium* sp. from Iranian *Taxus baccata* (yew) plant that showed antimitotic effects on human breast

Table 9.1 Compounds originated from fungal endophytes of medicinal plants and their bioactive properties

Plant	Plant part(s)	Fungal endophyte	Bioactive compound(s)	Biological activity	Reference
Juniperus polycarpos K. Koch var. *seravschanica* (Kom.) Kitam	Fruits	*Penicillium canescens*	1,2,3,5,6-pentahydroxy-8 methylxanthone 1,3,5,6- tetrahydroxy-8-methylxanthone 1,6-dihydroxy-3-methoxy-8-methylxanthone	Inhibitory activity against α-glucosidase	Malik et al. (2020)
Ocimum sanctum	Twigs and leaves	*Alternaria tenuissima, Colletotricum gloesporioides, Diaporthe* sp. and *Trichoderma* sp.	NM	α-glucosidase inhibition	Pavithra et al. (2014)
Combretum dolichopetalum	Leaves	*Nigrospora oryzae*	(S)-(+)-2-cis-4-trans-abscisic acid 7′-hydroxy-abscisic acid 4-des-hydroxyl altersolanol A	Antidiabetic activity	Uzor et al. (2017)
Piper longum L.	Leaves	*Periconia* sp.	Piperine (5-(3, 4-methylenedioxyphenyl)-1-piperidinopent-2, 4-dien-1-one)	Anti-tuberculosis activity	Verma et al. (2011)
Glycyrrhiza glabra	Roots, stems, and leaves	*Fusarium solani*	3,6,9-trihydroxy-7-methoxy-4,4-dimethyl-3,4-dihydro-1H-benzo[g] isochromene-5,10-dione Fusarubin 3-O-methylfusarubin Javanicin	Anti-tuberculosis activity	Shah et al. (2017)
Morinda citrofolia Linn.	Fruits and leaves	*Stemphylium solani, Leptosphaerulina australis*, and *Xylaria* sp.	NM	Anticancer activity	Wu et al. (2015)
Huperzia serrata	Roots, stems, leaves	*Colletotrichum gloeosporioides*	Huperzine A (HupA)	Alzheimer's disease	Shu et al. (2014)

NM Not Mentioned

cancer cell line MCF-7. Ma et al. (2018) suggested that Cytochalasin-H produced from mangrove-derived endophytic fungus *Phomopsis* sp. can be used as a potential chemopreventive drug for the treatment of lung cancer. In 2020, Majoumouo et al. highlighted the anticancer activity and apoptosis-inducing effects of extracts of *Fusarium oxyporum* isolated from medicinal plant *Terminalia catappa* against HeLa cells.

9.2.1.3 Alzheimer's Disease

Alzheimer's disease (AD) is a neurodegenerative disease characterized by progressive loss of cognitive ability such as memory, language, executive, visuospatial function, and substantial behavioral issues (Pilipenko et al. 2019). It is the most common form of dementia and accounts for 60–80% of cases (Duong et al. 2017). Cholinergic dysfunction is one of the primary causes of deterioration of learning and memory processing in AD patients (Fodale et al. 2006). Acetylcholine (Ach) is a cholinergic neurotransmitter that has a significant impact on motor neurons involved in memory formation (Khan et al. 2018), and during the progression of AD, there is a progressive decline in Ach (Banks 2012). Acetylcholinesterase (AChE) is one of the key enzymes in the cholinergic nervous system which hydrolyzes ACh. Progression of AD is also characterized by an increase in AChE (García-Ayllón et al. 2011). Therefore, one of the strategies for treating AD includes inhibiting the activity of this enzyme with particular inhibitors (Sun et al. 2019). Huperzine A (HupA) an anticholinesterase, is one of the most effective drugs for the treatment of AD. *Huperzia serrata* is a traditional Chinese medicinal herb that has been used to cure illness, bruising, cramps, hemoptysis, mental illness, and snakebite for hundreds of years. The endophytic fungus strain *Colletotrichum boninense* HS7-1 was isolated from *H. serrata* that produced HupA (Cui et al. 2021). An endophytic fungus, *Nigrospora oryzae* (GL15), isolated from the medicinal plant *Tinospora cordifolia*, was also reported to show inhibitory and antioxidant action against acetylcholinesterase (AChE) (Vig et al. 2022).

9.2.1.4 AIDS

The human immunodeficiency virus (HIV) continues to be a major public health concern around the world. Highly active antiretroviral therapy (HAART), the current treatment strategy for HIV infection, has reduced AIDS-related deaths, but still it can be an expensive regime for underdeveloped and developing countries where drug supplies are scarce and often not well tolerated, especially in people undergoing long-term treatment. The current medication also has limits in terms of the development of multidrug resistance, necessitating the development of novel anti-HIV chemicals derived from plants as a potential alternative in the fight against HIV (Palshetkar et al. 2020). Fungal endophytes are a potential source of various bioactive compounds that target HIV-1 cell surface receptor attachment proteins. One such protein is chemokine receptor CCR-5, which is the main HIV co-receptor. Yang et al. (2007) reported a CCR-5 inhibitor, Scheme 213766, from an endophytic fungus, *Chaetomium globosum* with a potent in-vitro activity. The other important protein that is inhibited for the treatment of HIV is HIV-1 protease. Vora et al.

(2021) reported the production of mulberroside C from endophyte *Phoma* sp. strain MaF04C isolated from *Morus alba*. The compound inhibited the HIV-1 protease enzyme.

9.2.1.5 Tuberculosis

The frequent use of antibiotics to treat common bacterial infections is leading to a critical issue of antibiotic resistance, indicating that our supply of effective antibiotics is diminishing. As antimicrobial resistance (AMR) grows, it enhances the pathogenicity and virulence of infectious microbes, making them more difficult to treat and control. This situation poses a global health threat due to the slow pace of new drug development. However, research has discovered unique bioactive metabolites in endophytic fungi, with novel chemical structures that could be crucial in combating AMR (Deshmukh et al. 2022).

Mycobacterium tuberculosis, the bacterium responsible for tuberculosis (TB), poses a significant global health threat due to its ability to cause severe and often fatal infections. The urgency for new drugs stems from the rise of drug-resistant TB strains, such as multidrug-resistant TB (MDR-TB) and extensively drug-resistant TB (XDR-TB), which do not respond to conventional treatments. Current TB therapy is lengthy and involves multiple antibiotics, leading to issues with patient adherence and increased risk of resistance. Additionally, the severe side effects of existing medications and the complexity of treating TB in HIV co-infected individuals further highlight the need for better treatment options (Heidary et al. 2022). As a result, natural sources of novel antimycobacterial agents are constantly being searched. Phomoxanthone A and B extracted from the endophytic fungus *Phomopsis* sp. BCC 1323 isolated from teak leaf, *Tectona grandis* L., exhibited action against *M. tuberculosis* (Isaka et al. 2001). In another report on *Phomopsis* sp. PSU-D15 isolated from *Garcinia dulcis* (Roxb.) Kurz., a compound Phomoenamide suppressed *M. tuberculosis* H37Ra (Rukachaisirikul et al. 2008). Growth of *M. tuberculosis* H37Rv was inhibited by tenuazonic acid of *Alternaria alternate* isolated from *Indigofera enneaphylla* L. (Sonaimuthu et al. 2011). Two novel benzopyrans, diaportheone A and B, were reported to be produced by *Diaporthe* sp. P133, isolated from *Pandanus amaryllifolius* leaves, and showed antimycobacterial activity against *M. tuberculosis* H(37)Rv (Bungihan et al. 2011).

9.2.2 Agricultural Applications

Endophytes residing within plants help manage environmental and biological stresses, enhancing plant productivity and sustainability. They achieve this through various methods, such as producing bioactive compounds that alter plant physiology and act as biocontrol agents against pathogens and pests (Mathur et al. 2022a). Additionally, endophytes stimulate plant growth and development by producing phytohormones and siderophores and engaging in nitrogen fixation and enzymatic activity that solubilizes minerals like phosphorus. They also provide natural resistance to soil toxins such as xenobiotics through degrading enzymes and promoting sustainable phytoremediation (Pimentel et al. 2011; Fouda et al. 2015; Lugtenberg et al. 2016). Recent research has highlighted the potential of endophytes in promoting plant growth and advancing sustainable agriculture techniques.

9.2.2.1 Plant Disease Management and Sustainable Agriculture

Endophytes typically establish mutually beneficial relationships with their host plants, providing defense against plant diseases. They offer direct protection by producing and releasing compounds that combat harmful bacteria, and they stimulate systemic host defense mechanisms, enhancing plant growth and development. Fungal endophytes have a diverse ancestry in producing unique bioactive chemicals, which have broad applications including disease and herbivore resistance. These bioactive compounds span various chemical classes, such as alkaloids, flavonoids, peptides, phenols, polyketones, quinols, steroids, and terpenoids (Hardoim et al. 2015). Additionally, scientists have found that certain endophytic fungi produce secondary metabolites that act as biocontrol agents, improving the host plants' ability to withstand diseases. Lahlali and Hijri (2010) identified *Trichoderma atroviridae*, an endophytic fungus, as an effective biocontrol agent against *Rhizoctonia solani* AG3, the pathogen responsible for black scurf in potatoes. Similarly, the biocontrol capacity of endophytic *Trichoderma* sp. against *Fusarium solani*, which causes root rot disease, was evaluated and showed significant suppression of *F. solani* growth and spore germination (Toghueo et al. 2016). Additionally, endophytic fungi *Lasiodiplodia pseudotheobromae*, isolated from medicinal plants in Wuhan Botanical Garden, China, demonstrated biocontrol activity against wheat powdery mildew (*Blumeria graminis* f.sp. *tritici*) (Xiang et al. 2016). Furthermore, secondary metabolites from endophytic fungi isolated from the leaves and petioles of medicinal plants *Vochysia divergens* and *Stryphnodendron adstringens* exhibited biocontrol activity against two citrus pathogens, *Colletotrichum abscissum* and *Phyllosticta citricarpa*, which cause post-bloom fruit drop and citrus black spot disease, respectively. These metabolites directly affected mycelium growth, conidia germination, and symptom development of the pathogenic fungi. Endophytes also inhibited the mycelial growth of *Fusarium graminearum*, the causal organism of fusarium head blight in maize (Iantas et al. 2021). These examples highlight the potential of endophytic fungi to serve as natural and effective biocontrol agents, offering an environmentally friendly alternative to chemical pesticides and advancing sustainable agricultural practices.

9.2.2.2 Fungal Endophytes as Plant Growth Enhancers

Through a variety of processes, endophytes can actively or passively support plant growth. According to Schulz and Boyle (2005), they secrete phytohormones to stimulate the growth of the host plant, although this does not seem to improve the plant's ability to absorb nutrients. Endophytes assist plant growth by engaging in nitrogen fixation, phosphate solubilization, synthesis of indole-3-acetic acid (IAA), and secretion of siderophores (Ahemad and Kibret 2014).

For instance, 15 isolated fungal endophytes from the genera *Penicillium*, *Alternaria*, and *Aspergillus* were obtained from healthy leaves of the medicinal plant *Ephedra pachyclada*. These endophytes exhibited varying abilities to produce ammonia and IAA, as well as phosphate solubilization. They stimulated plant growth, increased biomass production, and improved nutrient uptake in maize (Khalil et al. 2021). Similarly, five medicinal plants—*Anthemis altissima*, *Matricaria parthenium*, *Cichorium intybus*, *Achillea millefolium*, and *Achillea filipendulina*—harbored endophytic fungi that significantly improved seed germination. Notably, among these, *Alternaria consortiale* produced the highest levels of indole-3-acetic acid-like compounds and exhibited a growth-promoting ability on maize (*Zea mays* L.) (Hatamzadeh et al. 2023). Recently, in 2024, Jigjiddorj et al. reported that endophytic fungal strains isolated from the medicinal plant *Cynomorium songaricum* (Rupr.) showed their ability to solubilize complex phosphorus and zinc minerals and produce IAA. These examples highlight the potential of endophytes to be used as biofertilizers to promote plant growth (Pandya et al. 2013).

Additionally, Barazani et al. (2007) found that the presence of the endophytic fungus *Sebacina vermifera* enhanced the growth of *Nicotiana attenuate* by reducing the synthesis of ethylene. The presence of the endophyte *Acremonium strictum* in the roots of *Ammophila arenaria* (Marram grass) led to an increase in plant biomass, achieved by the endophyte's ability to act as an antagonist against root-feeding nematodes, as observed in a study by Hol et al. (2007). Moreover, Mathur et al. (2022b) reported that compounds from a partially purified fraction of *Aspergillus niger* (isolate 29) influenced the super-pathway of gibberellin and isoflavonoid biosynthesis, resulting in improved seed germination and enhanced plant growth in various cash crops. More recently, Mathur et al. (2024) demonstrated that endophytic *Aspergillus niger*, isolated from *Albizia lebbeck*, facilitated growth and enhanced physiological features in *Vigna* species by producing specific bioactive compounds. From a future perspective, endophytic fungi hold great potential in the agricultural sector as eco-friendly biofertilizers, enhancing plant growth performance, boosting plant production, and improving soil quality and fertility.

9.3 Conclusion

The exploration of endophytic fungi associated with medicinal plants has unveiled a wealth of bioactive compounds with significant pharmacological and agricultural potential. These compounds exhibit various bioactivities, including anticancer, antidiabetic, antiviral, antibacterial, antifungal, and antioxidant properties. The research presented in this chapter underscores the crucial role of endophytic fungi in drug discovery and their potential as a source of novel therapeutic agents. Notably, compounds derived from these fungi have shown promising results in combating diseases such as cancer, diabetes, Alzheimer's, HIV, tuberculosis, and other microbial infections. In addition to their pharmacological applications, endophytic fungi have demonstrated considerable potential in agriculture. They contribute to sustainable agriculture by enhancing plant growth, providing natural resistance to diseases, and improving overall plant health. Endophytes such as *Penicillium*, *Aspergillus*, and *Trichoderma* species have shown the ability to produce bioactive compounds that protect plants from harmful pathogens and pests and stimulate plant growth through the production of phytohormones and other beneficial metabolites.

The examples provided highlight the potential of endophytic fungi to yield effective and economically viable alternatives to traditional synthetic drugs and chemical fertilizers, addressing the growing concerns of drug resistance and the need for new treatments. Despite these promising findings, there is still a need for more standardized clinical trials to validate fully the therapeutic potential of these fungal metabolites. Future research should focus on the isolation and characterization of new endophytic fungi from unexplored medicinal plants, as well as the elucidation of their mechanisms of action. This continued effort will not only enhance our understanding of plant–microbe interactions but also contribute to the development of innovative treatments for various human ailments and sustainable agricultural practices. In summary, endophytic fungi represent a valuable and largely untapped resource in the realm of natural product drug discovery and sustainable agriculture. Their ability to produce bioactive compounds offers significant promise for the development of new medicines and agricultural applications, highlighting the importance of preserving and studying the biodiversity of medicinal plants and their microbial allies.

References

Ahemad M, Kibret M (2014) Mechanisms and applications of plant growth promoting rhizobacteria: current perspective. J King Saud Univ Sci 26(1):1–20. https://doi.org/10.1016/j.jksus.2013.05.001

Banday MZ, Sameer AS, Nissar S (2020) Pathophysiology of diabetes: an overview. Avicenna J Med 10(4):174–188. https://doi.org/10.4103/ajm.ajm_53_20

Banks WA (2012) Drug delivery to the brain in Alzheimer's disease: consideration of the blood-brain barrier. Adv Drug Deliv Rev 64:629–639

Barazani O, von Dahl CC, Baldwin IT (2007) *Sebacina vermifera* promotes the growth and fitness of *Nicotiana attenuata* by inhibiting ethylene signaling. Plant Physiol 144(2):1223–1232

Bray F, Laversanne M, Sung H, Ferlay J, Siegel RL, Soerjomataram I, Jemal A (2024) Global cancer statistics 2022: GLOBOCAN estimates of incidence and mortality worldwide for 36 cancers in 185 countries. CA Cancer J Clin 74:229–263. https://doi.org/10.3322/caac.21834

Bungihan ME, Tan MA, Kitajima M, Kogure N, Franzblau SG, Dela Cruz TE, Takayama H, Nonato MG (2011) Bioactive metabolites of *Diaporthe* sp. P133, an endophytic fungus isolated from *Pandanus amaryllifolius*. J Nat Med 65(3–4):606–609. https://doi.org/10.1007/s11418-011-0518-x

Clay K (1988) Fungal endophytes of grasses: a defensive mutualism between plants and fungi. Ecology 69(1):10–16

Clay K, Schardl CL (2002) Evolutionary origins and ecological consequences of endophyte symbiosis with grasses. Am Nat 160(S42):S99–S127

Cui JL, Guo TT, Ren ZX, Zhang NS, Wang ML (2015) Diversity and antioxidant activity of culturable endophytic fungi from alpine plants of *Rhodiola crenulata*, *R. angusta*, and *R. sachalinensis*. PLoS One 10(3):e0118204

Cui L, Noushahi HA, Zhang Y, Liu J, Cosoveanu A, Liu Y, Yan L, Zhang J, Shu S (2021) Endophytic fungal community of *Huperzia serrata*: Diversity and relevance to the production of Huperzine A by the plant host. Molecules 26(4):892

De Bary A (1866) Morphologie und physiologie der pilze, flechten und myxomyceten. W. Engelmann, Leipzig

Deshmukh SK, Dufossé L, Chhipa H, Saxena S, Mahajan GB, Gupta MK (2022) Fungal endophytes: a potential source of antibacterial compounds. J Fungi (Basel) 8(2):164. https://doi.org/10.3390/jof8020164

Duong S, Patel T, Chang F (2017) Dementia: what pharmacists need to know. Can Pharm J (Ott) 150(2):118–129

Fodale V, Quattrone D, Trecroci C, Caminiti V, Santamaria LB (2006) Alzheimer's disease and anaesthesia: implications for the central cholinergic system. Br J Anaesth 97(4):445–452

Fouda AH, Hassan SED, Eid AM, Ewais EED (2015) Biotechnological applications of fungal endophytes associated with medicinal plant *Asclepias sinaica* (Bioss.). Ann Agric Sci 60(1):95–104

García-Ayllón MS, Small DH, Avila J, Sáez-Valero J (2011) Revisiting the role of acetylcholinesterase in Alzheimer's disease: cross-talk with P-tau and β-amyloid. Front Mol Neurosci 4:22. https://doi.org/10.3389/fnmol.2011.00022

Guo B, Wang Y, Sun X, Tang K (2008) Bioactive natural products from endophytes: a review. Appl Biochem Microbiol 44(2):136–142. https://doi.org/10.1134/S0003683808020026

Gutierrez RMP, Gonzalez AMN, Ramirez AM (2012) Compounds derived from endophytes: a review of phytochemistry and pharmacology. Curr Med Chem 19(18):2992–3030

Hardoim PR, van Overbeek LS, Berg G, Pirttilä AM, Compant S, Campisano A, Döring M, Sessitsch A (2015) The hidden world within plants: ecological and evolutionary considerations for defining functioning of microbial endophytes. Microbiol Mol Biol Rev 79(3):293–320

Harper JK, Arif AM, Ford EJ, Strobel GA, Porco JA Jr, Tomer DP, Oneill KL, Heider EM, Grant DM (2003) Pestacin: a 1, 3-dihydro isobenzofuran from *Pestalotiopsis microspora* possessing antioxidant and antimycotic activities. Tetrahedron 59(14):2471–2476

Hatamzadeh S, Rahnama K, White JF, Oghaz NA, Nasrollahnejad S, Hemmati K (2023) Investigation of some endophytic fungi from five medicinal plants with growth promoting ability on maize (*Zea mays* L.). J Appl Microbiol 134(1). https://doi.org/10.1093/jambio/lxac015

Heidary M, Shirani M, Moradi M, Goudarzi M, Pouriran R, Rezaeian T, Khoshnood S (2022) Tuberculosis challenges: resistance, co-infection, diagnosis, and treatment. Eur J Microbiol Immunol (Bp) 12(1):1–17. https://doi.org/10.1556/1886.2021.00021

Hol WG, de la Pena E, Moens M, Cook R (2007) Interaction between a fungal endophyte and root herbivores of *Ammophila arenaria*. Basic Appl Ecol 8(6):500–509

Iantas J, Savi DC, Schibelbein RDS, Noriler SA, Assad BM, Dilarri G, Ferreira H, Rohr J, Thorson JS, Shaaban KA, Glienke C (2021) Endophytes of Brazilian medicinal plants with activity against phytopathogens. Front Microbiol 12:714750

Isaka M, Jaturapat A, Rukseree K, Danwisetkanjana K, Tanticharoen M, Thebtaranonth Y (2001) Phomoxanthones A and B, novel xanthone dimers from the endophytic fungus *Phomopsis* species. J Nat Prod 64(8):1015–1018

Jalgaonwala RE, Mohite BV, Mahajan RT (2011) A review: natural products from plant associated endophytic fungi. J Microbiol Biotechnol Res 1(2):21–32

Jia M, Chen L, Xin HL, Zheng CJ, Rahman K, Han T, Qin LP (2016) A friendly relationship between endophytic fungi and medicinal plants: a systematic review. Front Microbiol 7:906

Jigjiddorj E-A, Maidarjav A, Byambasuren B, Nyamgerel D (2024) Identification, antimicrobial and plant growth promoting activities of endophytic fungi associated with *Cynomorium songaricum* Rupr., a traditional medicinal plant in Mongolia. Diversity 16(2):122

Kasaei A, Mobini-Dehkordi M, Mahjoubi F, Saffar B (2017) Isolation of taxol-producing endophytic fungi from Iranian yew through novel molecular approach and their effects on human breast cancer cell line. Curr Microbiol 74(6):702–709. https://doi.org/10.1007/s00284-017-1231-0

Khalil AMA, Hassan SE-D, Alsharif SM, Eid AM, Ewais EE-D, Azab E, Gobouri AA, Elkelish A, Fouda A (2021) Isolation and characterization of fungal endophytes isolated from medicinal plant *Ephedra pachyclada* as plant growth-promoting. Biomol Ther 11(2):140. https://doi.org/10.3390/biom11020140

Khan H, Marya, Amin S, Kamal MA, Patel S (2018) Flavonoids as acetylcholinesterase inhibitors: current therapeutic standing and future prospects. Biomed Pharmacother 101:860–870. https://doi.org/10.1016/j.biopha.2018.03.007

Khan R, Naqvi STQ, Fatima N, Muhammad SA (2019) Study of antidiabetic activities of endophytic fungi isolated from plants. PAB 8(2):1287–1295

Kogel KH, Franken P, Hückelhoven R (2006) Endophyte or parasite–what decides? Curr Opin Plant Biol 9(4):358–363

Kousar R, Naeem M, Jamaludin MI, Arshad A, Shamsuri AN, Ansari N, Akhtar S, Hazafa A, Uddin J, Khan A, Al-Harrasi A (2022) Exploring the anticancer activities of novel bioactive compounds derived from endophytic fungi: mechanisms of action, current challenges and future perspectives. Am J Cancer Res 12(7):2897–2919

Kumar V, Prasher IB (2024) Phytochemical analysis and antioxidant activity of endophytic fungi isolated from *Dillenia indica* Linn. Appl Biochem Biotechnol 196:332–349. https://doi.org/10.1007/s12010-023-04498-7

Lahlali R, Hijri M (2010) Screening, identification and evaluation of potential biocontrol fungal endophytes against *Rhizoctonia solani* AG3 on potato plants. FEMS Microbiol Lett 311:152–159

Ludwig-Müller J (2015) Plants and endophytes: equal partners in secondary metabolite production? Biotechnol Lett 37:1325–1334

Lugtenberg BJJ, Caradus JR, Johnson LJ (2016) Fungal endophytes for sustainable crop production. FEMS Microbiol Ecol 92(12):fiw194

Ma Y, Wu X, Xiu Z, Liu X, Huang B, Hu L, Liu J, Zhou Z, Tang X (2018) Cytochalasin H isolated from mangrove-derived endophytic fungus induces apoptosis and inhibits migration in lung cancer cells. Oncol Rep 39(6):2899–2905. https://doi.org/10.3892/or.2018.6347

Majoumouo MS, Tincho MB, Kouipou Toghueo RM, Morris T, Hiss DC, Boyom FF, Mandal C (2020) Cytotoxicity potential of endophytic fungi extracts from *Terminalia catappa* against human cervical cancer cells. J Toxicol 2020:8871152. https://doi.org/10.1155/2020/8871152

Malik A, Ardalani H, Anam S, McNair LM, Kromphardt KJK, Frandsen RJN, Franzyk H, Staerk D, Kongstad KT (2020) Antidiabetic xanthones with α-glucosidase inhibitory activities from an endophytic *Penicillium canescens*. Fitoterapia 142:104522. https://doi.org/10.1016/j.fitote.2020.104522

Mathur P, Chaturvedi P, Sharma C, Bhatnagar P (2022a) Improved seed germination and plant growth mediated by compounds synthesized by endophytic *Aspergillus Niger* (isolate 29) isolated from *Albizia lebbeck* (L.) Benth. 3 Biotech 12:271. https://doi.org/10.1007/s13205-022-03332-x

Mathur P, Mehtani P, Sharma C, Bhatnagar P (2022b) Impact of pertinent endophytic fungal communities and their secondary metabolites on crops. In: Singh J, Sharma D (eds) Microbial resource technologies for sustainable development. Elsevier, pp 147–156. https://doi.org/10.1016/B978-0-323-90590-9.00002-X

Mathur P, Agrawal P, Ghosh U, Sharma C, Bhatnagar P, Chaturvedi P (2024) Growth promotion ability of endophytic *Aspergillus Niger* on different species of *Vigna*. Vegetos 37:192–201. https://doi.org/10.1007/s42535-023-00570-z

Mattoo AJ, Nonzom S (2021) Endophytic fungi: understanding complex cross-talks. Symbiosis 83:237–264

Mendes R, Garbeva P, Raaijmakers JM (2013) The rhizosphere microbiome: significance of plant beneficial, plant pathogenic, and human pathogenic microorganisms. FEMS Microbiol Rev 37(5):634–663

Nadeem M, Ram M, Alam P, Ahmad MM, Mohammad A, Al-Qurainy F, Abdin MZ (2012) *Fusarium solani*, P1, a new endophytic podophyllotoxin-producing fungus from roots of *Podophyllum hexandrum*. Afr J Microbiol Res 6(10):2493–2499

Nair SS, Kavrekar V, Mishra A (2013) In vitro studies on alpha amylase and alpha glucosidase inhibitory activities of selected plant extracts. Eur J Exp Biol 3(1):128–132

Palem PP, Kuriakose GC, Jayabaskaran C (2015) An endophytic fungus, *Talaromyces radicus*, isolated from *Catharanthus roseus*, produces vincristine and vinblastine, which induce apoptotic cell death. PLoS One 10(12):e0144476. https://doi.org/10.1371/journal.pone.0144476. eCollection 2015

Palshetkar A, Pathare N, Jadhav N, Pawar M, Wadhwani A, Kulkarni S, Singh KK (2020) In vitro anti-HIV activity of some Indian medicinal plant extracts. BMC Complement Med Ther 20(1):69

Pandya M, Kumar GN, Rajkumar S (2013) Invasion of rhizobial infection thread by non-rhizobia for colonization of *Vigna radiata* root nodules. FEMS Microbiol Lett 348(1):58–65. https://doi.org/10.1111/1574-6968.12245

Pavithra N, Sathish L, Suneel Kumar A, Venkatarathanamma V, Pushpalatha H, Bhanuprakash Reddy G, Ananda K (2014) In-vitro studies on α-Amylase, α-Glucosidase and aldose reductase inhibitors found in endophytic fungi isolated from *Ocimum sanctum*. Curr Enzym Inhib 10(2):129–136. https://doi.org/10.2174/1573408011666141205233252

Petrini O (1991) Fungal endophytes in tree leaves. In: Andrews JH, Hirano SS (eds) Microbial ecology of leaves. Springer, New York, pp 179–197

Philippot L, Raaijmakers JM, Lemanceau P, van der Putten WH (2013) Going back to the roots: the microbial ecology of the rhizosphere. Nat Rev Microbiol 11(11):789–799

Pilipenko V, Narbute K, Pupure J, Rumaks J, Jansone B, Klusa V (2019) Neuroprotective action of diazepam at very low and moderate doses in Alzheimer's disease model rats. Neuropharmacology 144:319–326

Pimentel MR, Molina G, Dionísio AP, Maróstica Junior MR, Pastore GM (2011) The use of endophytes to obtain bioactive compounds and their application in biotransformation process. Biotechnol Res Int 2011(1):576286

Qiao W, Ling F, Yu L, Huang Y, Wang T (2017) Enhancing taxol production in a novel endophytic fungus, *Aspergillus aculeatinus* Tax-6, isolated from *Taxus chinensis* var. *mairei*. Fungal Biol 121(12):1037–1044

Ran X, Zhang G, Li S, Wang J (2017) Characterization and antitumor activity of camptothecin from endophytic fungus *Fusarium solani* isolated from *Camptotheca acuminate*. Afr Health Sci 17(2):566–574

Rodriguez RJ, White JF Jr, Arnold AE, Redman RS (2009) Fungal endophytes: diversity and functional roles. New Phytol 182(2):314–330

Rukachaisirikul V, Sommart U, Phongpaichit S, Sakayaroj J, Kirtikara K (2008) Metabolites from the endophytic fungus *Phomopsis* sp. PSU-D15. Phytochemistry 69(3):783–787

Saikkonen K, Ion D, Gyllenberg M (2002) The persistence of vertically transmitted fungi in grass metapopulations. Proc Royal Soc B 269(1498):1397–1403

Schulz B, Boyle C (2005) The endophytic continuum. Mycol Res 109:661–686

Schulz B, Boyle C, Draeger S, Römmert AK, Krohn K (2002) Endophytic fungi: a source of novel biologically active secondary metabolites. Mycol Res 106(9):996–1004

Shah A, Rather MA, Hassan QP, Aga MA, Mushtaq S, Shah AM, Hussain A, Baba SA, Ahmad Z (2017) Discovery of anti-microbial and anti-tubercular molecules from *Fusarium solani*: an endophyte of *Glycyrrhiza glabra*. J Appl Microbiol 122(5):1168–1176

Shu S, Zhao X, Wang W, Zhang G, Cosoveanu A, Ahn Y, Wang M (2014) Identification of a novel endophytic fungus from *Huperzia serrata* which produces huperzine A. World J Microbiol Biotechnol 30:3101–3109. https://doi.org/10.1007/s11274-014-1737-6

Singh B, Kaur A (2016) Antidiabetic potential of a peptide isolated from an endophytic *Aspergillus awamori*. J Appl Microbiol 120(2):301–311

Sonaimuthu V, Parihar S, Thakur JP, Luqman S, Saikia D, Chanotiya CS, Jhonpaul M, Negi AS (2011) Tenuazonic acid: a promising antitubercular principle from *Alternaria alternate*. Microbiol Res 2(2):e17. https://doi.org/10.4081/mr.2011.e17

Spiteller P (2015) Chemical ecology of fungi. Nat Prod Rep 32(7):971–993

Stierle A, Strobel G, Stierle D (1993) Taxol and taxane production by *Taxomyces andreanae*, an endophytic fungus of Pacific yew. Science 260(5105):214–216. https://doi.org/10.1126/science.8097061

Sun K, Bai Y, Zhao R, Guo Z, Su X, Li P, Yang P (2019) Neuroprotective effects of matrine on scopolamine-induced amnesia via inhibition of AChE/BuChE and oxidative stress. Metab Brain Dis 34(1):173–181

Tan RX, Zou WX (2001) Endophytes: a rich source of functional metabolites. Nat Prod Rep 18(4):448–459

Toghueo RMK, Eke P, Zabalgogeazcoa Í, de Aldana BRV, Nana LW, Boyom FF (2016) Biocontrol and growth enhancement potential of two endophytic *Trichoderma* spp. from *Terminalia catappa* against the causative agent of Common Bean Root Rot (*Fusarium solani*). Biol Control 96:8–20

Turner TR, James EK, Poole PS (2013) The plant microbiome. Genome Biol 14(6):209. https://doi.org/10.1186/gb-2013-14-6-209

Uzor PF, Osadebe PO, Nwodo NJ (2017) Antidiabetic activity of extract and compounds from an endophytic fungus *Nigrospora oryzae*. Drug Res (Stuttg) 67(5):308–311. https://doi.org/10.1055/s-0042-122777

Verma VC, Lobkovsky E, Gange AC, Singh SK, Prakash S (2011) Piperine production by endophytic fungus *Periconia* sp. isolated from *Piper longum* L. J Antibiot 64(6):427–431

Vig R, Bhadra F, Gupta SK, Sairam K, Vasundhara M (2022) Neuroprotective effects of quercetin produced by an endophytic fungus *Nigrospora oryzae* isolated from *Tinospora cordifolia*. J Appl Microbiol 132(1):365–380

Vora J, Velhal S, Sinha S, Patel V, Shrivastava N (2021) Bioactive phytocompound mulberroside C and endophytes of *Morus alba* as potential inhibitors of HIV-1 replication: a mechanistic evaluation. HIV Med 22(8):690–704. https://doi.org/10.1111/hiv.13116

Watts D, Palombo EA, Jaimes Castillo A, Zaferanloo B (2023) Endophytes in agriculture: potential to improve yields and tolerances of agricultural crops. Microorganisms 11(5):1276. https://doi.org/10.3390/microorganisms11051276

Wen J, Okyere SK, Wang S, Wang J, Xie L, Ran Y, Hu Y (2022) Endophytic fungi: an effective alternative source of plant-derived bioactive compounds for pharmacological studies. J Fungi (Basel) 8(2):205. https://doi.org/10.3390/jof8020205

WHO (2009) Global health risks: mortality and burden of disease attributable to selected major risks. World Health Organization

Wu Y, Girmay S, da Silva VM, Perry B, Hu X, Tan GT (2015) The role of endophytic fungi in the anticancer activity of *Morinda citrifolia* Linn. (Noni). Evid Based Complement Alternat Med 2015:393960

Xiang L, Gong S, Yang L, Hao J, Xue M, Zeng F, Zhang X, Shi W, Wang H, Yu D (2016) Biocontrol potential of endophytic fungi in medicinal plants from Wuhan Botanical Garden in China. Biol Control 94:47–55. https://doi.org/10.1016/j.biocontrol.2015.12.002

Yadav G, Meena M (2021) Bioprospecting of endophytes in medicinal plants of Thar Desert: an attractive resource for biopharmaceuticals. Biotechnol Rep 30:e00629. https://doi.org/10.1016/j.btre.2021.e00629

Yadav M, Yadav A, Yadav JP (2014) In vitro antioxidant activity and total phenolic content of endophytic fungi isolated from *Eugenia jambolana* Lam. Asian Pac J Trop Med 7:S256–S261

Yan Z, Huang C, Guo H, Zheng S, He J, Lin J, Long Y (2020) Isobenzofuranone monomer and dimer derivatives from the mangrove endophytic fungus *Epicoccum nigrum* SCNU-F0002 possess α-glucosidase inhibitory and antioxidant activity. Bioorg Chem 94:103407

Yang SW, Mierzwa R, Terracciano J, Patel M, Gullo V, Wagner N, Baroudy B, Puar M, Chan TM, Chu M (2007) Sch 213766, a novel chemokine receptor CCR-5 inhibitor from *Chaetomium globosum*. J Antibiot 60:524–528. https://doi.org/10.1038/ja.2007.67

Zhang HW, Song YC, Tan RX (2006) Biology and chemistry of endophytes. Nat Prod Rep 23(5):753–771

Potential Role of Endophytes in Sustainable Agriculture

10

Aryadeep Roychoudhury, Alen D' Rozario, and Ritika Gupta

Abstract

The ever-growing demand for the implementation of sustainable practices in agriculture, which will be suitable for providing solutions to improve agricultural productivity, has led to the exploration of innovative approaches for achieving sustainable agriculture. This strictly prohibits relying exclusively on fertilizers and pesticides to enhance productivity. The practice of high-input agriculture that relies heavily on agrochemical usage is generally regarded to be harmful to the natural environment and is not considered economically viable. In addition, the changing climate, along with the chemical applications (of fertilizers and pesticides), seriously affect the soil health, pest resistance, and growth patterns of the crops. Therefore, microorganisms of the rhizosphere are being studied to promote crop growth along with control of the growth of the pathogen and resistance against the pathogens. Thus, studies on endophytes are considered as a possible alternative to some of the agrochemicals, thus reducing their harmful impact on the environment. Endophytes are the microorganisms that reside and colonize around the plant tissues including colonization of fungi, bacteria, and mycorrhiza. Endophytes specifically help in promoting plant growth by the utilization of various systematic methods, including biological nitrogen fixation, phosphate solubilization, phytohormone production, tolerance against stress, etc. and also develop resistance against pathogen by the release of specific secondary metabolites. The endophytes can also be regarded as bioinoculants since the establishment of symbiotic relationships can improve crop production. For

A. Roychoudhury (✉)
Discipline of Life Sciences, School of Sciences, Indira Gandhi National Open University, New Delhi, India

A. D' Rozario · R. Gupta
Department of Biotechnology, St. Xavier's College (Autonomous), Kolkata, West Bengal, India

M. Meena et al. (eds.), *Metabolic Sustainability of Endophytes*, Microorganisms for Sustainability 52, https://doi.org/10.1007/978-981-96-4004-1_10

sustainable agriculture to prosper, a considerable number of fungal and bacterial endophytes are being applied as soil and plant inoculants to boost plant yield without exerting unfavorable consequences on the productivity of the plant. Hence, endophytes can also be used as biofertilizers unlike the chemical fertilizers. The endophytes have proven to be effective in rendering better tolerance against both biotic and abiotic stress. Therefore, the endophytic microorganisms can serve as one of the most important components to maximize agricultural benefits. This chapter deals with the immense role played by the endophytes in improving crop productivity and sustainable agriculture.

Keywords

endophytic bacteria · endophytic fungi · mycorrhizal endophyte · stress tolerance · secondary metabolites · sustainable agriculture

10.1 Introduction

Mutualism between the endophytes and the plants is a type of relationship which is surmised to be ubiquitous among the higher plants. The endophytes either contain fungal or bacterial species residing within the leaves, flowers, stems, and fruits of the plants (Watts et al. 2023a). This relationship between plants and endophytes ranges from symbiosis to mild pathogenicity; the plant supplies carbohydrates essential for the growth of the endophyte, which in turn produces secondary metabolites that enhance plant tolerance to various stress, including salinity and nutrient deficiency (Deshmukh et al. 2015). Due to increasing population and fast growth in industrialization and urbanization, environmental damage is also increasing (Sansanwal et al. 2017). The chemical fertilizers are enormously used; on one hand, it is being used to fortify crop production to a larger extent, while on the other hand, it is also damaging the environment with harmful impacts on humans and animals. Thus, as a response to the adverse effects of prolonged chemical fertilizer use, organic farming has emerged as an alternative to meet the growing demand for safe food, to ensure long-term sustainability, and to address concerns about environmental pollution. Endophytic bacteria, residing within plant cells, offer the advantage of directly benefiting the plants. As a result, they are widely utilized to enhance microbial processes and improve nutrient availability, enabling plants to efficiently assimilate nutrients for their growth (Singh and Pandey 2012).These microbes contribute to soil fertility by fixing nitrogen, producing essential plant growth-promoting substances, and solubilizing insoluble potassium and phosphates. Utilizing such endophytic bacteria holds significant potential for developing agricultural solutions to enhance crop performance under stress conditions such as cold, drought, or heavy metal pollution. Not only endophytic bacteria, the endophytic fungi also take part in sustainable agriculture. *Metarhizium anisopliae* and *Beauveria bassiana* are the best characterized and most commonly used entomopathogenic fungi that assist in biological control programs for the control of arthropods that act as pests or disease vectors. Therefore, these endophytic fungi can protect plants and function as antagonists of plant pathogens through various mechanisms, including metabolite production, competition, and parasitism, ultimately leading to a decrease in pathogenic activities (Singh and Pandey 2012).

Abiotic stress includes various conditions such as drought, salinity, cold, and heat that hamper optimum agricultural yield (Verma et al. 2021). Erratic climatic conditions, irregular rainfall, heat waves, and rise in global temperature can significantly impact crop growth and yield. Drought is considered as one of the most severe forms of stress, affecting growth and productivity of the plants and disrupting the physiological and metabolic processes (Chaves et al. 2009). Some of the major impacts due to drought stress include changes in leaf morphology, nutrient absorption, photosynthetic efficiency, and seed germination. However, the impacts on plants can depend upon the duration of exposure of the plant towards stress. During short exposure, the plant system increases water use efficiency by reducing the stomatal aperture and the transpiration rate (Li et al. 2019); due to long-term exposure, the chloroplast is disrupted, thus affecting photochemical activities (Zhou et al. 2016). Severe drought stresses can also lead to osmotic stresses in the root; in saline environment, the osmotic pressure of the soil solution exceeds the osmotic capacity, thereby reducing water uptake by the plants while increasing the uptake of Na^+ and Cl^- (Munns et al. 2020). Salinity can be another difficult challenge for the plants that can reduce seedling growth and photosynthetic activity and increase ion toxicity. Similarly, lower amount of rainfall and high temperature both play a crucial role in increasing soil salinity. The extent of salinity is determined by the concentration and the exposure time. Some of the major impacts of salinity stress on the plant include reduction in leaf expansion, ultimately resulting in complete inhibition of cell division and stomatal closure, while prolonged exposure can cause premature leaf senescence, reducing photosynthetic activity and eventually leading to plant death (Cramer and Nowak 1992).

Under such stressful conditions, plant evokes a series of reaction mechanisms such as induction of stress responsive genes, regulation of functional proteins, and corresponding responses mainly in the chloroplast, mitochondria, and peroxisomes. The plants also induce the production of certain stress hormones and reactive oxygen species (ROS) that regulate to maintain physiology and normal functioning of the plants (Choudhury et al. 2013). To mitigate the harmful impacts of stress and their effects on the growth and productivity, the utilization of the beneficial microorganisms is often sought to be used as they play a crucial role in providing the required growth and also to combat the stressful conditions. Hence, in this chapter, we discuss how the microbial (bacterial, fungal, mycorrhizal) endophytes help in the process of sustainable agriculture by combating the different types of environmental challenges (Fig. 10.1).

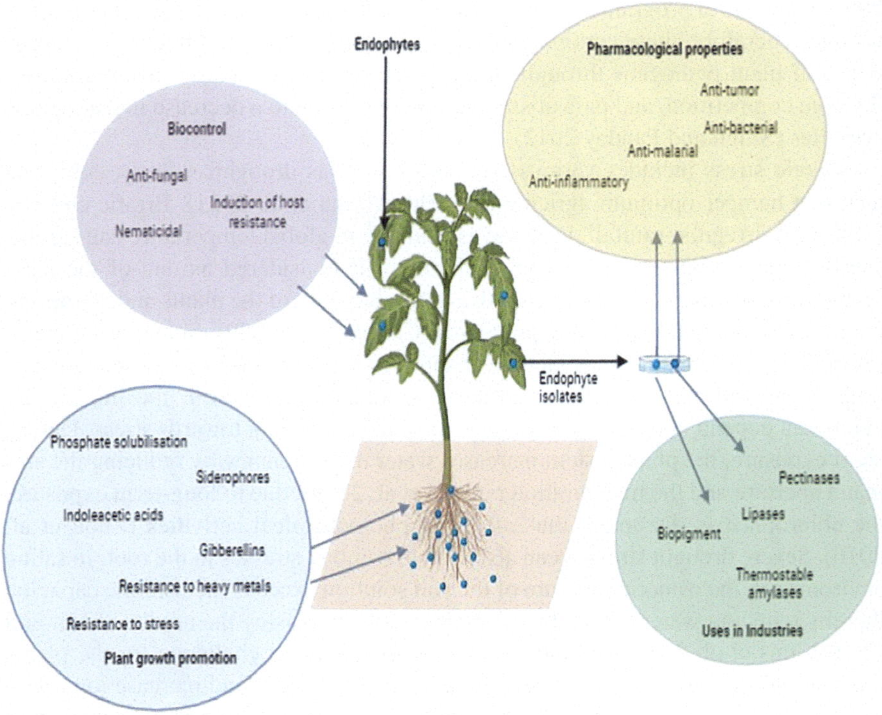

Fig. 10.1 Endophyte colonization with their different properties and uses. (Modified from BIORENDER site)

10.2 Different Types of Endophytes and Their Characteristics

10.2.1 Bacterial Endophytes

Bacterial endophytes support and enhance plant growth and development and include Proteobacteria, Actinobacteria, and Firmicutes. Some of the common characteristics of the these bacterial endophytes includes its ability to synthesize plant hormone such as indole-3-acetic acid (IAA), to solubilize phosphates, to secrete siderophores and to confer resistance to the plant against abiotic and biotic stress (Kandel et al. 2017). Additionally, some of the bacterial endophytes also carry genes that are required for biological nitrogen fixation (BNF), thus enabling them to convert nitrogen gas (N_2) into the available forms of nitrogen, like ammonium and nitrate, within the host plant (Bhattacharjee et al. 2008; Santi et al. 2013). The use of bacterial endophytes in agriculture can greatly reduce the use of chemical fertilizers. It has also been seen that a significant amount of N used in agriculture is lost in the environment in lakes, rivers, and the marine systems causing the growth of algae, thus exerting serious impacts on the water ecosystems. Greater and elevated concentration of N, in the form of ammonia and ammonium, leads to cases of eutrophication and in the form of nitric acid vapor causes reduction in the air quality and visibility (Kandel et al. 2017). Therefore, the natural symbiosis of the bacterial endophytes with the plants can significantly reduce the need for fertilizers for plant growth and therefore help in maintaining sustainable agriculture. As previously written, bacterial endophytes can synthesize phytohormones such as indole-3-acetic acid (IAA), cytokinins, and gibberellins that can not only promote plant growth but also regulate the internal hormone level of the plant. The produced IAA within the host plant enhances the development of lateral and adventitious roots, improving nutrient access and boosting root exudation (Spaepen and Vanderleyden 2011; Gamalero and Glick 2011). The secreted organic compounds such as siderophores in iron-deficient environmental conditions enable the plants to chelate iron from the environment, which are otherwise harmful to the plants (Ahmed and Holmström 2014). Lastly, the bacterial endophytes can also confer resistance and tolerance to host plants against the biotic and abiotic stress by the release of certain antimicrobial compounds, thus modulating the plant resistance response (Friesen et al. 2011; Mercado-Blanco and Lugtenberg 2014). Strains of *Bacillus*, *Enterobacter*, *Pseudomonas*, *Streptomyces*, etc. have proven to be successful in combating stress conditions like drought, heat, and salinity in various crop plants. Therefore, the symbiosis of the bacterial endophytes with plants not only help in resistance and tolerance against the different types of environmental stresses but also increase plant biomass and height which significantly assist in the process of sustainable agriculture.

10.2.2 Fungal Endophytes

Fungal endophytes provide various benefits, serving as biofertilizers to promote plant growth, biological control agents against pathogens and pests, and inducers of tolerance to abiotic stress. These roles significantly contribute to developing innovative strategies for sustainable agriculture (Poveda et al. 2022).

One of the main benefits of enhancing plant growth by endophytic fungi constitute increased agricultural output which in turn increases the access of nutrients such as nitrogen, phosphorus, potassium, zinc (Zn), and iron, and also leads to the production of several plant hormones. Fungal endophytes can alter plant growth by directly interfering on their hormonal pathways, for example using wheat plants (*Triticum aestivum*) and the endophytic fungus *Fusarium oxysporum*, it was reported that plant growth promotion can be observed through increased aerial and root biomass, enhanced root and shoot length, as well as improved vigor index and germination rates. It was also linked to increased levels of plant hormones such as auxins, gibberellins, and salicylic acid, along with decreased levels of abscisic acid (ABA) and jasmonic acid (JA) (Poveda et al. 2022). The fungal endophytes also offer protection to the plants against some of the hostile environmental conditions such as heat, salinity, and drought (Malarvizhi et al. 2023). Unusual and abrupt changes in the climatic conditions cause decrease in crop production at a global scale, for instance climatic alteration causes decrease in productivity of rice in many regions due to reduced water availability and soil salinization (Arora 2019). Therefore, the fungal endophytes provide tolerance to the rice plants; for example a rice endophyte *Penicillium simplicissimum* was identified to be tolerant to above 10% salt concentration, which explains that the endophyte has significant role in increasing the salt stress tolerance in their hosts (Potshangbam et al. 2017). In recent researches, it was found that the endophytic fungus *Penicillium minioluteum* could increase the yield of flavonoids, daidzein, and genistein in soybean found in salinity stress, thus positively influencing the growth pattern of the host plant along with shoot length and chlorophyll content. The fungal endophyte *Piriformospora indica* can also be used as a biofertilizer, bioprotector, bioregulator, and biotization agent, acting as a protectant of plant from the pathogens (Gill et al. 2016). Thus, the endophytic fungi has established itself for providing significant benefits to hosts by offering an innovative and viable solution to address the challenges of enhancing crop productivity and protecting them from pests, while minimizing the environmental impact.

10.2.3 Mycorrhizal Endophytes

Both sustainable agriculture and organic farming depend on applying beneficial microorganisms to mitigate the negative impacts on crop production and crop growth and tolerance against several kinds of stresses. Arbuscular mycorrhizal fungi (AMF) are being used by symbiotically associating them with the plants (Kalamulla et al. 2022). Though mycorrhizal association is being favored by most of the plant species, plants belonging to Brassicaceae, Cyperaceae, Juncaceae, etc. inhibit the mycorrhizal colonization (Singh et al. 2010; Lekberg et al. 2014). Colonization around the plants is divided into endomycorrhizae and ectomycorrhizae. The symbiotic relationship is formed between the plant and colonizing microbes by a collection of sequenced chemical signals transferred between the plant and the colonizing microbes and the mycorrhizae. (Bowles et al. 2016). The soil quality is based on the physical and chemical properties along with the variety, distribution, and action of the soil biota; thus, soil health means the ability of the soil for serving as a living ecosystem. AMF maintains the soil quality and health by interfering three main factors which include soil structure, plant physiology, and ecological interactions (Fall et al. 2022). In agriculture, the soil structure maintenance is the main prerequisite. AMF forms soil microaggregates, which are water-stable and are formed by secreting hydrophobic glycoprotein, glomalin, which acts as a long-term binding agent. Glomalin is deposited between the outer hyphal walls and adjacent soil particles by AMF for forming microaggregates and then macroaggregates, which therefore form the root of soil aggregation (Li et al. 2022). AMF can also respond to biotic and abiotic stress that the plants are being exposed to, where the pathogens act as biotic agents of stress, whereas drought, salinity, flooding, and heavy metals act as abiotic stress factors. Salinity causes decrease in water absorption in plants; AMF employs a variety of morphological, biochemical and physiological strategies for overcoming such condition by increasing water absorption; for instance, AMF enhance water and nutrient uptake by increasing root volume through the expansion of root length, surface area, and projected area (Evelin et al. 2019; Yadav et al. 2017). Various researches have also shown that the ROS production increases under salinity stress, thus causing damages to biomolecules such as lipids and proteins; AMF neutralize the damage by increasing the function of the antioxidant enzymes found in the plants (Evelin et al. 2019). AMF are also responsible for the improvement of plant nutrients and integration with other beneficial organisms leads to morphological, biochemical, and physiological changes in plants. This helps in pathogen biocontrol by creating a competition for space, infection sites, nutrients, and photosynthesis. For example, AMF competed with the main plant pathogenic nematode, *Meloidogyne incognita*, for the colony formation in the root nodules of the plant *Prunus persica* and also restricted gall formation (Chen et al. 2018).

10.3 Mode of Association of the Endophytes

10.3.1 Bacterial Endophytes

The association mechanism is based on the complex interaction between the host plant and the endophytic bacteria which naturally initiate from the roots and require identification of proper compounds in the root exudates by the endophytic bacteria. The production of root exudates by plants is required for interaction with the beneficial bacteria because of their own ecological advantage. Thus, the endophytic bacteria enters the plants through the root zone, but the aerial parts of the plants which include stems, leaves, flowers, and the cotyledons can also be used by them (Afzal et al. 2019). Thus, the sequential steps include:

1. *Rhizosphere colonization by the bacteria*: Occupying a place and getting nutrients from that place is a competitive task for bacteria. Only the bacterial species having the ability to colonize the plant rhizosphere competitively and thus survive in such environment can affect plant growth and development. Bacterial rhizosphere population can range between 10^7 and 10^9 cfu/g of the rhizosphere soil. They adhere to the surface of roots and form a string of cells which finally colonize the entire root surface along with some rhizodermal cells and establish the micro-colonies or biofilms by the bacteria (Compant et al. 2010a). Bacteria colonize the plant rhizosphere and the rhizoplane competitively, providing beneficial effects to the host plant.
2. *Root colonization by the endophytic bacteria*: Bacterial endophytes enter inside the plant root after establishing themselves in the rhizosphere. They also colonize themselves within the plant roots with subpopulations ranging from 10^5 to 10^7 cfu/g fresh weight. This involves bacterial adhesion to cell surface structures mediated by polysaccharides, pili, and bacterial adhesins. Sometimes the bacteria may reach root entry sites which are present on the root surface and it causes emergence of lateral root, causing wounds using the type IV pili, mediating twitching mobility (Hori and Matsumoto 2010; Böhm et al. 2007). Penetration process can also be both passive and active. Passive penetration occurs at the

opening present at the root tips, root emergence areas, or those sites that are created by deleterious organisms. Active penetration occurs by attachment and proliferation machinery, which involves the presence of polysaccharide, twitching motility, and the quorum sensing. Quorum sensing can influence bacterial movement and endophytic colonization inside the host plants.

3. *Systemic colonization of aerial plant tissues by the endophytic bacteria*: The endophytic bacteria spread systemically to colonize aboveground tissues after entry into the hosts and can establish stem and leaf population densities within the range between 10^3 and 10^4 cfu/g fw under natural conditions (Compant et al. 2010a). Next, bacterial flagella and plant transpiration stream support bacterial movement inside plants. Migration occurs along intracellular spaces for which the cell wall degrading enzymes like cellulases and pectinases are required. Movement through xylem elements occurs through perforated plates that allow movement of bacteria through large pores without requiring cell wall degrading enzymes. Therefore, the endophytic bacteria mostly colonize the leaf tissues from the plant roots, similar to the phytopathogenic bacteria. The bacterial endophytes can gain entry into the leaves from the phyllosphere via leaf stomata (Senthilkumar et al. 2011).

10.3.2 Fungal and Mycorrhizal Endophytes

As the method of association and colonization of the fungi and mycorrhizae are similar, we can club them together while describing their colonization process. Thus, the steps that are involved in the colonization of the fungal endophytes along with the mycorrhizal endophytes around the plant roots include the preliminary steps involving communication or signaling between the endophytic fungi and the rhizospheric root exudates of the plant, leading to the adhesion of the endophytic organism to the surface of the host plant. The colonization process constitutes the migration of the endophytes to the plant surfaces as a response to chemotactic transfers between the endophyte and plant root exudates followed by attachment (Omomowo and Babalola 2019). The attachment occurs by some structural secretory biomolecules/organelles, including fimbriae, pili, flagella. and the exopolysaccharide. Reports suggest that the initial attachment and colonization are facilitated by exopolysaccharide metabolites. The next complex step involves the colonization of the plant tissues by the endophytes, relating various mechanisms by which the endophytes penetrate the host tissue after establishing themselves on the epiphytic surfaces of a plant. Similar to the bacterial endophytes, the fungal endophytes exhibit active and passive penetration mechanisms by which they enter into the host tissue via the opening present in the root zones or aerial parts of the plant such as the stems, flowers, cotyledons, and leaves (Hardoim et al. 2008). The proliferation, penetration, and attachment are mediated by various bioactive secretory compounds such as lipopolysaccharides, exopolysaccharides, lytic enzymes, and lysozymes. The compatibility between the fungal endophyte and plant host must be present for the colonization to be successful. Additionally, effective signaling and communication between the endophytic fungi must be present along with the bioactive metabolites of the plant (Brader et al. 2014; Compant et al. 2010b). Since the successful colonization depends on various factors such as the genotype of the host plant, the extent of nutrient limitations, and biotic and abiotic factors, the endophytic microbial strains, which are efficiently adapted in terms of these factors, show a better chance of success.

10.4 Role of Endophytes in Growth and Development to Support Sustainable Agriculture

Endophytes help to promote growth of crops by various mechanisms. They interact with host plants symbiotically, i.e., they live on host plants and derive nutrients and shelter as well as help host plants to promote growth by secreting various phytohormones like abscisic acid (ABA), jasmonic acid (JA), gibberellic acid (GA), along with various metabolites. Endophytes like mycorrhizal fungi provide various macronutrients like nitrogen and phosphorous from soil to host plant, promoting growth. Bacterial endophytes like symbiotic nitrogen-fixing bacteria fix atmospheric nitrogen to ammonia and offer it to the host plant.

Bacterial endophytes help plant growth directly or indirectly. Direct growth mechanism is promoted by bacterial endophytes providing potassium, nitrogen, phosphorous, and iron to the host plant and secreting various kinds of growth-promoting hormones to the host plant. Indirect growth mechanism is promoted by bacterial endophytes producing various kinds of metabolites, hydrogen cyanide (HCN), and antibiotic against the pathogens (Rana et al. 2023).

10.5 Various Plant Growth-Promoting Mechanisms

1. *Nutrient fixation*: Plant growth-promoting endophytes (PGPEs) increase the concentration of chemical nutrients and enhance their accessibility by locking and fixing these nutrients for plant development and growth. Some nutrient-fixing bacteria are also found to be associated with maize (Gonzalez-Mendoza et al. 2022).
2. *Nitrogen fixation*: Nitrogen is a beneficial nutrient as it is the main component of amino acids. It is the main component of proteins and nucleic acids. Microorganisms that have the ability for biological nitrogen fixation reduces atmospheric nitrogen to ammonia (Rana et al. 2023). Symbiotic plant growth-promoting rhizobacteria (PGPRs) including *Rhizobium* sp., *Azoarcus* sp., *Beijerinckia* sp., and *Pantoea agglomerans* can fix atmospheric nitrogen to ammonia (Gonzalez-Mendoza et al. 2022). *Rhizobium tropici* is a type of endophytic rhizobacteria that is isolated from infecting nodules of leguminous plants (Rana et al. 2023). *Rhizobium* sp. contains genes like *nod* and *nif* which are responsible for nodule formation. Such factors help in nodule formation in root and inside nodules, they divide and produce an enzyme nitrogenase responsible

for reducing atmospheric nitrogen to ammonia. The nitrogenase is an oxygen-sensitive enzyme and is protected by producing leghemoglobin inside nodule. There are also some non-rhizobial endophytes, including *Azoarcus* sp., *Achromobacter* sp., *Burkholderia* sp., and *Gluconacetobacter* sp. (Rana et al. 2023). *Azoarcus* sp. secretes bacterial endoglucanase enzyme, which helps this bacterium to interact with host plant through the root hairs and root tips (Santoyoa et al. 2016). Researchers found that the endophytic *Bacillus* strain from pearl millet has antifungal and plant growth-promoting abilities. Non-leguminous plant production increases because of the nitrogen fixing ability of the endophytic diazotrophs (Rana et al. 2023). Besides bacteria, there are some endophytic fungi which also help to transfer nitrogen to plant and promote growth. Studies show that the fungus *Metarhizium robertsii* infects and kills the larva of *Galleria mellonella*, following which nitrogen from dead larva is transferred to the bean plants, thereby promoting growth (Baron and Rigobelo 2022).

3. *Phosphate solubilization*: We know that phosphorus is an essential and beneficial macronutrient for plant growth and development. Phosphate is the main component of ATP, also called as energy currency of cell. It helps to control various metabolic pathways in plant cell, viz., cell division, photosynthesis, respiration, nutrient uptake, etc. Large amount of phosphorous is present in inorganic form in soil. As this phosphorous is insoluble in nature, plant cannot use it for its growth and development (Rana et al. 2023). Various phosphate-solubilizing bacteria help for solubilization and mineralization of phosphorus so that plant can easily uptake the soluble form of phosphorous for their growth and development. Phosphate-solubilizing PGPRs include various bacteria like *Arthrobacter* sp., *Bacillus* sp., *Beijerinckia* sp., *Burkholderia* sp., *Enterobacter*, and *Microbacterium* (Gonzalez-Mendoza et al. 2022). These bacteria have the ability to solubilize inorganic phosphates via the production of siderophores, organic acids, hydroxyl PGPB, etc. (Rana et al. 2023).

   Mycorrhizal fungi also help in plant growth by increasing the uptake of phosphorous from soil by the host plant. Arbuscular mycorrhizal (AM) plant absorbs orthophosphate by root hairs and root epidermis through the mycorrhizal pathway with the help of external AM hyphae. This external hypha helps to absorb the phosphorus and passes it through the internal part of roots and releases into the apoplast so that the plant can easily derive phosphorous for its growth (Sally et al. 2003). From further study, it was found that *Trichoderma harzianum* helps phosphate solubilization in tomato plant (Rana et al. 2023). Researchers found that in maize plant, the endophytic fungus *Aspergillus sydowii* helps in the accumulation of higher amount of phosphorous in plant tissues to promote growth (Baron and Rigobelo 2022).

4. *Potassium solubility*: Potassium is one of the important nutrients for plant growth and development. Plant cannot use potassium in its insoluble form. So, various endophytes like potassium-solubilizing microorganisms help in solubilizing potassium so that plant can easily use it. Potassium-solubilizing endophytes include *Acidithiobacillus* sp., *Bacillus edaphicus*, *Pseudomonas* sp., and *Paenibacillus* sp. (Gonzalez-Mendoza et al. 2022). Potassium-soluble bacteria

have the ability for solubilizing potassium, aluminum, and silicon from insoluble minerals like mica, etc., which contain potassium (Rana et al. 2023). From various researches, it was found that some fungi also have the capability for potassium solubilization. These fungi belong to Ascomycota and they are isolated from various parts of Asia and South America. Potassium-solubilizing fungi include *Torulaspora globosa*, *Aspergillus fumigatus*, and *Aspergillus niger* (Velázquez et al. 2016).

5. *Siderophore production*: Siderophore is a type of organic, low molecular weight, iron-chelating compound. Interaction of endophytic microbe with the host plant helps to produce siderophore by microbes. This compound has iron-scavenging activity (Shaffique et al. 2022). There are various receptors found on bacterial membrane for binding of siderophore complex. In siderophore complex, Fe^{3+} is reduced to Fe^{2+}, which is released from siderophore into the cell by various gating mechanism. The siderophore helps to increase the concentration of metal so that the plant can easily use it for its growth and developmental purposes (Rana et al. 2023). Siderophore production process is induced by various endophytic bacteria that include *Pseudomonas* sp. and *Streptomyces* sp.. They help to produce soluble iron-chelating complex so that plant can easily use it (Gonzalez-Mendoza et al. 2022). Scientists found that *Serratia marcescens* and *Bradyrhizobium* sp. help in plant growth by the production of siderophore (Rana et al. 2023).
6. *Phytohormone production*: Various endophytes associated with the host plant produce hormones like GA, cytokinin, indole-3-acetic acid (IAA), and ABA. Some endophytic bacteria help to produce IAA that induces cell division and differentiation in plant. IAA also helps in formation of roots so that surface area of lateral roots increase. As a result, plant can absorb a vast amount of nutrients from soil. IAA also helps in pigment formation in plant so that photosynthesis is stimulated (Rana et al. 2023). Endophytic bacteria that help in the production of phytohormones include *Pantoea agglomerans*, *Rhodospirillum rubrum*, and *Bacillus subtilis* (Gonzalez-Mendoza et al. 2022). In tomato, it was observed that IAA is produced by *Trichoderma harzianum* that significantly stimulates plant growth (Gonzalez-Mendoza et al. 2022). Endophytic fungi also help in the production of phytohormones. For IAA biosynthesis, endophytic fungi use the precursor, L-tryptophan. Further research showed that the fungus *Phoma glomerata* helps in the production of IAA in plants (Baron and Rigobelo 2022). Gibberellin is also a type of phytohormone which helps in seed germination, stem elongation, sexual processes, etc. Researchers found that *Gibberella fujikuroi* is a type of endophytic fungi that help in the production of GA in plant. Further studies showed that the endophytic fungi like *Aspergillus fumigatus* and *Cladosporium sphaerospermum* also help in the production of GA in plant (Baron and Rigobelo 2022).

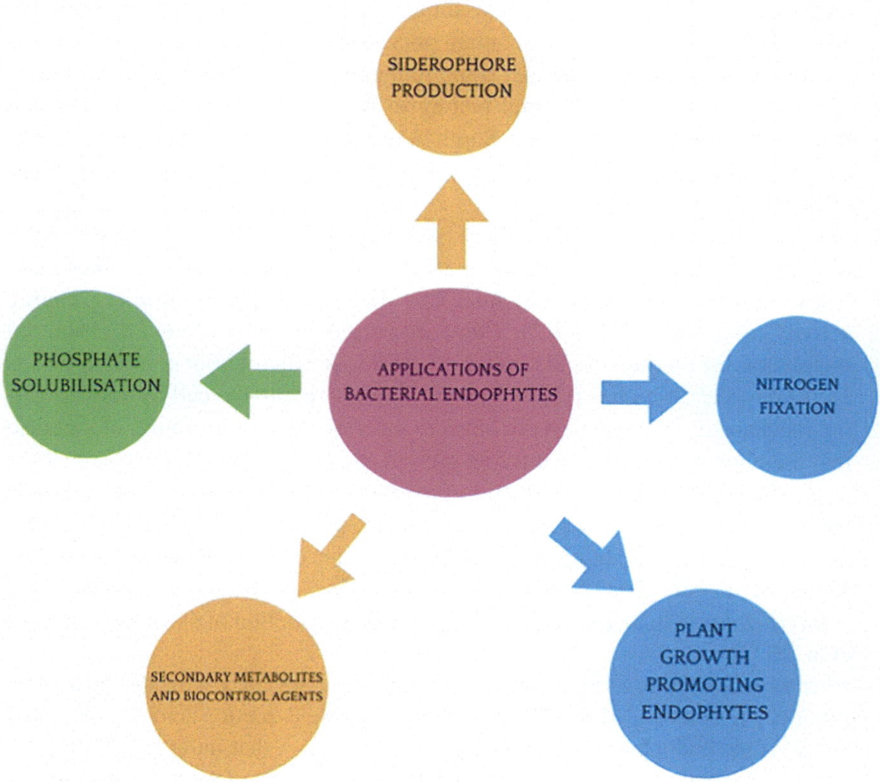

Fig. 10.2 Applications of bacterial endophytes

Figure 10.2 represents the various applications of endophytes in plants.

10.6 Role of Endophytes to Protect Plant from Abiotic and Biotic Stress

Endophytic microbial community lives with plant species. Under normal environmental conditions, they obtain nutrition from plant and help in plant growth and development. However, when environmental condition is not favorable, these endophytic microorganisms help plants to survive under stress condition. Abiotic stress occurs due to environmental factors like temperature extremes, osmotic stress, cold, salinity, oxygen deficiency, etc. Biotic stress is caused by living agents like pathogens, insects, etc. Endophytes help plant species to overcome stress and enhance tolerance of plants to such unfavorable environmental conditions. These endophytes secrete various hormones like ABA, salicylic acid, jasmonic acid, etc. that help plants to overcome stress. They also release various secondary metabolites like antimicrobial agents, antibiotics, etc., which also help to build up resistance against various diseases in plants. Thus, endophytes are efficient to elicit immune response in plant species.

10.6.1 Acclimatization to Abiotic Stress

Abiotic stress condition, as already mentioned, is related to environmental factors. Drought, salinity, heavy metal toxicity, temperature, or pH extremes, etc. have an impact on the microbial endophytes, which reside in plant roots. The associated microbes help the plants to acclimatize during these unfavorable environmental situations like heat stress, drought stress, etc. (Chen et al. 2022).

1. *Heat stress*: Temperature or heat is an important environmental factor for plant growth and development. We know that at excessive high temperature, protein and enzyme denaturation occurs and they become nonfunctional, arresting metabolic activity. As a result, plant growth does not take place and plants gradually die due to absence of metabolism. The same situation arises during cold stress. Therefore, optimum temperature is required for plant growth and development. Membrane fluidity changes during extreme temperature, thereby changing interaction among macromolecules (Kumar et al. 2019). Plant growth is affected due to inhibition in photosynthesis, cell division, enzyme functioning, etc. Microbial endophytes secrete various anti-stress products like heat shock proteins so that they can maintain protein and nucleic acid structure, membrane fluidity, etc. at high or low temperature (Gonzalez-Mendoza et al. 2022). Researchers have found some microbial endophytes that acclimatize plants to heat stress. These include *Pseudomonas cedrina*, *Brevundimonas terrae*, and *Arthrobacter nicotianae (*Kumar and Verma 2018). *Burkholderia phytofirmans* has been found to help the plants to tolerate low temperature so that plants can survive in such temperature (Ait Barka et al. 2006). *Dichanthelium lanuginosum*, a grass species is able to survive at high temperature due to the presence of *Curvularia protuberata* (Hata et al. 2002). In wheat plant, some fungal endophytes help to tolerate high temperature. As a result, grain size increases in the next generation of wheat (Kannahi and Senbagam 2014).
2. *Drought stress*: Drought stress reduces plant growth, productivity, and development (Kumar et al. 2019). Drought stress mainly occurs due to scarcity of water in soil or high transpiration rate. Hence, there is a limited supply of water to roots (Anjum et al. 2011). During drought stress, plant growth rate decreases due to decrease in photosynthesis and germination rate, and reactive oxygen species accumulate in plants along with disruption in membrane integrity (Cohen et al. 2008). During drought stress, photosynthesis level decreases in plant due to reduction in chlorophyll content. Proline accumulation also occurs during drought stress (Begum et al. 2022). Endophytic organisms help plants to acclimatize drought via several mechanisms. Endophytes help to balance the cellular water homeostasis during drought condition (Gonzalez-Mendoza et al. 2022). Endophytic bacteria that help plants to overcome drought stress are *Acinetobacter*

sp., *Azospirillum* sp., *Bacillus* sp., *Arthrobacter* sp., *Flavobacterium* sp., etc. (Vidal et al. 2022). Drought stress is also overcome due to increased biosynthesis of the plant hormone, ABA, which has a role in stomatal closing mechanism. Endophytic organisms help to increase the production of ABA in plants during drought stress (Bloemberg and Lugtenberg 2001). *Azospirillum brasilense* sp. has been shown to increase ABA level in *Arabidopsis thaliana* seedling (Choi et al. 2008). It has been observed that the growth of seedlings of *Pinus tabulaeformis* is enhanced due to the inoculation of the endophyte, *Phoma* sp., under drought condition (Verma et al. 2021). Another fungal endophyte *Ampelomyces* sp. was shown to increase tolerance of tomato plant in drought condition (Verma et al. 2021).

3. *Salinity stress*: Salinity stress is one of the most important abiotic factors that affects plant growth and development. This occurs due to the accumulation of water-soluble salt in soil. As higher amount of salt accumulates in soil, fertility of soil decreases due to death of some beneficial microorganisms at high salt concentration. Growth of plants ceases as physical and chemical properties of the soil changes (Kumar et al. 2019). Due to salinity stress, plant cannot uptake water from soil sufficiently. As a result, the rate of photosynthesis, nutrient uptake, and growth of plants decreases (Gonzalez-Mendoza et al. 2022). Some salt-tolerant endophytic bacteria help plants to acclimatize high salt concentration by various mechanisms. These endophytic organisms produce 1-amino cyclopropane 1 carboxylate (ACC) deaminase to reduce salinity stress (Win et al. 2018). Such organisms include genera like *Pseudomonas* sp., *Gracilibacillus* sp., *Cronobacter* sp., *Staphylococcus* sp., and *Halomonas* sp. (Win et al. 2018). Some fungal endophytes also play an important role in increasing salinity stress in plants. The fungal endophyte *Penicillium* sp. increases salinity stress in tomato plant (Verma et al. 2021).
4. *Oxidative stress*: Due to higher salinity and osmotic stress, reactive oxygen species (ROS) are generated in plants that include hydrogen peroxide, superoxide radicals, and hydroxyl radicals. They lead to oxidative damage to the cells (Gonzalez-Mendoza et al. 2022). ROS-scavenging enzymes like superoxide dismutase (SOD), catalase, glutathione reductase (GR), monodehydroascorbate reductase (MDHAR), etc. (Kumar et al. 2019) act as antioxidants and reduce the effects of ROS in plants. Endophytic organisms possess such ROS-scavenging enzymes. *Piriformospora indica* helps barley plant to increase tolerance to oxidative stress (Choi et al. 2008), while *Trichoderma longibrachiatum* helps wheat plants to survive during oxidative stress (Verma et al. 2021).
5. *Heavy metal stress*: Heavy metal stress reduces plant growth and development as soil fertility and crop production decrease due to the presence of heavy metals. In acidic soil, manganese and aluminum toxicity takes place thereby decreasing crop production (Kumar et al. 2019). Biological processes like photosynthesis, respiration, and nutrient uptake are inhibited. Mechanisms like phytovolatilization, efflux, membrane impermeability to metal ions, enzymatic detoxification, metal complexation, etc. (Gonzalez-Mendoza et al. 2022) serve to reduce cellular toxicity due to heavy metal stress. *Rhizobacteria* sp. and *Firmicutes* sp. help

plant growth during heavy metal stress (Chathurdevi and Gowrie 2016). In cadmium-stressed soil, the endophyte *Exophiala pisciphila* assisted in the growth of maize plant by activating the antioxidant enzymes (Chathurdevi and Gowrie 2016). *Gigaspora* sp. and *Pseudomonas* sp. enhanced heavy metal tolerance of plants by changing ACC content so that the ethylene level was increased (Ardanov et al. 2011).

6. *Nutrient stress*: Various endophytic organisms promote plant growth in nutrient-deficient conditions by several mechanisms that include IAA and GA production, siderophore production, phosphate solubilization, and cytokinin production (Kumar et al. 2019). In wheat plant, the endophytic organisms *Azotobacter chroococcum* and *Piriformospora indica* increased the uptake of Zn in nutrient-deficient condition (Abadi and Sepehri 2015). Mycorrhizal fungi assisted phosphate uptake in phosphorous-deficient plants, promoting their growth and development (Gonzalez-Mendoza et al. 2022). Some studies also explained that endophytes help in nutrient cycling by degradation of organic compounds like lignin and cellulose (Andreas et al. 2012).

10.6.2 Acclimatization to Biotic Stress

Biotic stress occurs due to attack by pathogens like bacteria, fungi, viruses, nematodes, insects, and viroids. Endophytic microorganisms act as antagonist against such pathogens (Gonzalez-Mendoza et al. 2022). When endophytes enter into plant hosts during growth, they also face several challenges inside. Due to coevolution, the genome of plant host and endophytic microorganism survive together. There is a complex mutual relationship between plant hosts and endophytes due to their coexistence and evolution. As a result of nonpathogenic complex relationship between plant host and endophytes, plant hosts can protect themselves from pathogenic microorganisms (Pathak et al. 2022). Following pathogen entry, several enzymes like chitinase, phenylalanine ammonia lyase, and peroxidase are activated in plants. Microbial endophytes produce various secondary metabolites like ACC-deaminase, beta-1,3-glucanase, and chitinase that help to neutralize pathogen by lysing their cell walls (Gonzalez-Mendoza et al. 2022). Some of these microbial endophytes include *Pseudomonas* sp., *Bacillus* sp., *Arthrobacter* sp., and *Serratia* sp. (Pathak et al. 2022). Endophytic microorganisms help to build up induced systematic resistance (ISR) in plants against pathogens. Several endophytic organisms help in the induction of ISR by secreting various phytohormones like salicylic acid, jasmonic acid, and ethylene, which help in controlling signaling pathway for ISR induction. ISR is induced by various secondary metabolites like antibiotics, siderophores, N-acyl homoserine lactone, volatile compounds, and acetoin (Choi et al. 2008). The resistance in potato plant against the pathogen *Pectobacterium atrosepticum* was generated due to the presence of the endophyte *Methylobacterium* sp. (Ardanov et al. 2011). Fungal endophytes also take part in developing ISR in plant host. They produce various secondary metabolites like steroids, alkaloids, peptides, terpenoids, phenols, and polyketones. These secreting secondary metabolites have inhibitory effect against pathogens (Choi et al. 2008). The fungal endophyte *Trichoderma harzianum* develops resistance in tomato plant against bacterial wilt disease caused by the pathogen *Ralstonia solanacearum* (Yan and Khan 2021). Arbuscular mycorrhizal fungi also act as an endophyte which develops resistance against the pathogen *Phoma medicaginis* causing alfalfa leaf spot (Li et al. 2021). Microbial endophytes also develop systemic acquired resistance in plants by triggering hypersensitive response in the infected part of plants by salicylic acid signaling pathway. *Bacillus subtilis* induces resistance in wheat plant against *Blumeria graminis*, which causes powdery mildew disease (Muthu Narayanan et al. 2022).

10.7 Applications of Fungal Endophytes

Figure 10.3 represents the different applications of fungal endophytes.

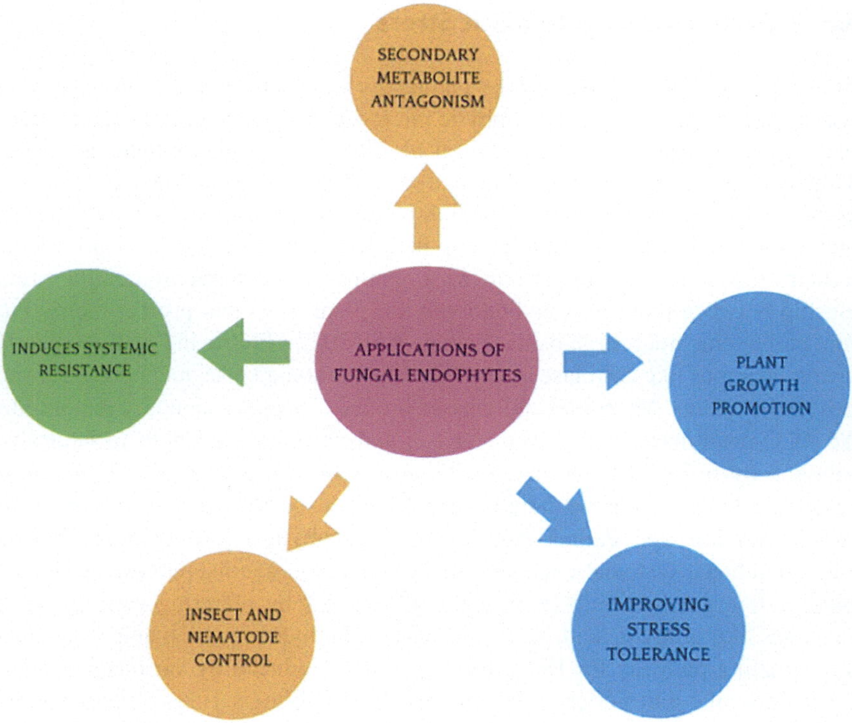

Fig. 10.3 Applications of fungal endophytes

10.8 Role of Endophytes in Increasing Yield

As the population density is expanding day by day, there is a probable chance of shortage of food in the world. Endophytes have a great role for increasing production of crops because they can promote plant growth and development by several mechanisms like nutrient fixation, secondary metabolite production, phytohormone production, and antimicrobial agent production. Therefore, endophytes have a major role in sustainable agriculture and crop production. Due to endophyte association, fresh and dry weight of plants increase as endophytes increase the amount of nutrient uptake and solubilization (Watts et al. 2023b). The endophyte *Bradyrhizobium japonicum* increases plant biomass and nodulation in soybean plant (Sansanwal et al. 2017). Endophytes increase shoot and root dry weight in *Glycine max* and *Glycine soja* (Long et al. 2008). In case of potato seedlings, there is an increase in seed germination rate and root and hypocotyl growth rate after inoculation with endophytic bacteria (Stajkovic et al. 2009). In green gram, the bacterial endophyte *Pseudomonas* sp. increased plant dry weight, nodule number, total chlorophyll content, seed protein, seed growth, and yield (Ahemad and Khan 2012). In case of maize, plant height, weight, and seed content was found to increase due to the presence of the bacterial endophytes *Pseudomonas putida* and *Pseudomonas fluorescens* (Dey et al. 2004). Inoculating *Lupinus luteus* with *Bradyrhizobium* sp. and *Pseudomonas* sp. increased the biomass, nitrogen content and plant yield under field conditions. (Dary et al. 2010). In wheat and soybean, *Pseudomonas* sp. increased plant productivity by improving nutrient uptake ability and enzyme activity (Sharma et al. 2011).

Table 10.1 provides the list of metabolites produced by endophytes for the control of plant diseases.

Table 10.1 Endophytes and their metabolites used in biological control of diseases

Metabolites/ compounds	Endophytic strain	Host plant	Properties/mechanisms
Antimicrobial compounds and cell wall degradation enzymes	*Streptomyces albidoflavus* OsiLf-2	Rice	Inducing plant resistance, lytic enzyme activity, and antimicrobial activity
Hydrolytic enzymes, protease, siderophores, and IAA	*Klebsiella pneumoniae* HR1	*Vigna mungo* L.	Inducing plant resistance and lytic enzyme activity, promoting plant growth
Eugenol, lauric acid, and caprylic acid	*Neopestalotiopsis* sp. and *Diaporthe* sp.	*Cinnamomum loureiroi*	Antimicrobial activity
Volatile substances	*Pseudomonas putida* BP25	Black pepper	Antimicrobial activity
Proteolytic enzymes and cellulase	*Phoma putaminum*, *Penicillium*, and *Myrmecridium schulzeri*	*Bauhinia forficata*	Lytic enzyme activity

10.9 Conclusion and Future Perspectives

Microbial endophytes, which are associated with plants, help in the growth and development of plants and increase crop yield by several mechanisms. They acclimatize plants to various unfavorable and stressed conditions and also help in improving resistance of host plants against various diseases. As these endophytes help in increasing productivity, they have a most important effect on sustainable agriculture. Endophytes actually work in eco-friendly manner because they improve plant growth, nutrient uptake, stress tolerance, and disease resistance naturally without using fertilizers and pesticides. Though many aspects are already discovered regarding endophyte and plant interactions, many roles of endophytes still need to be discovered in future for further improvement in agriculture. Inoculation methods of endophytes need to be standardized to derive the maximum efficacy of endophytes, their effects upon human health, and various processes to scale up their production. In future, these endophytes will be more extensively used for developing stress-tolerant plant species and various nutrient-enriched eco-friendly fertilizers and pesticides. Increasing productivity for supply of food without harming environment and soil fertility, using endophytes is very much effective and safe process. These endophytes are nontoxic, eco-friendly, and cheap. Farmers can easily apply these endophytes in their fields as a substituent of fertilizers and pesticides for agricultural production. Endophytes secrete various plant hormones, secondary metabolites, and enzymes, which are highly effective as biocontrol agents. Endophytes also reduce environmental pollution by reducing the concentration of various pollutants and toxic heavy metals. Hence, endophytes have a lot of potentiality in future as positive regulators of sustainable agriculture.

References

Abadi VAJM, Sepehri M (2015) Effect of *Piriformospora indica* and *Azotobacter chroococcum* on mitigation of zinc deficiency stress in wheat (*Triticum aestivum* L). Symbiosis 69:9–19

Afzal I, Shinwari ZK, Sikandar S, Shahzad S (2019) Plant beneficial endophytic bacteria: mechanisms, diversity, host range and genetic determinants. Microbiol Res 221:36–49. https://doi.org/10.1016/j.micres.2019.02.001

Ahemad M, Khan MS (2012) Ecological assessment of biotoxicity of pesticides towards plant growth promoting activities of pea (*Pisum sativum*)-specific rhizobium sp. strain MRP1. Emir J Food Agric 24:334–343

Ahmed E, Holmström SJM (2014) Siderophores in environmental research: roles and applications. Microb Biotechnol 7:196–208. https://doi.org/10.1111/1751-7915.12117

Ait Barka E, Nowak J, Clément C (2006) Enhancement of chilling resistance of inoculated grapevine plantlets with a plant growth-promoting rhizobacterium, *Burkholderia phytofirmans* strain PsJN. Appl Environ Microbiol 72:7246–7252

Andreas T, Christophe C, Essaïd AB (2012) Physiological and molecular changes in plants at low temperatures. Planta 235:1091–1105. https://doi.org/10.1007/s00425-012-1641-y

Anjum SA, Wang LC, Farooq M, Hussain M, Xue LL, Zou CM (2011) Brassinolide application improves the drought tolerance in maize through modulation of enzymatic antioxidants and leaf gas exchange. J Agron Crop Sci 197:177–185. https://doi.org/10.1111/j.1439-037X.2010.00459.x

Ardanov P, Ovcharenko L, Zaets I, Kozyrovska N, Pirttilä AM (2011) Endophytic bacteria enhancing growth and disease resistance of potato (*Solanum tuberosum* L.). Biol Control 56:43–49. https://doi.org/10.1016/j.biocontrol.2010.09.014
Arora NK (2019) Impact of climate change on agriculture production and its sustainable solutions. Environ Sustain 2(2):95–96. https://doi.org/10.1007/s42398-019-00078-w
Baron NC, Rigobelo EC (2022) Endophytic fungi: a tool for plant growth promotion and sustainable agriculture. Mycology 13:39
Begum N, Ahanger MA, Su Y, Lei Y, Mustafa NSA, Ahmad P et al (2022) Improved drought tolerance by AMF inoculation in maize involves physiological and biochemical implications. Plan Theory 11(18):2437. https://doi.org/10.3390/plants11182437
Bhattacharjee RB, Singh A, Mukhopadhyay SN (2008) Use of nitrogen-fixing bacteria as biofertiliser for non-legumes: prospects and challenges. Appl Microbiol Biotechnol 80:199–209. https://doi.org/10.1007/s00253-008-1567-2
Bloemberg GV, Lugtenberg BJ (2001) Molecular basis of plant growth promotion and biocontrol by rhizobacteria. Curr Opin Plant Biol 4(4):343–350
Böhm M, Hurek T, Reinhold-Hurek B (2007) Twitching motility is essential for endophytic rice colonization by the N_2-fixing endophyte *Azoarcus* sp. strain BH72. Mol Plant Microbe Interact 20:526–533
Bowles TM, Barrios-Masias FH, Carlisle EA, Cavagnaro TR, Jackson LE (2016) Effects of arbuscular mycorrhizae on tomato yield, nutrient uptake, water relations, and soil carbon dynamics under deficit irrigation in field conditions. Sci Total Environ 566–567:1223–1234. https://doi.org/10.1016/j.scitotenv.2016.05.178. Epub 2016 Jun 5. PMID: 27266519
Brader G, Compant S, Mitter B, Trognitz F, Sessitsch A (2014) Metabolic potential of endophytic bacteria. Curr Opin Biotechnol 27:30–37. https://doi.org/10.1016/j.copbio.2013.09.012
Chathurdevi G, Gowrie SU (2016) Endophytic fungi isolated from medicinal plant—a source of potential bioactive metabolites. Int J Curr Pharm Res 8:50–56
Chaves MM, Flexas J, Pinheiro C (2009) Photosynthesis under drought and salt stress: regulation mechanisms from whole plant to cell. Ann Bot 103(4):551–560. https://doi.org/10.1093/aob/mcn125
Chen M, Arato M, Borghi L, Nouri E, Reinhardt D (2018) Beneficial services of arbuscular mycorrhizal fungi—from ecology to application. Front Plant Sci 9:e1270
Chen C, Zhong C, Gao X, Tan C, Baj H, Ning K (2022) Root-associated microbiota: the multifaceted hubs associated with environmental factors, growth status and accumulation of secondary metabolites. Environ Microbiome 17(1):23. https://doi.org/10.1186/s40793-022-00418-0
Choi O, Kim J, Kim JG, Jeong Y, Moon JS, Park CS, Hwang I (2008) Pyrroloquinoline quinone is a plant growth promotion factor produced by *Pseudomonas fluorescens* B16. Plant Physiol 146:657–668
Choudhury S, Panda P, Sahoo L, Panda SK (2013) Reactive oxygen species signaling in plants under abiotic stress. Plant Signal Behav 8:4. https://doi.org/10.4161/psb.23681
Cohen AC, Bottini R, Piccoli P (2008) *Azospirillum brasilense* sp 245 produces ABA in chemically-defined culture medium and increases ABA content in Arabidopsis plants. Plant Growth Regul 54:9
Compant S, Clément C, Sessitsch A (2010a) Plant growth-promoting bacteria in the rhizo-and endosphere of plants: their role, colonization, mechanisms involved and prospects for utilization. Soil Biol Biochem 42:669–678
Compant S, Clément C, Sessitsch A (2010b) Plant growth-promoting bacteria in the rhizo- and endosphere of plants: their role, colonization, mechanisms involved and prospects for utilization. Soil Biol Biochem 42(5):669–678. https://doi.org/10.1016/j.soilbio.2009.11.024
Cramer GR, Nowak RS (1992) Supplemental manganese improves the relative growth, net assimilation and photosynthetic rates of salt-stressed barley. Physiol Plant 84(4):600–605
Dary M, Chamber-Perez MA, Palomares AJ, Pajuelo E (2010) "In situ" phytostabilisation of heavy metal polluted soils using *Lupinus luteus* inoculated with metal resistant plant-growth promoting rhizobacteria. J Hazard Mater 177:323–330

Deshmukh S, Verekar S, Bhave S (2015) Endophytic fungi: a reservoir of antibacterials. Front Microbiol 5:715. https://doi.org/10.3389/fmicb.2014.00715

Dey R, Pal KK, Bhatt DM, Chauhan SM (2004) Growth promotion and yield enhancement of peanut *Arachis hypogaea* L. by application of plant growth-promoting rhizobacteria. Microbiol Res 159(4):371–394

Evelin H, Devi TS, Gupta S, Kapoor R (2019) Mitigation of salinity stress in plants by arbuscular mycorrhizal symbiosis: current understanding and new challenges. Front Plant Sci 10:470

Fall AF, Nakabonge G, Ssekandi J, Founoune-Mboup H, Apori SO, Ndiaye A, Badji A, Ngom K (2022) Roles of arbuscular mycorrhizal fungi on soil fertility: contribution in the improvement of physical, chemical and biological properties of the soil. Front Fungal Biol 3:1–11

Friesen ML, Porter SS, Stark SC, Von Wettberg EJ, Sachs JL (2011) Microbially mediated plant functional Traits-2011.pdf. Annu Rev Ecol Evol Syst 42:23–46. https://doi.org/10.1146/annurev-ecolsys-102710-145039

Gamalero E, Glick BR (2011) Mechanisms used by plant growth-promoting bacteria. In: Maheshwari DKK (ed) Bacteria in agrobiology: plant nutrient management. Springer, Berlin/Heidelberg, pp 17–47

Gill S, Gill R, Trivedi DK, Anjum NA, Sharma KK, Ansari MW, Ansari AA, Johri AK, Pereira R, Prasad E, Varma A, Tuteja N (2016) *Piriformospora indica*: potential and significance in plant stress tolerance. Front Microbiol 7:332. https://doi.org/10.3389/fmicb.2016.00332

Gonzalez-Mendoza VM, de la Torre M, Rocha J (2022) Plant-growth promoting endophytic bacteria and their role for maize acclimation to abiotic stress; Submitted: 28 November 2022 Reviewed: 03 January 2023 Published: 17 February 2023; https://doi.org/10.5772/intechopen.109798

Hardoim PR, van Overbeek LS, Elsas JD (2008) Properties of bacterial endophytes and their proposed role in plant growth. Trends Microbiol 16(10):463–471. https://doi.org/10.1016/j.tim.2008.07.008. Epub 2008 Sep 12

Hata K, Atari R, Sone K (2002) Isolation of endophytic fungi from leaves of *Pasania edulis* and their within-leaf distributions. Mycoscience 43(5):369–373

Hori K, Matsumoto S (2010) Bacterial adhesion: from mechanism to control. Biochem Eng J 48:424–434

Kalamulla R, Karunarathna SC, Tibpromma S, Galappaththi MCA, Suwannarach N, Stephenson SL, Asad S, Salem ZS, Yapa N (2022) Arbuscular Mycorrhizal fungi in sustainable agriculture. Sustain For 14(19):12250. https://doi.org/10.3390/su141912250

Kandel S, Joubert P, Doty S (2017) Bacterial endophyte colonization and distribution within plants. Microorganisms 5(4):77. https://doi.org/10.3390/microorganisms504007

Kannahi M, Senbagam N (2014) Studies on siderophore production by microbial isolates obtained from rhizosphere soil and its antibacterial activity. J Chem Pharm Res 6(4):1142–1145

Kumar A, Verma JP (2018) Does plant-microbe interaction confer stress tolerance in plants: a review? Microbiol Res 207:41–52. https://doi.org/10.1016/j.micres.2017.11.004

Kumar V, Prasad R, Kumar M, Choudhary DK (eds) (2019) Microbiome in plant health and disease challenges and opportunities. Springer, Singapore

Lekberg Y, Rosendahl S, Olsson PA (2014) The fungal perspective of arbuscular mycorrhizal colonization in "nonmycorrhizal" plants. New Phytol 4:1399–1403

Li P et al (2019) Physiological and transcriptome analyses reveal short-term responses and formation of memory under drought stress in rice. Front Genet 10:55

Li Y, Duan T, Nan Z, Li Y (2021) Arbuscular Mycorrhizal fungus alleviates alfalfa leaf spots caused by *Phoma medicaginis* revealed by RNA-Seq analysis. J Appl Microbiol 130:547–560. [CrossRef] [PubMed]

Li Y, Xu J, Hu J, Zhang T, Wu X, Yang Y (2022) Arbuscular mycorrhizal fungi and glomalin play a crucial role in soil aggregate stability in Pb—contaminated soil. Int J Environ Res Public Health 19:5029

Long HH, Schmidt DD, Baldwin IT (2008) Native bacterial endophytes promote host growth in a species-specific manner; phytohormone manipulations do not result in common growth responses. PLoS One 3(7):2702–2709

Malarvizhi K, Murali TS, Kumaresan V (2023) Fungal endophytes of crop plants: diversity, stress tolerance and biocontrol potential. Egypt J Biol Pest Control 33:67. https://doi.org/10.1186/s41938-023-00711-1

Mercado-Blanco J, Lugtenberg B (2014) Biotechnological applications of bacterial endophytes. Curr Biotechnol 3:60–75. https://doi.org/10.2174/22115501113026660038

Munns R, Passioura JB, Colmer TD, Byrt CS (2020) Osmotic adjustment and energy limitations to plant growth in saline soil. New Phytol 225(3):1091–1096. https://doi.org/10.1111/nph.15862

Muthu Narayanan M, Ahmad N, Shivanand P, Metali F (2022) The role of endophytes in combating fungal- and bacterial-induced stress in plants. Molecules 27:6549

Omomowo OI, Babalola OO (2019) Bacterial and fungal endophytes: tiny giants with immense beneficial potential for plant growth and sustainable agricultural productivity. Microorganisms 7(11):481. https://doi.org/10.3390/microorganisms7110481

Pathak P, Rai VK, Can H, Singh SK, Kumar D, Bhardwaj N, Roychowdhury R, Basilio LC, de Azevedo K, Verma H, Kumar A (2022) Plant-endophyte interaction during biotic stress management. Plants (Basel) 11(17):2203

Potshangbam M, Devi SI, Sahoo D, Strobel GA (2017) Functional characterization of endophytic fungal community associated with *Oryza sativa* L. and *Zea mays* L. Front Microbiol 8:325

Poveda J, Baptista P, Sacristán S, Velasco P (2022) Editorial: beneficial effects of fungal endophytes in major agricultural crops. Front Plant Sci 13:1061112. https://doi.org/10.3389/fpls.2022.1061112

Rana N, Rathore A, Ghabru A, Chauhan S (2023) Endophytes: role and applications in sustainable agriculture. Pharm Innov Int J 12(3):139–151. https://doi.org/10.22271/tpi.2023.v12.i3b.19528

Sally E, Smith F, Smith A, Jakobsen I (2003) Mycorrhizal fungi can dominate phosphate supply to plants irrespective of growth responses. Plant Physiol 133(1):16–20. https://doi.org/10.1104/pp.103.024380

Sansanwal R, Ahlawat U, Priyanka, Wati L (2017) Role of endophytes in agriculture. Chem Sci Rev Lett 6:2397–2407

Santi C, Bogusz D, Franche C (2013) Biological nitrogen fixation in non-legume plants. Ann Bot 111:743–767. https://doi.org/10.1093/aob/mct048

Santoyoa G, Moreno-Hagelsiebb G, Orozco-Mosquedac M d C, Glickc BR (2016) Plant growth-promoting bacterial endophytes. Microbiol Res 183:92

Senthilkumar M, Anandham R, Madhaiyan M, Venkateswaran V, Sa T (2011) Endophytic bacteria: perspectives and applications in agricultural crop production. In: Bacteria in agrobiology: crop ecosystems. Springer, pp 61–96

Shaffique S, Khan MA, Wani SH, Pande A, Imran M, Kang S-M, Rahim W, Khan SA, Bhatta D, Kwon E-H, Lee I-J (2022) A review on the role of endophytes and plant growth promoting rhizobacteria in mitigating heat stress in plants. Microorganisms 10:1286

Sharma SK, Johri BN, Ramesh A, Joshi OP, Prasad SVS (2011) Selection of plant growth promoting *Pseudomonas* spp. that enhanced productivity of soybean-wheat cropping system in Central India. J Microbiol Biotechnol 21:1127–1142

Singh A, Pandey J (2012) Metal contamination and health risk from consumption of organically grown vegetables influenced by atmospheric deposition in a seasonally dry tropical region of India. Bull Environ Contamin Toxicol 89:384–389

Singh SR, Singh U, Chaubey AK, Bhat M (2010) Mycorrhizal fungi for sustainable agriculture-a review. Agric Rev 31(2):79–92

Spaepen S, Vanderleyden J (2011) Auxin and plant-microbe interactions. Cold Spring Harb Perspect Biol 3:1–13. https://doi.org/10.1101/cshperspect.a001438

Stajkovic O, De-Meyer S, Miličić B, Willems A, Delic D (2009) Isolation and characterization of endophytic non-rhizobial bacteria from root nodules of alfalfa (*Medicago sativa* L.). Bot Serb 33:107–114

Velázquez E, Silva LR, Ramírez-Bahena M-H, Peix A (2016) Diversity of potassium-solubilizing microorganisms and their interactions with plants. In: Potassium solubilizing microorganisms for sustainable agriculture. Springer, New Delhi, pp 99–110

Verma H, Kumar D, Kumar V, Kumari M, Singh SK, Sharma VK, Droby S, Santoyo G, White JF, Kumar A (2021) The potential application of endophytes in management of stress from drought and salinity in crop plants. Microorganisms 9(8):1729. https://doi.org/10.3390/microorganisms9081729

Vidal C, Gonazalez F, Santander C, Perez R, Gallardo V, Santos C et al (2022) Management of rhizosphere microbiota and plant production under drought stress: a comprehensive review. Plan Theory 11(18):2437. https://doi.org/10.3390/plants11182437

Watts D, Palombo EA, Jaimes Castillo A, Zaferanloo B (2023a) Endophytes in agriculture: potential to improve yields and tolerances of agricultural crops. Microorganisms 11(5):1276. https://doi.org/10.3390/microorganisms11051276. PMID: 37317250; PMCID: PMC10221004

Watts D, Palombo EA, Castillo AJ, Zaferanloo B (2023b) Endophytes in agriculture: potential to improve yields and tolerances of agricultural crops. Microorganisms 11(5):1276

Win KT, Tanaka F, Okazaki K, Ohwaki Y (2018) The ACC deaminase expressing endophyte *Pseudomonas* spp. enhances NaCl stress tolerance by reducing stress-related ethylene production, resulting in improved growth, photosynthesinthetic performance, and ionic balance in tomato plants. Plant Physiol Biochem 127:599–607. https://doi.org/10.1016/j.plaphy.2018.04.038

Yadav RS, Mahatma MK, Thirumalaisamy PP, Meena HN, Bhaduri D, Arora S, Panwar J (2017) Arbuscular Mycorrhizal Fungi (AMF) for sustainable soil and plant health in salt-affected soils. In: Arora S, Singh AK, Singh YP (eds) Bioremediation of salt affected soils: an Indian perspective. Springer International Publishing AG, Cham, pp 133–156

Yan L, Khan RAA (2021) Biological control of bacterial wilt in tomato through the metabolites produced by the biocontrol fungus, *Trichoderma harzianum*. Egypt J Biol Pest Control 31:5. [CrossRef]

Zhou X, Zhou L, Nie Y, Fu Y, Du Z, Shao J et al (2016) Similar responses of soil carbon storage to drought and irrigation in terrestrial ecosystems but with contrasting mechanisms: a meta-analysis. Agric Ecosyst Environ 228:70–81. https://doi.org/10.1016/j.agee.2016.04.030

Endophytic Fungi of Rice and Their Role in Sustainable Agriculture

11

Purnamsree Gogoi, Madhu Kamle, Vidhi Jain, Namita Ashish Singh, and Pradeep Kumar

Abstract

To address the global food demand, enhanced agricultural productivity has been attained through the application of endophytic microbes. Rice (*Oryza sativa*) serves as a staple food grain for a significant portion of the global population, contributing to over 20% of the daily calorie intake. The deepwater rice (DWR) variety is distinguished by its ability to elongate under deep flooding and endure anaerobic conditions resulting from rising water levels. Few crop plants exhibit the capacity to thrive under as diverse agroclimatic conditions as rice. The exploration and comprehension of rice-associated endophytic fungi bear considerable implications for sustainable agricultural practices. Fungal endophytes isolated from diverse rice varieties play a pivotal role in sustainable agriculture by their

Authors Purnamsree Gogoi and Madhu Kamle have equally contributed to this chapter.

P. Gogoi
Applied Microbiology Laboratory, Department of Forestry, North Eastern Regional Institute of Science and Technology, Nirjuli, Arunachal Pradesh, India

M. Kamle
Applied Microbiology Laboratory, Department of Forestry, North Eastern Regional Institute of Science and Technology, Nirjuli, Arunachal Pradesh, India

Department of Biochemistry, University of Lucknow, Lucknow, India

V. Jain · N. A. Singh
Applied Microbiology and Food Safety Laboratory, Department of Microbiology, Mohanlal Sukhadia University, Udaipur, Rajasthan, India

P. Kumar (✉)
Applied Microbiology Lab, Department of Botany, University of Lucknow, Lucknow, Uttar Pradesh, India

College of Life Science & Biotechnology, Korea University, Seoul, Republic of Korea

M. Meena et al. (eds.), *Metabolic Sustainability of Endophytes*, Microorganisms for Sustainability 52, https://doi.org/10.1007/978-981-96-4004-1_11

nitrogen fixation capability, plant growth-promoting activity, biocontrol potential, biofertilization, and management of biotic and abiotic stresses, among other functions. Given the manifold beneficial effects offered by rice fungal endophytes, numerous inquiries have been undertaken to elucidate plant–fungus symbiotic relationships. There are limited studies available on fungal endophytes associated with deepwater ecosystems. However, as elaborated in this review, the utilization of endophytic fungi holds significant potential as biofertilizers, bioinoculants, and biocontrol agents in sustainable agriculture.

Keywords

Endophytic fungi · Deepwater rice · Biocontrol activity · Biofertilizers · Sustainable agriculture

11.1 Introduction

With the introduction of the Green Revolution in agriculture, it has become possible to increase global food production. However, productivity levels remain low compared to the world's fast-growing population (Aamir et al. 2020). The growing global demand for food has been met through large-scale exploitation of resources. However, this has led to undesirable consequences in the ecosystem, such as soil salinity, extreme temperatures, groundwater pollution, nutrient depletion, soil degradation, and the emergence of new pests and pathogens. These significant changes have prompted scientists and researchers to seek modern, alternative, safe, and eco-friendly farming methods. Further, plant disease management strategies require synthetic chemicals, which harm the environment and human health. To overcome such undesired conditions associated with the soil ecosystem, we must explore natural resources for agricultural productivity to support increasing food demand. We must prioritize ecological sustainability and a clean, green environment. Utilizing beneficial microbial populations to enhance agriculture productivity is a recognized and innovative approach to sustainable farming.

11.1.1 Endophytes

The term "endophyte" comes from the Greek words "Endos," meaning inside, and "Phyton," meaning plant (Schulz and Boyle 2006). It was coined by de Bary in the year 1866 to distinguish living organisms inside the plant tissue from those on the plant's surface (Arora et al. 2014). Endophytes are defined as microorganisms that live inside the host tissue without causing any harm or producing symptoms in the host plant (Laskar et al. 2012). Endophytes may be bacteria and fungi, which are different from phytopathogens. They do not cause diseases and harm to host species. Endophytes are found in various plant tissue types, such as seeds and ovules, fruits, stems, roots, leaves, tubers, buds, xylem, rachis, and bark (Zhang et al.

2006). Several new endophytes may present in plants, indicating that endophytic microorganisms are essential to microbial biodiversity and antimicrobial properties (Zhang et al. 2009). The relationship between endophytes and plants is influenced by the genes of both organisms and environmental conditions. Endophytes enter the plant through the root tips and spread throughout the roots, stem, and leaves of the plant (Mano and Morisaki 2008). Endophytes play a crucial role in the plant's physiological functions, such as improving resistance to diseases and stress, as well as enhancing productivity (Laskar et al. 2012). Both climatic and soil-related factors contribute equally to the behavior and effects of endophytes on plants (Kandel et al. 2017). Under different conditions, they also help regulate plant growth-promoting hormones such as auxin, gibberellins, and cytokines (Egamberdieva et al. 2017). It is demonstrated that endophyte-infected plants often grow faster than noninfected ones because of the production of phytohormones, cytokines, and other plant growth promotion (PGP) substances (Tan and Zou 2001). The beneficial effects of endophytes are gaining importance with the potentiality of novel antimicrobial activity, role in plant growth promotion and sustainable agriculture. With the development of modern tools and techniques of molecular biology, it has now become possible to explore interactions of endophytes with their host and other pathogenic organisms and to establish the precise and correct identification of these microorganisms.

11.1.2 Endophytic Fungi

Fungal endophytes are well known to contribute to and enhance plant health benefits, enabling the adaptation of the host plant to biotic and abiotic stress. Endophytic fungi produce a variety of bioactive compounds that help protect host plants from pests and diseases. These compounds also play a crucial role in the fungi's interactions with their plant hosts (Lugtenberg et al. 2016). The functions of endophytic fungi in plant communities are diverse and can be influenced by factors such as the host plant, interactions with other microorganisms, and environmental conditions (Pili et al. 2016). Research has shown that endophytic fungi can benefit crops by promoting biomass production and enhancing resistance to various types of stress, both biotic and abiotic (Hamilton and Bauerle 2012). One significant contribution of fungal endophytes to agriculture is their ability to act as biocontrol agents (BCAs) against plant pathogens, insects, nematodes, and pests (Potshangbam et al. 2017). Additionally, fungal endophytes can induce systemic resistance in plants, which is an important mechanism for plant health (Arnold and Herre 2003). Fungal endophytes have the potential to be used as biofertilizers. Rhizosphere soil and endophytic fungi can enhance plant productivity, and endophytes can survive in the host plant for long periods. Therefore, fungi can be used in integrated pest management or organic farming systems for sustainable crop production (Leewijit et al. 2016).

11.1.3 Rice Plant

Rice (*Oryza sativa*) is a nutritious cereal crop primarily used for human consumption. It serves as a staple food grain for a substantial segment of the global population, contributing over 20% of the daily calorie intake (Ray et al. 2013). It is documented that Asian nations collectively account for approximately 90% of the worldwide rice production. The extensive geographical spread of rice cultivation has fostered the emergence of a wide array of varietal types (Khush 1997). *O. sativa* has many cultivars or ecotypes adapted to various environmental conditions. Rice is grown on all continents except Antarctica. There are hardly any crop plants that grow under as diverse agroclimatic conditions as rice. More than 40,000 varieties have been reported worldwide (Khush 1997).

The deepwater rice (DWR) variety is characterized by its ability to elongate under deep flooding, and the elongation is mainly the extension of internodes (Vergara et al. 1976). DWR can tolerate anaerobic conditions caused by rising water, utilizing accelerated stem extension that maintains the apical part above water (Baruah et al. 2006). It is recognized as a crucial food source for more than 100 million rural inhabitants residing in flood-prone regions (Catling 1993). Verma et al. (2001) stated that a poor understanding of DWR and its agro-system is an enormous loss in agricultural planning. As a result, it underestimates its importance as a staple food crop. The study on deepwater rice is significantly less by scientists and researchers for the reason that there may be difficulty in duplicating growing conditions and the more significant risks of flood damage (Vergara et al. 1976; Baruah et al. 2006).

Exploration of endophytic fungi is one of the significant unexploited resources that have potential applications in plant disease management strategies and several economically sound applications in future agriculture systems. Exploring and understanding endophytic fungi from different rice varieties significantly impacts sustainable agricultural practices. Tian et al. (2004) suggested that the chemical composition in soil, rice cultivars, and plant physiological status could influence endophytic microbial communities of rice plants. Laskar et al. (2012) identified *Cladosporium* sp., *Fusarium* sp., *Aspergillus* sp., *Penicillium* sp., *and Helminthosporium* sp. Wetland rice fields harbor more endophytes than upland ones, as mentioned in the study of fungal endophytes. By considering the capabilities of endophytic fungi, farmers can potentially reduce their reliance on chemical fertilizers and pesticides, leading to environment-friendly farming practices. The rice crops have unique adaptations to survive under diverse agroclimatic conditions, making them an exciting source for isolating endophytic fungi.

11.2 Role of Rice-Associated Endophytic Fungi in Sustainable Agriculture

Endophytic fungi play a crucial role in sustainable agriculture. They have gained increasing recognition for their significant contributions to the overall health of the rice plant, their ability to improve nutrient uptake, and their role in enhancing the

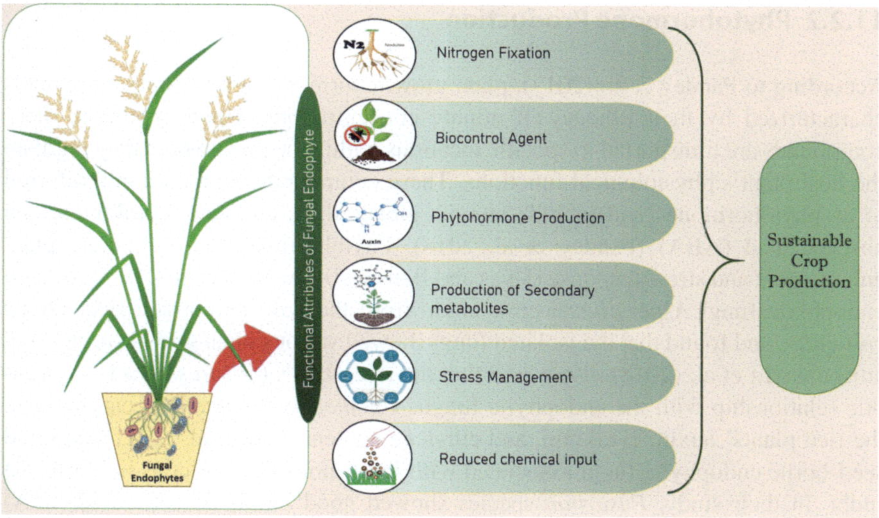

Fig. 11.1 Role of fungal endophytes in sustainable agriculture

plant's resilience to various stresses. By forming symbiotic relationships with the rice plant, these endophytic fungi effectively enhance rice productivity, while also supporting and promoting eco-friendly agricultural practices (Fig. 11.1).

11.2.1 Nitrogen Fixation

The process of nitrogen (N) fixation involves the conversion of atmospheric nitrogen into ammonia, and it is commonly utilized as a method of biofertilization by endophytic microorganisms to facilitate plant growth (Gaiero et al. 2013). Nitrogen (N) is an essential element for optimal plant growth, despite its abundant presence in the atmosphere (78%). The availability of nitrogen to plants can be enhanced through the application of endophytic microbes, which have the capability to convert free nitrogen into ammonia through biological nitrogen fixation using a complex enzyme system known as nitrogenase (Pandey et al. 2017, 2018). Research conducted by Yang et al. (2014) demonstrated that the symbiosis of rice with the fungal endophyte *Phomopsis liquidambari* led to improved rice yield. They expounded that the fungal endophyte *P. liquidambari* could enhance nitrogen (N) uptake and metabolism of rice plants under hydroponic conditions (Yang et al. 2014). Furthermore, research by Airin et al. (2023) proposed that isolated endophytic fungi *Aspergillus welwitschiae* from the wild halophytic rice, *Oryza coarctata*, could serve as a potential eco-friendly biofertilizer to enhance rice yield in highly saline coastal regions. The fungal endophytes exhibited various plant growth-promoting activities along with nitrogen fixation capability at high salinity.

11.2.2 Phytohormone Production

According to Pandey et al. (2017), plant growth hormones are organic compounds characterized by their efficacy in minute concentrations. Upon synthesis, these compounds are transported to specific locations within the plant where they regulate the host plant's physiological functions. These compounds are commonly referred to as phytohormones and include auxins, gibberellins, ethylene, cytokinins, and abscisic acid (ABA) (Pandey et al. 2018). Phytohormone helps regulate plant enlargement and stress response (Vyas and Bansal 2018). Airin et al. (2023) isolated endophytic fungi *Aspergillus welwitschiae* from the wild halophytic rice, *Oryza coarctata*, and found that the isolated fungi showed various plant growth-promoting attributes. Li et al. (2018) discovered that the rice plant's (*Oryza sativa* L.) symbiotic relationship with the endophytic fungus *Phomopsis liquidambar* can regulate the rice plants' auxin, cytokinin, and ethylene content. Roy et al. (2021) explored seed-borne endophytic fungi associated with indigenous rice varieties in Northeast India. In their study, *Fusarium* species showed good indole-3-acetic acid (IAA) production and several bioactive compounds; hence, they suggested that seed-borne endophyte can be used as a bioinoculant for crop improvement in future programs.

11.2.3 Biocontrol Agents

In crop productivity, pathogenic organisms that affect plant health are a major threat. Instead of using synthetic chemicals on crop fields, biocontrol agent is an effective and environment-friendly method against disease-causing pathogens (Vyas and Bansal 2018). Biocontrol agents (BCAs) are living organisms capable of inhibiting, antagonizing, and suppressing plant pathogens. Numerous endophytes have exhibited biocontrol potential against diverse plant pathogens, owing to their systematic distribution within plants and superior competence in colonizing plant tissues compared to phytopathogens (Rai et al. 2007; Pandey et al. 2018). According to Ryan et al. (2008), endophytes have demonstrated efficacy in combating plant disease development through de novo synthesis of novel compounds. Endophytic microbes employ various mechanisms as biocontrol agents, including the synthesis of chelators to effectively sequester vital elements required for growth, rendering them unavailable to phytopathogens; the production of antimicrobial compounds and cell wall lysing enzymes; the synthesis of phytohormones; and the ability to induce systemic resistance, among others (Weyens et al. 2009). The following list enumerates the biocontrol activity of fungal endophytes isolated from rice plants (Table 11.1).

11.2.4 Fungal Endophytes as a Source of Secondary Metabolites

Secondary metabolites are organic compounds not primarily involved in the growth, development, and reproduction of any organisms, but they often play an essential

Table 11.1 Rice-associated endophytic fungi as biocontrol agents

Sl. No.	Fungal endophytes	Host plant/ tissue	Biocontrol activity	Reference
1.	*Streptomyces* sp., *Chaetomium globosum*, *Penicillium chrysogenum*, *Fusarium oxysporum*, and *Cladosporium cladosporioides*	Leaves and roots of rice plants	Antifungal activity against *Rhizoctonia solani*, *Macrophomina phaseolina*, *Nigrospora oryzae*, *Phoma sorghina*, and *Alternaria alternata*	Naik et al. (2009)
2.	*Fusarium* sp. and *Trichoderma* sp.	Rice rhizosphere	Biocontrol activity against the rice root-knot nematode *Meloidogyne graminicola*	Le et al. (2009)
3.	*Antennariella placitae*	Rice plant	Antifungal activity against *Ustilaginoidea virens* strains by reducing the severity of false smut disease on susceptible rice plants	Andargie et al. (2017)
4.	*Fusarium*, *Sarocladium*, *Aspergillus* and *Penicillium*	Rice plant	Biocontrol activity against *Pythium ultimum*, *Scletorium oryzae*, *Rhizoctonia solani*, *Pyricularia oryzae*	Potshangbam et al. (2017)
5.	*Trichoderma* sp. 1, *Trichoderma* sp. 2, and *Chaetomium* sp.	Leaves, stems, and roots of rice plants	Controlled brown spot disease of rice by suppressing the growth of *Bipolaris oryzae*	Priyadarshani et al. (2018)
6.	*Paecilomyces tenuis* EF1, *Talaromyces pinophilus* EF2, *Nigospora oryzae* EF4, *Trichoderma longibrachiatum* EF5 and EF7, and *Aspergillus terreus* EF6	Leaves of rice plant	Antifungal activity against *Magnaporthe grisea*, *Macrophomina phaseolina*, *Rhizoctonia solani*, *Fusarium oxysporum*, and *Colletotrichum falcatum*	Sornakili et al. (2020)

role in the ecological interactions of fungi with their hosts or other microorganisms (Bills and Gloer 2016). Fungal endophytes produce various secondary metabolites, including alkaloids, terpenoids, polyketides, peptides, and phenolic compounds (Mousa and Raizada 2013; Meena et al. 2022). The chemical diversity of these metabolites makes them valuable for various applications in the pharmaceuticals, agriculture, and industry sectors (Anamika et al. 2018). Secondary metabolites are produced either for signaling or defense or in the process of establishing their interaction with the host plant, which can directly influence the attack of the pathogen (Baron and Rigobelo 2022). The diversity of secondary metabolites produced by fungal endophytes is vast, and different species or strains may synthesize unique compounds (Bills and Gloer 2016) (Table 11.2).

Table 11.2 Rice-associated fungal endophytes as source of secondary metabolite

Sl. No.	Secondary metabolites	Source species	Role in sustainable agriculture	References
1	6-isoprenylindole-3-carboxylic acid (indole-derivative)	*Colletotrichum* sp.	Antagonistic against pathogenic bacteria as well as fungi	Lu et al. (2000)
2	Alipathic organic acids (e.g., lactic, acetic, hydroxyl-methyl valeric acids, ethyl-hexyl acetate, salbutamol sulfate, dodecanone, octadeconoic, and butanoic acids), aromatic nitroamino compounds (Benzyl glycinate), and volatile metabolites (e.g., 2,3 butanediol dinitrate, pentadecane, tetramethyl hexahydro napthaleone, 1-iodo 2 methylundecane	*Paecilomyces tenuis* EF1, *Talaromyces pinophilus* EF2, *Nigrospora sphaerica* EF3, *Nigrospora oryzae* EF4, *Trichoderma longibrachiatum* EF5, and EF7, *Aspergillus terreus* EF6	Induced antagonistic activity against phytopathogens such as *Magnaporthe grisea*, *Rhizoctonia solani*, and *Macrophomina phaseolina*. The result paved the way to improve rice productivity without using pesticides and herbicides	Sornakili et al. (2020)
3	Phthalic acid, Pyrrolo (1,2-A), Pyrazine 1,4-dione, hexahydro-3-(2-met), Phenol 2,4-Bis (1,1 dimethyl ethyl), 1H-Inden-1-One, 2–3 dihydro-5,6 dimethoxy-3 methyl, 1 Propene, 3-(2-cyclopentenyl)-2-methyl-1,1 diphenyl, Phosgene Oxime, 1H-Benzotriazole, Benzaldehyde, 4,5-di benzyloxy-2-chloro, phenylacetic acid, 4-acetoxy-5-benzyloxy-2-fluoro	*Fusarium* sp. AM15 *Fusarium* sp. AM16 *Fusarium* sp. AM6 *Fusarium* sp. AM25	Antifungal activity against *Magnaporthe grisea*. The isolated fungal endophyte could be explored as bioinoculants for crop improvement in future research programs	Roy et al. (2021)

11.2.5 Protection Against Abiotic and Biotic Stress

Endophytic fungi present a promising alternative for safeguarding plants against both biotic and abiotic stresses (Baron and Rigobelo 2022). These fungi exhibit the capability to mitigate abiotic stresses, including drought, extreme temperatures, salinity, and toxic heavy metals (Aly et al. 2011; Yadav and Meena 2021). Abiotic stresses exert adverse effects on plant morphology and physiology through the genetic regulation of cellular pathways, leading to various dysfunctions (Egamberdieva et al. 2017). The heightened production of reactive oxygen species (ROS) is a common consequence of all stresses, stemming from imbalances in metabolic processes (Ganie et al. 2022). Notably, endophytic fungi associated with upland rice play a pivotal role in enhancing drought tolerance by detoxifying ROS through the augmentation of superoxide dismutase, peroxidase, and catalase activities in inoculated seedlings (Pang et al. 2020). Phytohormones indirectly facilitate plant resilience by modulating their response to abiotic stress (Baron and Rigobelo 2022). For example, abscisic acid (ABA) regulates stomatal closure to curtail excessive water loss and modulates stress response-related genes. The symbiosis with endophytic fungi leads to a reduction in ABA levels (Khan et al. 2015). Sodhi and Saxena (2023) investigated seed-borne endophytic fungi *Nigrospora oryzae* under drought and salinity tolerance. The findings indicated that the ability of *N. oryzae* to produce IAA effectively mitigates the impact of stress conditions, enabling the rice plant to sustain itself. Waqas et al. (2015) found that japonica rice treated with *Paecilomyces formosus* LWL1 produced phytohormones abscisic acid and jasmonic acid (JA), which induced heat stress tolerance. Mohd et al. (2017) focused on colonizing endophytic fungi *Piriformospora indica* on rice roots under arsenic toxicity. Their results suggested that *P. indica* from arsenic toxicity by three different mechanisms, viz. reducing the availability of free arsenic in the plant environment, biotransformation of the toxic arsenic salts insoluble particle matter and modulating the antioxidative system of the host cell.

For biotic stress protection, endophytic fungi activate induced systematic resistance (ISR), which produces metabolites against pathogens; moreover, parasitism or competition can inhibit the entry and growth of pathogens (Baron and Rigobelo 2022; Yadav and Meena 2022). The action mechanisms of endophytes in inducing resistance may encompass increased synthesis of phytoalexins and pathogenesis-related (PR) proteins, cell wall thickening via deposition of lignin and glucans, and augmented cuticle thickness. These mechanisms are accountable for impeding the penetration and development of pathogens in the host plant (Fontana et al. 2021). Microbiome-produced secondary metabolites play a substantial role in rice defense mechanisms (Ganie et al. 2022). Priming with endophyte *P. indica* can suppress rice root herbivory by the water weevil (*Lissorhoptrus oryzophilus*) and improve plant tolerance to the insect by stimulating root and shoot growth. This process is mediated by the gibberellic acid (GA) and jasmonic acid (JA) signaling pathway (Cosme et al. 2016). Srivastava et al. (2016) noted that the inoculation of SN13 enhances the rice immune response against *Rhizoctonia solani* by modulating levels of reactive oxygen species (ROS) through the production of ROS quenchers (arabitol, proline, mannitol, and phospholipases), production of non-metabolizable rare sugars such as turanose, activation of the defense response through the MAPK signaling pathway, and production of alkaloids (terpenes and quinazoline) while maintaining cell wall integrity.

11.2.6 Fungal Endophytes as Biofertilizers

It has been proven that the application of chemical fertilizers leads to the deterioration of crop productivity and soil fertility in agricultural systems (Aamir et al. 2020). Biofertilizers, which include bacteria, algae, fungi, etc., can fix atmospheric nitrogen (Aamir et al. 2020). Endophytic fungi can act as biofertilizers and are known to increase resistance to biotic and abiotic stress. A biofertilizer based on *Trichoderma* showed a significant increase in plant height, photosynthetic rate, chlorophyll a and b content, and stomatal conductance (Doni et al. 2018). Biofertilizer formulations can be made from endophytic fungi, reducing the use of mineral fertilizers in agroecosystems (Poveda et al. 2021). The effect of *Trichoderma* as a biofertilizer on the rice farming system showed improvements in growth, physiological traits, and yield of rice under the System of Rice Intensification (SRI) method (Aamir et al. 2020). The application of endophytic fungi associated with mangroves could promote the positive growth of black rice (*Oryza sativa* L. cultivar 'Cempolreng') (Boike et al. 2018). Fungal isolates such as *Aspergillus niger* and *Penicillium chrysogenum* from rice showed mycofertilizer potentials for the growth of rice crops. These isolates could also improve soil fertility and health (David et al. 2023).

11.3 Future Perspective and Concluding Remarks

In the context of a sustainable agroecosystem, the utilization of endophytic fungi represents an economically feasible, compelling, and environment-friendly approach. The endophytic fungi have shown many beneficial impacts on host plants and contribute significantly to sustainable agriculture. They have been documented for promoting plant growth through various functional attributes, viz., an increase in nutrient availability to the plants through fixation and solubilization of nutrients in the soil and the production of phytohormones. Besides plant growth promotion, endophytic fungi can increase plant resistance to different biotic and abiotic stresses. We can conclude that the successful use of endophytes would make crop production more eco-friendly and sustainable.

There has been an ever-increasing interest in finding eco-friendly and safe approaches to enhancing agricultural productivity. Fungal endophytes are essential components of sustainable agriculture. Considering the various beneficial effects of fungal endophytes, several investigations have been carried out to gain knowledge of the plant–fungus symbiotic relationship. In recent years, genetically modified endophytes have also been applied in sustainable agriculture (Vyas and Bansal 2018).

Despite the potential performance of fungal endophytes on other crop plants, it is remarkable that this group of fungi has gained little attention in the case of rice plants. Over the past decades, few publications have been available about endophytic microbiota communities in rice. Rice-associated endophytic fungi have been explored in a few countries, like Italy, China, Malaysia, and India (Pili et al. 2016). Moreover, no studies have investigated the endophytic fungi community associated with rice in deepwater ecosystems. Therefore, a more comprehensive study is required to explain endophytic fungi in deepwater rice, the total endophytic microbiota in deepwater ecosystems, and their phylogenetic relationship.

As stated in this review, endophytic fungi are helpful because they offer numerous direct and indirect benefits to crop plants. It is now possible to consider that no synthetic molecule or chemical has such a great range of positive impacts as these endophytic microorganisms. Hence, endophytic fungi have excellent potential as biofertilizers, biostimulation and/or biocontrol agents in sustainable agriculture.

Acknowledgments The authors would like to acknowledge higher authorities of their respective departments, institutes, and universities for their cooperation and support during their studies.

Author Contribution Conceptualization: PK & MK; Original draft writing: PG; Formatting and editing, critical revision: PK, MK, VJ & NAS.

Conflicts of Interests The authors declare no conflict of interest.

Funding No external funding was received for this research work.Data AvailabilityNo data were used for the research described in the article.

References

Aamir M, Rai KK, Zehra A, Kumar S, Yadav M, Shukla V, Upadhyay RS (2020) Fungal endophytes: classification, diversity, ecological role, and their relevance in sustainable agriculture. In: Microbial endophytes. Woodhead Publishing, pp 291–323

Airin AA, Arafat MI, Begum RA, Islam MR, Seraj ZI (2023) Plant growth-promoting endophytic fungi of the wild halophytic rice Oryza coarctata. Ann Microbiol 73(1):36

Aly AH, Debbab A, Proksch P (2011) Fungal endophytes: unique plant inhabitants with great promises. Appl Microbiol Biotechnol 90:1829–1845

Anamika, Joshi S, Sahgal M, Sahu S, Prakash A (2018) Fungal endophytes and their secondary metabolites: role in sustainable agriculture. In: Fungi and their role in sustainable development: current perspectives, pp 121–146

Andargie M, Congyi Z, Yun Y, Li J (2017) Identification and evaluation of potential biocontrol fungal endophytes against Ustilagonoidea virens on rice plants. World J Microbiol Biotechnol 33:1–10

Arnold AE, Herre EA (2003) Canopy cover and leaf age affect colonization by tropical fungal endophytes: ecological pattern and process in Theobroma cacao (Malvaceae). Mycologia 95(3):388–398

Arora S, Patel PN, Vanza MJ, Rao GG (2014) Isolation and characterization of endophytic bacteria colonizing halophyte and other salt tolerant plant species from coastal Gujarat. Afr J Microbiol Res 8(17):1779–1788

Baron NC, Rigobelo EC (2022) Endophytic fungi: a tool for plant growth promotion and sustainable agriculture. Mycology 13(1):39–55

Baruah KK, Rajkhowa SC, Das K (2006) Physiological analysis of growth, yield development and grain quality of some deep-water rice (Oryza sativa L.) Cultivars. J Agron Crop Sci 192(3):228–232

Bills GF, Gloer JB (2016) Biologically active secondary metabolites from the fungi. Microbiol Spectrum 4(6):4–6

Boike ST, Nadilla F, Baiduri N, Mardina V (2018) In vitro screening of endophytic fungi associated with mangroveas biofertilizer on the growth of black rice (Oryza sativa L. "Cempo Ireng"). In: IOP conference series. Materials science and engineering (Vol. 420, No. 1). IOP Publishing

Catling D (1993) Rice in deep water. Springer

Cosme M, Lu J, Erb M, Stout MJ, Franken P, Wurst S (2016) A fungal endophyte helps plants to tolerate root herbivory through changes in gibberellin and jasmonate signaling. New Phytol 211(3):1065–1076

David OM, Olawusi AC, Oluwole OA, Adeola PO, Odeyemi AT (2023) Isolation, molecular characterization and application of Aspergillus niger and Penicillium chrysogenum with Biofertilizer potentials to enhance rice growth. Trop J Nat Prod Res 7(4):2790

Doni F, Zain CRCM, Isahak A, Fathurrahman F, Anhar A, Mohamad WNAW et al (2018) A simple, efficient, and farmer-friendly Trichoderma-based biofertilizer evaluated with the SRI Rice Management System. Org Agric 8:207–223

Egamberdieva D, Wirth SJ, Alqarawi AA, Abd Allah EF, Hashem A (2017) Phytohormones and beneficial microbes: essential components for plants to balance stress and fitness. Front Microbiol 8:2104

Fontana DC, de Paula S, Torres AG, de Souza VHM, Pascholati SF, Schmidt D, Dourado Neto D (2021) Endophytic fungi: biological control and induced resistance to phytopathogens and abiotic stresses. Pathogens 10(5):570

Gaiero JR, McCall CA, Thompson KA, Day NJ, Best AS, Dunfield KE (2013) Inside the root microbiome: bacterial root endophytes and plant growth promotion. Am J Bot 100(9):1738–1750

Ganie SA, Bhat JA, Devoto A (2022) The influence of endophytes on rice fitness under environmental stresses. Plant Mol Biol 109, 447–467.

Hamilton CE, Bauerle TL (2012) A new currency for mutualism? Fungal endophytes alter antioxidant activity in hosts responding to drought. Fungal Divers 54:39–49

Kandel SL, Firrincieli A, Joubert PM, Okubara PA, Leston ND, McGeorge KM (2017) An in vitro study of bio-control and plant growth promotion potential of salicaceae endophytes. Front Microbiol 8:386. https://doi.org/10.3389/FMICB.2017.00386/BIBTEX

Khan AL, Hussain J, Al-Harrasi A, Al-Rawahi A, Lee IJ (2015) Endophytic fungi: resource for gibberellins and crop abiotic stress resistance. Crit Rev Biotechnol 35(1):62–74

Khush GS (1997) Origin, dispersal, cultivation and variation of rice. Plant Mol Biol 35(1):25–34

Laskar F, Nevita T, Sharma G (2012) Isolation and identification of endophytes from different cultivars of rice (*Oryza sativa* L.) under wetland and upland conditions in South Assam. J Pure Appl Microbiol 6:357–362

Le HTT, Padgham JL, Sikora RA (2009) Biological control of the rice root-knot nematode *Meloidogyne graminicola* on rice, using endophytic and rhizosphere fungi. Int J Pest Manag 55(1):31–36. https://doi.org/10.1080/09670870802450235

Leewijit T, Pongnak W, Soytong K, Poeaim S (2016) Isolation of soil and endophytic fungi from rice (Oryza sativa L.). Int J Agric Technol 12(7.2):2191–2202

Li X, Zhou J, Xu RS, Meng MY, Yu X, Dai CC (2018) Auxin, cytokinin, and ethylene involved in rice N availability improvement caused by endophyte Phomopsis liquidambari. J Plant Growth Regul 37:128–143

Lugtenberg BJ, Caradus JR, Johnson LJ (2016) Fungal endophytes for sustainable crop production. FEMS Microbiol Ecol 92(12):fiw194

Lu H, Zou WX, Meng JC, Hu J, Tan RX (2000) New bioactive metabolites produced by Colletotrichum sp., an endophytic fungus in Artemisia annua. Plant Sci. 151(1):67-73

Mano H, Morisaki H (2008) Endophytic bacteria in the rice plant. Microbes Environ 23(2):109–117

Meena M, Yadav G, Sonigra P, Nagda A, Mehta T, Swapnil P, Harish, Marwal A (2022) Role of elicitors to initiate the induction of systemic resistance in plants to biotic stress. Plant Stress 5:100103. https://doi.org/10.1016/j.stress.2022.100103

Mohd S, Shukla J, Kushwaha AS, Mandrah K, Shankar J, Arjaria N et al (2017) Endophytic fungi Piriformospora indica mediated protection of host from arsenic toxicity. Front Microbiol 8:754

Mousa WK, Raizada MN (2013) The diversity of anti-microbial secondary metabolites produced by fungal endophytes: an interdisciplinary perspective. Front Microbiol 4:65

Naik BS, Shashikala J, Krishnamurthy YL (2009) Study on the diversity of endophytic communities from rice (*Oryza sativa* L.) and their antagonistic activities in vitro. Microbiol Res. 164(3):290-6

Pandey PK, Singh MC, Singh SS, Kumar AK, Pathak MM, Shakywar RC, Pandey AK (2017) Inside the plants: endophytic bacteria and their functional attributes for plant growth promotion. Int J Curr Microbiol Appl Sci 6(2):11–21

Pandey PK, Singh S, Singh MC, Singh AK, Yadav SK, Pandey AK, Heisnam P (2018) Diversity, ecology, and conservation of fungal and bacterial endophytes. In: Microbial resource conservation: conventional to modern approaches, pp 393–430

Pang Z, Zhao Y, Xu P, Yu D (2020) Microbial diversity of upland rice roots and their influence on rice growth and drought tolerance. Microorganisms 8(9):1329

Pili NN, França SC, Kyndt T, Makumba BA, Skilton R, Höfte M et al (2016) Analysis of fungal endophytes associated with rice roots from irrigated and upland ecosystems in Kenya. Plant Soil 405:371–380

Potshangbam M, Devi SI, Sahoo D, Strobel GA (2017) Functional characterization of endophytic fungal community associated with Oryza sativa L. and Zea mays L. Front Microbiol 8:325

Poveda J, Eugui D, Abril-Urías P, Velasco P (2021) Endophytic fungi as direct plant growth promoters for sustainable agricultural production. Symbiosis 85(1):1–19

Priyadarshani CDN, Deshappriya N, Sandamali TGI (2018) Effect of fungal endophytes of rice variety Ld 368 on growth and brown spot disease incidence of rice. Trop Plant Res 5(3):292–302

Rai R, Dash PK, Prasanna BM, Singh A (2007) Endophytic bacterial flora in the stem tissue of a tropical maize (Zea mays L.) genotype: isolation, identification and enumeration. World J Microbiol Biotechnol 23:853–858

Ray DK, Mueller ND, West PC, Foley JA (2013) Yield trends are insufficient to double global crop production by 2050. PLoS One 8(6):e66428

Roy S, Chiranjib M, Talukdar R, Wary S, Tayung K (2021) Seed-borne endophytic fungi associated with some indigenous rice varieties of North East India and their growth promotion and antifungal potential. Indian J Agric Res. 55(5):603-8

Ryan RP, Germaine K, Franks A, Ryan DJ, Dowling DN (2008) Bacterial endophytes: recent developments and applications. FEMS Microbiol Lett 278(1):1–9

Schulz B, Boyle C (2006) What are endophytes? In: Microbial root endophytes. Springer Berlin Heidelberg, Berlin/Heidelberg, pp 1–13

Sodhi GK, Saxena S (2023) Plant growth-promoting endophyte Nigrospora oryzae mitigates abiotic stress in rice (Oryza sativa L.). FEMS Microbiol Ecol 99(9):fiad094

Sornakili A, Thankappan S, Sridharan AP, Nithya P, Uthandi S (2020) Antagonistic fungal endophytes and their metabolite-mediated interactions against phytopathogens in rice. Physiol Mol Plant Pathol 112:101525

Srivastava S, Bist V, Srivastava S, Singh PC, Trivedi PK, Asif MH et al (2016) Unraveling aspects of Bacillus amyloliquefaciens mediated enhanced production of rice under biotic stress of Rhizoctonia solani. Front Plant Sci 7:587

Tan RX, Zou WX (2001) Endophytes: a rich source of functional metabolites. Nat Prod Rep 18(4):448–459

Tian XL, Cao LX, Tan HM, Zeng QG, Jia YY, Han WQ, Zhou SN (2004) Study on the communities of endophytic fungi and endophytic actinomycetes from rice and their antipathogenic activities in vitro. World J Microbiol Biotechnol 20:303–309

Vergara BS, Jackson B, De Datta SK (1976) Deep water rice and its response to deep water stress. Clim Rice 1:301–331

Verma SC, Ladha JK, Tripathi AK (2001) Evaluation of plant growth promoting and colonization ability of endophytic diazotrophs from deep water rice. J Biotechnol 91(2–3):127–141

Vyas P, Bansal A (2018) Fungal endophytes: role in sustainable agriculture. In: Fungi and their role in sustainable development: current perspectives, pp 107–120. https://doi.org/10.1007/978-981-13-0393-7_7

Waqas M, Khan AL, Shahzad R, Ullah I, Khan AR, Lee IJ (2015) Mutualistic fungal endophytes produce phytohormones and organic acids that promote japonica rice plant growth under prolonged heat stress. J Zhejiang Univ Sci B 16(12):1011

Weyens N, van der Lelie D, Taghavi S, Newman L, Vangronsveld J (2009) Exploiting plant–microbe partnerships to improve biomass production and remediation. Trends Biotechnol 27(10):591–598

Yadav G, Meena M (2021) Bioprospecting of endophytes in medicinal plants of Thar Desert: An attractive resource for biopharmaceuticals. Biotechnol Rep 30:e00629. https://doi.org/10.1016/j.btre.2021.e00629

Yadav G, Meena M (2022) Biological control of plant diseases by endophytes. In: Shah MP, Deka D (eds) Endophytic association: what, why and how (volume: developments in applied microbiology and biotechnology). Academic/Elsevier, pp 119–135. https://doi.org/10.1016/B978-0-323-91245-7.00007-9

Yang B, Ma HY, Wang XM, Jia Y, Hu J, Li X, Dai CC (2014) Improvement of nitrogen accumulation and metabolism in rice (Oryza sativa L.) by the endophyte Phomopsis liquidambari. Plant Physiol Biochem 82:172–182

Zhang HW, Song YC, Tan RX (2006) Biology and chemistry of endophytes. Nat Prod Rep 23(5):753. https://doi.org/10.1039/b609472b

Zhang P, Zhou PP, Yu LJ (2009) An endophytic taxol-producing fungus from Taxus media, Cladosporium cladosporioides MD2. Curr Microbiol 59:227–232

Plant Endophytic Fungi: Communication Mechanism to Deal with Biotic Stress

12

Ankush Yadav, Rupesh Bhardwaj, Pritee Kumari, Laishram Amarjit Singh, Prashant Swapnil, and Mukesh Meena

12.1 Introduction

Fungi, eukaryotic and heterotrophic organisms, exhibit diverse morphologies, including elongated hyphae and budding yeast cells, reproducing through both meiotic and mitotic processes. This evolutionary diversity has resulted in fungi becoming one of the most species-rich groups among eukaryotes. Despite their vast biodiversity, our understanding of fungal distribution and diversity, especially for endophytic fungi, remains limited due to their complex lifestyles and interactions with other organisms. Endophytic fungi are a fascinating group of microorganisms that live within the tissues of plants without causing apparent harm. The term "endophytic fungi" refers to those fungal species that inhabit the internal tissues of plants throughout all or part of their life cycle (Wen et al. 2022). Unlike pathogenic fungi, which cause disease, endophytic fungi form mutualistic relationships with their host plants, often providing various benefits that enhance plant growth and survival (Seth et al. 2023; Yadav and Meena 2021, 2022). These fungi establish a symbiotic relationship with plants, wherein both parties derive benefits. They interact with plants through vertical and horizontal transmissions. Vertical transmission occurs when fungi pass from parent plants to offspring via seeds, ensuring stable, inherited benefits like stress tolerance and nutrient uptake. Horizontal transmission involves the spread of fungi between plants through environmental means such as spore

A. Yadav · R. Bhardwaj · P. Kumari · L. A. Singh · P. Swapnil (✉)
Cyanobacterial and Plant Stress Biology Lab, School of Basic Science, Department of Botany, Central University of Punjab, Bathinda, Punjab, India
e-mail: prashant.swapnil@cup.edu.in

M. Meena (✉)
Laboratory of Phytopathology and Microbial Biotechnology, Department of Botany, Mohanlal Sukhadia University, Udaipur, Rajasthan, India
e-mail: drmukeshmeena321@mlsu.ac.in

M. Meena et al. (eds.), *Metabolic Sustainability of Endophytes*, Microorganisms for Sustainability 52, https://doi.org/10.1007/978-981-96-4004-1_12

dispersal or insect vectors, fostering genetic diversity and adaptability (Aharwal et al. 2021). This method allows plants to acquire new beneficial traits in response to environmental changes. They colonize the intercellular spaces and vascular tissues of their host plants, gaining access to nutrients and a protected environment. In return, they contribute to the plant's health by improving nutrient uptake, enhancing stress resistance, and producing bioactive compounds that can deter pathogens and pests. This symbiotic relationship is crucial for the adaptation of plants to diverse and challenging environments. One of the remarkable features of endophytic fungi is their ability to produce a wide array of metabolites. These include phytohormones, which regulate plant growth and development, as well as secondary metabolites with antimicrobial, antifungal, and insecticidal properties (Jha et al. 2023). Volatile organic compounds (VOCs) produced by endophytic fungi can also enhance plant defenses and stress tolerance (Ling et al. 2024). Additionally, they improve the absorption of vital nutrients like nitrogen, magnesium, zinc, sulfur, and phosphorus from the soil, further supporting plant growth and survival (Chaudhary et al. 2022). Research on endophytic fungi continues to uncover their profound impact on plant biology and ecosystem functioning. Endophytic fungi can also enhance plant growth, protect against pathogens, and improve tolerance to abiotic stresses such as drought and salinity. This knowledge has significant implications for sustainable agriculture, as harnessing the benefits of endophytic fungi can reduce the need for chemical fertilizers and pesticides (Watts et al. 2023; Meena et al. 2023). By promoting plant health naturally, endophytic fungi contribute to more resilient agricultural systems. Future research aims to explore the full potential of these fungi, including their applications in crop improvement, biocontrol, and environmental conservation. As we deepen our understanding of endophytic fungi and their interactions with plants, we open new avenues for innovative and sustainable approaches to agriculture and plant science.

12.1.1 Current Knowledge of Endophytic Fungi

Endophytic fungi, residing within plant tissues without causing apparent harm, are a diverse group of microorganisms that play significant roles in plant health and development. The diversity of endophytic fungi is vast, encompassing various species that inhabit different plant parts, including roots, stems, leaves, and seeds (Toppo et al. 2022). This diversity reflects the fungi's ability to adapt to a wide range of host plants and environmental conditions. The earliest fossil records of endophytic fungi date back to the Devonian period, approximately 400 million years ago, suggesting their long-standing presence and evolutionary significance (Ogbe et al. 2020). These ancient associations indicate that endophytic fungi have played a crucial role in plant evolution, contributing to the success and diversification of land plants. Gene compatibility between endophytic fungi and their host plants is a crucial factor determining the success of their symbiotic relationship. Specific genes in both fungi and plants are responsible for the recognition and establishment of mutualistic interactions (Priyashantha et al. 2023; Chandran et al. 2021). For example, the *sym* (symbiosis) genes in plants and corresponding *effector* genes in fungi facilitate the colonization process (Ho-Plágaro and García-Garrido 2022). These genes code for proteins that are involved in signaling pathways, enabling the fungi to suppress plant immune responses and establish a stable symbiotic relationship. Such genetic compatibilities allow endophytic fungi to colonize plant tissues effectively, facilitating nutrient exchange and enhancing plant stress tolerance. The presence and activity of endophytic fungi are influenced by various factors, including geographical conditions, climate, soil nutrients, and plant species (Rigobelo and Baron 2021). Geographical location and climate affect the composition and abundance of endophytic fungal communities, as different fungi have varying requirements for temperature, humidity, and soil conditions (Fan et al. 2023). Fungi in tropical regions exhibit higher diversity due to favorable environmental conditions. Nutrient availability in the soil also plays a pivotal role, as it can influence the growth and metabolic activity of endophytic fungi, thereby affecting their symbiotic functions. Additionally, plant species and their specific root exudates can influence the selection of particular endophytic fungi, which shaping the fungal community composition (Broeckling et al. 2008; Kumar et al. 2022). The role of endophytic fungi in plants can be compared to that of beneficial gut microbes in humans (Akram et al. 2023). Just as gut microbes aid in digestion, nutrient absorption, and immune function, endophytic fungi enhance nutrient uptake, improve stress tolerance, and bolster plant immunity (Fig. 12.1). Both systems involve complex interactions

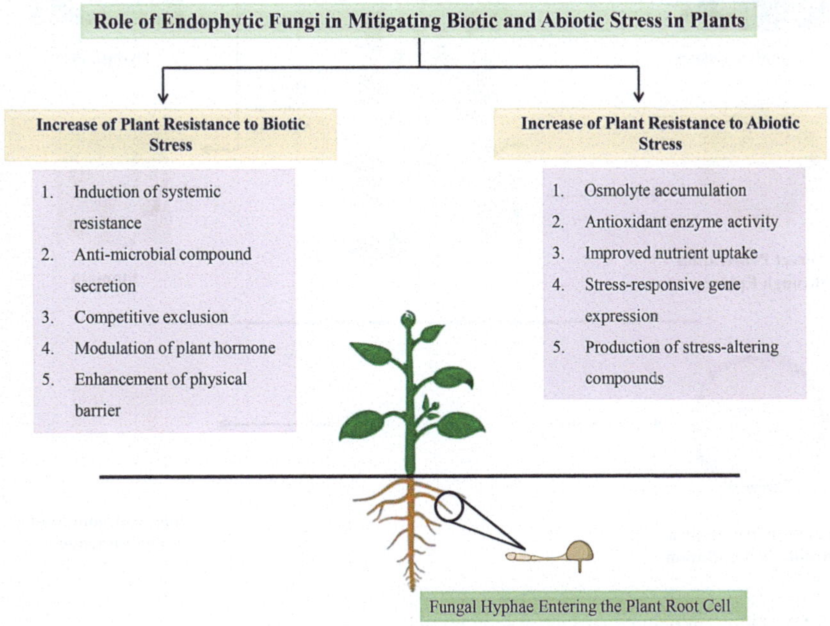

Fig. 12.1 Role of endophytic fungi in mitigating biotic and abiotic stress in plants

where microorganisms and their hosts benefit from each other. In plants, endophytic fungi produce various metabolites, such as phytohormones (e.g., indole-3-acetic acid [IAA]) and secondary metabolites (e.g., alkaloids, terpenoids), that promote growth and defense mechanisms (Singh and Kumar 2023). Similarly, human gut microbes produce essential vitamins, short-chain fatty acids, and other metabolites that contribute to overall health.

12.2 Entry of Fungi into Plant Tissue

The entry of endophytic fungi into plant tissues is a complex process that involves multiple pathways and sophisticated mechanisms to establish a successful symbiotic relationship. Different pathways of entry include direct penetration of the plant epidermis, entry through natural openings such as stomata or lenticels, and invasion through wounds or root tips (Chaudhary et al. 2022) (Fig. 12.2). Arbuscular mycorrhizal fungi (AMF) typically enter plant roots through direct penetration of the epidermal cells, forming specialized structures known as arbuscules within the root cortical cells (Wahab et al. 2023). This process involves the expression of a set of conserved genes known as the common symbiosis signaling pathway (CSSP) genes,

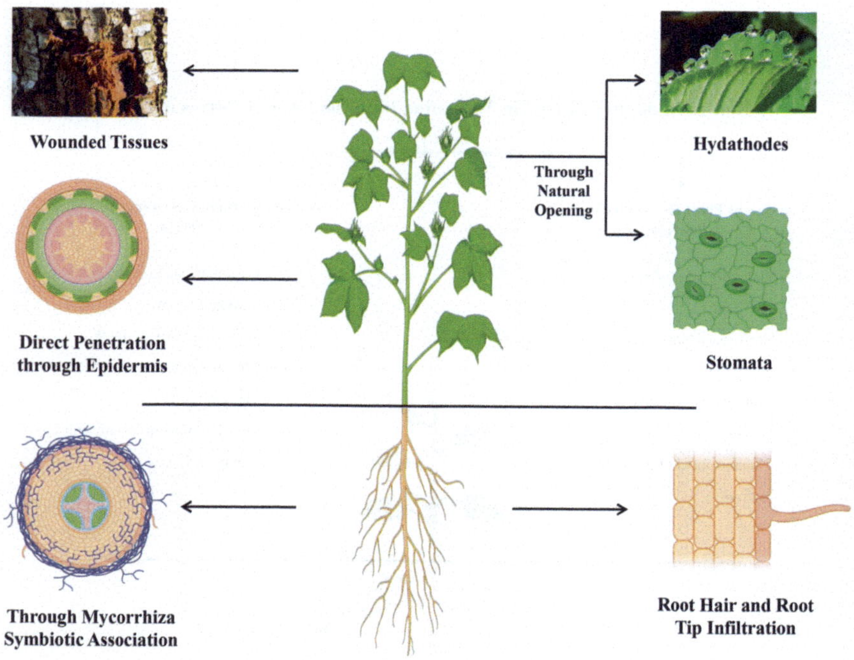

Fig. 12.2 Different pathways of entry used by fungi to enter plants

which are essential for the establishment and maintenance of the symbiotic relationship (Ho-Plágaro and García-Garrido 2022).

One of the critical steps in the entry of fungi into plant tissues is the overcoming of physical barriers presented by the plant cell wall. Endophytic fungi employ various strategies to breach these barriers. The production of cell wall degrading enzymes (CWDEs), such as cellulases, hemicellulases, and pectinases, is a common mechanism used by fungi to break down the plant cell wall components (Kubicek et al. 2014). For example, *Piriformospora indica* produces cellulases that degrade the cellulose component of the plant cell wall, facilitating its entry into the root tissues (Wan et al. 2021). Additionally, some fungi secrete cutinases that degrade the cuticle layer, allowing for easier penetration through the epidermis.

Upon entering the plant tissues, endophytic fungi often envelop their hyphae in a plant-derived membrane, known as the periarbuscular membrane (PAM) in the case of AMF (Johri et al. 2015). This membrane separates the fungal hyphae from the plant cytoplasm, maintaining a distinct interface where nutrient exchange occurs without direct cytoplasmic contact. This encapsulation helps in avoiding the activation of the plant's immune response, allowing the fungi to reside within the plant tissues without eliciting a strong defense reaction. The interaction between endophytic fungi and plant cells is also mediated by the expression of specific extracellular matrix (ECM) genes (Alam et al. 2021). These genes encode proteins that facilitate the attachment of fungal hyphae to the plant cell walls and the formation of a stable interface. The expression of ECM genes in *Trichoderma harzianum* has been shown to be critical for its colonization of maize roots, aiding in the establishment of a mutualistic relationship (Saravanakumar et al. 2017).

Studies on the model legume *Medicago truncatula* have revealed that the CSSP genes, such as *DMI1*, *DMI2*, and *DMI3*, are crucial for the formation of arbuscules by AMF (Hogg et al. 2005). These genes encode proteins involved in calcium signaling and nuclear responses that are essential for symbiosis. Moreover, the secretion of effector proteins by fungi can modulate plant immune responses, promoting symbiotic compatibility. *Laccaria bicolor*, an ectomycorrhizal fungus, secretes small secreted proteins (SSPs) that suppress plant defense mechanisms, facilitating its entry and colonization of poplar roots (Cope et al. 2019).

12.2.1 Plant Immune Response to Foreign Invasion

Plants have evolved complex immune systems to defend against a wide array of pathogens, including bacteria, fungi, viruses, and nematodes. The plant immune system is generally divided into two main layers: pattern-triggered immunity (PTI) and effector-triggered immunity (ETI). The first layer of defense involves the recognition of microbe-associated molecular patterns (MAMPs) and pathogen-associated molecular patterns (PAMPs) by pattern recognition receptors (PRRs) located on the plant cell surface (Mogensen 2009) (Fig. 12.3). MAMPs and PAMPs are conserved molecular signatures found in many pathogens. PTI involves several responses, including the production of reactive oxygen species (ROS), activation of

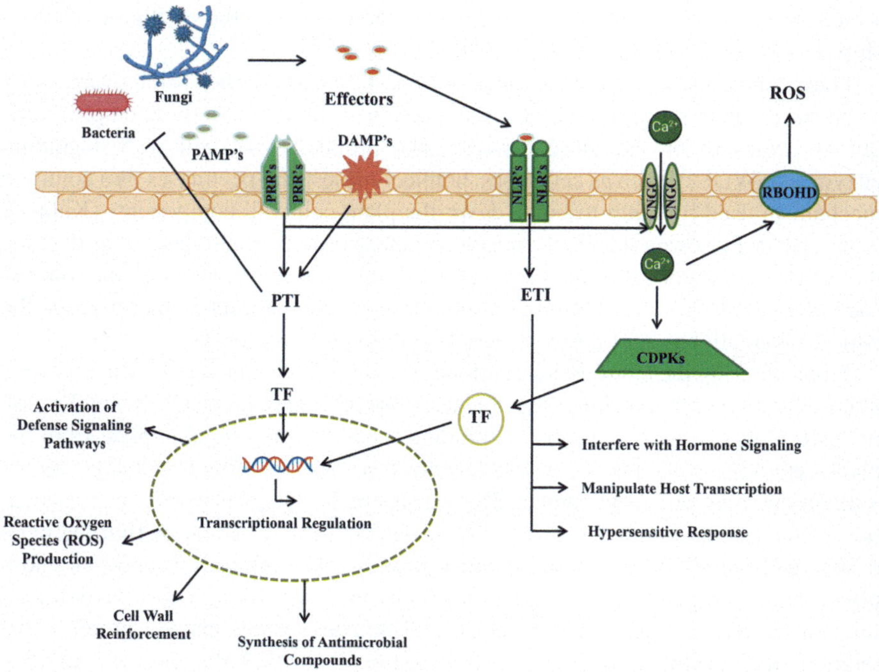

Fig. 12.3 Plants perceive pathogen-associated molecular patterns (PAMPs) through pattern recognition receptors (PRRs) on their cell surface, triggering a cascade of immune responses. Upon recognition, these receptors activate signaling pathways that lead to the rapid production of reactive oxygen species (ROS) and an influx of calcium ions into the cytosol. Calcium signaling is mediated by cyclic nucleotide-gated channels (CNGCs), which facilitate the activation of downstream proteins such as calcium-dependent protein kinases (CDPKs). These early signaling events result in the transcriptional activation of defense-related genes and the production of antimicrobial compounds. Abbreviations: pathogen-associated molecular patterns (PAMPs), pattern recognition receptors (PRRs), reactive oxygen species (ROS), cyclic nucleotide-gated channels (CNGCs), calcium-dependent protein kinases (CDPKs), leucine-rich repeat receptors (NLRs), effector-triggered immunity (ETI), plant innate immunity (PTI), transcription factor (TF), respiratory burst oxidase homologue D (RBOHD)

mitogen-activated protein kinases (MAPKs), and transcriptional reprogramming to produce defense-related proteins. Bacterial flagellin recognized by the PRR FLS2 in *Arabidopsis* triggers a robust immune response. Similarly, chitin recognition by the LysM receptor kinase CERK1 in rice leads to the activation of plant innate immunity (PTI) (Noman et al. 2019). In addition to MAMPs and PAMPs, plants recognize damage-associated molecular patterns (DAMPs), which are endogenous signals released upon cell damage. DAMPs, such as oligogalacturonides, cellodextrins, and ATP, act as signals to amplify the immune response (Hou et al. 2019). Oligogalacturonides, derived from the breakdown of pectin in the plant cell wall, can trigger defense responses by binding to specific receptors. Cellodextrins, produced from cellulose degradation, also serve as DAMPs that signal the presence of damage and initiate defense mechanisms. Scientific findings have highlighted various examples of DAMP and PAMP recognition, leading to enhanced immunity in different plants. In tomatoes, the receptor-like kinase (RLK) EIX2 recognizes the fungal protein ethylene (ET)-inducing xylanase, triggering PTI (Ron and Avni 2004). Also, in rice, PRR XA21 detects a conserved microbial signature in *Xanthomonas oryzae*, leading to strong immune responses that prevent bacterial blight (Kumar et al. 2020).

Despite these robust defenses, some pathogens secrete effector proteins that can suppress PTI and promote infection. These effectors are recognized by intracellular nucleotide-binding leucine-rich repeat receptors (NLRs), leading to the activation of effector-triggered immunity (ETI). ETI is often more robust and specific than PTI and is characterized by localized cell death known as the hypersensitive response (HR), which restricts pathogen growth (Chen et al. 2022a, b). Effectors can be broadly categorized based on their function, such as those that inhibit host proteases, manipulate host transcription, or interfere with hormone signaling. Other than that, one of the crucial components of the plant immune signaling network is calcium signaling. Cyclic nucleotide-gated channels (CNGCs) play a vital role in mediating cytosolic calcium influx upon pathogen detection. When pattern recognition receptors (PRRs) recognize pathogen-associated molecular patterns (PAMPs) or damage-associated molecular patterns (DAMPs), they trigger an influx of Ca^{2+} into the cytosol from both external and internal stores. This rise in cytosolic Ca^{2+} is sensed by various Ca^{2+}-binding proteins, such as calmodulin (CaM) and calcium-dependent protein kinases (CDPKs), which then activate downstream signaling pathways (Wang and Luan 2024). CDPKs can phosphorylate target proteins that regulate defense gene expression and metabolic responses. The production of reactive oxygen species (ROS) in the apoplast is another early response to pathogen recognition. Nicotinamide adenine dinucleotide phosphate oxidases (NADPH oxidases), such as respiratory burst oxidase homologue D (RBOHD) in *Arabidopsis*, are activated upon pathogen detection, leading to a burst of ROS (Camejo et al. 2016). These ROS function as signaling molecules to further amplify the defense response, as well as having direct antimicrobial effects. Additionally, ROS can activate mitogen-activated protein kinase (MAPK) cascades, which are crucial for transmitting the immune signal from the cell surface to the nucleus. Upon activation by ROS or other signaling molecules, MAPKs phosphorylate various transcription factors and other regulatory proteins, leading to changes in gene expression (Son et al. 2013). In *Arabidopsis*, the MAPK cascade involving MPK3 and MPK6 is activated by the perception of flagellin and is essential for the transcriptional activation of numerous defense-related genes. Nitric oxide (NO) is another important signaling molecule in plant defense. NO can modulate the activity of various proteins through S-nitrosylation and interacts with ROS to form reactive nitrogen species (RNS), which further contribute to the defense response. NO signaling is involved in the regulation of stomatal closure, a defense mechanism that prevents pathogen entry through stomata (Khan et al. 2023). Phytohormones such as salicylic acid (SA), jasmonic acid (JA), and ethylene (ET) play critical roles in the regulation of plant immune responses. SA is primarily associated with defense against biotrophic pathogens and the establishment of systemic acquired resistance (SAR).

A transcription factor NPR1 plays a central role in the SA signaling pathway (Backer et al. 2019; Meena et al. 2022). Upon SA accumulation, NPR1 translocates to the nucleus and interacts with TGA transcription factors to induce the expression of *PR* genes. These genes encode proteins with antimicrobial properties that enhance the plant's resistance to pathogens. On other hand, JA and ET are involved in defense against necrotrophic pathogens and herbivorous insects. ERF transcription factors are involved in the ET signaling pathway, regulating genes associated with defense against necrotrophic pathogens. In the JA signaling pathway, the MYC transcription factors are critical regulators of defense gene expression (Lorenzo et al. 2004). For example, MYC2 controls the expression of genes involved in the biosynthesis of secondary metabolites such as alkaloids and terpenoids, which have defensive roles. Moreover, the interaction between JA and ET signaling pathways is mediated by the transcription factor ORA59, which integrates signals from both pathways to regulate defense responses (Chen et al. 2019). The interplay between these hormones fine-tunes the plant's immune response to different types of pathogens. *Arabidopsis* mutant *npr1*, which is defective in SA signaling, shows enhanced susceptibility to biotrophic pathogens, highlighting the crucial role of SA in plant defense. Callose deposition at the site of pathogen attack is another defense mechanism in plants. Callose is a β-1,3-glucan polymer that is deposited at the cell wall to reinforce it and prevent pathogen penetration (German et al. 2022). This process is regulated by various signaling pathways, including those involving ROS, Ca^{2+}, and MAPKs. In *Arabidopsis*, the callose synthase gene *GSL5* is required for callose deposition during pathogen infection (Wawrzynska et al. 2010).

12.3 Microbial Receptors and Signaling Pathways

12.3.1 Molecular Dialogue Between Endophytic Fungi and Plants

12.3.1.1 Rhizosphere Attraction and Root Exudates

Plant roots secrete a diverse array of exudates that play a crucial role in shaping the microbial community in the rhizosphere. These exudates act as signals and nutrients, attracting endophytic fungi and facilitating their colonization. The key components of root exudates include flavonoids, strigolactones, and other signaling molecules (Steinkellner et al. 2007). Strigolactones are crucial in attracting arbuscular mycorrhizal fungi (AMF), by inducing hyphal branching and promoting fungal growth towards the roots. These exudates create a conducive environment that encourages the proliferation of endophytic fungi in the rhizosphere, establishing the foundation for a symbiotic relationship (Liao et al. 2018).

12.3.1.2 Common Symbiosis Signaling Pathway (CSSP)

Once in the vicinity of the plant roots, endophytic fungi engage in a sophisticated molecular dialogue with the plant to establish symbiosis. This process is mediated by the common symbiosis signaling pathway (CSSP), which is a shared signaling mechanism used by plants to interact with different types of symbiotic fungi, including AMF and certain endophytic fungi (Fig. 12.4).

12.3.2 Perception of Fungal Signals

The process begins with the perception of fungal signals by plant receptors. Fungi release signaling molecules, such as mycorrhizal lipochitooligosaccharides (Myc-LCOs) or other fungal-derived signals, which are recognized by LysM receptor-like

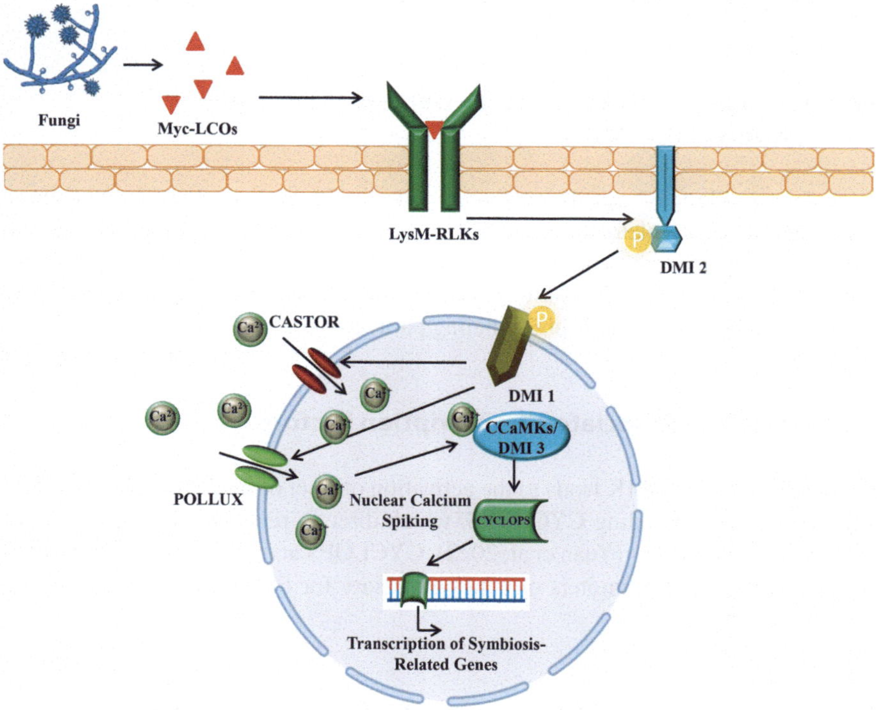

Fig. 12.4 Endophytic fungi initiate symbiosis with plant roots through the common symbiosis signaling pathway (CSSP). Fungal signaling molecules, such as Myc-LCOs, are recognized by LysM receptor-like kinases (LysM-RLKs) on the plant root surface. This recognition triggers nuclear calcium spiking, mediated by CASTOR and POLLUX ion channels, leading to the activation of calcium- and calmodulin-dependent protein kinase (CCaMK). CCaMK subsequently activates transcription factors like CYCLOPS, driving the expression of symbiosis-related genes. Additionally, DMI proteins play crucial roles in generating calcium signals and facilitating the structural modifications necessary for fungal entry. Abbreviations: Common Symbiosis Signaling Pathway (CSSP), LysM Receptor-Like Kinases (LysM-RLKs), Mycorrhizal Lipochitooligosaccharides (Myc-LCOs), Calcium- and Calmodulin-Dependent Protein Kinase (CCaMK), Doesn't Make Infections (DMI)

kinases (LysM-RLKs) on the plant root surface (Girardin et al. 2019). This recognition is crucial for distinguishing beneficial fungi from pathogens.

12.3.3 Nuclear Calcium Spiking

Upon recognition of fungal signals, the CSSP is activated, leading to a series of intracellular signaling events. One of the earliest responses is the induction of calcium spiking within the nuclear envelope. This calcium signaling is facilitated by a complex of proteins, including the nuclear-localized ion channels CASTOR and POLLUX, which are crucial for the generation of calcium oscillations (Charpentier et al. 2008). These oscillations serve as a secondary messenger system, relaying the symbiotic signal from the cell membrane to the nucleus.

12.3.4 Calcium- and Calmodulin-Dependent Protein Kinases (CCaMKs)

The calcium spiking is sensed by calcium- and calmodulin-dependent protein kinase (CCaMK), a key regulator in the CSSP. CCaMK decodes the calcium signals and activates a downstream signaling cascade by phosphorylating specific target proteins (Miller et al. 2013). This kinase is central to the symbiotic signaling pathway and acts as a hub where multiple signaling pathways converge.

12.3.5 Nuclear-Associated Transcription Factors

The activation of CCaMK leads to the activation of a set of nuclear-associated transcription factors, including CYCLOPS, which directly regulates the expression of symbiosis-related genes (Yuan et al. 2022). CYCLOPS acts as a transcriptional activator, binding to the promoters of genes necessary for fungal accommodation and nutrient exchange.

12.3.6 Involvement of DMI (Doesn't Make Infections) Proteins

DMI proteins, including DMI1, DMI2, and DMI3, play pivotal roles in the CSSP. DMI1 is involved in the generation of nuclear calcium spiking, DMI2 acts as a receptor-like kinase on the plasma membrane, and DMI3 (which is a CCaMK) is crucial for interpreting calcium signals and initiating downstream responses (Mitra et al. 2004). These proteins are integral to the successful establishment of symbiosis.

12.3.7 Formation of the Pre-penetration Apparatus (PPA)

As a result of CSSP activation, plants undergo structural modifications to facilitate fungal entry. The pre-penetration apparatus (PPA) is formed in the host plant cells, guiding the fungal hyphae to the inner cortical cells where symbiosis is established (Hull et al. 2021). This apparatus ensures that the fungal hyphae remain enclosed in a plant-derived membrane, preventing direct contact with the plant cytoplasm, which is essential for maintaining the symbiotic relationship without triggering immune responses.

12.3.8 Regulation of Symbiosis-Related Gene Expression

The CSSP ultimately leads to the activation of specific genes involved in the symbiotic interaction. These genes encode proteins necessary for the formation of symbiotic structures, such as arbuscules in the case of AMF, and for the exchange of nutrients between the plant and the fungi (Skiada et al. 2020). The expression of these genes is tightly regulated and is dependent on the precise decoding of the CSSP signals.

12.4 The Resilience of Endophytic Fungi Against Plant Immune Responses

Endophytic fungi have evolved sophisticated mechanisms to evade plant immune responses, particularly MAMP-triggered immunity (MTI), which is one of the primary layers of plant defense. The initial recognition of endophytes by the host plant occurs through pattern recognition receptors (PRRs) on the plant cell surface that detect microbe-associated molecular patterns (MAMPs) such as chitin, glucans, and other fungal cell wall components. One of the key strategies involves the use of LysM domain-containing effectors, such as Ecp6 from *Cladosporium fulvum* and Slp1 from *Magnaporthe oryzae* (Kombrink and Thomma 2013). These effectors bind to chitin fragments, a common fungal MAMP recognized by plant immune receptors like chitin elicitor receptor kinase 1 (CERK1) in *Arabidopsis* and OsCEBiP in rice, effectively sequestering these fragments and preventing their recognition by the plant's pattern recognition receptors (PRRs) (Hu et al. 2021). By shielding chitin from detection, endophytic fungi circumvent the activation of downstream immune responses, such as reactive oxygen species (ROS) production and mitogen-activated protein kinase (MAPK) cascades, which are crucial components of MTI. Following MAMP recognition, PRRs such as flagellin-sensing 2 (FLS2) and elongation factor Tu receptor (EFR) initiate signaling cascades by interacting with co-receptors like BRI1-associated kinase 1 (BAK1) (Zhang and Zhou 2010). This leads to the phosphorylation and activation of downstream kinases, such as *Botrytis*-induced kinase 1 (BIK1), which propagate the immune signal towards MAPK pathways. MAPKs, including MPK3 and MPK6, then phosphorylate a range of transcription factors, such as WRKY33, that are responsible for upregulating defense genes involved in antimicrobial compound production and reinforcing the cell wall through callose deposition (Lal et al. 2018). However, endophytic fungi can suppress this pathway by producing specific effectors that target these key components. Effector AVR-Pii from *Magnaporthe oryzae* can interfere with the phosphorylation of MAPKs, effectively blocking the downstream transcriptional activation of immune-related genes (Singh et al. 2016). At the root level, endophytic fungi like *Piriformospora indica* alter the plant's immune signaling to facilitate colonization. Upon initial contact, fungal effectors are delivered into the plant cells, where they modulate host gene expression by interacting with transcription factors or other regulatory proteins. *Piriformospora indica* secretes a protein known as FGB1, which suppresses the plant's ROS burst and modifies the expression of genes associated with salicylic acid (SA) signaling (Zhang et al. 2020). By downregulating SA biosynthesis and signaling, these fungi diminish the plant's defensive responses, promoting a more compatible interaction.

Additionally, endophytes manipulate phytohormone signaling pathways to modulate immune responses. *Fusarium oxysporum* secretes effector proteins that can interact with the jasmonic acid (JA) and ethylene (ET) signaling pathways, which are antagonistic to SA signaling (Ding et al. 2022). This manipulation shifts the plant's immune response from SA-mediated defenses, which are generally detrimental to fungal colonization, towards JA/ET pathways that are more permissive to fungal growth. Effector FvSnod1 from *Fusarium verticillioides* interacts with plant receptor-like kinases (RLKs) to suppress immune signaling, enhancing fungal colonization by avoiding the activation of canonical defense responses (Tang et al. 2017). Another sophisticated mechanism involves the direct modulation of the plant's transcriptional machinery. Endophytic fungi can secrete small RNAs (sRNAs) that are taken up by plant cells and act as molecular decoys or inhibitors of plant immune components. sRNAs from *Botrytis cinerea* have been shown to hijack the plant's RNA interference (RNAi) machinery, leading to the silencing of defense-related genes, including those involved in ROS production and MAPK signaling (Wang et al. 2017). By delivering these sRNAs into the plant, endophytes can effectively reprogram the host's immune responses to favor colonization. Moreover, endophytic fungi can inhibit MTI by altering the dynamics of Ca^{2+} signaling, which is one of the earliest events in plant immune response. Fungal effectors can interfere with the Ca^{2+} influx and downstream Ca^{2+}-binding proteins such as calmodulins (CaMs) and calcium-dependent protein kinases (CDPKs), which are essential for the activation of defense responses (Ranty et al. 2006). For example, the effector Avr4 from *Cladosporium fulvum* not only binds to chitin to prevent PRR activation but also disrupts the Ca^{2+} signaling pathway, further dampening the plant's immune response (Van Esse et al. 2007). A critical aspect of immune suppression by endophytes is the avoidance of ROS burst, a common plant response to pathogen attack. Many endophytic fungi produce antioxidants and ROS-scavenging enzymes that neutralize the oxidative burst, thus preventing the amplification of defense signals (Javed et al. 2022). During colonization, endophytes may secrete catalases (CATs) and superoxide dismutases (SODs) that degrade hydrogen peroxide, a key signaling molecule in plant defense. This reduction in ROS levels not only diminishes direct antimicrobial effects but also disrupts the signaling network required for full activation of defense genes. *Piriformospora indica* secretes a protein known as FGB1, which suppresses the plant's ROS burst and modifies the expression of genes associated with salicylic acid (SA) signaling (Jwa and Hwang 2017). By downregulating SA biosynthesis and signaling, these fungi diminish the plant's defensive responses, promoting a more compatible interaction.

By targeting multiple nodes in the immune signaling network from the initial recognition by PRRs to the downstream transcriptional and hormonal responses, endophytes fine-tune the plant's immune system to facilitate colonization while avoiding detection and triggering of full-scale immune responses. This intricate interplay of fungal effectors and plant receptors exemplifies the complexity of plant–microbe interactions at the molecular level, highlighting the delicate balance between immune evasion and mutualistic association that characterizes endophytic colonization.

12.4.1 Other Factors that Help the Endophytic Fungi to Help Colonization

12.4.1.1 Apoplastic Proteins

The apoplast, the space outside the plasma membrane that includes the cell wall and intercellular spaces, is a critical site of interaction between plants and endophytic fungi. Apoplastic proteins, such as expansins, hydrolases, and effectors secreted by fungi, contribute to the remodeling of plant cell walls and suppress immune responses, thus facilitating fungal colonization (Doehlemann and Hemetsberger 2013).

12.4.1.2 Expansins and Cell Wall-Modifying Enzymes

Expansins are proteins secreted by fungi that loosen the plant cell wall by disrupting the hydrogen bonds between cellulose fibers (Cosgrove 2015). This activity does not break the wall structurally but allows the fungal hyphae to penetrate more easily. *Piriformospora indica* secretes cell wall-modifying enzymes, such as cellulases and pectinases, which break down plant cell wall components, thus enabling easier access to internal plant tissues (Li et al. 2023). These enzymes also play a role in the controlled degradation of cell walls to avoid triggering strong defense responses that would occur with extensive damage.

12.4.1.3 Apoplastic Effectors

Fungal apoplastic effectors, such as *Magnaporthe oryzae* effectors Avr-Piz-t and Slp1, play crucial roles in modulating the plant's immune responses (Wei et al. 2023). These effectors are secreted into the apoplast, where they can bind to or inhibit plant proteins involved in defense signaling. For example, Slp1 competes with plant chitin-binding proteins, effectively sequestering chitin fragments and preventing their recognition by plant PRRs like CERK1 (Hu et al. 2021). This avoidance of immune detection enables the fungus to persist and colonize plant tissues without activating full-scale immune responses.

12.4.1.4 Protease Inhibitors and Host Defense Suppression

Endophytic fungi can also secrete protease inhibitors into the apoplast, which target and inhibit plant proteases involved in defense. *Fusarium oxysporum* produces a protease inhibitor that binds to serine proteases in the plant apoplast, preventing them from degrading fungal effectors and allowing these effectors to accumulate and function effectively in suppressing host defenses (Wang et al. 2020). This activity is essential for creating a favorable environment for fungal colonization and avoiding triggering defense responses such as ROS production.

12.4.1.5 Apoplastic Nucleotides

Nucleotides, such as ATP and ADP, present in the apoplast act as signaling molecules that can influence plant–fungal interactions (Tanaka et al. 2010). Apoplastic nucleotides are often released during cell damage or stress and act as damage-associated molecular patterns (DAMPs) that activate immune responses (Tanaka et al. 2014). However, endophytic fungi have evolved mechanisms to modulate these nucleotide signals to promote colonization.

12.4.1.6 Apyrases and Ectonucleotidases

Enzymes like apyrases and ectonucleotidases secreted by endophytic fungi hydrolyze extracellular ATP (eATP) and ADP, reducing their concentration in the apoplast (Matthus et al. 2023). eATP is a potent DAMP that triggers immune signaling through PRRs, leading to the activation of defense responses. By degrading eATP, these enzymes effectively suppress DAMP-triggered immunity (DTI), allowing the fungi to escape immune recognition. *Laccaria bicolor*, an ectomycorrhizal fungus, secretes apyrases that reduce eATP levels in the apoplast, facilitating colonization by avoiding the activation of plant defenses (Cope et al. 2019).

12.4.1.7 Nucleotide Binding and Signaling Disruption

Certain apoplastic proteins produced by fungi can bind to nucleotides like ATP, ADP, and AMP, disrupting their signaling roles in plant defense. For example, fungal ecto-ATPases can bind and hydrolyze eATP in the apoplast, thereby modulating calcium signaling cascades that would otherwise lead to immune activation (Kesten et al. 2023).

12.5 Role of Endophytic Fungi in Host Plant

12.5.1 Fungal Endophytes and Their Effect on Parasitic Fungus

Endophytic fungi exhibit remarkable antagonistic activities against parasitic fungi through the production of various bioactive compounds and enzymes. *Ampelomyces*, a well-known endophytic fungus, acts as a mycoparasite specifically targeting powdery mildew fungi, such as *Erysiphe* spp. and *Sphaerotheca* spp. (Németh et al. 2021). This fungus invades the parasitic fungal structures, including hyphae and conidiophores, disrupting their cellular integrity and leading to the collapse of the pathogen. Endophytic fungi secrete a variety of bioactive compounds that inhibit the growth of parasitic fungi and induce systemic responses in the host plant. Among these compounds are glucosides, cytochalasans, azaphilones, terpenoids, non-ribosomal peptides, macrolide polyketides, benzenoids, and lactones. Glucosides, such as phenolic glucosides produced by *Pestalotiopsis* spp., have been shown to interfere with the cellular processes of parasitic fungi by disrupting cell wall integrity and inhibiting fungal respiration (Grabka et al. 2022). Cytochalasans, which are produced by *Xylaria* spp., act on the actin cytoskeleton of parasitic fungi, leading to the disruption of hyphal growth and development (Watchaputi et al. 2023). Azaphilones, found in *Talaromyces* spp., possess strong antifungal properties by targeting fungal cell membranes and inhibiting spore germination (Bhat et al. 2023). Terpenoids, such as those produced by *Trichoderma harzianum*, are known to induce systemic resistance in plants and inhibit the growth of parasitic fungi like *Fusarium* spp. through membrane disruption and interference with fungal signaling pathways (Yao et al. 2023). Non-ribosomal peptides, such as gliotoxin from *Trichoderma virens*, disrupt the cellular machinery of parasitic fungi by inducing oxidative stress and damaging cellular components (Bulgari et al. 2020). Macrolide polyketides, including those secreted by *Penicillium* spp., exhibit strong antifungal activities by targeting mitochondrial function, ultimately leading to fungal cell death (Wang et al. 2023). Benzenoids, which include phenolic compounds produced by *Phomopsis* spp., act as potent inhibitors of fungal enzymes crucial for parasitic fungal growth, such as chitinases and glucanases (Brzezinska and Jankiewicz 2012). These benzenoids interfere with the structural integrity of the fungal cell wall, making parasitic fungi more vulnerable to environmental stresses and plant immune responses. Lactones, such as those derived from *Aspergillus* spp., disrupt fungal cell membrane integrity and inhibit spore germination, effectively reducing the spread of parasitic fungi (Conrado et al. 2022). Exoenzymes also play a critical role in the antagonistic interactions between endophytic fungi and parasitic fungi. *Trichoderma* species secrete a suite of enzymes, including chitinases, glucanases, and proteases, which degrade the cell walls of parasitic fungi like *Rhizoctonia solani* and *Sclerotinia sclerotiorum* (Tyśkiewicz et al. 2022). The enzymatic degradation of the parasitic fungal cell wall not only suppresses pathogen growth but also releases fragments that can act as elicitors, triggering the host plant's immune responses. *Nigrospora oryzae* inhibits the growth of *Colletotrichum* spp. by secreting azaphilones that disrupt fungal cell membranes (Rai et al. 2021). *Chaetomium globosum* produces chaetoglobosins, cytochalasans that impair actin polymerization in parasitic fungi, hampering their ability to invade host tissues (Chen et al. 2020). *Penicillium citrinum* secretes citrinin, a polyketide that interferes with fungal respiration, thereby inhibiting the growth of pathogens like *Botrytis cinerea* (Zhang et al. 2021). *Fusarium solani* produces fusaric acid, a benzenoid that targets mitochondrial function in parasitic fungi, reducing their viability (Perincherry et al. 2019). *Aspergillus fumigatus* secrete fumagillin, a lactone compound that disrupts the metabolic processes of parasitic fungi, leading to growth inhibition. *Alternaria alternata* secretes alternariol, a mycotoxin that inhibits the growth of parasitic fungi by interfering with DNA replication and transcription (Grover and Lawrence 2017). *Phoma* spp. secretes phomalactone, which has been shown to inhibit spore germination and hyphal growth in pathogens like *Fusarium oxysporum* (Rai et al. 2022). *Beauveria bassiana* secretes beauvericin, a non-ribosomal peptide that targets ion channels in fungal cells, causing cell membrane disruption and subsequent cell death (Ávila-Hernández et al. 2020). Some other examples of endophytic fungi are shown in Table 12.1.

12.5.2 Stress Tolerance: Temperature, Drought, and Salinity Stress

Endophytic fungi play a significant role in helping plants overcome various abiotic stresses, including temperature extremes, salinity, and drought. These fungi enhance the resilience of their host plants by modulating stress-responsive pathways, producing protective metabolites, and activating stress-related proteins and enzymes. The mechanisms by which endophytic fungi assist plants under these adverse conditions often involve the upregulation of specific molecular proteins, such as heat shock proteins (HSPs) and other molecular chaperones, which help in maintaining cellular homeostasis (Morales-Vargas et al. 2024). During heat stress, endophytes like *Piriformospora indica* and *Trichoderma harzianum* have been shown to enhance the expression of heat shock proteins (HSPs) in their host plants (Ameen et al. 2024). *P. indica* triggers the upregulation of HSP70 and HSP90, which are crucial for protein folding and repair, thereby protecting plants such as wheat and rice from heat-induced damage. Additionally, endophytic fungi produce antioxidant enzymes like superoxide dismutase (SOD) and catalase (CAT), which reduce oxidative damage caused by heat stress by scavenging reactive oxygen species (ROS) (Zandi and Schnug 2022). In cold stress, *Penicillium resedanum* improve the cold tolerance of host plants by regulating the expression of cold-responsive (*COR*) genes, including C-repeat binding factors (CBFs) and dehydration-responsive element-binding proteins (DREBs) (Kamran et al. 2022). These transcription factors activate the expression of cold-responsive (*COR*) genes, which play a vital role in maintaining membrane fluidity and stabilizing cellular structures at low temperatures.

Piriformospora indica has been shown to alleviate salt stress in rice by upregulating sodium/proton antiporters like Salt Overly Sensitive 1 (SOS1) and High-Affinity Potassium Transporter 1 (HKT1), which help in the extrusion of excess sodium ions and maintain potassium levels within the cells (Jogawat et al. 2013). This ion-regulating mechanism reduces the toxic effects of high salt concentrations, allowing the plant to maintain normal metabolic functions. Furthermore, endophytic fungi enhance the production of compatible solutes like proline, glycine betaine, and trehalose, which act as osmoprotectants and stabilize proteins and membranes under saline conditions. *Trichoderma* spp., known for their salinity

Table 12.1 Examples of endophytic fungi, the compounds they secrete, their host plants, the parasitic fungi they inhibit, and the mode of action

Endophytic fungi	Compound secreted	Host plant	Parasitic fungus inhibited	Mode of action	References
Ampelomyces quisqualis	Not specified	Tomato	*Pseudoidium neolycopersici*	Mycoparasitism, invades fungal structures and disrupts cell integrity	Németh et al. (2021)
Trichoderma harzianum	Terpenoids, chitinases	Tomato	*Fusarium oxysporum*	Cell wall degradation, induces systemic resistance	Harman et al. (2019)
Pestalotiopsis spp.	Phenolic glucosides	*Camellia sinensis*	*Botrytis cinerea*	Disrupts cell wall integrity, inhibits respiration	Wu et al. (2022)
Xylaria spp.	Cytochalasans	Tea	*Colletotrichum* spp.	Disrupts actin cytoskeleton, impedes hyphal growth	Bills et al. (2012)
Talaromyces spp.	Azaphilones	Rice	*Magnaporthe oryzae*	Targets cell membranes, inhibits spore germination	Franck et al. (2013)
Penicillium spp.	Macrolide polyketides	Maize	*Aspergillus flavus*	Inhibits mitochondrial function, causing cell death	Muhammad et al. (2024)
Trichoderma virens	Gliotoxin (non-ribosomal peptide)	Maize	*Rhizoctonia solani*	Induces oxidative stress, damages cellular components	Howell (2003)
Chaetomium globosum	Chaetoglobosins	Wheat	*Bipolaris sorokiniana*	Inhibits actin polymerization, hampers invasion capability	Darshan et al. (2021)
Aspergillus fumigatus	Fumagillin	Peanut	*Aspergillus niger*	Disrupts metabolic processes, inhibits growth	Guruceaga et al. (2019)
Beauveria bassiana	Beauvericin (non-ribosomal peptide)	Cotton	*Verticillium dahliae*	Disrupts ion channels, causes cell membrane disruption	Wang et al. (2021)
Fusarium solani	Fusaric acid	Soybean	*Phytophthora sojae*	Targets mitochondrial function, reduces fungal viability	Żelechowski et al. (2021)
Alternaria alternata	Alternariol	Tomato	*Sclerotinia sclerotiorum*	Inhibits DNA replication and transcription	Philip et al. (2024)
Phoma spp.	Phomalactone	Tomato	*Fusarium oxysporum*	Inhibits spore germination, hyphal growth	Srinivas et al. (2019)

tolerance-promoting abilities, induce the expression of antioxidant enzymes such as SOD, CAT, and glutathione reductase (GR), which mitigate the oxidative stress associated with high salinity (Zhang et al. 2019). By regulating the cellular redox state and protecting key cellular components from salt-induced damage, these fungi enable plants to sustain growth and productivity in saline environments.

Trichoderma harzianum improves drought tolerance in crops like maize and tomato by increasing the expression of aquaporins, which are proteins that facilitate water transport across cell membranes (Yu et al. 2023). *Fusarium solani* induces the production of abscisic acid (ABA), a phytohormone that plays a pivotal role in the plant's drought response. ABA regulates stomatal closure, thereby reducing water loss through transpiration (Lim et al. 2015). The endophyte also upregulates the expression of drought-responsive genes, including those encoding late embryogenesis abundant (LEA) proteins, which protect cellular structures during dehydration by acting as molecular shields (Aziz et al. 2021). The accumulation of osmolytes, such as proline and soluble sugars, is another critical drought-tolerance mechanism facilitated by endophytic fungi. These compounds help maintain cellular osmotic balance and protect proteins and membranes from desiccation.

12.5.3 Enhancement of Nutrient Uptake and Plant Growth Promotion

Endophytic fungi play a crucial role in enhancing nutrient uptake and promoting plant growth through a variety of biochemical and molecular mechanisms. These fungi form intimate associations with their host plants, allowing for improved access to essential nutrients such as nitrogen, phosphorus, and micronutrients that are often limited in the soil (Sarkar et al. 2021). One of the primary ways endophytic fungi enhance nutrient uptake is through the secretion of organic acids, siderophores, and extracellular enzymes that solubilize and mobilize nutrients from the soil, making them more available to the plant roots (Chen et al. 2022a, b). *Piriformospora indica* has been shown to enhance phosphorus uptake in barley by secreting phosphatases and organic acids that solubilize insoluble phosphate compounds in the soil (Wu et al. 2018). *P. indica* also has been applied in crop improvement programs, showing positive effects on various crops, including wheat, maize, and rice, by enhancing their nutrient status and growth under nutrient-limited conditions. *Serendipita vermifera* increases nitrogen uptake in plants through the production of ammonia-lyase enzymes that facilitate nitrogen assimilation pathways (Ray et al. 2019). In orchids, *S. vermifera* enhances nitrogen uptake, leading to improved growth and flowering, which has significant implications for both natural ecosystems and horticultural industries (Ray et al. 2018). Additionally, *Rhizophagus irregularis* and *Funneliformis mosseae*, which are closely related to arbuscular mycorrhizal fungi, improve nutrient uptake by forming extensive hyphal networks that increase the surface area for nutrient absorption (Deja-Sikora et al. 2023). These fungi release glomalin-related soil proteins, which stabilize soil structure, enhance nutrient retention, and improve plant access to water and nutrients.

Endophytic fungi also promote plant growth by producing plant growth-promoting hormones such as indole-3-acetic acid (IAA), gibberellins, and cytokinins. *Colletotrichum tofieldiae* synthesizes IAA, which enhances root growth and branching, thus increasing the plant's ability to explore the soil for nutrients (Hiruma et al. 2016). This interaction is particularly beneficial under conditions of nutrient stress. Recent studies have identified specific metabolites and biochemicals produced by endophytic fungi that directly influence plant growth and nutrient uptake. Production of strigolactones by *Gigaspora margarita* not only stimulates fungal colonization but also promotes lateral root formation and enhances nutrient uptake efficiency in host plants like tomatoes (Mitra et al. 2021). These strigolactones act as signaling molecules that modulate root architecture, improving the plant's nutrient foraging capacity. *Penicillium* spp. secrete gibberellins that significantly increase plant growth by modulating the expression of growth-regulating genes in the host, leading to enhanced cell elongation and division (Keswani et al. 2022). Ethylene is a stress hormone that can inhibit plant growth under adverse conditions. *Trichoderma harzianum* produces 1-aminocyclopropane-1-carboxylate (ACC) deaminase, which degrades ACC, the precursor of ethylene, thereby reducing ethylene synthesis in the plant (Rauf et al. 2021). This reduction in ethylene levels leads to improved root growth and better overall plant vigor, especially under stress conditions like drought and salinity. In agricultural applications, these are being explored for their potential to enhance crop productivity. Inoculating soybeans with *Fusarium oxysporum* strain Fo47 and peppers with *Trichoderma asperellum* significantly enhances nutrient uptake, resulting in increased plant growth, biomass, and improved fruit quality (De Lamo and Takken 2020; Sood et al. 2020).

12.6 Conclusion

Endophytic fungi play a pivotal role in enhancing plant health through diverse mechanisms, including nutrient uptake, stress tolerance, and protection against pathogens. Their symbiotic relationship with plants is underpinned by a complex interplay of genetic compatibility, signaling pathways, and biochemical interactions that facilitate fungal colonization and mutual benefit. These endophytes produce an array of metabolites, such as phytohormones, secondary metabolites, and enzymes, that not only promote plant growth but also confer resistance to biotic and abiotic stresses. They enhance nutrient uptake by secreting organic acids, phosphatases, and other extracellular enzymes that solubilize nutrients like phosphorus and nitrogen, making them readily available to plants. Plant growth-promoting hormones, such as indole-3-acetic acid (IAA), which optimize root architecture for better soil exploration, can be produced by these endophytes. The production of strigolactones and gibberellins further enhances root development and nutrient foraging, contributing to improved plant growth and resilience.

In terms of biotic stress management, endophytic fungi secrete bioactive compounds, including glucosides, cytochalasans, terpenoids, and azaphilones, which inhibit the growth of parasitic fungi, bacteria, and viruses. They act through diverse mechanisms such as disrupting cell walls, interfering with cellular respiration, and modulating plant immune responses. Exoenzymes like chitinases and glucanases play a crucial role in degrading the cell walls of pathogens, thereby enhancing the plant's defense system. They also play a critical role in mitigating abiotic stresses such as temperature extremes, salinity, and drought. They upregulate heat shock proteins (HSPs), enhance the expression of cold-responsive genes, and modulate antioxidant enzymes to reduce oxidative stress. By regulating ion transporters and enhancing osmolyte production, these fungi help plants maintain cellular homeostasis under saline and drought conditions, ensuring sustained growth and productivity.

Acknowledgments The authors would like to extend their sincere appreciation to the funding agency Science and Engineering Research Board (SERB), State University Research Excellence (SURE) & Empowerment and Equity Opportunities for Excellence in Science, New Delhi, India. All the authors acknowledge their host institute for infrastructure support.

Conflict of Interest The authors declare no conflict of interest.

Funding This research was supported by the funding agency Science and Engineering Research Board (SERB), State University Research Excellence (SURE) (File Number: SUR/2022/005216) and Empowerment and Equity Opportunities for Excellence in Science (File Number: EEQ/2022/000455), New Delhi, India. The research was also supported by the Ministry of Education and the State Project Directorate-Rashtriya Uchchatar Shiksha Abhiyaan (SPD-RUSA), Rajasthan, through the RUSA 2.0 project, as well as the funding agency DST-FIST (GP193).

Data Availability Not applicable.

References

Aharwal RP, Kumar S, Sandhu SS (2021) Endophytic mycoflora: antibacterial secondary metabolites and their therapeutic potential. Curr Pharmacol Rep 7(4):150–170. https://doi.org/10.1007/s40495-021-00261-w

Akram S, Ahmed A, He P, He P, Liu Y, Wu Y, Munir S, He Y (2023) Uniting the role of endophytic fungi against plant pathogens and their interaction. J Fungi 9(1):72. https://doi.org/10.3390/jof9010072

Alam B, Lǐ J, Gě Q, Khan MA, Gōng J, Mehmood S, Yuán Y, Gǒng W (2021) Endophytic fungi: from symbiosis to secondary metabolite communications or vice versa? Front Plant Sci 12. https://doi.org/10.3389/fpls.2021.791033

Ameen M, Mahmood A, Sahkoor A, Zia MA, Ullah MS (2024) The role of endophytes to combat abiotic stress in plants. Plant Stress 12:100435. https://doi.org/10.1016/j.stress.2024.100435

Ávila-Hernández JG, Carrillo-Inungaray ML, De La Cruz-Quiroz R, Wong-Paz JE, Muñiz-Márquez DB, Parra R, Aguilar CN, Aguilar-Zárate P (2020) *Beauveria bassiana* secondary metabolites: a review inside their production systems, biosynthesis, and bioactivities. Mex J Biotechnol 5(4):1–33. https://doi.org/10.29267/mxjb.2020.5.4.1

Aziz MA, Sabeem M, Mullath SK, Brini F, Masmoudi K (2021) Plant Group II LEA proteins: intrinsically disordered structure for multiple functions in response to environmental stresses. Biomol Ther 11(11):1662. https://doi.org/10.3390/biom11111662

Backer R, Naidoo S, Van Den Berg N (2019) The Nonexpressor of Pathogenesis-Related Genes 1 (NPR1) and related family: mechanistic insights in plant disease resistance. Front Plant Sci 10. https://doi.org/10.3389/fpls.2019.00102

Bhat MP, Rudrappa M, Hugar A, Gunagambhire PV, Kumar RS, Nayaka S, Almansour AI, Perumal K (2023) In-vitro investigation on the biological activities of squalene derived from the soil fungus *Talaromyces pinophilus*. Heliyon 9(11):e21461. https://doi.org/10.1016/j.heliyon.2023.e21461

Bills GF, González-Menéndez V, Martín J, Platas G, Fournier J, Peršoh D, Stadler M (2012) *Hypoxylon pulicicidum* sp. nov. (Ascomycota, Xylariales), a pantropical insecticide-producing endophyte. PLoS ONE 7(10):e46687. https://doi.org/10.1371/journal.pone.0046687

Broeckling CD, Broz AK, Bergelson J, Manter DK, Vivanco JM (2008) Root exudates regulate soil fungal community composition and diversity. Appl Environ Microbiol 74(3):738–744. https://doi.org/10.1128/aem.02188-07

Brzezinska MS, Jankiewicz U (2012) Production of antifungal Chitinase by *Aspergillus niger* LOCK 62 and its potential role in the biological control. Curr Microbiol 65(6):666–672. https://doi.org/10.1007/s00284-012-0208-2

Bulgari D, Fiorini L, Gianoncelli A, Bertuzzi M, Gobbi E (2020) Enlightening gliotoxin biological system in agriculturally relevant *Trichoderma* spp. Front Microbiol 11. https://doi.org/10.3389/fmicb.2020.00200

Camejo D, Guzmán-Cedeño Á, Moreno A (2016) Reactive oxygen species, essential molecules, during plant–pathogen interactions. Plant Physiol Biochem 103:10–23. https://doi.org/10.1016/j.plaphy.2016.02.035

Chandran H, Meena M, Swapnil P (2021) Plant growth-promoting rhizobacteria as a green alternative for sustainable agriculture. Sustainability 13(19):10986. https://doi.org/10.3390/su131910986

Charpentier M, Bredemeier R, Wanner G, Takeda N, Schleiff E, Parniske M (2008) *Lotus japonicus* CASTOR and POLLUX are ion channels essential for perinuclear calcium spiking in legume root endosymbiosis. Plant Cell 20(12):3467–3479. https://doi.org/10.1105/tpc.108.063255

Chaudhary P, Agri U, Chaudhary A, Kumar A, Kumar G (2022) Endophytes and their potential in biotic stress management and crop production. Front Microbiol 13. https://doi.org/10.3389/fmicb.2022.933017

Chen X, Wang D, Fang X, Chen X, Mao Y (2019) Plant specialized metabolism regulated by jasmonate signaling. Plant Cell Physiol 60(12):2638–2647. https://doi.org/10.1093/pcp/pcz161

Chen J, Zhang W, Guo Q, Yu W, Zhang Y, He B (2020) Bioactivities and future perspectives of chaetoglobosins. Evid Based Complement Alternat Med 2020:1–10. https://doi.org/10.1155/2020/8574084

Chen Z, White JF, Malik K, Chen H, Jin Y, Yao X, Wei X, Li C, Nan Z (2022a) Soil nutrient dynamics relate to Epichloë endophyte mutualism and nitrogen turnover in a low nitrogen environment. Soil Biol Biochem 174:108832. https://doi.org/10.1016/j.soilbio.2022.108832

Chen J, Zhang X, Rathjen JP, Dodds PN (2022b) Direct recognition of pathogen effectors by plant NLR immune receptors and downstream signalling. Essays Biochem 66(5):471–483. https://doi.org/10.1042/ebc20210072

Conrado R, Gomes TC, Roque GSC, De Souza AO (2022) Overview of bioactive fungal secondary metabolites: cytotoxic and antimicrobial compounds. Antibiotics 11(11):1604. https://doi.org/10.3390/antibiotics11111604

Cope K, Bascaules A, Irving T, Venkateshwaran M, Maeda J, Garcia K, Rush T, Ma C, Labbé J, Jawdy S, Steigerwald E, Setzke J, Fung E, Schnell K, Wang Y, Schleif N, Bücking H, Strauss S, Maillet F, Ané J (2019) The ectomycorrhizal fungus *Laccaria bicolor* produces lipochitooligosaccharides and uses the common symbiosis pathway to colonize *Populus* roots. Plant Cell 31(10):2386–2410. https://doi.org/10.1105/tpc.18.00676

Cosgrove DJ (2015) Plant expansins: diversity and interactions with plant cell walls. Curr Opin Plant Biol 25:162–172. https://doi.org/10.1016/j.pbi.2015.05.014

Darshan K, Aggarwal R, Bashyal BM, Singh J, Kundu A, Yadav S, Saharan MS (2021) Antifungal metabolite profiling in *Chaetomium globosum* potential strain Cg2 effective against *Bipolaris sorokiniana*. Indian J Agric Sci 91(5). https://doi.org/10.56093/ijas.v91i5.113103

De Lamo FJ, Takken FLW (2020) Biocontrol by *Fusarium oxysporum* using endophyte-mediated resistance. Front Plant Sci 11. https://doi.org/10.3389/fpls.2020.00037

Deja-Sikora E, Werner K, Hrynkiewicz K (2023) AMF species do matter: *Rhizophagus irregularis* and *Funneliformis mosseae* affect healthy and PVY-infected *Solanum tuberosum* L. in a different way. Front Microbiol 14. https://doi.org/10.3389/fmicb.2023.1127278

Ding L, Li Y, Wu Y, Li T, Geng R, Cao J, Zhang W, Tan X (2022) Plant disease resistance-related signaling pathways: recent progress and future prospects. Int J Mol Sci 23(24):16200. https://doi.org/10.3390/ijms232416200

Doehlemann G, Hemetsberger C (2013) Apoplastic immunity and its suppression by filamentous plant pathogens. New Phytol 198(4):1001–1016. https://doi.org/10.1111/nph.12277

Fan L, Li Y, Wang X, Leng F, Li S, Zhu N, Chen K, Wang Y (2023) Culturable endophytic fungi community structure isolated from *Codonopsis pilosula* roots and effect of season and geographic location on their structures. BMC Microbiol 23(1):132. https://doi.org/10.1186/s12866-023-02848-3

Franck WL, Gokce E, Oh Y, Muddiman DC, Dean RA (2013) Temporal analysis of the *Magnaporthe oryzae* proteome during conidial germination and cyclic AMP (CAMP)-mediated appressorium formation. Mol Cell Proteomics 12(8):2249–2265. https://doi.org/10.1074/mcp.m112.025874

German L, Yeshvekar R, Benitez-Alfonso Y (2022) Callose metabolism and the regulation of cell walls and plasmodesmata during plant mutualistic and pathogenic interactions. Plant Cell Environ 46(2):391–404. https://doi.org/10.1111/pce.14510

Girardin A, Wang T, Ding Y, Keller J, Buendia L, Gaston M, Ribeyre C, Gasciolli V, Auriac M, Vernié T, Bendahmane A, Ried MK, Parniske M, Morel P, Vandenbussche M, Schorderet M, Reinhardt D, Delaux P, Bono J, Lefebvre B (2019) LCO receptors involved in arbuscular mycorrhiza are functional for rhizobia perception in legumes. Curr Biol 29(24):4249–4259.e5. https://doi.org/10.1016/j.cub.2019.11.038

Grabka R, D'Entremont TW, Adams SJ, Walker AK, Tanney JB, Abbasi PA, Ali S (2022) Fungal endophytes and their role in agricultural plant protection against pests and pathogens. Plan Theory 11(3):384. https://doi.org/10.3390/plants11030384

Grover S, Lawrence C (2017) The *Alternaria alternata* mycotoxin alternariol suppresses lipopolysaccharide-induced inflammation. Int J Mol Sci 18(7):1577. https://doi.org/10.3390/ijms18071577

Guruceaga X, Perez-Cuesta U, De Cerio AA, Gonzalez O, Alonso RM, Hernando FL, Ramirez-Garcia A, Rementeria A (2019) Fumagillin, a mycotoxin of *Aspergillus fumigatus*: biosynthesis, biological activities, detection, and applications. Toxins 12(1):7. https://doi.org/10.3390/toxins12010007

Harman G, Doni F, Khadka R, Uphoff N (2019) Endophytic strains of *Trichoderma* increase plants' photosynthetic capability. J Appl Microbiol 130(2):529–546. https://doi.org/10.1111/jam.14368

Hiruma K, Gerlach N, Sacristán S, Nakano RT, Hacquard S, Kracher B, Neumann U, Ramirez D, Bucher M, O'Connell R, Schulze-Lefert P (2016) Root endophyte *Colletotrichum tofieldiae* confers plant fitness benefits that are phosphate status dependent. Cell 165(2):464–474. https://doi.org/10.1016/j.cell.2016.02.028

Hogg BV, Cullimore JV, Ranjeva R, Bono J (2005) The DMI1 and DMI2 early symbiotic genes of *Medicago truncatula* are required for a high-affinity nodulation factor-binding site associated to a particulate fraction of roots. Plant Physiol 140(1):365–373. https://doi.org/10.1104/pp.105.068981

Ho-Plágaro T, García-Garrido JM (2022) Molecular regulation of arbuscular mycorrhizal symbiosis. Int J Mol Sci 23(11):5960. https://doi.org/10.3390/ijms23115960

Hou S, Liu Z, Shen H, Wu D (2019) Damage-associated molecular pattern-triggered immunity in plants. Front Plant Sci 10. https://doi.org/10.3389/fpls.2019.00646

Howell CR (2003) Mechanisms employed by *Trichoderma* species in the biological control of plant diseases: the history and evolution of current concepts. Plant Dis 87(1):4–10. https://doi.org/10.1094/pdis.2003.87.1.4

Hu S, Li J, Dhar N, Li J, Chen J, Jian W, Dai X, Yang X (2021) Lysin motif (LySM) proteins: interlinking manipulation of plant immunity and fungi. Int J Mol Sci 22(6):3114. https://doi.org/10.3390/ijms22063114

Hull R, Choi J, Paszkowski U (2021) Conditioning plants for arbuscular mycorrhizal symbiosis through DWARF14-LIKE signalling. Curr Opin Plant Biol 62:102071. https://doi.org/10.1016/j.pbi.2021.102071

Javed J, Rauf M, Arif M, Hamayun M, Gul H, Ud-Din A, Ud-Din J, Sohail M, Rahman MM, Lee I (2022) Endophytic fungal consortia enhance basal drought-tolerance in *Moringa oleifera* by upregulating the antioxidant enzyme (APX) through heat shock factors. Antioxidants 11(9):1669. https://doi.org/10.3390/antiox11091669

Jha P, Kaur T, Chhabra I, Panja A, Paul S, Kumar V, Malik T (2023) Endophytic fungi: hidden treasure chest of antimicrobial metabolites interrelationship of endophytes and metabolites. Front Microbiol 14. https://doi.org/10.3389/fmicb.2023.1227830

Jogawat A, Saha S, Bakshi M, Dayaman V, Kumar M, Dua M, Varma A, Oelmüller R, Tuteja N, Johri AK (2013) *Piriformospora indica* rescues growth diminution of rice seedlings during high salt stress. Plant Signal Behav 8(10):e26891. https://doi.org/10.4161/psb.26891

Johri AK, Oelmüller R, Dua M, Yadav V, Kumar M, Tuteja N, Varma A, Bonfante P, Persson BL, Stroud RM (2015) Fungal association and utilization of phosphate by plants: success, limitations, and future prospects. Front Microbiol 6. https://doi.org/10.3389/fmicb.2015.00984

Jwa N, Hwang BK (2017) Convergent evolution of pathogen effectors toward reactive oxygen species signaling networks in plants. Front Plant Sci 8. https://doi.org/10.3389/fpls.2017.01687

Kamran M, Imran QM, Ahmed MB, Falak N, Khatoon A, Yun B (2022) Endophyte-mediated stress tolerance in plants: a sustainable strategy to enhance resilience and assist crop improvement. Cells 11(20):3292. https://doi.org/10.3390/cells11203292

Kesten C, Leitner V, Dora S, Sims JW, Dindas J, Zipfel C, De Moraes CM, Sanchez-Rodriguez C (2023) Soil-borne fungi alter the apoplastic purinergic signaling in plants by deregulating the homeostasis of extracellular ATP and its metabolite adenosine. elife 12. https://doi.org/10.7554/elife.92913

Keswani C, Singh SP, García-Estrada C, Mezaache-Aichour S, Glare TR, Borriss R, Rajput VD, Minkina TM, Ortiz A, Sansinenea E (2022) Biosynthesis and beneficial effects of microbial gibberellins on crops for sustainable agriculture. J Appl Microbiol 132(3):1597–1615. https://doi.org/10.1111/jam.15348

Khan M, Ali S, Azzawi TNIA, Yun B (2023) Nitric oxide acts as a key signaling molecule in plant development under stressful conditions. Int J Mol Sci 24(5):4782. https://doi.org/10.3390/ijms24054782

Kombrink A, Thomma BPHJ (2013) LYSM effectors: secreted proteins supporting fungal life. PLoS Pathog 9(12):e1003769. https://doi.org/10.1371/journal.ppat.1003769

Kubicek CP, Starr TL, Glass NL (2014) Plant cell wall–degrading enzymes and their secretion in plant-pathogenic fungi. Annu Rev Phytopathol 52(1):427–451. https://doi.org/10.1146/annurev-phyto-102313-045831

Kumar A, Kumar R, Sengupta D, Das SN, Pandey MK, Bohra A, Sharma NK, Sinha P, Sk H, Ghazi IA, Laha GS, Sundaram RM (2020) Deployment of genetic and genomic tools toward gaining a better understanding of rice-*Xanthomonas oryzae* pv. *oryzae* interactions for development of durable bacterial blight resistant rice. Front Plant Sci 11. https://doi.org/10.3389/fpls.2020.01152

Kumar R, Swapnil P, Meena M, Selpair S, Yadav BG (2022) Plant growth-promoting rhizobacteria (PGPR): approaches to alleviate abiotic stresses for enhancement of growth and development of medicinal plants. Sustain For 14(23):15514. https://doi.org/10.3390/su142315514

Lal NK, Nagalakshmi U, Hurlburt NK, Flores R, Bak A, Sone P, Ma X, Song G, Walley J, Shan L, He P, Casteel C, Fisher AJ, Dinesh-Kumar SP (2018) The receptor-like cytoplasmic kinase

BIK1 localizes to the nucleus and regulates defense hormone expression during plant innate immunity. Cell Host Microbe 23(4):485–497.e5. https://doi.org/10.1016/j.chom.2018.03.010

Li L, Feng Y, Qi F, Hao R (2023) Research progress of *Piriformospora indica* in improving plant growth and stress resistance to plant. J Fungi 9(10):965. https://doi.org/10.3390/jof9100965

Liao D, Wang S, Cui M, Liu J, Chen A, Xu G (2018) Phytohormones regulate the development of arbuscular mycorrhizal symbiosis. Int J Mol Sci 19(10):3146. https://doi.org/10.3390/ijms19103146

Lim C, Baek W, Jung J, Kim J, Lee S (2015) Function of ABA in stomatal defense against biotic and drought stresses. Int J Mol Sci 16(12):15251–15270. https://doi.org/10.3390/ijms160715251

Ling L, Feng L, Li Y, Yue R, Wang Y, Zhou Y (2024) Endophytic fungi volatile organic compounds as crucial biocontrol agents used for controlling fruit and vegetable postharvest diseases. J Fungi 10(5):332. https://doi.org/10.3390/jof10050332

Lorenzo O, Chico JM, Saénchez-Serrano JJ, Solano R (2004) JASMONATE-INSENSITIVE1 encodes a MYC transcription factor essential to discriminate between different jasmonate-regulated defense responses in *Arabidopsis*[W]. Plant Cell 16(7):1938–1950. https://doi.org/10.1105/tpc.022319

Matthus E, Ning Y, Shafiq F, Davies JM (2023) Phosphate-deprivation and damage signalling by extracellular ATP. Front Plant Sci 13. https://doi.org/10.3389/fpls.2022.1098146

Meena M, Yadav G, Sonigra P, Nagda A, Mehta T, Swapnil P, Harish, Marwal A (2022) Role of elicitors to initiate the induction of systemic resistance in plants to biotic stress. Plant Stress 5:100103. https://doi.org/10.1016/j.stress.2022.100103

Meena M, Yadav G, Sonigra P, Nagda A, Mehta T, Swapnil P, Harish, Marwal A, Kumar S (2023) Multifarious responses of forest soil microbial community toward climate change. Microb Ecol 86:49–74. https://doi.org/10.1007/s00248-022-02051-3

Miller JB, Pratap A, Miyahara A, Zhou L, Bornemann S, Morris RJ, Oldroyd GE (2013) Calcium/calmodulin-dependent protein kinase is negatively and positively regulated by calcium, providing a mechanism for decoding calcium responses during symbiosis signaling. Plant Cell 25(12):5053–5066. https://doi.org/10.1105/tpc.113.116921

Mitra RM, Gleason CA, Edwards A, Hadfield J, Downie JA, Oldroyd GED, Long SR (2004) A Ca^{2+}/calmodulin-dependent protein kinase required for symbiotic nodule development: gene identification by transcript-based cloning. Proc Natl Acad Sci 101(13):4701–4705. https://doi.org/10.1073/pnas.0400595101

Mitra D, Rad KV, Chaudhary P, Ruparelia J, Sagarika MS, Boutaj H, Mohapatra PKD, Panneerselvam P (2021) Involvement of strigolactone hormone in root development, influence and interaction with mycorrhizal fungi in plant: mini-review. Curr Res Microbial Sci 2:100026. https://doi.org/10.1016/j.crmicr.2021.100026

Mogensen TH (2009) Pathogen recognition and inflammatory signaling in innate immune defenses. Clin Microbiol Rev 22(2):240–273. https://doi.org/10.1128/cmr.00046-08

Morales-Vargas AT, López-Ramírez V, Álvarez-Mejía C, Vázquez-Martínez J (2024) Endophytic fungi for crops adaptation to abiotic stresses. Microorganisms 12(7):1357. https://doi.org/10.3390/microorganisms12071357

Muhammad M, Ahmad J, Basit A, Mohamed HI, Khan A, Kamel EA (2024) Antimicrobial activity of *Penicillium* species metabolites. In: Elsevier eBooks, pp 369–383. https://doi.org/10.1016/b978-0-323-95241-5.00004-6

Németh MZ, Mizuno Y, Kobayashi H, Seress D, Shishido N, Kimura Y, Takamatsu S, Suzuki T, Takikawa Y, Kakutani K, Matsuda Y, Kiss L, Nonomura T (2021) *Ampelomyces* strains isolated from diverse powdery mildew hosts in Japan: their phylogeny and mycoparasitic activity, including timing and quantifying mycoparasitism of *Pseudoidium neolycopersici* on tomato. PLoS One 16(5):e0251444. https://doi.org/10.1371/journal.pone.0251444

Noman A, Aqeel M, Lou Y (2019) PRRs and NB-LRRs: from signal perception to activation of plant innate immunity. Int J Mol Sci 20(8):1882. https://doi.org/10.3390/ijms20081882

Ogbe AA, Finnie JF, Van Staden J (2020) The role of endophytes in secondary metabolites accumulation in medicinal plants under abiotic stress. S Afr J Bot 134:126–134. https://doi.org/10.1016/j.sajb.2020.06.023

Perincherry L, Lalak-Kańczugowska J, Stępień Ł (2019) Fusarium-produced mycotoxins in plant-pathogen interactions. Toxins 11(11):664. https://doi.org/10.3390/toxins11110664

Philip B, Behiry SI, Salem MZM, Amer MA, El-Samra IA, Abdelkhalek A, Heflish A (2024) *Trichoderma afroharzianum* TRI07 metabolites inhibit *Alternaria alternata* growth and induce tomato defense-related enzymes. Sci Rep 14(1):1874. https://doi.org/10.1038/s41598-024-52301-2

Priyashantha AKH, Dai D, Bhat DJ, Stephenson SL, Promputtha I, Kaushik P, Tibpromma S, Karunarathna SC (2023) Plant–fungi interactions: where it goes? Biology 12(6):809. https://doi.org/10.3390/biology12060809

Rai N, Keshri PK, Verma A, Kamble SC, Mishra P, Barik S, Singh SK, Gautam V (2021) Plant associated fungal endophytes as a source of natural bioactive compounds. Mycology 12(3):139–159. https://doi.org/10.1080/21501203.2020.1870579

Rai M, Zimowska B, Gade A, Ingle P (2022) Promising antimicrobials from *Phoma* spp.: progress and prospects. AMB Express 12(1):60. https://doi.org/10.1186/s13568-022-01404-y

Ranty B, Aldon D, Galaud J (2006) Plant calmodulins and calmodulin-related proteins. Plant Signal Behav 1(3):96–104. https://doi.org/10.4161/psb.1.3.2998

Rauf M, Awais M, Ud-Din A, Ali K, Gul H, Rahman MM, Hamayun M, Arif M (2021) Molecular mechanisms of the 1-aminocyclopropane-1-carboxylic acid (ACC) deaminase producing *Trichoderma asperellum* MAP 1 in enhancing wheat tolerance to waterlogging stress. Front Plant Sci 11. https://doi.org/10.3389/fpls.2020.614971

Ray P, Guo Y, Kolape J, Craven KD (2018) Non-targeted colonization by the endomycorrhizal fungus, *Serendipita vermifera*, in three weeds typically co-occurring with switchgrass. Front Plant Sci 8. https://doi.org/10.3389/fpls.2017.02236

Ray P, Abraham PE, Guo Y, Giannone RJ, Engle NL, Yang ZK, Jacobson D, Hettich RL, Tschaplinski TJ, Craven KD (2019) Scavenging organic nitrogen and remodelling lipid metabolism are key survival strategies adopted by the endophytic fungi, *Serendipita vermifera* and *Serendipita bescii* to alleviate nitrogen and phosphorous starvation in vitro. Environ Microbiol Rep 11(4):548–557. https://doi.org/10.1111/1758-2229.12757

Rigobelo EC, Baron NC (2021) Endophytic fungi: a tool for plant growth promotion and sustainable agriculture. Mycology 13(1):39–55. https://doi.org/10.1080/21501203.2021.1945699

Ron M, Avni A (2004) The receptor for the fungal elicitor ethylene-inducing xylanase is a member of a resistance-like gene family in tomato. Plant Cell 16(6):1604–1615. https://doi.org/10.1105/tpc.022475

Saravanakumar K, Li Y, Yu C, Wang Q, Wang M, Sun J, Gao J, Chen J (2017) Effect of *Trichoderma harzianum* on maize rhizosphere microbiome and biocontrol of Fusarium stalk rot. Sci Rep 7(1):1771. https://doi.org/10.1038/s41598-017-01680-w

Sarkar S, Dey A, Kumar V, Batiha GE, El-Esawi MA, Tomczyk M, Ray P (2021) Fungal endophyte: an interactive endosymbiont with the capability of modulating host physiology in myriad ways. Front Plant Sci 12. https://doi.org/10.3389/fpls.2021.701800

Seth K, Vyas P, Deora S, Gupta AK, Meena M, Swapnil P, Harish. (2023) Understanding plant-plant growth-promoting rhizobacteria (PGPR) interactions for inducing plant defense. In: Swapnil P, Meena M, Harish, Marwal A, Vijayalakshmi S, Zehra A (eds) Plant-microbe interaction—recent advances in molecular and biochemical approaches, vol 2. Academic/Elsevier, pp 201–226. https://doi.org/10.1016/B978-0-323-91876-3.00010-5

Singh VK, Kumar A (2023) Secondary metabolites from endophytic fungi: production, methods of analysis, and diverse pharmaceutical potential. Symbiosis 90(2):111–125. https://doi.org/10.1007/s13199-023-00925-9

Singh R, Dangol S, Chen Y, Choi J, Cho Y, Lee J, Choi M, Jwa N (2016) *Magnaporthe oryzae* effector AVR-Pii helps to establish compatibility by inhibition of the rice NADP-malic enzyme resulting in disruption of oxidative burst and host innate immunity. Mol Cells *39*(5):426–438. https://doi.org/10.14348/molcells.2016.0094

Skiada V, Avramidou M, Bonfante P, Genre A, Papadopoulou KK (2020) An endophytic *Fusarium*–legume association is partially dependent on the common symbiotic signalling pathway. New Phytol 226(5):1429–1444. https://doi.org/10.1111/nph.16457

Son Y, Kim S, Chung H, Pae H (2013) Reactive oxygen species in the activation of MAP kinases. In: Methods in enzymology on CD-ROM/methods in enzymology, pp 27–48. https://doi.org/10.1016/b978-0-12-405881-1.00002-1

Sood M, Kapoor D, Kumar V, Sheteiwy MS, Ramakrishnan M, Landi M, Araniti F, Sharma A (2020) *Trichoderma*: the "Secrets" of a multitalented biocontrol agent. Plan Theory 9(6):762. https://doi.org/10.3390/plants9060762

Srinivas C, Devi DN, Murthy KN, Mohan CD, Lakshmeesha T, Singh B, Kalagatur NK, Niranjana S, Hashem A, Alqarawi AA, Tabassum B, Abd-Allah EF, Nayaka SC, Srivastava RK (2019) *Fusarium oxysporum* f. sp. *lycopersici* causal agent of vascular wilt disease of tomato: biology to diversity–a review. Saudi J Biol Sci 26(7):1315–1324. https://doi.org/10.1016/j.sjbs.2019.06.002

Steinkellner S, Lendzemo V, Langer I, Schweiger P, Khaosaad T, Toussaint J, Vierheilig H (2007) Flavonoids and strigolactones in root exudates as signals in symbiotic and pathogenic plant-fungus interactions. Molecules 12(7):1290–1306. https://doi.org/10.3390/12071290

Tanaka K, Gilroy S, Jones AM, Stacey G (2010) Extracellular ATP signaling in plants. Trends Cell Biol 20(10):601–608. https://doi.org/10.1016/j.tcb.2010.07.005

Tanaka K, Choi J, Cao Y, Stacey G (2014) Extracellular ATP acts as a damage-associated molecular pattern (DAMP) signal in plants. Front Plant Sci 5. https://doi.org/10.3389/fpls.2014.00446

Tang D, Wang G, Zhou J (2017) Receptor kinases in plant-pathogen interactions: more than pattern recognition. Plant Cell 29(4):618–637. https://doi.org/10.1105/tpc.16.00891

Toppo P, Subba R, Roy K, Mukherjee S, Mathur P (2022) Elucidating the strategies for isolation of endophytic fungi and their functional attributes for the regulation of plant growth and resilience to stress. J Plant Growth Regul 42(3):1342–1363. https://doi.org/10.1007/s00344-022-10638-w

Tyśkiewicz R, Nowak A, Ozimek E, Jaroszuk-Ściseł J (2022) *Trichoderma*: the current status of its application in agriculture for the biocontrol of fungal phytopathogens and stimulation of plant growth. Int J Mol Sci 23(4):2329. https://doi.org/10.3390/ijms23042329

Van Esse HP, Bolton MD, Stergiopoulos I, De Wit PJGM, Thomma BPHJ (2007) The chitin-binding *Cladosporium fulvum* effector protein Avr4 is a virulence factor. Mol Plant-Microbe Interact 20(9):1092–1101. https://doi.org/10.1094/mpmi-20-9-1092

Wahab A, Muhammad M, Munir A, Abdi G, Zaman W, Ayaz A, Khizar C, Reddy SPP (2023) Role of arbuscular mycorrhizal fungi in regulating growth, enhancing productivity, and potentially influencing ecosystems under abiotic and biotic stresses. Plan Theory 12(17):3102. https://doi.org/10.3390/plants12173102

Wan J, He M, Hou Q, Zou L, Yang Y, Wei Y, Chen X (2021) Cell wall associated immunity in plants. Stress Biol 1(1):3. https://doi.org/10.1007/s44154-021-00003-4

Wang C, Luan S (2024) Calcium homeostasis and signaling in plant immunity. Curr Opin Plant Biol 77:102485. https://doi.org/10.1016/j.pbi.2023.102485

Wang M, Weiberg A, Dellota E, Yamane D, Jin H (2017) *Botrytis* small RNA Bc-siR37 suppresses plant defense genes by cross-kingdom RNAi. RNA Biol 14(4):421–428. https://doi.org/10.1080/15476286.2017.1291112

Wang Y, Wang Y, Wang Y (2020) Apoplastic proteases: powerful weapons against pathogen infection in plants. Plant Commun 1(4):100085. https://doi.org/10.1016/j.xplc.2020.100085

Wang H, Peng H, Li W, Cheng P, Gong M (2021) The toxins of *Beauveria bassiana* and the strategies to improve their virulence to insects. Front Microbiol 12. https://doi.org/10.3389/fmicb.2021.705343

Wang L, Lu H, Jiang Y (2023) Natural polyketides act as promising antifungal agents. Biomol Ther 13(11):1572. https://doi.org/10.3390/biom13111572

Watchaputi K, Jayasekara LACB, Ratanakhanokchai K, Soontorngun N (2023) Inhibition of cell cycle-dependent hyphal and biofilm formation by a novel cytochalasin 19,20-epoxycytochalasin Q in *Candida albicans*. Sci Rep 13(1):9724. https://doi.org/10.1038/s41598-023-36191-4

Watts D, Palombo EA, Castillo AJ, Zaferanloo B (2023) Endophytes in agriculture: potential to improve yields and tolerances of agricultural crops. Microorganisms 11(5):1276. https://doi.org/10.3390/microorganisms11051276

Wawrzynska A, Rodibaugh NL, Innes RW (2010) Synergistic activation of defense responses in *Arabidopsis* by simultaneous loss of the GSL5 callose synthase and the EDR1 protein kinase. Mol Plant-Microbe Interact 23(5):578–584. https://doi.org/10.1094/mpmi-23-5-0578

Wei Y, Liang S, Zhu X, Liu X, Lin F (2023) Recent advances in effector research of *Magnaporthe oryzae*. Biomol Ther 13(11):1650. https://doi.org/10.3390/biom13111650

Wen J, Okyere SK, Wang S, Wang J, Xie L, Ran Y, Hu Y (2022) Endophytic fungi: an effective alternative source of plant-derived bioactive compounds for pharmacological studies. J Fungi 8(2):205. https://doi.org/10.3390/jof8020205

Wu M, Wei Q, Xu L, Li H, Oelmüller R, Zhang W (2018) *Piriformospora indica* enhances phosphorus absorption by stimulating acid phosphatase activities and organic acid accumulation in Brassica napus. Plant Soil 432(1–2):333–344. https://doi.org/10.1007/s11104-018-3795-2

Wu C, Wang Y, Yang Y (2022) Pestalotiopsis diversity: species, dispositions, secondary metabolites, and bioactivities. Molecules 27(22):8088. https://doi.org/10.3390/molecules27228088

Yadav G, Meena M (2021) Bioprospecting of endophytes in medicinal plants of Thar Desert: an attractive resource for biopharmaceuticals. Biotechnol Rep 30:e00629. https://doi.org/10.1016/j.btre.2021.e00629

Yadav G, Meena M (2022) Biological control of plant diseases by endophytes. In: Shah MP, Deka D (eds) Endophytic association: what, why and how (Volume: developments in applied microbiology and biotechnology). Academic Press, Elsevier, pp 119–135. https://doi.org/10.1016/B978-0-323-91245-7.00007-9

Yao X, Guo H, Zhang K, Zhao M, Ruan J, Chen J (2023) *Trichoderma* and its role in biological control of plant fungal and nematode disease. Front Microbiol 14. https://doi.org/10.3389/fmicb.2023.1160551

Yu C, Jiang X, Xu H, Ding G (2023) *Trichoderma longibrachiatum* inoculation improves drought resistance and growth of *Pinus massoniana* seedlings through regulating physiological responses and soil microbial community. J Fungi 9(7):694. https://doi.org/10.3390/jof9070694

Yuan P, Luo F, Gleason C, Poovaiah BW (2022) Calcium/calmodulin-mediated microbial symbiotic interactions in plants. Front Plant Sci 13. https://doi.org/10.3389/fpls.2022.984909

Zandi P, Schnug E (2022) Reactive oxygen species, antioxidant responses and implications from a microbial modulation perspective. Biology 11(2):155. https://doi.org/10.3390/biology11020155

Żelechowski M, Molcan T, Bilska K, Myszczyński K, Olszewski J, Karpiesiuk K, Wyrębek J, Kulik T (2021) Patterns of diversity of fusarium fungi contaminating soybean grains. Toxins 13(12):884. https://doi.org/10.3390/toxins13120884

Zhang J, Zhou J (2010) Plant immunity triggered by microbial molecular signatures. Mol Plant 3(5):783–793. https://doi.org/10.1093/mp/ssq035

Zhang F, Wang Y, Liu C, Chen F, Ge H, Tian F, Yang T, Ma K, Zhang Y (2019) *Trichoderma harzianum* mitigates salt stress in cucumber via multiple responses. Ecotoxicol Environ Saf 170:436–445. https://doi.org/10.1016/j.ecoenv.2018.11.084

Zhang J, Coaker G, Zhou J, Dong X (2020) Plant immune mechanisms: from reductionistic to holistic points of view. Mol Plant 13(10):1358–1378. https://doi.org/10.1016/j.molp.2020.09.007

Zhang H, Ahima J, Yang Q, Zhao L, Zhang X, Zheng X (2021) A review on citrinin: its occurrence, risk implications, analytical techniques, biosynthesis, physiochemical properties and control. Food Res Int 141:110075. https://doi.org/10.1016/j.foodres.2020.110075

Adaptation Potential of Endophytic Fungi to Produce Secondary Metabolites

13

A. Usha Raja Nanthini

Abstract

Endophytes, primarily fungi, inhabit plant tissues without causing visible harm and have evolved alongside plants for over 400 million years. This long coevolution has transformed them from pathogens to mutualistic partners, enabling them to produce specialized metabolites that enhance plant survival and adaptation. Endophytes exhibit remarkable diversity and are found in a wide range of climates and plant species. Their interactions with plants involve complex chemical signaling, particularly through root exudates, which facilitate microbial colonization and competition. These signals enable endophytes to establish themselves in plant tissues, where they contribute to growth and stress responses by producing bioactive compounds. By tapping into plant metabolic pathways, endophytic fungi enhance resistance to biotic and abiotic stresses. They promote plant growth through the production of phytohormones and improved nutrient uptake. Additionally, they bolster disease tolerance via competition, antibiotic production, and mycoparasitism. Endophytes display adaptability through genetic mechanisms, allowing them to alternate between parasitic, saprotrophic, and endophytic lifestyles. Recent studies have highlighted diverse endophytic fungal strains in mushrooms that produce a variety of bioactive metabolites, indicating a need for revised classification. This chapter explores the adaptation potential of endophytic fungi in producing secondary metabolites.

A. Usha Raja Nanthini (✉)
Department of Biotechnology, Mother Teresa Women's University,
Kodaikanal, Tamilnadu, India

M. Meena et al. (eds.), *Metabolic Sustainability of Endophytes*, Microorganisms for Sustainability 52, https://doi.org/10.1007/978-981-96-4004-1_13

13.1 Introduction

Microorganisms, though invisible to the naked eye, are essential players in the intricate web of life on Earth. These microscopic entities drive critical processes that sustain ecosystems, regulate climate, and promote human health and well-being. Their influence is profound, as they underpin nutrient cycling, decomposition, and the overall stability of various ecological systems. Plants, in particular, serve as significant reservoirs for diverse microorganisms, including bacteria and fungi, collectively known as plant microbiomes (Dhyani et al. 2019). This symbiotic relationship is vital for maintaining ecosystem equilibrium. Among these microorganisms, endophytic fungi inhabit the internal and external tissues of plants, establishing mutualistic connections without harming their hosts (Aly et al. 2011).

Endophytic fungi possess a remarkable capacity for adaptation, particularly in their ability to synthesize secondary metabolites—compounds not directly involved in their growth or reproduction but crucial for environmental interactions. These metabolites encompass a variety of chemical classes, including alkaloids, terpenoids, and polyketides, enabling endophytes to thrive across diverse ecological niches and engage with a range of plant species. The production of secondary metabolites also plays a defensive role, protecting host plants from herbivores, pathogens, and competing flora. This adaptive trait enhances the resilience and fitness of both the fungi and their host plants. Furthermore, endophytic fungi can modulate their metabolite production in response to environmental stresses such as drought, salinity, and temperature extremes. Their genetic flexibility allows them to respond dynamically to shifting environmental conditions and host signals, making this adaptability critical for their survival and effectiveness as endophytes.

13.2 Endophytes

Endophytes are microorganisms that inhabit living plant tissues without causing visible disease symptoms, coexisting with their hosts for at least part of their life cycle (Bacon and White 2000). These organisms are particularly abundant in roots and can be transmitted vertically—from parent plants to seeds—and horizontally, through sexual reproduction or other forms of infection (Malik et al. 2023). Endophytes encompass a variety of microorganisms, including fungi, bacteria, and actinomycetes, with fungi being the most prevalent (Staniek et al. 2008). They belong to diverse phyla, predominantly Ascomycota (89%), followed by Basidiomycota (9%) and Mucoromycota (2%) (Rana et al. 2019). Fungal endophytes are particularly versatile, colonizing various plant structures, including leaf surfaces (epiphytes), leaf tissues (foliar endophytes), bark (bark endophytes), and wood (xylem endophytes and decomposers) (Stone et al. 2004). Fungal endophytes can be classified into two primary groups based on their life history, ecological roles, and phylogenetic relationships: Clavicipitaceous endophytes (C-endophytes) and non-Clavicipitaceous endophytes (NC-endophytes).

Clavicipitaceous endophytes, commonly referred to as grass endophytes or Class 1 endophytes, primarily consist of the Clavicipitaceae family within the Hypocreales order of Ascomycota. This group includes 27 identified genera, such as *Balansia, Epichloë, Balansiopsis, Atkinsonella, Echinodothis, Myriogenospora, Parepichloe,* and *Neotyphodium*, which are associated with specific grasses (Poaceae) and thrive in cooler climates. These endophytes form mutualistic relationships with their hosts, colonizing intercellular spaces in aboveground plant tissues and propagating both vertically through seed infection and horizontally (Hardoim et al. 2015; De Silva et al. 2016; Santangelo et al. 2015).

Non-Clavicipitaceous endophytes, on the other hand, represent a highly diverse polyphyletic group with less clearly defined taxonomy. These endophytes are found globally, across a range of habitats and host plants—including nonvascular, vascular, and woody species (Suryanarayanan et al. 2000; Mohali et al. 2005; Šraj-Kržič et al. 2006; Selim et al. 2017). The majority belong to Ascomycota or Basidiomycota, with genera such as *Fusarium, Colletotrichum, Xylaria,* and *Phomopsis* prevalent in terrestrial ecosystems (De Silva et al. 2016; Jayawardena et al. 2016). These endophytes can reproduce both sexually and asexually, dispersing through spores or conidia. Unlike some endophytes that rely on their host for survival, non-Clavicipitaceous endophytes can remain dormant, responding to chemical signals from plants under stress, such as injury or environmental changes (Mishra et al. 2021a, b). Their colonization patterns vary, with NC-endophytes often extensively invading roots and the rhizosphere, while their presence in aerial parts of plants is typically more localized.

Endophytic fungi are remarkably diverse, with estimates suggesting there may be up to three million species globally (Hawksworth and Lücking 2017; Manganyi and Ateba 2020). They have been found in various climatic zones, from Arctic tundras to tropical forests, with the highest diversity observed in tropical and temperate rainforests. It is estimated that around 30,000 land plant species host these fungi, often containing multiple endophytic species (Arnold 2008). The relationships between endophytic fungi and their host plants can vary widely, ranging from mutualistic to antagonistic interactions. In grass-endophyte systems, these fungi can act as herbivore antagonists and promote plant growth (Clay 1990), while forest endophytes may display mutualistic antagonism against insects and pathogens (Faeth 2002). The structure and diversity of endophyte communities are influenced by various abiotic and biotic factors, including plant–microbe interactions, biogeography, and soil conditions (Deng and Cao 2017). This complexity reflects the richness of the molecular compounds produced by endophytes as they adapt to their environments.

13.3 Evolutionary Adaptation

Endophytes have been identified in nearly every plant species studied, from the Arctic to tropical rainforests (Tan and Zou 2001). Fossil evidence suggests that the relationship between plants and endophytes may date back millions of years, coinciding with the emergence of higher plants (Gutierrez et al. 2012). It is believed that endophytes evolved from parasitic or pathogenic fungi associated with both grasses and woody plants (Carroll 1986, 1991, 1992). Over time, these fungi likely reduced their pathogenicity and extended their latency periods, transitioning into beneficial endophytes (Petrini et al. 1992). The coevolution of endophytic fungi and their host plants fosters the development of specialized secondary metabolites tailored to their interactions. This evolutionary relationship enhances the adaptive potential and survival strategies of both parties. As endophytes and plants evolve together, a balance is struck between fungal activity and plant response, resulting in a stable mutualism over time (Mariño et al. 2007).

Research indicates that endophytic fungi are heterotrophic organisms with complex life cycles that engage in symbiotic relationships with their plant hosts. Fossil records show that these associations have existed for at least 400 million years, playing significant roles in the ecology, growth, fitness, and evolution of both groups (Krings et al. 2007). As plants transitioned from aquatic to terrestrial environments, they faced challenges such as nutrient-poor soils, high carbon dioxide levels, fluctuating temperatures, and limited water availability. Fungal endophytes have been instrumental in helping plants adapt to these stressors, enhancing their survival through genetic adaptations and the production of secondary metabolites (Arora and Ramawat 2017).

Endophytic fungi exhibit genetic diversity influenced by their prolonged associations with host plants, which can include the incorporation of plant genetic material into their genomes. Recent studies reveal that endophytic fungi and their host plants utilize distinct yet complementary metabolic pathways to produce bioactive compounds (Jennewein et al. 2001). For instance, *Fusarium oxysporum* is known to synthesize plant-derived lignan podophyllotoxin, a compound that exemplifies this interaction (Germaine et al. 2004). Another example is the endophytic fungus *Trametes hirsuta*, isolated from the rhizomes of *Podophyllum hexandrum*. This fungus also produces the podophyllotoxin and other related compounds (Puri et al. 2006). Subsequent studies have identified various fungal endophytes from other podophyllotoxin-producing plants that also generate this compound and its analogs (Eyberger et al. 2006; Kour et al. 2008; Kusari et al. 2009b). The consistent production of lignans by these fungi supports the idea that microorganisms have evolved the capacity to compete for survival alongside plants by acquiring the mechanisms to synthesize similar compounds and tolerate high concentrations of plant secondary metabolites (Kusari et al. 2009b).

Most plants host one or more fungal endophytes, particularly woody species that can contain hundreds of different endophytic fungi. These fungi are widely distributed across various geographical and climatic regions, showcasing remarkable species diversity. Their abundance and ecological functions indicate a high level of adaptation to different environments and selective pressures (Rodriguez et al. 2009). Overall, the colonization of host tissues by endophytic fungi enhances plant tolerance to specific environmental stresses, contributing significantly to their survival. Less adaptable microorganisms may struggle to thrive when resources are scarce, resulting in competition for survival. Organisms have developed two main strategies: one involves producing allelochemicals that inhibit competitors, while the other involves synthesizing compounds that facilitate symbiotic relationships with host plants. Such symbiotic interactions enable both partners to endure and reproduce in challenging environments (Macías-Rubalcava et al. 2014). This symbiotic relationship between endophytes and their plant hosts fosters a sustained coevolutionary process within biotic communities (Ji et al. 2009). Notably, genetic changes associated with host specificity have been observed in plant pathogens, highlighting the adaptive processes occurring during host-pathogen coevolution. For instance, while *Colletotrichum graminicola* and *Colletotrichum sublineolum* are genetically similar, they have evolved to infect specific hosts—maize and sorghum, respectively—demonstrating the nuanced adaptations driven by their interactions with host plants (Buiate et al. 2017).

13.4 Adaptation by Host-Specific Interaction

The colonization of plants by endophytes involves a complex interplay of chemical signals exchanged between plants, endophytes, and other microorganisms (Gaiero et al. 2013; Hardoim et al. 2015). Root exudates play a crucial role in this process, releasing compounds that attract specific soil-borne or rhizosphere microorganisms, ultimately benefiting the host plant (Francis et al. 2010; Rosenblueth and Martı'nez-Romero 2006; Kloepper et al. 2004; Philippot et al. 2013; Compant et al. 2010). However, the mechanisms behind this selective recruitment remain poorly understood. Root exudates contain various chemical signals and nutrients—such as carbohydrates, organic acids, and amino acids—that help microorganisms identify suitable hosts (Fellbaum et al. 2012; Rudrappa et al. 2008; Rosenblueth and Martı'nez-Romero 2006). Successful root colonization by microbes depends on several factors, including the expression of relevant genes, chemotactic responses to exudates, competitive abilities against other microorganisms, resistance to plant defenses, and the establishment of a niche within the plant tissue (Bais et al. 2006; Gaiero et al. 2013; Compant et al. 2010; Rosenblueth and Martı'nez-Romero 2006; Shidore et al. 2012). In fact, root exudates can induce the expression of genes necessary for endophytic colonization (Shidore et al. 2012). Once inside the root cells, microbes can be "digested" by host plants for nutrients through a process known as rhizophagy, which also promotes the formation of extensive root hairs in microbe-associated plants. As a result, endophyte-infected plants generally exhibit enhanced survival, growth rates, and flowering compared to those without such associations.

Endophytes typically inhabit healthy plant tissues, residing in areas such as roots, stems, and petioles (Ma et al. 2013; Deepthi et al. 2018). Most bacterial endophytes enter through the roots but can also access plants via stomata, wounds, or even flowers and fruits (Compant et al. 2010; Hardoim et al. 2015). Additionally, compounds exuded from stems and leaves may attract further microbial colonization (Compant et al. 2010). Ultimately, both exudates and the plant's immune system help select for microbes that have developed strategies for effective colonization (Lareen et al. 2016). Chemical signals initially guide microbial colonization of the rhizoplane and other surfaces, allowing them to penetrate the plant tissues (Hardoim et al. 2015). Once established, endophytes can move through vascular tissues, influenced by the availability of plant resources (Johnston-Monje and Raizada 2011; Zakria et al. 2007). Some studies suggest that leaf-associated microbial endophytes often arise from the migration of root endophytes rather than from surface colonization (Bringel and Coue'e 2015; Hartley et al. 2015).

Endophytic fungi have developed specialized mechanisms to penetrate and thrive within plant tissues. They secrete exoenzymes that facilitate colonization and proliferation in the apoplastic fluid of their hosts (Chandra 2012). By synthesizing compounds that enhance their ability to colonize, endophytes can regulate metabolic processes in a balanced manner (Chandra 2012). The production of secondary metabolites often arises from metabolic communication between endophytes and their host plants (Priti et al. 2009). Over time, these mutualistic relationships result in the production of bioactive compounds as by-products of the interaction. Additionally, genetic material can be exchanged due to the long-term cohabitation between hosts and endophytes (Wang and Dai 2011; Nadeem et al. 2012).

Endophytes have also evolved mechanisms to detoxify the bioactive chemicals produced by their host plants, enabling them to effectively colonize plant tissues (Wang and Dai 2011). They can transform harmful metabolites into beneficial compounds with therapeutic properties (Zikmundova et al. 2002; Saunders and Kohn 2009), utilizing biotransformation enzymes to modify various molecules (Chen et al. 2016). For example, thioridazine has been stereoselectively biotransformed by several endophytic fungi, including *Glomerella cingulata*, *Phomopsis* sp., *Diaporthe phaseolorum*, and *Aspergillus fumigatus* (Borges et al. 2007). Research indicates that endophytes may enhance the fitness of their host plants by altering their physiology or influencing interactions with external factors such as pathogens and herbivores. Numerous studies have reported improvements in plant biomass, photosynthetic rates, nutrient uptake, and other growth parameters under both controlled and natural conditions (Carroll 1988). These interactions contribute positively to plant sustainability, biodiversity, and ecological stability (Rosier et al. 2016; Bai et al. 2017; Sasse et al. 2017).

Endophytes exhibit host specificity, often residing within particular plant species or groups, with some species restricted to a single host type (Holliday 1998). This specificity can be classified into establishment specificity—where a fungus proliferates within a particular host species—and expression specificity—where a fungus can colonize multiple hosts, although its structure is limited by the number of available plant taxa (Balkrishna et al. 2022). The first report of endophytic mycelium was made in 1898 in the seeds of *Latium teinutentuin* (Waghunde et al. 2021). With around 300,000 plant species known to harbor one or more endophytes—often comprising one to one hundred strains depending on the host system (Gao et al. 2018)—these organisms can be categorized as obligate, opportunistic, or facultative endophytes. Obligate endophytes spend most or all of their lives within plants, while opportunistic endophytes thrive externally and enter plants under favorable conditions. Facultative endophytes, which include the majority of endophytes, engage in an ambiguous mutualistic relationship with their host (Hardoim et al. 2015; Plett et al. 2011).

Endophyte–plant interactions typically fall under three symbiotic types: mutualism, commensalism, and parasitism. Most plant–endophyte associations are either commensal or mutualistic, although some pathogenic microbes can also inhabit the endosphere (Wani et al. 2015; Brader et al. 2017; Khare et al. 2018; Rodriguez et al. 2009; Yadav and Meena 2022). Generally, these interactions are viewed as mutualistic symbiosis, where both partners benefit. Endophytes provide essential services to host plants, including the production of phytohormones and secondary metabolites, increased nutrient availability, and enhanced stress tolerance, all in exchange for fixed carbon from the plant (Field et al. 2015; Wani et al. 2015; Martinez-Klimova et al. 2017; Rabiey et al. 2019; Wang et al. 2019; Slama et al. 2021; Yadav and Meena 2021). The intricate relationships between endophytes and their host plants are vital for plant development, survival, and defense mechanisms. Achieving a mutualistic relationship requires a delicate balance between host defenses and fungal virulence. An imbalance can lead to disease, as increased virulence can suppress the plant's defenses (Chandra 2012). This is illustrated in Fig. 13.1.

Additionally, the production of secondary metabolites by endophytes significantly enhances host plant protection against pathogens and herbivores. For instance, the endophyte *Acremonium*, which lacks a sexual reproduction stage, can only propagate through its host plant's seeds. Evidence suggests that the frequency of *Acremonium* infections in host populations increases when the plants harboring these endophytes exhibit improved survival and reproductive success (Clay 1993). While fungi can colonize intercellular or intracellular areas of plants, they tend to establish extensive colonization more readily in roots than in aerial tissues. Successful reproduction of endophytes in aboveground organs often depends on the availability of the host's apoplastic fluid (Schulz et al. 2002). Fungal interactions with their hosts can be categorized into three types based on their effects: beneficial symbiotic (mutualistic endophytes), non-beneficial commensal (virulent endophytes), and pathogenic (virulent pathogens). These associations can either enhance, have no effect on, or reduce host fitness (Kogel et al. 2006).

Despite the importance of these interactions, much remains to be understood about the dynamics of endophyte–host relationships in natural habitats. It has been suggested that the relationship is characterized by a finely tuned balance between fungal virulence and plant defenses (Saikkonen et al. 1998; Schardl et al. 1991; Schulz et al. 2002). A mutualistic relationship may emerge when endophytic colonization leads to higher phenolic defenses compared to pathogenic infections (Schulz et al. 2002). Imbalances in nutrient exchange and genetic factors can shift the interaction from mutualistic to parasitic (Kogel et al. 2006). Additionally, miscommunication can inadvertently trigger disease processes under stress conditions (Rodriguez and Redman 2008). Evidence suggests that culturable endophytes associated with woody plant leaves are more likely to originate from external sources (horizontal transmission) rather than from vertical transmission (Arnold and Lutzoni 2007). Overall, endophytic fungi tend to exhibit antagonistic behavior, while those colonizing roots demonstrate more mutualistic traits (Schardl et al. 1991; Schulz

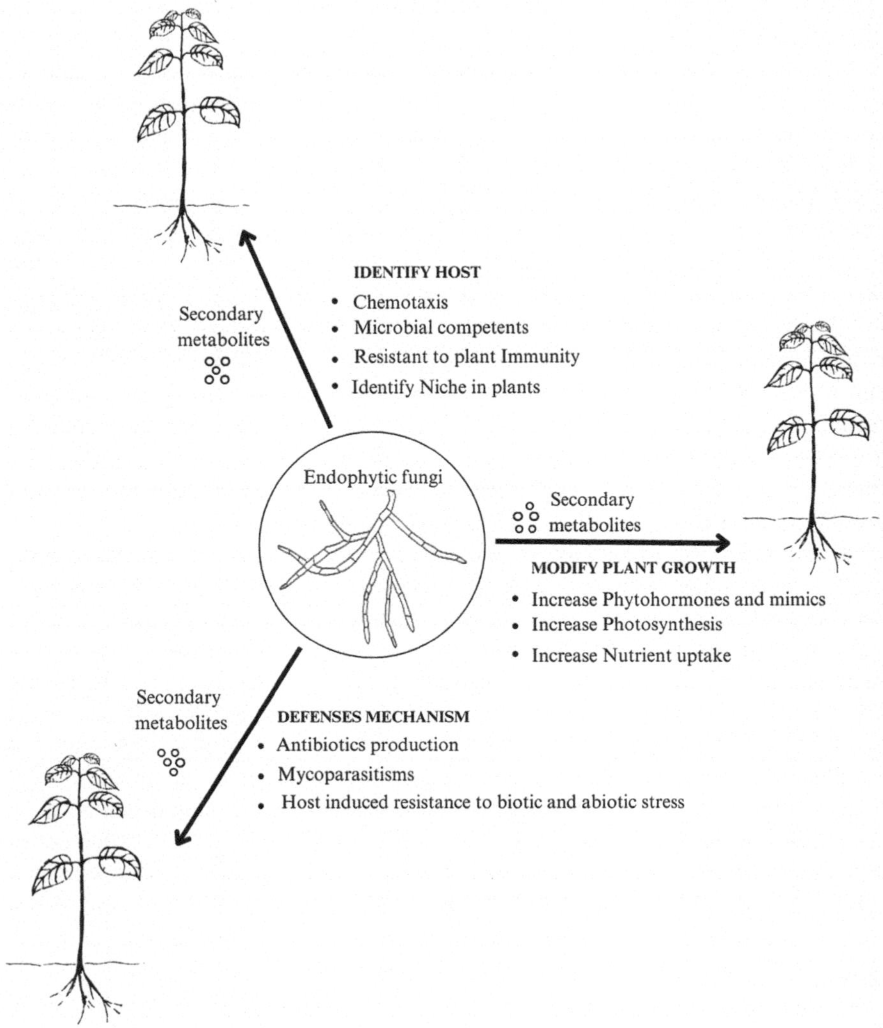

Fig. 13.1 Diverse mechanisms for secondary metabolite synthesis in endophytic fungi

et al. 2002). They produce exoenzymes essential for penetrating and colonizing their hosts, utilizing the apoplastic medium as a growth environment. Further investigation into the molecular and chemical signals governing this intricate communication between endophytes and their hosts is warranted. Table 13.1 illustrates the bioactive secondary metabolites produced by endophytic fungi in response to host-specific interactions.

Table 13.1 Secondary metabolites produced by endophytic fungi in response to host-specific interactions and their mode of action

Bioactive secondary metabolites	Mode of action	References
Tyrosol and farnesol	Quorum sensing (QS) signal molecules in yeasts	Chen et al. (2004), Yajima (2014)
Diterpene momilactone A	Defense compounds secreted by noninfected rice can be metabolized and detoxified by the endophyte *Magnaporthe grisea*. This ability to detoxify the diterpene momilactone A may be a strategy for the endophyte to overcome the toxicity of this defense compound, facilitating successful colonization of the plant	Hasegawa et al. (2010), Toyomasu et al. (2008), Kato-Noguchi et al. (2008)
Phytoalexins including camalexin, an indolic compound derived from tryptophan	Phytopathogenic fungi metabolize camalexin into less toxic products	Balachandar et al. (2007), Aguilar et al. (2015)
Myc factors—classes of chitin oligomers (oligosaccharides), chitooligosaccharides (COs), and lipochitooligosaccharides (LCOs)	Produced in higher quantities when spore germination is induced by strigolactones (plant hormones)	Gough and Be'card (2016), Gough and Cullimore (2011), Maillet et al. (2011), Genre et al. (2013)
Citric acid and fumaric acid	Increase the ability of beneficial microorganisms to attract to the roots of infected cucumbers and facilitate biofilm development.	Liu et al. (2014)
Camptothecin	*Fusarium solani* relies on the host plant's strictosidine synthase enzyme to synthesize camptothecin. To counteract the toxicity of camptothecin, *F. solani* modifies the amino acid sequence of its own topoisomerase I, enabling it to develop self-resistance	Kusari et al. (2011)
Podophyllotoxin and other related aryl tetralin lignans	Produced by *Trametes hirsuta* from *Podophyllum hexandrum* and by acquiring the necessary machinery for biosynthesizing similar compounds and developing tolerance to elevated levels of plant secondary metabolites they have evolved the ability to compete and thrive alongside plants	Kusari et al. (2009a, b, c), Puri et al. (2006)
Vindoline	*Curvularia* sp. and *Choanephora infundibulifera* from *Catharanthus roseus* leaves increased the content by 229–403% in endophyte-inoculated plants by upregulating the expression of genes involved in the terpene indole-alkaloid biosynthetic pathway	Pandey et al. (2016)

(continued)

Table 13.1 (continued)

Bioactive secondary metabolites	Mode of action	References
Hypericin and emodin	*Thielavia subthermophilia*, from the stems of *Hypericum perforatum*, lack the *hyp-1* gene. This absence suggests that distinct biosynthetic pathways or molecular mechanisms are at work in the endophyte compared to its host plant	Kusari et al. (2009a, b, c)
Artemisinin	The incorporation of the endophytic fungus *Colletotrichum* sp. into plant tissue cultures of *Artemisia annua* increased the production of the sesquiterpene lactone endoperoxide by upregulating the expression of plant genes associated with artemisinin biosynthesis	Wang et al. (2001), Li et al. (2012)
Tetraspanin membrane protein	Produced by *Magnaporthe grisea* and *Botrytis cinerea* appressorium (infectious structure of fungi) to penetrate host plant surface tissue, additional to physical pressure. These fungi causes an oxidative burst by producing reactive oxygen species (ROS)	van Kan (2006), Garcia-Arenal and Sagasta (1980), Choquer et al. (2007), Mbengue et al. (2016)
Pep1, Pit2, Cmu1, Tin2, and See1	Effector proteins secreted by *Fusarium oxysporum* into the xylem during the colonization for virulence and help overcome plant physiological adaptations, facilitating fungal infection and dissemination	Gawehns et al. (2015)
Reactive species	The host plant suppresses these factors to create more favorable conditions for symbiotic association with *Serendipita indica*	Hua et al. (2017)

13.5 Adaptation by Metabolic Diversity

Primary metabolites, including lipids, proteins, carbohydrates, and amino acids, are essential for growth and development, serving as the building blocks for an organism's metabolism. These primary metabolites also play a crucial role as precursors for the synthesis of secondary metabolites, which, while not vital for basic life processes, are essential for plants to adapt to their environments (Bourgaud et al. 2001). Similarly, bioactive secondary metabolites produced by microorganisms serve as defensive agents against external threats (Bérdy 2005; Kusari and Spiteller 2012). Both endophytes and their host plants utilize these precursors in their distinct biosynthetic pathways for secondary metabolites. Research by Kirby and Keasling (2009) and Deepika et al. (2016) suggests that endophytes can mimic host pathways to synthesize their secondary metabolites. Studies employing blocking mutants and radiolabeling techniques have examined the biosynthesis of specific phytochemicals, including ergot alkaloids, aflatoxins, and lovastatin (Keller et al. 2005; Meena et al. 2023).

Previously thought to have negligible physiological effects, secondary metabolites are now recognized for their roles in enhancing plant immunity (Soliman et al. 2015), controlling diseases (Terhonen et al. 2016), facilitating nutrient uptake (Hiruma et al. 2016), and improving resilience to abiotic stress (Khan et al. 2013). Plants that produce toxic alkaloids in conjunction with endophytes exhibit enhanced resistance to both biotic and abiotic stressors (Carroll 1988; Chagas et al. 2018). Additionally, coculturing with endophytic elicitors has proven effective in boosting plant secondary metabolite production and resistance (Li and Tao 2009; Meena et al. 2022). Endophytic fungi are classified based on their biological interactions with plant systems. Group I endophytes transfer genetic elements vertically into plant systems, while Group II endophytes generally assist in managing external stresses. Group III endophytes acquire genes from other fungal species through horizontal gene transfer, leading to the production of bioactive compounds. The specific secondary metabolites produced by endophytes vary depending on the plant's condition and age, spanning a range of structural classes, including terpenoids, steroids, flavonoids, alkaloids, phenols, and peptides (Calhoun et al. 1992; Bhupenchandra et al. 2024).

Reports suggest that endophytes produce bioactive secondary metabolites that bolster host plant resilience against herbivores and inhibit microbial diseases (Gutierrez et al. 2012; Tan and Zou 2001). For instance, fungal endophytes in ryegrass (*Lolium perenne*) are associated with increased levels of hydroxycinnamic acids and glycosylated flavonoids, including chlorogenic acid, which enhance the plant's antioxidant activity (Ludwig-Müller 2015; Qawasmeh et al. 2012). Additionally, various antimicrobial metabolites produced by plant-associated microorganisms, such as ergot alkaloids and aflatoxins, play a vital role in competition within the host plant's microenvironment (Spiteller 2015; Zhang et al. 2006; Chauhan et al. 2023). By reducing competition, these secondary metabolites help endophytes thrive in challenging conditions, sometimes even disadvantaging related species. The mutualistic relationships between plants and endophytic fungi lead to the production of important bioactive compounds that improve stress tolerance. Under abiotic stress conditions such as drought and salinity, endophytes can regulate the levels of antioxidant enzymes (catalase [CAT], peroxidase [POD], superoxide dismutase [SOD]) to alleviate stress-induced cellular damage.

Endophytic fungi also show potential as biocontrol agents through the secretion of enzymes (cellulase, amylase, and protease) and secondary metabolites, which could reduce reliance on synthetic pesticides (Yadav et al. 2010). They influence host metabolic processes, enhancing drought and metal tolerance, growth, and nutrient uptake (Poveda and Baptista 2021; Cui et al. 2021). For example, *Penicillium* sp. can solubilize phosphate under saline conditions, making it available to plants, while also reducing disease severity through increased jasmonic acid and antioxidant enzyme concentrations (Patel et al. 2021). In return for the benefits provided, plants offer spatial structure, moisture protection, and nutrients to their endophytes.

Host plants may also supply essential substances for the endophyte's life cycle and self-defense (Rudgers et al. 2004).

The production of bioactive compounds by endophytes requires a diverse array of plant species, components, and environmental conditions (Giauque and Hawkes 2013; Guevara-Araya et al. 2020; Harrison and Griffn 2020). Environmental factors significantly influence endophytic communities, including cultivation history, climate, and seasonal changes (Correa-Galeote et al. 2018; Hosseyni-Moghaddam and Soltani 2014; Wang et al. 2019). Endophytic fungi can upregulate plant genes associated with phytoalexin biosynthesis, thereby enhancing plant defenses (Li et al. 2012). This suggests that endophytes may have evolved to produce secondary metabolites similar to those of their host plants, potentially influencing the biosynthesis of plant metabolites (Kusari et al. 2013). These metabolites are synthesized through various pathways, such as the shikimate pathway (alkaloids, flavonoids) and the (Tricarboxylic acid (TCA)) cycle (isoprenoids, polyketides, terpenoids) (Tohge et al. 2013; Meena et al. 2019). Research has shown that endophytic fungi from medicinal plants can produce secondary metabolites (Tejesvi et al. 2007). This indicates that endophytes serve as a reservoir of novel compounds with potential applications in pharmaceuticals and agrochemicals, exhibiting properties such as antimicrobial, antiviral, antifungal, anticancer, and antioxidant effects (Calhoun et al. 1992). For instance, huperzine A, an alkaloid isolated from the Chinese plant *Huperzia serrata*, is also produced in low yields by endophytic fungi like *Shiraia* sp. (Zhu et al. 2010; Zhao et al. 2013). Enhancements in huperzine production have been observed with the addition of elicitors such as host plant extracts and L-lysine, a precursor for alkaloid biosynthesis (Yan et al. 2014). This suggests that plant-derived substances and small molecules regulate huperzine production (Fernanda et al. 2017) (Table 13.2).

13.6 Adaptation by Enhancing Growth

Endophytic fungi enhance host plant growth by producing and supplying essential plant hormones, including auxins, cytokinins, and gibberellins. Research has identified mechanisms similar to those found in higher plants, demonstrating that both plants and fungi can synthesize gibberellins (Kumar et al. 2013). In stressful conditions, endophytes secrete a variety of secondary metabolites into plant cells, integrating them into the plant's stress response pathways (Nanda et al. 2019). These stress-responsive compounds, such as gibberellins, cytokinins, salicylic acid, and indole-3-acetic acid (IAA), promote growth and facilitate various physiological changes within the plant (Shekhawat et al. 2021; Shaffique et al. 2022). By enhancing germination success and growth rates, endophytes improve the host's overall fitness and competitiveness. They boost the plant's ability to absorb nutrients and increase the production of growth hormones like indole-3-acetic acid, indole-3-acetonitrile, and cytokinins (Hartley and Gange 2009; Tan and Zou 2001; Zhang et al. 2006).

Table 13.2 Secondary metabolites produced by endophytic fungi by adapting metabolic diversity

Bioactive secondary metabolites	Endophyte	Host plant	References
Alkaloids and nitrogen-containing heterocyclic compounds			
Variecolorin Grand alkaloid E-7 Dioxopiperazine alkaloid	*Eurotium rubrum*	*Hibiscus tiliaceus*	Li et al. (2008)
Camptothecin, 9-methoxycamptothecin	*Fusarium solani*	*Apodytes dimidiate*	Shweta et al. (2010)
9-methoxy-camptothecin Camptothecin 10-hydroxy-camptothecin	*Phomopsis* sp., *Fomitopsis* sp., and *Alternaria alternata*	*Miquelia dentata*	Shweta et al. (2013)
Azaphilone alkaloids Chaetomugilides A–C Chaetoviridin E	*Chaetomium globosum*	*Ginkgo biloba*	Li et al. (2013)
Mycoleptodiscin B	*Mycoleptodiscus* sp.	*Desmotes incomparabilis*	Ortega et al. (2013)
Citriquinochroman	*Penicillium citrinum*	*Ceratonia siliqua*	El-Neketi et al. (2013)
Pestalactam A Pestalactam B	*Pestalotiopsis* sp.	Plant host can vary depending on the geographical location and fungal species involved	Davis et al. (2010)
Aspochalasin D Aspochalasin J	*Trichoderma gamsii*	*Panax notoginseng*	Ding et al. (2012)
Camptothecin	*Fusarium solani*	*Camptotheca acuminate*	Ran et al. (2017)
Piperine	*Colletotrichum gloeosporioides*	*Piper nigrum*	Chithra et al. (2014)
Camptothecin	*Entrophospora infrequens* *Neurospora* sp.	*Nothapodytes foetida*	Amna et al. (2006)
Deoxytryptoquialanone Tryptoquialanine A Tryptoquialanine C 15-dimethyl-2-epifumiquinazoline A Citrinadin A Ceoxycitrinadin A	*Penicillium citrinum* *Penicillium digitatum*	*Citrus limon*	Costa et al. (2019), Huang et al. (2007)
Vincristine	*Fusarium oxysporum*	*Catharanthus roseus*	Fernanda et al. (2017)
Chaetoglobosin A	*Fusarium oxysporum* *Chaetonium globosum*	*Catharanthus roseus* *Maytenus hookeri*	Zhang et al. (2002)
Chaetoglobosin U Chaetoglosins C, E and F Penochalasin A	*Chaetonium globosum*	*Imperata cylindrical*	Ding et al. (2006)

(continued)

Table 13.2 (continued)

Bioactive secondary metabolites	Endophyte	Host plant	References
Alanditrypinone Alantryphenone Alantrypinene Alantryleunone	*Eupenicillium* sp.	*Murraya paniculata*	Proenca Barros and Rodrigues-Filho (2005)
Huperzine A	*Shiraia* sp.	*Huperzia serrata*	Zhu et al. (2010), Zhao et al. (2013)
Coumarins			
(3R*,4S*)-6,8-dihydroxy-3,4,7-trimethylisocoumarin	*Penicillium* sp.	*Bruguiera sexangula* Linn	Han et al. (2009)
5-methyl-8-(3-methylbut-2-enyl) furanocoumarin	*Penicillium* sp.	Mangrove	Huang et al. (2012)
Arundinone B	*Microsphaeropsis arundinis*	*Ulmus macrocarpa*	Luo et al. (2013)
Pestalustaine A and B Sesquiterpene Coumarin derivatives	*Pestalotiopsis adusta*	*Sinopodophyllum hexandrum*	Xiao et al. (2018)
Lactones			
Cytospolide P	*Cytospora* sp.	*Ilex canariensis*	Lu et al. (2011)
Macrolides	*Colletotrichum* sp.	*Pandanus amaryllifolius*	De Vries et al. (2016)
Cytosporone B and C Dothiorelone C	*Chaetomium globosum*	*Ginkgo biloba*	Bungihan et al. (2013), Kumar et al. (2017)
Aspergillide B1	*Aspergillus terreus*	*Glycine max*	El-Hawary et al. (2021)
Oocydin A	*Serratia marcescens*	*Rhyncholacis penicillate*	Strobel et al. (2003)
Brefeldin-A	*Paecilomyces* sp. *Aspergillus clavatus*	*Taxus mairei* *Torreya grandis*	Cuconati et al. (1998), Wang et al. (2002)
Phomanolide	*Phoma* sp.	*Aconitum vilmorinianum*	Liu et al. (2019a, b)
Graphislactone	*Cephalosporium* sp.	*Trachelospermum jasminoides*	Joseph and Priya (2011)
Cytochalasin U Cytochalasin	*Geniculosporium* sp.	Red alga *Polysiphonia* sp.	Krohn et al. (2005)
Graphislactones A, G, and H	*Cephalosporium acremonium*	*Trachelospermum jasminoides*	Zhang et al. (2005)
Nucleoside			
Cordycepin	*Cordyceps militaris* *Irpex lacteus* *Cordyceps ninchukispora*	*Tetheella fuctuosa* *Ochropacha duplaris* *Tethea ocularis* *Cordyceps hawkesii* *Beilschmiedia erythrophloia*	Kryukov et al. (2020), Liu et al. (2022), Chang et al. (2017)

(continued)

Table 13.2 (continued)

Bioactive secondary metabolites	Endophyte	Host plant	References
Peptides			
Pullularin A Pullularin B Verticillin D	*Bionectria ochroleuca*	*Sonneratia caseolaris*	Ebrahim et al. (2012)
Beauvericin	*Aspergillus terreus*	*Bruguiera gymnorhiza*	Deng et al. (2013)
Beauvericin	*Epicoccum nigrum*	*Entada abyssinica*	Dzoyem et al. (2017)
Coronamycins	*Streptomyces* sp.	*Monstera* sp.	Ezra et al. (2004)
Peroxides			
Merulin A Merulin C	XG8D	*Xylocarpus granatum*	Chokpaiboon et al. (2010)
Nor-sesquiterpene peroxides Talaperoxide B Talaperoxide D	*Talaromyces favus*	*Sonneratia apetala*	Li et al. (2011)
Polyketides			
Clavatol Patulin	*Aspergillus clavatonanicus*	*Taxus mairei*	Zhang et al. (2008)
Xyloketals B	*Xylaria* sp.	*Avicennia marina*	Zhou et al. (2018)
Chaetomugilin A and D	*Chaetomium globosum* *Phomopsis* sp.	*Ginkgo biloba*	Kumar et al. (2017)
3α-Hydroxy-3, 5-dihydromonacolin L	*Aspergillus terreus*	*Glycine max*	El-Hawary et al. (2021)
Guignasulphide	*Guignardia* sp.	*Hopea hainanensis*	Wang et al. (2010)
Chaetoglobosin	*Chaetomium globosum*	*Curcuma wenyujin*	Wang et al. (2012)
Alternariol Alternariol 9-methyl ether	*Alternaria alternate* PGL-3	*Punica granatum*	Abou El-kassem et al. (2019)
Gliovictin Gliovictin acetate	*Penicillium janczewskii*	*Prumnopitys andina*	Schmeda-Hirschmann et al. (2005)
Asperfumoid Aspernigrin A Aspernigerin	*Aspergillus fumigatus* *Cladosporium herbarum* *Aspergillus niger*	*Cynodon dactylon*	Liu et al. (2004), Ye et al. (2005), Shen et al. (2006a, b)
Griseofulvin	*Xylaria* sp. *Penicillium griseofulvum*	*Abies holophylla* *Mentha pulegium* L.	Macías-Rubalcava and Sánchez-Fernández (2017), Reshma et al. (2019)

(continued)

Table 13.2 (continued)

Bioactive secondary metabolites	Endophyte	Host plant	References
Tetrahydroaltersolanol C Alterporriol Q	*Alternaria* sp.	*Sacrophyton* sp. soft coral	Zheng et al. (2012)
Chrysogenamide A	*Penicillium citrinum* *Penicillium digitatum*	*Citrus limon*	Costa et al. (2019), Huang et al. (2007)
Cryptocin	*Cryptosporiopsis* cf. *quercina*	*Tripterygium wilfordii*	Li et al. (2000)
Myrothecines A, B, and C	*Myrothecium roridum*	*Trachelospermum jasminoides* *Artemisia annua*	Shen et al. (2006a, b)
Polyol			
Aureonitol	*Daldinia eschscholtzii*	*Musa paradisiaca*	Victor et al. (2020)
Pestalpolyol I	*Pestalotiopsis clavispora*	*Rhizophora harrisonii*	Hemphill et al. (2016)
Pyrans and pyrones			
Benzopyran (2R*, 4R*)-3,4-dihydro-4-methoxy-2-methyl-2H-1benzopyran-5-ol	*Nodulisporium* sp.	*Aquilaria sinensis*	Wu et al. (2010)
Naphtho-gamma-pyrone	*Aspergillus tubingensis*	*Oryza sativa*	Huang et al. (2011b)
Nigerapyrones B, D, and E Congeners Asnipyrone A	*Aspergillus niger*	*Avicennia marina*	Liu et al. (2011)
Quinones			
Anthraquinones Alterporriol L Alterporriol K	*Alternaria* sp.	*Aegiceras corniculatum*	Huang et al. (2011a)
9-dehydroxyeurotinone Variecolorin G 12-demethyl-12-oxo-eurotechinulin B Emodin Some other anthraquinone molecule	*Eurotium rubrum*	*Hibiscus tiliaceus*	Yan et al. (2012)
Altersolanol A	*Stemphylium globuliferum*	*Mentha pulegium*	Teiten et al. (2013)
Peniprequinolone	*Penicillium janczewskii*	*Prumnopitys andina*	Schmeda-Hirschmann et al. (2005)
Steroids			
Penicisteroid A Penicisteroid B	*Penicillium chrysogenum*	*Laurencia*	Gao et al. (2011)

(continued)

Table 13.2 (continued)

Bioactive secondary metabolites	Endophyte	Host plant	References
Jesterone Hydroxyjesterone	*Pestalotiopsis jester*	*Fragraea bodenii*	Li and Strobel (2001)
Phytoecdysteroids	*Aspergillus terreus*	*Bruguiera gymnorhiza*	Deng et al. (2013)
Solamargine	*Aspergillus favus*	*Solanum nigrum*	El-Hawary et al. (2016)
Ergosterol Cerevesterol	2L-5	*Ocimum basilicum*	Haque et al. (2005)
(20S,22S)-4a-homo-22-hydroxy-4-oxaergasta-7,24(28)-dien-3-one	*Gliocladium* sp.	*Taxus chinensis*	Zhang et al. (2002)
3b-Hydroxyergosta-4,22-diene Ergosterol 3b-hydroxy-5a,8a-epi-dioxyergosta-6,22-diene	*Aspergillus fumigatus*	*Cynodon dactylon*	Liu et al. (2004)
Diterpenes			
Taxol Taxane	*Taxomyces andreanae*	*Taxus brevifolia*	Stierle et al. (1993)
Sphaeropsidin A Sphaeropsidin D	*Paraconiothyrium* sp. *Smardaea* sp.	*Ceratodon purpureus*	Shiono et al. (2011)
Taxol	*Alternaria* *Fusarium* *Aspergillus niger*	*Salacia oblonga*	Roopa et al. (2015)
Sordaricin	*Xylaria* sp.	*Garcinia dulcis*	Digra and Nonzom (2023)
Taxol	*Metarhizium anisopliae* *Taxomyces andreanae*	*Taxus baccata*	Das et al. (2022)
Triterpenes			
(+)-(3S,6S,7R,8S)-periconone A, as well as (−)-(1R,4R,6S,7S)-2-caren-4,8-olide.	*Periconia* sp.	*Annona muricate*	Ge et al. (2011)
Azadirachtin	*Eupenicillium parvum*	*Azadirachta indica*	Kusari and Spiteller (2012)
Sesquiterpenes			
Aciphyllene	*Muscodor albus*	*Cinnamomum zeylanicum*	Atmosukarto et al. (2005)

(continued)

Table 13.2 (continued)

Bioactive secondary metabolites	Endophyte	Host plant	References
1(10) E,5E-germacradiene-3,11-diol 1(10) E,5E-germacradiene-2,11-diol 1(10) E,5E-germacradiene-11-diol	*Streptomyces griseus*	*Kandelia candel*	Guan et al. (2005)
Phenols			
Pyrogallol	ZA161	*Albizia zygia*	Ibrahim et al. (2021)
Furans			
Pestacin (1,3-dihydroisobenzofuran) Isopestacin (isobenzofuranone)	*Pestalotiopsis microspore*	*Terminalia morobensis*	Pandey et al. (2014) Toghueo and Boyom (2019)
2-methyl furan	*Muscodor albus*	*Cinnamomum zeylanicum*	Atmosukarto et al. (2005)

Beneficial microbes can also modulate ethanol levels by enzymatically degrading its precursor, 1-aminocyclopropane-1-carboxylic acid (ACC). The enzyme ACC deaminase cleaves ACC into ammonia and 2-oxobutanoate, reducing plant stress and promoting growth, thereby enhancing drought tolerance (Zolla et al. 2013; Yang et al. 2019; Glick 2005; Stearns et al. 2012; Sziderics et al. 2007; Chandran et al. 2021; Kumar et al. 2024). Many endophytic fungi are known for their phytohormone production, synthesizing auxins, cytokinins, and gibberellic acid, often in conjunction with siderophores (Mishra et al. 2016; Tochhawng et al. 2019; Abdalla et al. 2020). Some fungi that naturally produce indole-3-acetic acid (IAA) can modulate auxin signaling, thereby influencing root development and enhancing plant fitness (Lareen et al. 2016; Duca et al. 2014; Bailly et al. 2014; Kim et al. 2011). For instance, the fungal strain *Serendipita indica* has been shown to promote the growth of Chinese cabbage (*Brassica campestris* subsp. *chinensis*) and *Arabidopsis* (Lee et al. 2011).

Fungal species from genera such as *Fusarium*, *Alternaria*, *Xylogone*, and *Didymella*, isolated from *Sophora flavescens*, have demonstrated significant IAA production, positively affecting primary root length in *Arabidopsis* (Turbat et al. 2020). Additionally, endophytic colonization by *Chaetomium globosum* has fostered cucumber growth through the production of hormones like zeatin, gibberellic acid, IAA, and indole butyric acid (IBA) (Tian et al. 2022). Beyond phytohormones, beneficial microorganisms also produce various plant growth-promoting compounds, including adenine, adenine ribosides, polyamines like spermidine and spermine, and volatile organic compounds such as acetoin and 2,3-butanediol (Hardoim et al. 2015).

13.7 Adaptation by Defense Mechanisms

Plants can distinguish between beneficial and pathogenic microbes through microbe-associated molecular patterns, which encompass microbial secondary metabolites. These compounds engage with transmembrane pattern recognition receptors, initiating pattern-triggered immunity (PTI). This response results in notable morphological and physiological changes in plant cells, including callose deposition, stomatal closure, and the activation of effectors and secondary metabolites (Fernanda et al. 2017). Both beneficial and pathogenic microorganisms can modulate plant immunity by influencing hormonal signaling pathways. They may induce or inhibit specific pathways through the production of phytohormones, phytohormone mimics, or effectors (Fernanda et al. 2017). In response, plants have evolved resistance genes (R genes) that encode proteins to recognize specific effectors. When these effectors are detected, plants activate effector-triggered immunity (ETI), resulting in similar changes to those observed in PTI, including cell death and the activation of defense signaling pathways. These responses lead to local hypersensitive responses and systemic acquired resistance (SAR), enhancing the plant's ability to resist both biotic and abiotic stress through a priming mechanism (Fernanda et al. 2017).

In response to abiotic stress, plants activate various pathways involving jasmonic acid, abscisic acid, and salicylic acid, which serve as defense signaling molecules. Since many fungal endophytes are derived from plant pathogenic fungi, they may also act as pathogens, prompting plants to enhance their defenses (Yung et al. 2021). Research has shown that endophytes can enhance plant growth and reproduction by interacting with ethylene-targeted transcription factors or by bolstering defense mechanisms, similar to the effects of ethylene and jasmonic acid (Di et al. 2016; Forni et al. 2017; Van der Ent et al. 2009; Yang et al. 2019). C-group endophytes, recognized as prototypical obligatory endophytes, release protective bioactive substances that help host plants resist herbivores and thrive in drought conditions (Poveda 2021). Many endophytic fungi produce biologically active secondary metabolites, including paclitaxel, camptothecin, and azadirachtin (Stierle et al. 1993; Eyberger et al. 2006; Kusari et al. 2008, 2009a, b, c, 2012a, b). Fungi employ three primary defense strategies against phytopathogens: competition for resources, antibiotic production, and mycoparasitism. These strategies enhance plant defenses and increase tolerance to pathogenic virulence factors. Table 13.3 illustrates the secondary metabolites produced by endophytic fungi as adaptation potential to tolerate biotic stress.

Given that global crop losses due to plant diseases are estimated at 16% (Fontana et al. 2021), endophytic fungi play a critical role in activating induced systemic resistance (ISR) and systemic acquired resistance (SAR) to combat biotic stress. Endophytes may exert their effects through mechanisms such as antibiosis, competition, direct parasitism, and host-induced resistance (Card et al. 2016). Antibiosis refers to the antagonistic effects of diffusible substances produced by endophytes (Fravel 1988; Baker 1968; Kong et al. 2010), while competition involves vying for nutrients for their own benefit (Schouten 2016). Direct parasitism occurs when a

Table 13.3 Secondary metabolites produced by endophytic fungi as adaptation potential to tolerate biotic stress

Bioactive compounds	Biological effects	References
Indole derivatives, indole-3-acetic acid and indole-3-ethanol, a sesquiterpene, and a diacetamide from the cultures of *Epichloe* and *Neotyphodium* sp.	Antifungal activity against *Cryphonectria parasitica*	Yue et al. (2000)
Ethyl 2,4-dihydroxy-5,6-dimethylbenzoate and phomopsilactone, isolated from *Pestalotia cassiae*, from *Cassia spectabilis*	Antifungal activity against the *C. cladosporioides* and *C. sphaerospermum*	Silva et al. (2005)
Gamma-aminobutyric acid (GABA)	Expression gets increased in its roots to prepare the host plant for biotic or abiotic stress in Chinese cabbage	Saini et al. (2024)
Artemisinin	Phytoalexin with allelopathic effects that protects plants against herbivores and microbial pathogens	Wang et al. (2001), Li et al. (2012)
Jesterone and hydroxyjesterone produced by *Pestalotiopsis jesteri* from *Fragaea bodenii*	Anti-oomycete metabolites against *Phytophthora* and *Pythium*	Li and Strobel (2001)
Indole-3-acetic acid (IAA); indole-3-ethanol (IEtOH); methylindole-3-carboxylate; indole-3-carboxaldehyde; diacetamide and cyclonerodiol, isolated from its fermentation culture of *Epichlöe festucae*	Inhibit *Cryptonectria parasitica*, a plant pathogen responsible of chestnut blight	Yue et al. (2000)
Colletotric acid produced by *Colletotrichum gloeosporides* (Penz.) Penz. & Sacc, from *Artemisia mongolica*	Antifungal against the plant pathogen *Helminthosporium sativum*	Zou et al. (2000)
Lolitrem B, an indole-diterpene neurotoxic mycotoxin produced by *Acremonium lolii* from *Lolium perenne* L.	Cattle intoxication	Berg et al. (2004)
Lolines (found only in plantendophyte), ergot alkaloids, and pyrrolopyrazine alkaloids (e.g., peramine), produced by *Acremonium coenophialum* from *Festuca arundinacea* Schreb.	Cattle intoxication	Justus et al. (1997)
Ergonovine, chanoclavine, ergovaline, ergocrystine, and ergocryptine produced by *Neotyphodium coenophialum* and *Neotyphodium lolii* from *Festuca arundinacea* (tall fescue) and *Lolium perenne* (perennial ryegrass)	Insect-defensive alkaloids produced by both the endophyte and the plant affect the development of the fall armyworm *Spodoptera frugiperda*	Salminen et al. (2005)
Lolitrem B and peramine	Affect sucking insects *Rhopalosiphum padi* (the bird cherry-oat aphid) and *Schizaphis graminum* (the greenbug) with species-dependent effects	Siegel et al. (1990)

(continued)

Table 13.3 (continued)

Bioactive compounds	Biological effects	References
5-hydroxybenzofurans produced by an unidentified endophyte from *gaultheria procumbes* L.	Affect *Choristoneura fumiferana* Clem (eastern spruce budworm)	Findlay et al. (1997)
Paclitaxel (Taxolentis) produced by endophytic fungus *Taxomyces andreanae* from the bark of the Pacific yew *Taxus brevifolia*	Antifungal activity against wood-decaying fungi	Stierle et al. (1993)

pathogen attaches hyphally and degrades host cell walls using lytic enzymes (Sharma et al. 2009), and host-induced resistance reflects the ability of endophytes to modulate plant immunity (Zipfel and Oldroyd 2017).

Many endophytic microorganisms produce a diverse array of secondary metabolites with various biological activities (Newman and Cragg 2015). These metabolites help organisms access scarce resources, survive in hostile environments, and inhibit competitors (Xie et al. 2019). To promote plant growth under stress, fungal endophytes produce siderophores, antibiotics, and phytohormones; they also mineralize nutrients and perform other beneficial functions (Yung et al. 2021). Some compounds may serve dual roles as both antibiotics and elicitors of plant-induced systemic resistance (Card et al. 2016). Siderophores facilitate iron uptake and phosphate solubilization, promoting plant growth while defending against pathogens (Chowdappa et al. 2020). Over 500 siderophores from various fungi have been identified (Chowdappa et al. 2020). Endophytes also enhance biomass formation and nitrogen uptake while degrading organic waste (Idbella et al. 2019). Recent studies have shown that fungi such as *Aspergillus fumigatus*, *Aspergillus niger*, *Curvularia*, and *Trichoderma* can solubilize and mobilize phosphorus, potassium, and zinc salts, thereby promoting plant growth and improving crop yields (Mehta et al. 2019; Haro and Benito 2019).

As endophytes reside within host plant tissues, they form barriers against phytopathogens that might colonize these plants (Moy et al. 2000). They are rich sources of secondary bioactive metabolites, including tannins, terpenoids, saponins, steroids, and phenolic acids (Gouda et al. 2016). Additionally, endophytes produce chemicals that inhibit the maturation of other antagonists and withstand biotic stresses from rhizospheric nematodes and insects (Khan et al. 2012). Key endophytes that enhance biotic stress tolerance while promoting host plant growth include species from *Trichoderma*, *Epicoccum*, *Aspergillus*, *Colletotrichum*, *Gliocladium*, *Fusarium*, *Petriella*, *Piriformospora*, and *Epichloe* (Ali et al. 2019a, b). Endophytes such as *Penicillium*, *Ascomycete*, *Verticillium*, and *Cladosporium* have been reported to produce triterpenoids, ginsenosides, and saponins, enhancing stress resistance and root development (Sahoo et al. 2017). Given their potential as biocontrol agents, these endophytes warrant further exploration. For example, *Epichloe* species have been shown to control herbivorous insects, providing their host plants with biotic resilience.

Endophytes colonizing roots, twigs, and stems, like *Trichoderma*, exhibit antibacterial, antifungal, and cytotoxic properties. Notably, secondary metabolites produced by endophytic fungi play a critical role in mutualistic relationships with host plants. Commonly identified metabolites include epichloe alkaloids, such as lolines (e.g., N-methylloline), ergot alkaloids (e.g., ergovaline), lolitrems (e.g., lolitrem B), and peramine. Ergot alkaloids are linked to plant defenses against both vertebrate and invertebrate herbivores, while lolitrems act specifically against vertebrates, and lolines and peramine target invertebrate herbivores (Saikkonen et al. 2013; Schardl et al. 2013a, b). The Clavicipitaceae family (Ascomycota) dominates fungal endophytes in temperate grasses, producing alkaloids that help host plants survive various herbivore attacks (Johnson et al. 2013). Numerous studies have demonstrated that plants infected with endophytes exhibit increased resistance to pathogens, supporting their potential as effective biological control agents (Schouten 2016; Wang et al. 2016; Perez et al. 2016; Gond et al. 2015). For instance, extracts from *Poa ampla* inoculated with *Neotyphodium typhonium* effectively repelled mosquito larvae, while extracts from non-inoculated plants had no effect (Ju et al. 1998). Endophyte colonization can trigger a reprogramming of host metabolism, favoring secondary metabolism and altering host development (Dupont et al. 2015). A notable endophytic functional trait is the ability to quench pathogen quorum sensing (QS) molecules (Kusari et al. 2015). Endophytes operate within a complex system, influenced by both the plant host and environmental factors.

For example, *Serendipita indica* colonizing banana (*Musa* spp.) has been shown to increase the activity of antioxidant enzymes such as superoxide dismutase (SOD), peroxidase (POD), catalase (CAT), and ascorbate peroxidase (APX), enhancing resistance to the pathogen *Fusarium oxysporum* f.sp. *cubense* (Cheng et al. 2020a, b). Additionally, *Penicillium* species associated with *Suaeda japonica* secrete gibberellins, illustrating how fungal endophytes can influence plant growth-promoting compounds under stress. The endophyte *Fusarium oxysporum* has demonstrated nematocidal activity against the root-knot nematode *Meloidogyne incognita* in tomato plants (Diedhiou et al. 2003). Furthermore, 11 *Fusarium* isolates from the roots of the great dwarf banana variety significantly reduced nematode populations per gram of root material and mitigated the severity of banana root diseases such as "Panama disease," caused by *Fusarium oxysporum* f.sp. *cubense* (Pocasangre et al. 2000).

Mutualistic symbionts, such as endophytes, play a crucial role in enhancing host plant adaptation to abiotic stress over the long term. These endophytes significantly increase the root and shoot biomass of colonized plants, facilitating drought stress tolerance through accelerated photosynthesis, enhanced energy release, and improved antioxidant capacity. Research has shown that endophytes can boost the generation of reactive oxygen species (ROS) while also producing a variety of antioxidant compounds, including phenolic acids, isobenzofuranones, mannitol, and other carbohydrates. This antioxidant capability contributes to the host's resilience against stressors such as drought and metal toxicity (Aly et al. 2011). These compounds can disrupt host cell membranes, allowing nutrients to leak out and be absorbed by fungal hyphae (Herrera-Carillo et al. 2009; White Jr and Torres 2010).

Mass spectrometry analyses have identified 45 differentially expressed proteins in *Serendipita indica* involved in ROS scavenging, photosynthesis, signal transduction, metabolism, and plant defense responses (Yadava et al. 2015). Notably, *S. indica* increases the foliar potassium (K^+)/sodium (Na^+) ratio, a key indicator of salt stress tolerance, alongside calcium (Ca^{2+}) accumulation that may influence stress signaling. Other studies have demonstrated that *S. indica* induces systemic responses to salt stress by modifying the physiological and proteomic responses of the host plant (Alikhani et al. 2013). Siderophores produced by endophytes help biodegrade biomass and recycle nutrients in the environment, enhancing the phyto-immobilization and availability of essential elements like zinc, nitrogen, and phosphorus (Yung et al. 2021). They can also lower ethylene levels by degrading 1-aminocyclopropane-1-carboxylate (ACC), thereby helping plants manage osmolytes and regulate membrane ion conductivity during stress. Through these mechanisms, endophytes enhance the plants' ability to withstand various abiotic stressors. Moreover, endophytes equipped with effective degradation pathways and metal sequestration systems significantly increase host plant tolerance to heavy metals, enabling survival in contaminated soils (Weyens et al. 2009).

Endophytes can also harbor viruses that contribute to plant heat tolerance, among other advantages (Lugtenberg et al. 2016; Schouten 2016). Thermophilic fungal endophytes isolated from the hot desert plant *Cullen plicatum* enhance drought and heat stress tolerance through ecophysiological mechanisms, leading to improved growth in host plants (Ali et al. 2019a, b). Additionally, these fungal endophytes have been found to improve the cold and chilling tolerance of rice plants (Barka et al. 2006; Redman et al. 2011). Under conditions of water scarcity, the beneficial effects of endophytes on host plants become even more pronounced, resulting in increased biomass, height, yield, and water content, along with enhanced antioxidant enzyme activity (Dastogeer 2018). In extreme environments characterized by salt and drought, endophytes from genera such as *Periconia*, *Neocamarosporium*, and *N. goegapense*, associated with salt lake plants, help *Hordeum vulgare* (barley) cope with stress, leading to improved biomass, shoot length, proline content, and antioxidant enzyme activity (Moghaddam et al. 2022). The rich reservoir of bioactive secondary metabolites produced by fungal endophytes underpins a symbiotic relationship that enables mutual survival in challenging environmental conditions.

13.8 Adaptation by Genetic Plasticity

Endophytes exhibit remarkable genetic plasticity, enabling them to thrive in diverse environments such as high-temperature regions, temperate forests, mangrove ecosystems, and tropical rainforests (Arnold 2008). Their ability to adapt to varying climatic conditions is largely attributed to the expression of specific genes that facilitate their colonization. Comparative genomic studies between endophytic isolates and their non-endophytic counterparts reveal essential traits necessary for establishing and maintaining plant–microbe interactions, highlighting the genetic foundations of niche adaptation (Lòpez-Fernàndez et al. 2015). The distinction between endophytes and pathogens often hinges on subtle genetic differences, emphasizing a narrow boundary in their colonization strategies (Kaul et al. 2016). Fungal endophytes adopt various lifestyles—parasitic, saprotrophic, necrotrophic, and endophytic—demonstrating their versatility. Their genomes contain an AT-rich block, a unique survival mechanism that houses effector genes and transposable elements, allowing for rapid genetic diversification and adaptability to host-specific challenges (Rouxel et al. 2011). For instance, *Serendipita indica* has been shown to harbor genes linked to both saprotrophic and biotrophic lifestyles, facilitating its interaction with host plants (Zuccaro et al. 2011).

Research on *Empetrum nigrum* highlights its capacity to produce metabolites that inhibit pathogens like *Xanthomonas albilineans*, *Fusarium verticillioides*, and *Ceratocystis paradoxa*, while simultaneously enhancing the biomass of sugarcane root systems (Fávaro et al. 2012). Additionally, endophytic fungi can modulate the plant's immune response, fostering beneficial interactions that may enhance growth and metabolite production (Yuan et al. 2019). This interaction boosts photosynthesis, enhances glycolysis and the TCA cycle, and promotes the biosynthesis of sesquiterpenoids. Endophytes also express phosphate transporter genes (PiPT), which facilitate phosphate absorption, and they can improve nitrogen and phosphorus availability in their symbiotic relationship with plants (Behie and Bidochka 2014). Under low-nitrogen conditions, endophytic colonization leads to the upregulation of genes involved in nitrogen uptake, resulting in increased levels of total nitrogen, amino acids, proteins, and free NH_4^+ (Yang et al. 2014).

Quantitative and differential analyses have shown significant fluctuations in metabolite levels, with some metabolites being upregulated or downregulated by more than 100-fold. Transcriptomic studies suggest that histone deacetylase (HDAC) inhibitors typically enhance the biosynthetic machinery, but the responses of secondary metabolites are more complex than mere increases in bioactive diversity (Keshri et al. 2021). A comparative analysis of over ten species within the Clavicipitaceae family has revealed that variations in the peripheral genes associated with alkaloid loci are responsible for the diversity of their secondary metabolites (Schardl et al. 2013a, b). The biosynthetic pathways of these metabolites are orchestrated by biosynthetic gene clusters (BGCs), which encode enzymatic reactions that produce a wide array of bioactive compounds, including alkaloids, terpenoids, flavonoids, and more (Schueffler and Anke 2014; Tawfike et al. 2019). The regulation of fungal BGCs occurs at both transcriptional and epigenetic levels (Keller 2019). Up to 50% of these clusters contain specific transcription factors that recognize motifs in gene promoters. Additionally, epigenetic mechanisms such as DNA methylation and histone acetylation play crucial roles in controlling gene expression, allowing for the exploration of the chemical diversity encoded in fungal endophyte genomes (Guo and Zou 2020).

Transcriptional changes in the endophyte genome can enhance plant stress tolerance. For example, the transcriptome of the heavy metal-tolerant *Exophiala pisciphila* revealed multiple glutathione-S-transferase genes associated with stress response (Shen et al. 2015a, b). Studies using SOLID-SAGE transcriptomic techniques have identified around 200 differentially expressed plant-associated genes in endophyte-infected versus endophyte-free *Festuca rubra*, suggesting that these genes may contribute to disease resistance, a hallmark of endophyte-infected fescue (Ambrose and Belanger 2012). Genomic analyses have illuminated critical genes involved in plant penetration and colonization, revealing significant differences among endophytes, nodule-forming symbionts, phytopathogens, and soil bacteria. Notably, certain traits, such as nitrogen fixation and protection against reactive oxygen species, serve as distinguishing features for endophytes compared to other microbial groups (Hardoim et al. 2015; Mitter et al. 2013; Amadou et al. 2008; Tisserant et al. 2013; Tian et al. 2021; Karpinets et al. 2014; Monteiro et al. 2012).

13.9 Fungal Endophytes from Algae

Fungal endophytes are fascinating organisms that have formed endosymbiotic associations with various marine algae, showcasing a rich diversity of bioactivities. These endophytes have been successfully isolated from green, red, and brown algae, highlighting their ecological significance (Suryanarayanan and Johnson 2014). Over 125 morphologically distinct endophytic fungi have been identified from a range of marine algal species, including *Codium*, *Gracilaria*, *Halimeda*, *Caulerpa*, *Gelidium*, *Chaetomorpha antennina*, *Ulva fasciata*, *Sargassum*, and *Turbinaria* (Mathan et al. 2013; Thirunavukkarasu et al. 2011). These fungi exhibit remarkable antimicrobial and enzymatic properties. For instance, *Chaetomium globosum*, isolated from the tissues of the marine red alga *Polysiphonia urceolata*, is known for producing a variety of secondary metabolites. Another interesting find is *Geniculosporium* sp., reported as an endophyte in *Polysiphonia* sp. (Krohn et al. 2005), showcasing the intricate relationships between fungi and algae. One particularly intriguing discovery is an unidentified endophytic fungus from *Sargassum* that exhibits significant antimicrobial activity (Yang et al. 2006). Furthermore, a novel compound, isorhodoptilometrin-1-methyl ether, along with seven known compounds, was isolated from *Aspergillus versicolor*, an endophyte of the marine alga *Halimeda opuntia* (Verma et al. 2017). These findings reveal that fungal endophytes engage in diverse interactions with algae, leading to the production of a variety of bioactive compounds.

13.10 Fungal Endophytes from Mushrooms

Fungal endophytes associated with mushrooms exhibit a relationship akin to the endophyte–plant association. Wild mushrooms often host endophytic fungi within their ecological niches, suggesting a complex symbiotic relationship that may require redefinition. Notably, some fungi isolated from mushrooms have also been found in plants, raising questions about the historical precedence of these relationships. For instance, the fruiting bodies of *Tricholoma matsutake* were found to harbor 13 fungal strains, 15 yeast strains, and 14 bacterial strains (Li et al. 2016). Additionally, 28 endophytic fungal strains were identified from *Cordyceps sinensis*

(Guo et al. 2020). Among these, *Penicillium herquei* stood out as the only strain sensitive to 5-aza-2-deoxycytidine, yielding a greater variety of metabolites than the others. This study also identified three α-pyrone derivatives produced in the presence of the compound. Further research revealed 22 endophytes in *C. hawkesii* Gray (Liu et al. 2022), focusing on cordycepin-producing fungi. Four significant endophytes were identified: *Irpex lacteus*, *Arthrinium arundinis*, *Arthrinium* sp., and *Bjerkandera* sp., with *Irpex lacteus* yielding the highest amounts of cordycepin. Zhang et al. (2021) detected 94 fungal genera in various samples of *C. militaris*, predominantly including *Cordyceps*, *Trichopaea*, *Isaria*, *Saitozyma*, and *Amanita*, while 89 other genera made up less than 0.25% of the total fungal community.

The fruiting bodies of the Russulaceae family also contain the endophyte *Calcarisporium arbuscular*, known for producing a range of secondary metabolites (Cheng et al. 2020a, b). Pei-Xin et al. (2014) isolated and identified 16 fungal strains from *Morchella semilibera* (formerly *Morchella crassipes*), which included 15 ascomycetes and one Zygomycota isolate, *Mortierella alpina*. This oleaginous fungus is notable for producing various glycerol-lipids, glycerophospholipids, and sphingolipids. Other endophytes isolated from *Morchella semilibera* included *Chaetomium* and *Lecanicillium fungicola* var. *aleophilum*. Overall, the endophytic fungi associated with mushrooms likely possess the metabolic pathways necessary for synthesizing bioactive secondary metabolites, similar to those found in their host mushrooms.

13.11 Conclusion

Endophytic fungi, with their profound evolutionary adaptations and metabolic diversity, play a pivotal role in plant health and ecosystem stability. Their ability to produce bioactive secondary metabolites and adapt to diverse environmental conditions underscores their importance in mutualistic plant interactions. By leveraging host pathways and responding to environmental stresses, these fungi enhance both their own survival and the resilience of their host plants. Moreover, fungal endophytes exhibit significant evolutionary adaptation by refining their interactions with host plants, often developing host-specific relationships that enhance their survival. They may imitate host metabolic pathways, utilizing these as biosynthetic routes for their secondary metabolites. This adaptation not only enhances their growth but also boosts their defense mechanisms against biotic and abiotic stresses. By improving their ability to tolerate environmental extremes and biotic pressures, endophytic fungi contribute to the resilience and fitness of their host plants. Additionally, fungal endophytes can also originate from algae and mushrooms, extending their diversity and adaptation strategies. Endophytes associated with algae and mushrooms demonstrate unique ecological roles and metabolic capabilities, contributing to a broader understanding of their functional potential and evolutionary adaptations.

References

Abdalla MA, Aro AO, Gado D, Passari AK, Mishra VK, Singh BP, McGaw LJ (2020) Isolation of endophytic fungi from South African plants, and screening for their antimicrobial and extracellular enzymatic activities and presence of type I polyketide synthases. S Afr J Bot 134:336–342. https://doi.org/10.1016/j.sajb.2020.03.021

Aguilar OM, Beker MP, Blanco FA, Zanetti ME, Calderoli PA (2015) Interaction between host and Rhizobial strains: affinities and coevolution. In: Biological nitrogen fixation. Wiley, pp 203–208. https://doi.org/10.1002/9781119053095

Ali A, Bilal S, Khan AL, Mabood F (2019a) Endophytic *Aureobasidium pullulans* BSS6 assisted developments in phytoremediation potentials of *Cucumis sativus* under Cd and Pb stress. J Plant Interact 14:303–313. https://doi.org/10.1080/17429145.2019.1633428

Ali AH, Radwan U, El-Zayat S, El-Sayed MA (2019b) The role of the endophytic fungus, *Thermomyces lanuginosus*, on mitigation of heat stress to its host desert plant *Cullen plicata*. Biologia Futura 70(1):1–7. https://doi.org/10.1556/019.70.2019.01

Alikhani M, Khatabi B, Sepehri M, Nekouei MK, Mardi M, Salekdeh GH (2013) A proteomics approach to study the molecular basis of enhanced salt tolerance in barley (*Hordeum vulgare* L.) conferred by the root mutualistic fungus *Piriformospora indica*. Mol BioSyst 9(6):1498–1510. https://doi.org/10.1039/c3mb70069k

Aly A, Debbab A, Proksch P (2011) Fungal endophytes: unique plant inhabitants with great promises. Appl Microbiol Biotechnol 90(6):1829–1845. https://doi.org/10.1007/s00253-011-3270-y

Amadou C, Pascal G, Mangenot S, Glew M, Bontemps C, Capela D, Carrère S, Cruveiller S, Dossat C, Lajus A, Marchetti M, Poinsot V, Rouy Z, Servin B, Saad M, Schenowitz C, Barbe V, Batut J, Médigue C, Masson-Boivin C (2008) Genome sequence of the β-rhizobium *Cupriavidus taiwanensis* and comparative genomics of rhizobia. Genome Res 18:1472–1483. https://doi.org/10.1101/gr.076448.108

Ambrose KV, Belanger FC (2012) SOLiD-SAGE of endophyte-infected red fescue reveals numerous effects on host transcriptome and an abundance of highly expressed fungal secreted proteins. PLoS One 7(12). https://doi.org/10.1371/journal.pone.0053214

Amna T, Puri SC, Verma V, Sharma JP, Khajuria RK, Musarrat J, Spiteller M, Qazi GN (2006) Bioreactor studies on the endophytic fungus *Entrophospora infrequens* for the production of an anticancer alkaloid camptothecin. Can J Microbiol 52(3):189–196. https://doi.org/10.1139/w05-122

Arnold A (2008) Hidden within our botanical richness, a treasure trove of fungal endophytes. Plant Press 32:13–15

Arnold AE, Lutzoni F (2007) Diversity and host range of foliar fungal endophytes: are tropical leaves biodiversity hotspots? Ecology 88(3):541–549. https://doi.org/10.1890/05-1459

Arora J, Ramawat KG (2017) An introduction to endophytes. In: Maheshwari D (ed) Endophytes: biology and biotechnology, Sustainable development and biodiversity, vol 15. Springer, Cham. https://doi.org/10.1007/978-3-319-66541-2_1

Atmosukarto I, Castillo U, Hess WM, Sears J, Strobel G (2005) Isolation and characterization of *Muscodor albus* I-41.3 s, a volatile antibiotic producing fungus. Plant Sci 169(5):854–861. https://doi.org/10.1016/j.plantsci.2005.06.002

Bacon CW, White JF (2000) Microbial endophytes. Marcel Deker Inc, New York

Bai Y, Kissoudis C, Yan Z, Visser RGF, Linden GVD (2017) Plant behaviour under combined stress: tomato responses to combined salinity and pathogen stress. Plant J 93(4):781–793. https://doi.org/10.1111/tpj.13800

Bailly A, Groenhagen U, Schulz S, Geisler M, Eberl L, Weisskopf L (2014) The inter-kingdom volatile signal indole promotes root development by interfering with auxin signalling. Plant J 80:758–771. https://doi.org/10.1111/tpj.12666

Bais HP, Weir TL, Perry LG, Gilroy S, Vivanco JM (2006) The role of root exudates in rhizosphere interactions with plants and other organisms. Annu Rev Plant Biol 57:233–266. https://doi.org/10.1146/annurev.arplant.57.032905.105159

Baker R (1968) Mechanism of biological control of soil borne pathogens. Annu Rev Phytopathol 6:263–294

Balachandar D, Raja P, Kumar K, Sundaram SP (2007) Non-rhizobial nodulation in legumes. Biotechnol Mol Biol Rev 20:49–57. https://doi.org/10.13140/2.1.1775.2960

Balkrishna A, Joshi R, Dhyani A, Arya V (2022) Endophytic fungi: a comprehensive review on their secondary metabolites, pharmacological interventions and host plant interactions. In: Endophytic fungi biodiversity, antimicrobial activity and ecological implications. Nova Science Publishers, pp 1–48. https://doi.org/10.52305/IKKF9948

Barka EA, Nowak J, Cle'ment C (2006) Enhancement of chilling resistance of inoculated grapevine plantlets with a plant growth-promoting rhizobacterium, Burkholderia phytofirmans strain PsJN. Appl Environ Microbiol 72:7246–7252. https://doi.org/10.1128/AEM.01047-06

Behie SW, Bidochka MJ (2014) Nutrient transfer in plant–fungal symbioses. Trends Plant Sci 19(11):734–740. https://doi.org/10.1016/j.tplants.2014.06.007

Bérdy J (2005) Bioactive microbial metabolites. J Antibiot 58(1):1–26. https://doi.org/10.1038/ja.2005.1

Berg G, Krechel A, Ditz M, Sikora RA, Ulrich A, Hallmann J (2004) Endophytic and ectophytic potato-associated bacterial communities differ in structure and antagonistic function against plant pathogenic fungi. J FEMS Microbiol Ecol 51:215. https://doi.org/10.1016/j.femsec.2004.08.006

Bhupenchandra I, Chongtham SK, Devi AG, Dutta P, Sahoo MR, Mohanty S, Kumar S, Choudhary AK, Devi EL, Sinyorita S, Devi SH, Mahanta M, Kumari A, Devi HL, Josmee RK, Pusparani A, Pathaw N, Gupta S, Meena M, Janmeda P, Seth CS, Arora A, Sahoo A, Swapnil P (2024) Unlocking the potential of arbuscular mycorrhizal fungi: exploring role in plant growth promotion, nutrient uptake mechanisms, biotic stress alleviation, and sustaining agricultural production systems. J Plant Growth Regul 43(9):1–39. https://doi.org/10.1007/s00344-024-11467-9

Borges KB, Borges W, Pupo M, Bonato PS (2007) Endophytic fungi as models for the stereoselective biotransformation of thioridazine. Appl Microbiol Biotechnol 77:669–674. https://doi.org/10.1007/s00253-007-1171-x

Bourgaud F, Gravot A, Milesi A, Gontier E (2001) Production of plant secondary metabolites: a historical perspective. Plant Sci 161–165:839–851. https://doi.org/10.1016/S0168-9452(01)00490-3

Brader G, Compant S, Vescio K, Mitter B, Trognitz F, Ma L-J, Sessitsch A (2017) Ecology and genomic insights on plant-pathogenic and -nonpathogenic endophytes. Annu Rev Phytopathol 55:61–83. https://doi.org/10.1146/annurev-phyto-080516-035641

Bringel F, Coue'e I (2015) Pivotal roles of phyllosphere microorganisms at the interface between plant functioning and atmospheric trace gas dynamics. Front Microbiol 6:486. https://doi.org/10.3389/fmicb.2015.00486

Buiate EAS, Xavier KV, Moore N, Torres MF, Farman ML, Schardl CL, Vaillancourt LJ (2017) A comparative genomic analysis of putative pathogenicity genes in the host-specific sibling species *Colletotrichum graminicola* and *Colletotrichum sublineola*. BMC Genomics 18:24. https://doi.org/10.1186/s12864-016-3457-9

Bungihan ME, Tan MA, Takayama HE, dela Cruz TE, Nonato MG (2013) A new macrolide isolated from the endophytic fungus *Colletotrichum* sp. Philipp Sci Lett 6(1):57–57

Calhoun LA, Findlay JA, Miller JD, Whitney NJ (1992) Metabolites toxic to spruce budworm from balsam for needle endophytes. Mycol Res 96(4):281–286. https://doi.org/10.1016/S0953-7562(09)80939-8

Card S, Johnson L, Teasdale S, Caradus J (2016) Deciphering endophyte behaviour: the link between endophyte biology and efficacious biological control agents. FEMS Microbiol Ecol 92:19. https://doi.org/10.1093/femsec/fiw114

Carroll GC (1986) The biology of endophytism in plants with particular reference to woody perennials. In: Fokkema NJ, Van den Heuvel J (eds) Microbiology of the phyllosphere. Cambridge University Press, pp 205–222

Carroll G (1988) Fungal endophytes in stems and leaves: from latent pathogen to mutualistic symbiont. Ecology 69(1):2–9. https://doi.org/10.2307/1943154

Carroll GC (1991) Beyond pest deterrence-alternative strategies and hidden costs of endophytic mutualisms in vascular plants. In: Andrews JH, Monano SS (eds) Microbial ecology of leaves. Springer, New York, pp 358–378. https://doi.org/10.1007/978-1-4612-3168-4_18

Carroll GC (1992) Fungal mutualism. In: Carroll GC, Wicklow DT (eds) The fungal community. Its organization and role in the ecosystem. Dekker, New York, pp 327–354

Chagas FO, Pessotti RD, Caraballo-Rodriguez AM, Pupo MT (2018) Chemical signaling involved in plant microbe interactions. Chem Soc Rev 47(5):1652–1704. https://doi.org/10.1039/C7CS00343A

Chandra S (2012) Endophytic fungi: novel sources of anticancer lead molecules. Appl Microbiol Biotechnol 95:47–59. https://doi.org/10.1007/s00253-012-4128-7

Chandran H, Meena M, Swapnil P (2021) Plant growth-promoting rhizobacteria as a green alternative for sustainable agriculture. Sustain For 13(19):10986. https://doi.org/10.3390/su131910986

Chang HS, Cheng MJ, Wu MD, Hing-Yuen C (2017) Secondary metabolites produced by an endophytic fungus *Cordyceps ninchukispora* from the seeds of *Beilschmiedia erythrophloia* Hayata. Phytochem Lett 22:179–184. https://doi.org/10.1016/j.phytol.2017.08.005

Chauhan P, Sharma N, Tapwal A, Kumar A, Verma GS, Meena M, Seth CS, Swapnil P (2023) Soil microbiome: diversity, benefits and interactions with plants. Sustain For 15:14643. https://doi.org/10.3390/su151914643

Chen H, Fujita M, Feng Q, Clardy J, Fink GR (2004) Tyrosol is a quorum-sensing molecule in Candida albicans. Proc Natl Acad Sci USA 101:5048–5052. https://doi.org/10.1073/pnas.0401416101

Chen L, Zhang QY, Jia M, Ming Q-L, Yue W, Rahman K, Qin L-P, Han T (2016) Endophytic fungi with antitumor activities: their occurrence and anticancer compounds. Crit Rev Microbiol 42(3):454–473. https://doi.org/10.3109/1040841X.2014.959892

Cheng C, Li D, Qi Q, Sun X (2020a) The root endophytic fungus *Serendipita indica* improves resistance of Banana to *Fusarium oxysporum* f. sp. *cubense* tropical race 4. Eur J Plant Pathol 156:87–100. https://doi.org/10.1007/s10658-019-01863-3

Cheng JT, Cao F, Cheng JT, Li Y-Q, Mao X-M (2020b) Genomic and transcriptomic survey of an endophytic fungus *Calcarisporium arbuscula* NRRL 3705 and potential overview of its secondary metabolites. BMC Genomics 21:1–13. https://doi.org/10.1186/s12864-020-06813-6

Chithra S, Jasim B, Sachidanandan P, Jyothis M, Radhakrishnan EK (2014) Piperine production by endophytic fungus *Colletotrichum gloeosporioides* isolated from *Piper nigrum*. Phytomedicine 21(4):534–540. https://doi.org/10.1016/j.phymed.2013.10.020

Chokpaiboon S, Sommit D, Teerawatananond T, Muangsin N, Bunyapaiboonsri T, Pudhom K (2010) Cytotoxic nor-chamigrane and chamigrane endoperoxides from a basidiomycetous fungus. J Nat Prod 73(5):1005–1007. https://doi.org/10.1021/np100103j

Choquer M, Fournier E, Kunz C, Levis C, Pradier J-M, Simon A, Viaud M (2007) *Botrytis cinerea* virulence factors: new insights into a necrotrophic and polyphageous pathogen. FEMS Microbiol Lett 277:1–10. https://doi.org/10.1111/j.1574-6968.2007.00930.x

Chowdappa S, Jagannath S, Konappa N, Udayashankar AC, Jogaiah S (2020) Detection and characterization of antibacterial siderophores secreted by endophytic fungi from *Cymbidium aloifolium*. Biomol Ther 10(10):1412. https://doi.org/10.3390/biom10101412

Clay K (1990) Fungal endophytes of grasses. In: Annual review of ecology and systematics, pp 275–297

Clay K (1993) The ecology and evolution of endophytes. Agric Ecosyst Environ 44(1–4):39–64. https://doi.org/10.1016/0167-8809(93)90038-Q

Compant S, Cle'ment C, Sessitsch A (2010) Plant growth-promoting bacteria in the rhizo- and endosphere of plants: their role, colonization, mechanisms involved and prospects for utilization. Soil Biol Biochem 42:669–678. https://doi.org/10.1016/j.soilbio.2009.11.024

Correa-Galeote D, Bedmar EJ, Arone GJ (2018) Maize endophytic bacterial diversity as affected by soil cultivation history. Front Microbiol 9:484. https://doi.org/10.3389/fmicb.2018.00484

Costa JH, Wassano CI, Angolini CFF, Scherlach K, Hertweck C, Fill TP (2019) Antifungal potential of secondary metabolites involved in the interaction between citrus pathogens. Sci Rep 9(1):1–11. https://doi.org/10.1038/s41598-019-55204-9

Cuconati A, Molla A, Wimmer E (1998) Brefeldin A inhibits cell-free, de novo synthesis of poliovirus. J Virol 72(8):6456–6464. https://doi.org/10.1128/JVI.72.8.6456-6464.1998

Cui R, Lu X, Chen X, Malik WA (2021) A novel raffinose biological pathway is observed by symbionts of cotton=*Verticillium dahliae* to improve salt tolerance genetically on cotton. J Agron Crop Sci 207:956–969. https://doi.org/10.1111/jac.12556

Das T, Dey A, Pandey DK, Panwar JS, Nandy S (2022) Fungal endophytes as biostimulants of secondary metabolism in plants: a sustainable agricultural practice for medicinal crops. In: New and future developments in microbial biotechnology and bioengineering. Elsevier, pp 283–314. https://doi.org/10.1016/B978-0-323-85163-3.00010-7

Dastogeer KMG (2018) Influence of fungal endophytes on plant physiology is more pronounced under stress than well-watered conditions: a meta-analysis. Planta 248:1403–1416. https://doi.org/10.1007/s00425-018-2982-y

Davis RA, Carroll AR, Andrews KT, Boyle GM, Tran TL, Healy PC, Kalaitzis JA, Shivas RG (2010) Pestalactams A–C: novel caprolactams from the endophytic fungus *Pestalotiopsis* sp. Org Biomol Chem 8(8):1785–1790. https://doi.org/10.1039/b924169h

De Silva NI, Lumyong S, Hyde KD, Bulgakov T, Phillips AJL, Yan JY (2016) Mycosphere essays 9: defining biotrophs and hemibiotrophs. Mycosphere 7(5):545–559. https://doi.org/10.5943/mycosphere/7/5/2

de Vries M, Wösten HAB, van der Lee TAJ (2016) Macrolides: a comprehensive review of their production, biosynthesis, and role in fungal physiology. Fungal Biol Rev 30(2):78–92

Deepika VB, Murali TS, Satyamoorthy K (2016) Modulation of genetic clusters for synthesis of bioactive molecules in fungal endophytes: a review. Microbiol Res 182:125–140. https://doi.org/10.1016/j.micres.2015.10.009

Deepthi VC, Sumathi S, Faisal M, Elyas KK (2018) Isolation and identification of endophytic fungi with anti-microbial activities from the leaves of *Elaeocarpus sphaericus* (gaertn.) K. Schum and *Myristica fragrans* houtt. Int J Pharm Sci Res. https://doi.org/10.13040/IJPSR.0975-8232.9(7).2783-91

Deng Z, Cao L (2017) Fungal endophytes and their interactions with plants in phytoremediation: a review. Chemosphere 168:1100–1106. https://doi.org/10.1016/j.chemosphere.2016.10.097

Deng CM, Liu SX, Huang CH, Pang J-Y, Lin Y-C (2013) Secondary metabolites of a mangrove endophytic fungus *Aspergillus terreus* (No. GX7-3B) from the South China Sea. Mar Drugs 11(7):2616–2624. https://doi.org/10.3390/md11072616

Dhyani A, Jain R, Pandey A (2019) Contribution of root associated microbial communities on soil quality of Oak and Pine forests in the Himalayan ecosystem. Trop Ecol 60(2):271–280. https://doi.org/10.1007/s42965-019-00031-2

Di X, Takken FLW, Tintor N (2016) How phytohormones shape interactions between plants and the soil-borne fungus *Fusarium oxysporum*. Front Plant Sci 7. https://doi.org/10.3389/fpls.2016.00170

Diedhiou PM, Hallmann J, Oerke EC, Dehne HW (2003) Effects of arbuscular mycorrhizal fungi and a non-pathogenic *Fusarium oxysporum* on *Meloidogyne incognita* infestation on tomato. Mycorrhiza 13:199–204

Digra S, Nonzom S (2023) An insight into endophytic antimicrobial compounds: an updated analysis. Plant Biotechnol Rep 14:1–31. https://doi.org/10.1007/s11816-023-00824-x

Ding G, Song YC, Chen YR, Xu C, Ge HM, Wang XT, Tan RX (2006) Chaetoglobosin U, a cytochalasan alkaloid from endophytic *Chaetomium globosum* IFB-E019. J Nat Prod 69:302. https://doi.org/10.1021/np050515+

Ding G, Wang H, Li L, Chen AJ (2012) Trichoderones A and B: two pentacyclic cytochalasans from the plant endophytic fungus *Trichoderma gamsii*. Eur J Org Chem 13:2516–2519. https://doi.org/10.1002/ejoc.201200053

Duca D, Lorv J, Patten CL, Rose D, Glick BR (2014) Indole-3-acetic acid in plant-microbe interactions. Antonie Van Leeuwenhoek 106:85–125. https://doi.org/10.1007/s10482-013-0095-y

Dupont PY, Eaton CJ, Wargent JJ (2015) Fungal endophyte infection of ryegrass reprograms host metabolism and alters development. New Phytol 208:1227–1240. https://doi.org/10.1111/nph.13614

Dzoyem JP, Melong R, Tsamo AT, Maffo T, Kapche DGWF, Ngadjui BT, McGaw LJ, Eloff JN (2017) Cytotoxicity, antioxidant and antibacterial activity of four compounds produced by an endophytic fungus *Epicoccum nigrum* associated with *Entada abyssinica*. Rev Bras 27:251–253. https://doi.org/10.1016/j.bjp.2016.08.011

Ebrahim W, Kjer J, El Amrani M, Wray V, Lin W, Ebel R, Lai D, Proksch P (2012) Pullularins E and F, two new peptides from the endophytic fungus *Bionectria ochroleuca* isolated from the mangrove plant *Sonneratia caseolaris*. Mar Drugs 10(5):1081–1091. https://doi.org/10.3390/md10051081

El-Hawary SS, Mohammed R, AbouZid SF, Bakeer W, Ebel R, Sayed AM, Rateb ME (2016) Solamargine production by a fungal endophyte of *Solanum nigrum*. J Appl Microbiol 120(4):900–911. https://doi.org/10.1111/jam.13077

El-Hawary SS, Mohammed R, Bahr HS, Attia EZ, El-Katatny MH, Abelyan N, Al-Sanea MM, Moawad AS, Abdelmohsen UR (2021) Soybean-associated endophytic fungi as potential source for anti-COVID-19 metabolites supported by docking analysis. J Appl Microbiol 131(3):1193–1211. https://doi.org/10.1111/jam.15031

El-Kassem LA, Hawas UW, El-Souda S (2019) Anti-HCV protease potential of endophytic fungi and cytotoxic activity. Biocatal Agric Biotechnol 19. https://doi.org/10.1016/j.bcab.2019.101170

El-Neketi M, Ebrahim W, Lin W, Gedara S, Badria F, Saad H-EA, Lai D, Proksch P (2013) Alkaloids and polyketides from *Penicillium citrinum*, an endophyte isolated from the Moroccan plant *Ceratonia siliqua*. J Nat Prod 76(6):1099–1104. https://doi.org/10.1021/np4001366

Eyberger AL, Dondapati R, Porter JR (2006) Endophyte fungal isolates from *Podophyllum peltatum* produce podophyllotoxin. J Nat Prod 69:1121–1124. https://doi.org/10.1021/np060174f

Ezra D, Catillo UF, Strobel GA, Hess WM, Porter H, Jensen JB, Condron MAM, Teplow DB, Sears J, Maranta M, Hunter M, Weber B, Yaver D (2004) Coronamycins, peptide antibiotics produced by a verticillate *Streptomyces* sp. (MSU-2110) endophytic on *Monstera* sp. Microbiology 150:785. https://doi.org/10.1099/mic.0.26645-0

Faeth SH (2002) Are endophytic fungi defensive plant mutualists? Oikos 98(1):25–36. https://doi.org/10.1034/j.1600-0706.2002.980103.x

Fávaro LCDL, Sebastianes FLDS, Araújo WL (2012) *Epicoccum nigrum* P16, a sugarcane endophyte, produces antifungal compounds and induces root growth. PLoS One 7(6):e36826. https://doi.org/10.1371/journal.pone.0036826

Fellbaum CR, Gachomo EW, Beesetty Y, Choudhari S, Strahan GD, Pfeffer PE, Kiers ET, Bücking H (2012) Carbon availability triggers fungal nitrogen uptake and transport in arbuscular mycorrhizal Symbiosis. Proc Natl Acad Sci USA 109:2666–2671. https://doi.org/10.1073/pnas.1118650109

Fernanda OC, Rita de CP, Andre's MCR, Monica TP (2017) Chemical signaling involved in plant–microbe interactions. Chem Soc Rev 47(5). https://doi.org/10.1039/c7cs00343a

Field KJ, Pressel S, Duckett JG, Rimington WR, Bidartondo MI (2015) Symbiotic options for the conquest of land. Trends Ecol Evol 30:477–486. https://doi.org/10.1016/j.tree.2015.05.007

Findlay JA, Buthelezi S, Li GQ, Seveck M, Miller JD (1997) Insect toxins from an endophytic fungus from wintergreen. J Nat Prod 60:1214. https://doi.org/10.1021/np970222j

Fontana DC, de Paula S, Torres AG, de Souza VHM, Pascholati SF, Schmidt D, Neto DD (2021) Endophytic fungi: biological control and induced resistance to phytopathogens and abiotic stresses. Pathogens 10(5):570. https://doi.org/10.3390/pathogens10050570

Forni C, Duca D, Glick BR (2017) Mechanisms of plant response to salt and drought stress and their alteration by rhizobacteria. Plant Soil 410(1–2):335–356. https://doi.org/10.1007/s11104-016-3007-x

Francis I, Holsters M, Vereecke D (2010) The Gram-positive side of plant–microbe interactions. Environ Microbiol 12(1):1–12. https://doi.org/10.1111/j.1462-2920.2009.01989.x

Fravel DR (1988) Role of antibiosis in the biocontrol of plant diseases. Annu Rev Phytopathol 26:75–91. https://doi.org/10.1146/annurev.py.26.090188.000451
Gaiero JR, McCall CA, Thompson KA, Day NJ, Best AS, Dunfield KE (2013) Inside the root microbiome: bacterial root endophytes and plant growth promotion. Am J Bot 100:1738–1750. https://doi.org/10.3732/ajb.1200572
Gao S-S, Li X-M, Li C-S, Proksch P, Wang B-G (2011) Penicisteroids a and B, antifungal and cytotoxic polyoxygenated steroids from the marine alga-derived endophytic fungus *Penicillium chrysogenum* QEN-24S. Bioorg Med Chem Lett 21(10):2894–2897. https://doi.org/10.1016/j.bmcl.2011.03.076
Gao H, Li G, Lou H-X (2018) Structural diversity and biological activities of novel secondary metabolites from endophytes. Molecules 23:646. https://doi.org/10.3390/molecules23030646
Garcia-Arenal F, Sagasta EM (1980) Scanning electron microscopy of *Botrytis cinerea* penetration of bean (*Phaseolus vulgaris*) hypocotyls. Phytopathology 99:37–42
Gawehns F, Ma LS, Bruning O, Houterman PM, Boeren S, Cornelissen BJC, Rep M, Takken FLW (2015) The effector repertoire of *fusarium oxysporum* determines the tomato xylem proteome composition following infection. Front Plant Sci 6:17. https://doi.org/10.3389/fpls.2015.00967
Ge HL, Zhang DW, Li L, Xie D, Zou J-H, Si Y-K, Dai J (2011) Two new terpenoids from endophytic fungus *Periconia* sp. F-31. Chem Pharm Bull 59(12):1541–1544. https://doi.org/10.1248/cpb.59.1541
Genre A, Chabaud M, Balzergue C, Bécard G, Rochange SF (2013) High phosphate reduces host ability to develop arbuscular mycorrhizal symbiosis without affecting root calcium spiking responses to the fungus. New Phytol 198:190–202. https://doi.org/10.3389/fpls.2013.00426
Germaine K, Keogh E, Garcia-Cabellos G, Borremans B, van der Lelie D, Barac T, Oeyen L, Vangronsveld J, Moore FP, Moore ERB, Campbell CD, Ryan D, Dowling DN (2004) Colonisation of poplar trees by *gfp* expressing bacterial endophytes. FEMS Microbiol Ecol 48(1):109–118. https://doi.org/10.1016/j.femsec.2003.12.009
Giauque H, Hawkes CV (2013) Climate affects symbiotic fungal endophyte diversity and performance. Am J Bot 100(7):1435–1444. https://doi.org/10.3732/ajb.1200568
Glick BR (2005) Modulation of plant ethylene levels by the bacterial enzyme ACC deaminase. FEMS Microbiol Lett 251:1–7. https://doi.org/10.1016/j.femsle.2005.07.030
Gond SK, Bergen MS, Torres MS, White JF Jr, Kharwar RN (2015) Effect of bacterial endophyte on expression of defense genes in Indian popcorn against *fusarium moniliforme*. Symbiosis 66:133–140. https://doi.org/10.1007/s13199-015-0348-9
Gouda S, Das G, Sen SK (2016) Endophytes: a treasure house of bioactive compounds of medicinal importance. Front Microbiol 7:1538. https://doi.org/10.3389/fmicb.2016.01538
Gough C, Be'card G (2016) Molecular mycorrhizal symbiosis. Wiley, pp 107–123
Gough C, Cullimore J (2011) Lipo-chitooligosaccharide signaling in endosymbiotic plant-microbe interactions. Mol Plant-Microbe Interact 24:867–878. https://doi.org/10.1094/MPMI-01-11-0019
Guan S, Grabley S, Groth I, Lin W, Christner A, Guo D, Sattler I (2005) Structure determination of germacrane-type sesquiterpene alcohols from an endophyte *Streptomyces griseus* subsp. Magn Reson Chem 43(1028):1028–1031. https://doi.org/10.1002/mrc.1710
Guevara-Araya M, Vilo C, Urzúa A, González-Teuber M (2020) Differences in community composition of endophytic fungi between above-and below-ground tissues of *Aristolochia chilensis* in an arid ecosystem. Rev Chil Hist Nat 93(1):1–9. https://doi.org/10.1186/s40693-020-00091-y
Guo Z, Zou ZM (2020) Discovery of new secondary metabolites by epigenetic regulation and NMR comparison from the plant endophytic fungus *Monosporascus eutypoides*. Molecules 25:4192. https://doi.org/10.3390/molecules25184192
Guo Z, Chen W, Dai G, Huang Y (2020) Cordycepin suppresses the migration and invasion of human liver cancer cells by downregulating the expression of CXCR4. Int J Mol Med 45:141–150. https://doi.org/10.3892/IJMM.2019.4391
Gutierrez RMP, Gonzalez AMN, Ramirez AM (2012) Compounds derived from endophytes: a review of phytochemistry and pharmacology. Curr Med Chem 19(18):2992–3030. https://doi.org/10.2174/092986712800672111

Han Z, Mei W, Zhao Y, Deng Y (2009) A new cytotoxic isocoumarin from endophytic fungus Penicillium sp. 091402 of the mangrove plant *Bruguiera sexangula*. Chem Nat Compd 45:805–807

Haque MA, Hossain MS, Rahman MZ (2005) Isolation of bioactive secondary metabolites from the endophytic fungus of *Ocimum basilicum*. J Pharm Sci 4(2):127–130. https://doi.org/10.3329/dujps.v4i2.215

Hardoim PR, Van Overbeek LS, Berg, Prittila AM, Compant S, Campisano A, Doring M, Sessitsch A (2015) The hidden world within plants: ecological and evolutionary considerations for defining functioning of microbial endophytes. Microbiol Mol Biol Rev 79(3):293–320. https://doi.org/10.1128/MMBR.00050-14

Haro R, Benito B (2019) The role of soil fungi in K+ plant nutrition. Int J Mol Sci 20:3169. https://doi.org/10.3390/ijms20133169

Harrison JG, Griffn EA (2020) The diversity and distribution of endophytes across biomes, plant phylogeny and host tissues: how far have we come and where do we go from here? Environ Microbiol 22(6):2107–2123. https://doi.org/10.1111/1462-2920.14968

Hartley SE, Gange AC (2009) Impacts of plant symbiotic fungi on insect herbivores: mutualism in a multitrophic context. Annu Rev Entomol 54:323–342. https://doi.org/10.1146/annurev.ento.54.110807.090614

Hartley SE, Eschen R, Horwood JM, Gange AC, Hill EM (2015) Infection by a foliar endophyte elicits novel arabidopside-based plant defence reactions in its host, *Cirsium arvense*. New Phytol 205:816–827. https://doi.org/10.1111/nph.13067

Hasegawa M, Mitsuhara I, Seo S, Imai T, Koga J, Okada K, Yamana H, Ohashi Y (2010) Phytoalexin accumulation in the interaction between Rice and the blast fungus. Mol Plant-Microbe Interact 23:1000–1011. https://doi.org/10.1094/MPMI-23-8-1000

Hawksworth DL, Lücking R (2017) Fungal diversity revisited: 2.2 to 3.8 million species. In: Heitman J, Howlett BJ, Crous PW, Stukenbrock EH, James TY, Gow NAR (eds) The fungal kingdom. ASM Press, Washington, DC, pp 79–95

Hemphill CFP, Daletos G, Liu Z, Lin W, Proksch P (2016) Polyketides from the mangrove-derived fungal endophyte *Pestalotiopsis clavispora*. Tetrahedron Lett 57(19):2078–2083. https://doi.org/10.1016/j.tetlet.2016.03.101

Herrera-Carillo Z, Torres MS, Singh AP, Vorsa N, Gianfagna T, Meyer W, White Jr JF (2009) Phenolic, flavonoid and antioxidant profiling for cool-season grasses with and without endophyte. Paper presented at the proceedings of the 18th annual Rutgers Turfgrass symposium, p 43

Hiruma K, Gerlach N, Sacristan S, Nakano RT, Bucher M, O'Connell RJ, Schulze-Lefert P (2016) Root endophyte *Colletotrichum tofeldiae* confers plant fitness benefits that are phosphate status dependent. Cell 165(2):464–474. https://doi.org/10.1016/j.cell.2016.02.028

Holliday P (1998) A dictionary of plant pathology, 2nd edn. Cambridge University Press

Hosseyni-Moghaddam MS, Soltani J (2014) Bioactivity of endophytic *Trichoderma* fungal species from the plant family Cupressaceae. Ann Microbiol 64(2):753–761. https://doi.org/10.1007/s13213-013-0710-1

Hua MD-S, Senthil Kumar R, Shyur L-F, Cheng Y-B, Tian Z, Oelmuller R, Yeh K-W (2017) Metabolomic compounds identified in *Piriformospora indica*-colonized Chinese cabbage roots delineate symbiotic functions of the interaction. Sci Rep 7(1):1–14. https://doi.org/10.1038/s41598-017-08715-2

Huang WY, Cai YZ, Xing J, Corke H, Sun M (2007) A potential antioxidant resource: endophytic fungi from medicinal plants. Econ Bot 61(1):14–30. https://doi.org/10.1663/0013-0001(2007)61[14:APAREF]2.0.CO;2

Huang CH, Pan JH, Chen B, Yu M, Huang H-B, Zhu X, Lu Y-J, She Z-G, Lin Y-C (2011a) Three bianthraquinone derivatives from the mangrove endophytic fungus *Alternaria* sp. ZJ9-6B from the South China Sea. Mar Drugs 9(5):832–843. https://doi.org/10.3390/md9050832

Huang HB, Xiao ZE, Feng XJ, Huang C-H, Zhu X, Ju J-H, Li M-F, Lin Y-C, Liu L, She Z-G (2011b) Cytotoxic naphtho-γ-pyrones from the mangrove endophytic fungus *Aspergillus tubingensis* (GX1-5E). Helv Chim Acta 94(9):1732–1740. https://doi.org/10.1002/hlca.201100050

Huang Z, Yang J, Cai X, She Z, Lin Y (2012) A new furanocoumarin from the mangrove endophytic fungus Penicillium sp.(ZH16). Nat Prod Res 26(14):1291–1295

Ibrahim M, Oyebanji E, Fowora M, Aiyeolemi A, Orabuchi C, Akinnawo B, Adekunle AA (2021) Extracts of endophytic fungi from leaves of selected Nigerian ethnomedicinal plants exhibited antioxidant activity. BMC Complement Med Ther 21(1):1–13. https://doi.org/10.1186/s12906-021-03269-3

Idbella M, Zotti M, Cesarano G, Fechtali T, Mazzoleni S, Bonanomi G (2019) Fungal endophytes affect plant response to leaf litter with contrasting chemical traits. Community Ecol 20:205–213. https://doi.org/10.1556/168.2019.20.2.10

Jayawardena RS, Hyde KD, Damm U, Cai L, Liu M, Li XH, Zhang W, Zhao WS, Yan YJ (2016) Notes on currently accepted species of *colletotrichum*. Mycosphere 7(8):1192–1260. https://doi.org/10.5943/mycosphere/si/2c/9

Jennewein S, Rithner CD, Williams RM, Croteau RB (2001) Taxol biosynthesis: taxane 13α-hydroxylase is a cytochrome P450-dependent monooxygenase. Proc Natl Acad Sci 98(24):13595–13600. https://doi.org/10.1073/pnas.25153939

Ji H, Li X, Zhang H (2009) Natural products and drug discovery. EMBO Rep 10(3):194–200. https://doi.org/10.1038/embor.2009.12

Johnson LJ, De Bonth ACM, Briggs LR, Caradus JR, Finch SC, Fleetwood DJ, Fletcher LR, Hume DE, Johnson RD, Popay AJ, Tapper BA, Simpson WR, Voisey CR, Card SD (2013) The exploitation of epichloae endophytes for agricultural benefit. Fungal Divers 60:171–188. https://doi.org/10.1007/s13225-013-0239-4

Johnston-Monje D, Raizada MN (2011) Conservation and diversity of seed associated endophytes in *Zea* across boundaries of evolution, ethnography and ecology. PLoS One 6:e20396. https://doi.org/10.1371/journal.pone.0020396

Joseph B, Priya RM (2011) Bioactive compounds from endophytes and their potential. Am J Biochem Mol Biol 1(3):291–309. https://doi.org/10.3923/ajbmb.2011.291.309

Ju Y, Sacalis JN, Still CC (1998) Bioactive favonoids from endophyte-infected blue grass (Poa ampla). J Agric Food Chem 46(9):3785–3788. https://doi.org/10.1021/jf980189m

Justus M, Witte L, Hartmann T (1997) Levels and tissue distribution of loline alkaloids in endophyte-infected *Festuca pratensis*. Phytochemistry 44:51. https://doi.org/10.1016/S0031-9422(96)00535-3

Karpinets TV, Park BH, Syed MH, Klotz MG, Uberbacher EC (2014) Metabolic environments and genomic features associated with pathogenic and mutualistic interactions between bacteria and plants. Mol Plant-Microbe Interact 27:664–677. https://doi.org/10.1094/MPMI-12-13-0368-R

Kato-Noguchi H, Ino T, Ota K (2008) Secretion of momilactone A from rice roots to the rhizosphere. J Plant Physiol 165:691–696. https://doi.org/10.1016/j.jplph.2007.07.018

Kaul S, Sharma T, Dhar KM (2016) "Omics" tools for better understanding the plant–endophyte interactions. Front Plant Sci 7:955. https://doi.org/10.3389/fpls.2016.00955

Keller NP (2019) Fungal secondary metabolism: regulation, function and drug discovery. Nat Rev Microbiol 17:167–180. https://doi.org/10.1038/s41579-018-0121-1

Keller NP, Turner G, Bennett JW (2005) Fungal secondary metabolism from biochemistry to genomics. Nat Rev Microbiol 3(12):937–947. https://doi.org/10.1038/nrmicro1286

Keshri PK, Rai N, Verma A, Kamble SC, Barik S, Mishra P, Singh SK, Salvi P, Gautam V (2021) Biological potential of bioactive metabolites derived from fungal endophytes associated with medicinal plants. Mycol Prog 20(5):577–594. https://doi.org/10.1007/s11557-021-01695-8

Khan AL, Hamayun M, Khan SA (2012) Pure culture of *Metarhizium anisopliae* LHL07 reprograms soybean to higher growth and mitigates salt stress. World J Microbiol Biotechnol 28(4):1483–1494. https://doi.org/10.1007/s11274-011-0950-9

Khan AL, Hussain J, Al-Harrasi A, Al-Rawahi A, Lee I-J (2013) Endophytic fungi: resource for gibberellins and crop abiotic stress resistance. Crit Rev Biotechnol 35(1):62–74. https://doi.org/10.3109/07388551.2013.800018

Khare E, Mishra J, Arora NK (2018) Multifaceted interactions between endophytes and plant: developments and prospects. Front Microbiol 9:2732. https://doi.org/10.3389/fmicb.2018.02732

Kim YC, Leveau J, Gardener BBM, Pierson EA, PiersonIII LS, Ryu C-M (2011) The multifactorial basis for plant health promotion by plant-associated bacteria. Appl Environ Microbiol 77:1548–1555. https://doi.org/10.1128/AEM.01867-10

Kirby J, Keasling JD (2009) Biosynthesis of plant isoprenoids: perspectives for microbial engineering. Annu Rev Plant Biol 60(1):335–355. https://doi.org/10.1146/annurev.arplant.043008.091955

Kloepper JW, Ryu CM, Zhang S (2004) Induced systemic resistance and promotion of plant growth by *Bacillus* spp. Phytopathology 94:1259–1266. https://doi.org/10.1094/PHYTO.2004.94.11.1259

Kogel K-H, Franken P, Hückelhoven R (2006) Endophyte or parasite – what decides? Curr Opin Plant Biol 9(4):358–363. https://doi.org/10.1016/j.pbi.2006.05.001

Kong KF, Schneper L, Mathee K (2010) Beta-lactam antibiotics: from antibiosis to resistance and bacteriology. APMIS 118:1–36. https://doi.org/10.1111/j.1600-0463.2009.02563.x

Kour A, Shawl AS, Rehman S (2008) Isolation and identification of an endophytic strain of *fusarium oxysporum* producing podophyllotoxin from *Juniperus recurve*. World J Microbiol Biotechnol 24:1115–1121. https://doi.org/10.1007/s11274-007-9582-5

Krings M, Taylor TN, Hass H, Kerp H, Dotzler N, Hermsen EJ (2007) Fungal endophytes in a 400-million-yr-old land plant: infection pathways, spatial distribution, and host responses. New Phytol 174:648–657. https://doi.org/10.1111/j.1469-8137.2007.02008.x

Krohn K, Dai J, Florke U, Aust H-J, Dräger S, Schulz B (2005) Botryane metabolites from the fungus *Geniculosporium* sp. isolated from the marine red alga *Polysiphonia*. J Nat Prod 68:400–405. https://doi.org/10.1021/np0498206

Kryukov VY, Kryukova NA, Tomilova OG, Vorontsova Y, Chertkova E, Pervushin AL, Slepneva I, Glupav VV, Yaroslavtseva ON (2020) Comparative analysis of the immune response of the wax moth *Galleria mellonella* after infection with the fungi *Cordyceps militaris* and *Metarhizium robertsii*. Microb Pathog 141. https://doi.org/10.1016/j.micpath.2020.103995

Kumar A, Patil D, Rajamohanan PR, Ahmad A (2013) Isolation, purification and characterization of vinblastine and vincristine from Endophytic fungus *Fusarium oxysporum* isolated from *Catharanthus roseus*. PLoS One 8(9):e71805. https://doi.org/10.1371/journal.pone.0071805

Kumar G, Chandra P, Choudhary M (2017) Endophytic fungi: a potential source of bioactive compounds. Chem Sci Rev Lett 6:2373–2381. https://doi.org/10.3390/jof8020205

Kumar GA, Kumar S, Bhardwaj R, Meena M, Swapnil P, Seth CS, Yadav A (2024) Recent advancements in multifaceted roles of flavonoids in plant–rhizomicrobiome interactions. Front Plant Sci 14:1297706. https://doi.org/10.3389/fpls.2023.1297706

Kusari S, Spiteller M (2012) Metabolomics of endophytic fungi producing associated plant secondary metabolites: progress, challenges and opportunities. Metabolomics 10:241–266. https://doi.org/10.5772/31596

Kusari S, Lamshöft M, Zühlke S, Spiteller M (2008) An endophytic fungus from *Hypericum perforatum* that produces hypericin. J Nat Prod 71:159–162. https://doi.org/10.1021/np070669k

Kusari S, Zuhlke S, Kosuth, Cellarova E, Spiteller M (2009a) Light-independent metabolomics of endophytic *Thielavia subthermophila* provides insight into microbial Hypericin biosynthesis. J Nat Prod 72:1825–1835. https://doi.org/10.1021/np9002977

Kusari S, Lamshoft M, Spiteller M (2009b) *Aspergillus fumigatus* Fresenius, an endophytic fungus from *Juniperus communis* L. Horstmann as a novel source of the anticancer pro-drug deoxypodophyllotoxin. J Appl Microbiol 107:1019–1030. https://doi.org/10.1111/j.1365-2672.2009.04285.x

Kusari S, Zuehlke S, Spiteller M (2009c) An endophytic fungus from *Camptotheca acuminata* that produces camptothecin and analogues. J Nat Prod 72:2–7. https://doi.org/10.1021/np800455b

Kusari S, Zuehlke S, Spiteller M (2011) Chemical ecology of endophytic fungi: origins of secondary metabolites. J Nat Prod 74:764–775. https://doi.org/10.1016/j.chembiol.2012.06.004

Kusari S, Hertweck C, Spiteller M (2012a) Chemical ecology of endophytic fungi: origins of secondary metabolites. Chem Biol 19:792–798. https://doi.org/10.1016/j.chembiol.2012.06.004

Kusari S, Verma VC, Lamshoeft M, Spiteller M (2012b) An endophytic fungus from *Azadirachta indica* a. Juss. That produces azadirachtin. World J Microbiol Biotechnol 28:1287–1294. https://doi.org/10.1007/s11274-011-0876-2

Kusari S, Pandey SP, Spiteller M (2013) Untapped mutualistic paradigms linking host plant and endophytic fungal production of similar bioactive secondary metabolites. Phytochemistry 91:81–87. https://doi.org/10.1016/j.phytochem.2012.07.021

Kusari P, Kusari S, Spiteller M, Kayser O (2015) Implications of endophyte-plant crosstalk in light of quorum responses for plant biotechnology. Appl Microbiol Biotechnol 99:5383–5390. https://doi.org/10.1007/s00253-015-6660-8

Lareen A, Burton F, Schafer P (2016) Plant root-microbe communication in shaping root microbiomes. Plant Mol Biol 90:575–587. https://doi.org/10.1007/s11103-015-0417-8

Lee YC, Johnson JM, Chien C-T, Sun C, Cai D, Lou B, Oelmüller R, Yeh K-W et al (2011) Growth promotion of Chinese cabbage and *Arabidopsis* by *Piriformospora indica* is not stimulated by mycelium-synthesized auxin. Mol Plant-Microbe Interact 24:421–431. https://doi.org/10.1094/MPMI-05-10-0110

Li JY, Strobel GA (2001) Jesterone and hydroxy-jesterone antioomycete cyclohexenone epoxides from the endophytic fungus *Pestalotiopsis jesteri*. Phytochemistry 57:261. https://doi.org/10.1016/S0031-9422(01)00021-8

Li YC, Tao WY (2009) Paclitaxel-producing fungal endophyte stimulates the accumulation of taxoids in suspension cultures of *Taxus cuspidate*. Sci Hortic 121:97–102. https://doi.org/10.1016/j.scienta.2009.01.016

Li JY, Strobel GA, Harper J, Lobkovsky E, Clardy J (2000) Cryptocin, a potent tetramic acid antimycotic from the endophytic fungus *Cryptosporiopsis* cf. *quercina*. Org Lett 23:767. https://doi.org/10.1021/ol000008d

Li D-L, Li X-M, Li T-G, Dang H-Y, Wang B-G (2008) Dioxopiperazine alkaloids produced by the marine mangrove derived endophytic fungus *Eurotium rubrum*. Helv Chim Acta 91(10):1888–1893. https://doi.org/10.1002/hlca.200890202

Li H, Huang H, Shao C, Huang H, Jiang J, Zhu X, Liu Y, Liu L, Lu Y, Li M, Lin Y, She Z (2011) Cytotoxic norsesquiterpene peroxides from the endophytic fungus *Talaromyces favus* isolated from the mangrove plant *Sonneratia apetala*. J Nat Prod 74(5):1230–1235. https://doi.org/10.1021/np200164k

Li J, Zhao GZ, Varma A (2012) An endophytic *Pseudonocardia* species induces the production of artemisinin in *Artemisia annua*. PLoS One 7:e51410. https://doi.org/10.1371/journal.pone.0051410

Li X, Tian Y, Yang SX, Zhang Y-m, Qin J-c (2013) Cytotoxic azaphilone alkaloids from *Chaetomium globosum* TY1. Bioorg Med Chem Lett 23(10):2945–2947. https://doi.org/10.1016/j.bmcl.2013.03.044

Li Q, Chen C, Penttinen P, Xiong C, Zheng L, Huang W (2016) Microbial diversity associated with *Tricholoma matsutake* fruiting bodies. Microbiology (Russian Fed) 85:531–539. https://doi.org/10.1134/S0026261716050106

Liu JY, Song YC, Zhang Z, Wang L, Guo ZJ, Zou WX, Tan RX (2004) *Aspergillus fumigatus* CY018, an endophytic fungus in *Cynodon dactylon* as a versatile producer of new and bioactive metabolites. J Biotechnol 114:279. https://doi.org/10.1016/j.jbiotec.2004.07.008

Liu D, Li XM, Meng L, Li C-S, Gao S-S, Shang Z, Proksch P, Huang C-G, Wang B-G (2011) Nigerapyrones A–H, α-Pyrone Derivatives from the Marine Mangrove-Derived Endophytic Fungus *Aspergillus niger* MA-132. J Nat Prod 74(8):1787–1791

Liu Y, Zhang N, Qiu M, Feng H, Jorge M, Vivanco J, Shen Q, Zhang R (2014) Enhanced rhizosphere colonization of beneficial *bacillus amyloliquefaciens* SQR9 by pathogen infection FEMS. Microbiol Lett 353:49–56. https://doi.org/10.1111/1574-6968.12406

Liu P, Zhang D, Shi R, Yang Z, Zhao F, Tian Y (2019a) Antimicrobial potential of endophytic fungi from *Astragalus chinensis*. 3 Biotech 9:1–9. https://doi.org/10.1007/s13205-019-1948-5

Liu SS, Jiang JX, Huang R, Wang Y-T, Jiang B-G, Zheng K-X, Wu S-H (2019b) A new antiviral 14-nordrimane sesquiterpenoid from an endophytic fungus *Phoma* sp. Phytochem Lett 29:75–78. https://doi.org/10.1016/j.phytol.2018.11.005

Liu Z, Leng G, Wen J, Deng G, Jiang J (2022) Cordycepin production by a novel endophytic fungus *Irpex lacteus* CHG05 isolated from *Cordyceps hawkesii* gray. Folia Microbiol (Praha) 10:851. https://doi.org/10.1007/s12223-022-00981-6

Lòpez-Fernàndez S, Sonego P, Moretto M, Pancher M, Engelen K, Pertot I, Campisano A (2015) Whole-genome comparative analysis of virulence genes unveils similarities and differences between endophytes and other symbiotic bacteria. Front Microbiol 6:419. https://doi.org/10.3389/fmicb.2015.00419

Lu S, Sun P, Li T, Kurtán T, Mándi A, Antus S, Krohn K, Draeger S, Schulz B, Yi Y, Li L, Zhang W et al (2011) Bioactive nonanolide derivatives isolated from the endophytic fungus *Cytospora* sp. J Org Chem 76(23):9699–9710. https://doi.org/10.1021/jo201755v

Ludwig-Müller J (2015) Plants and endophytes: equal partners in secondary metabolite production? Biotechnol Lett 37(7):1325–1334. https://doi.org/10.1007/s10529-015-1814-4

Lugtenberg BJJ, Caradus JR, Johnson LJ (2016) Fungal endophytes for sustainable crop production. FEMS Microbiol Ecol 92:17. https://doi.org/10.1093/femsec/fiw194

Luo J, Liu X, Li E, Guo L, Chee Y (2013) Arundinols A–C and arundinones A and B from the plant endophytic fungus *Microsphaeropsis arundinis*. J Nat Prod 76(1):107–112. https://doi.org/10.1021/np300806a

Ma L, Cao YH, Cheng MH, Huang Y, Mo MH, Wang Y, Yang JZ, Yang FX (2013) Phylogenetic diversity of bacterial endophytes of *Panax notoginseng* with antagonistic characteristics towards pathogens of root-rot disease complex. Antonie Van Leeuwenhoek 103:299–312. https://doi.org/10.1007/s10482-012-9810-3

Macías-Rubalcava ML, Sánchez-Fernández RE (2017) Secondary metabolites of endophytic *Xylaria* species with potential applications in medicine and agriculture. World J Microbiol Biotechnol 33:1–22. https://doi.org/10.1007/s11274-016-2174-5

Macías-Rubalcava ML, Ruiz-Velasco Sobrino ME, Melendez-Gonzalez C, Hernández-Ortega S (2014) Naphthoquinone spiroketals and organic extracts from the Endophytic fungus *Edenia gomezpompae* as potential herbicides. J Agric Food Chem 62(16):3553–3562. https://doi.org/10.1021/jf500965k

Maillet F, Poinsot V, Andre O et al (2011) Fungal lipochitooligosaccharide symbiotic signals in arbuscular mycorrhiza. Nature 469:58–63. https://doi.org/10.1038/nature09622

Malik S, Kishore S, Dutta AK, Dhasmana A (2023) Sustainable agriculture approach through endophytes. In: Endophytic association: what, why and how. https://doi.org/10.1016/B978-0-323-91245-7.00019-5

Manganyi MC, Ateba CN (2020) Untapped potentials of endophytic fungi: a review of novel bioactive compounds with biological applications. Microorganisms 8:1934. https://doi.org/10.3390/microorganisms8121934

Mariño CG, Cabrera R, Reina M, Gonzalez-Coloma A (2007) Fungal endophytes and their role in plant protection. Curr Org Chem 11(8):707–720. https://doi.org/10.2174/1385272077805

Martinez-Klimova E, Rodríguez-Peña K, Sánchez S (2017) Endophytes as sources of antibiotics. Biochem Pharmacol 134:1–17. https://doi.org/10.1016/j.bcp.2016.10.010

Mathan S, Subramanian V, Nagamony S, Ganapathy K (2013) Isolation of endophytic fungi from marine algae and its bioactivity. Int J Res Pharm Sci 4(1):45–49

Mbengue M, Navaud O, Peyraud R, Barascud M, Badet T, Vincent R, Barbacci A, Raffaele S (2016) Emerging trends in molecular interactions between plants and the broad host range fungal pathogens *Botrytis cinerea* and *Sclerotinia sclerotiorum*. Front Plant Sci 7:9. https://doi.org/10.3389/fpls.2016.00422

Meena H, Hnamte S, Siddhardha B (2019) Secondary metabolites from endophytic fungi: chemical diversity and application. In: Advances in endophytic fungal research. Fungal biology, pp 145–169. https://doi.org/10.1007/978-3-030-03589-1_7

Meena M, Yadav G, Sonigra P, Nagda A, Mehta T, Swapnil P, Harish MA (2022) Role of elicitors to initiate the induction of systemic resistance in plants to biotic stress. Plant Stress 5:100103. https://doi.org/10.1016/j.stress.2022.100103

Meena M, Yadav G, Sonigra P, Nagda A, Mehta T, Swapnil P, Harish MA, Kumar S (2023) Multifarious responses of forest soil microbial community toward climate change. Microb Ecol 86:49–74. https://doi.org/10.1007/s00248-022-02051-3

Mehta P, Sharma R, Putatunda C, Walia A (2019) Endophytic fungi: role in phosphate solubilization. In: Singh BP (ed) Advances in endophytic fungal research: present status and future challenges. Springer International Publishing, Cham, pp 183–209. https://doi.org/10.1007/978-3-030-03589-1_9

Mishra VK, Singh G, Passari AK, Yadav MK, Gupta VK, Singh BP (2016) Distribution and antimicrobial potential of endophytic fungi associated with ethnomedicinal plant *Melastoma malabathricum* L. J Environ Biol 37(2):229–237

Mishra S, Bhattacharjee A, Sharma S (2021a) An ecological insight into the multifaceted world of plant-endophyte association. Crit Rev Plant Sci 40(2):127–146. https://doi.org/10.1080/07352689.2021.1901044

Mishra Y, Sharma L, Dhiman M, Sharma MM (2021b) Endophytic fungal diversity of selected medicinal plants and their bio-potential applications. In: Fungi bio-prospects in sustainable agriculture, environment and nano-technology. Elsevier, pp 227–283. https://doi.org/10.1016/B978-0-12-821394-0.00010-X

Mitter B, Petric A, Shin MW, Chain PSG, Hauberg-Lotte L, Reinhold-Hurek B, Nowak J, Sessitsch A (2013) Comparative genome analysis of *Burkholderia phytofirmans* PsJN reveals a wide spectrum of endophytic lifestyles based on interaction strategies with host plants. Front Plant Sci 4:120. https://doi.org/10.3389/fpls.2013.00120

Moghaddam MS, Safaie N, Rahimlou S, Hagh-Doust N (2022) Inducing tolerance to abiotic stress in *Hordeum vulgare* L. by halotolerant endophytic fungi associated with salt lake plants. Front Microbiol 13:906365. https://doi.org/10.3389/fmicb.2022.906365

Mohali S, Burgess T, Wingfeld M (2005) Diversity and host association of the tropical tree endophyte *Lasiodiplodia theobromae* revealed using simple sequence repeat markers. For Pathol 35:385–396. https://doi.org/10.1111/j.1439-0329.2005.00418.x

Monteiro RA, Balsanelli E, Tuleski T et al (2012) Genomic comparison of the endophyte *Herbaspirillum seropedicae* SmR1 and the phytopathogen *Herbaspirillum rubrisubalbicans* M1 by suppressive subtractive hybridization and partial genome sequencing. FEMS Microbiol Ecol 80:441–451. https://doi.org/10.1111/j.1574-6941.2012.01309.x

Moy M, Belanger F, Duncan R, Freehoff R, Leary A, Meyer C, Sullivan W, White JFJ, James F (2000) Identification of epiphyllous mycelial nets on leaves of grasses infected by clavicipitaceous endophytes. Symbiosis. http://hdl.handle.net/10222/77715

Nadeem M, Ram M, Alam P et al (2012) Fusarium solani, P1, a new endophytic podophyllotoxin-producing fungus from roots of *Podophyllum hexandrum*. Afr J Microbiol Res 6(10):2493–2499. https://doi.org/10.5897/AJMR11.1596

Nanda S, Mohanty B, Joshi RK (2019) Endophyte-mediated host stress tolerance as a means for crop improvement. In: Reference series in phytochemistry, pp 677–701. https://doi.org/10.1007/978-3-319-76900-4_28-1

Newman DJ, Cragg GM (2015) Endophytic and epiphytic microbes as "sources" of bioactive agents. Front Chem 3:1–13. https://doi.org/10.3389/fchem.2015.00034

Ortega HE, Graupner PR, Asai Y et al (2013) Mycoleptodiscins A and B, cytotoxic alkaloids from the endophytic fungus *Mycoleptodiscus* sp. F0194. J Nat Prod 76(4):741–744. https://doi.org/10.1021/np300792t

Pandey PK, Singh S, Yadav RNS, Singh AK, Singh MCK (2014) Fungal endophytes: promising tools for pharmaceutical science. Int J Pharm Sci Rev Res 25(2):128–138

Pandey SS, Singh S, Babu CSV, Shankerv K, Srivastava NK, Shukla AK, Kalra A (2016) Fungal endophytes of *Catharanthus roseus* enhance vindoline content by modulating structural and regulatory genes related to terpenoid indole alkaloid biosynthesis. Sci Rep 6. https://doi.org/10.1038/srep26583

Patel S, Parekh V, Patel K, Jha S (2021) Plant growth-promoting activities of *Penicillium* sp. NAUSF2 ameliorate Vigna radiata salinity stress in phosphate-defcient saline soil. Appl Biochem Microbiol 57:500–507. https://doi.org/10.1134/S000368382104013X

Pei-xin H, Wei L, Xin-sheng H, Lu-jing G (2014) Diversity of endophytic fungi in ascocarps of *Morchella crassipes*. J Zhengzhou Univ Light Ind (Natural Sci Ed) 29:1–6

Perez LI, Gundel PE, Omacini (2016) Can the defensive mutualism between grasses and fungal endophytes protect non-symbiotic neighbours from soil pathogens? Plant Soil 405:289–298. https://doi.org/10.1007/s11104-015-2568-4

Petrini O, Sieber TH, Toti L, Viret O (1992) Ecology, metabolite production, and substrate utilization in endophytic fungi. Nat Toxins 1:185–196. https://doi.org/10.1002/nt.2620010306

Philippot L, Raaijmakers JM, Lemanceau P, van der Putten WH (2013) Going back to the roots: the microbial ecology of the rhizosphere. Nat Rev Microbiol 11:789–799. https://doi.org/10.1038/nrmicro3109

Plett JM, Kemppainen M, Kale SD, Kohler A, League V, Brun A, Pardo AG, Martin F (2011) A secreted effector protein of Laccaria bicolor is required for symbiosis development. Curr Biol 21:1197–1203. https://doi.org/10.1016/j.cub.2011.05.033

Pocasangre L, Sikora RA, Vilich V, Schuster R-P (2000) Survey of banana endophytic fungi from Central America and screening for biological control of the burrowing nematode (*Radopholus similis*). INFOMUSA 9:3. https://doi.org/10.17660/ActaHortic.2000.531.47

Poveda J (2021) Beneficial effects of microbial volatile organic compounds (MVOCs) in plants. Appl Soil Ecol 168:104118. https://doi.org/10.1016/j.apsoil.2021.104118

Poveda J, Baptista P (2021) Filamentous fungi as biocontrol agents in olive (*Olea europaea* L.) diseases: mycorrhizal and endophytic fungi. Crop Prot 146:105672. https://doi.org/10.1016/j.cropro.2021.105672

Priti V, Ramesha BT, Singh S, Ravikanrth G, Ganeshiah GN, Suryanarayanan TS, Uma Shaanker R (2009) How promising are endophytic fungi as alternative sources of plant secondary metabolites? Curr Sci 97(4):477–478

Proenca Barros FA, Rodrigues-Filho E (2005) Four spiroquinazoline alkaloids from *Eupenicillium* sp. isolated as an endophytic fungus from leaves of *Murraya paniculata* (Rutaceae). Biochem Syst Ecol 33:257. https://doi.org/10.1016/j.bse.2004.09.002

Puri SC, Nazir A, Chawla R et al (2006) The endophytic fungus *Trametes hirsuta* as a novel alternative source of podophyllotoxin and related aryl tetralin lignans. J Biotechnol 122:494–510. https://doi.org/10.1016/j.jbiotec.2005.10.015

Qawasmeh A, Obied HK, Raman A, Wheatley W (2012) Influence of fungal endophyte infection on phenolic content and antioxidant activity in grasses: interaction between *Lolium perenne* and different strains of *Neotyphodium lolii*. J Agric Food Chem 60(13):3381–3388. https://doi.org/10.1021/jf204105k

Rabiey M, Hailey LE, Roy SR, Grenz K, Al-Zadjali MAS, Barret GA, Jackson RW (2019) Endophytes vs. tree pathogens and pests: can they be used as biological control agents to improve tree health? Eur J Plant Pathol 155:711–729. https://doi.org/10.1007/s10658-019-01814-y

Ran X, Zhang G, Li S, Wang J (2017) Characterization and antitumor activity of camptothecin from endophytic fungus *Fusarium solani* isolated from *Camptotheca acuminate*. Afr Health Sci 17(2):566–574. https://doi.org/10.4314/ahs.v17i2.34

Rana KL, Kour D, Sheikh I, Dhiman A, Yadav N, Yadav AN, Rastegari AA, Singh K, Saxena AK (2019) Endophytic fungi: biodiversity, ecological significance, and potential industrial applications. In: Yadav AN et al (eds) Recent advancement in white biotechnology through fungi, fungal biology. https://doi.org/10.1007/978-3-030-10480-1_1

Redman RS, Kim YO, Woodward CJ, Greer C, Espino L, Doty SL, Rodriguez RJ (2011) Increased fitness of Rice plants to abiotic stress via habitat adapted symbiosis: a strategy for mitigating impacts of climate change. PLoS One 6:e14823. https://doi.org/10.1371/journal.pone.0014823

Reshma J, Vinaya C, Linu M (2019) Agricultural applications of endophytic microfora. In: Seed endophytes, pp 385–403. https://doi.org/10.1007/978-3-030-10504-4_18

Rodriguez R, Redman RJ (2008) More than 400 million years of evolution and some plants still can't make it on their own: plant stress tolerance via fungal symbiosis. J Exp Bot 59(5):1109–1114. https://doi.org/10.1093/jxb/erm342

Rodriguez RJ, White JF Jr, Arnold AE, Redman RS (2009) Fungal endophytes: diversity and functional roles. New Phytol 182(2):314–330. https://doi.org/10.1111/j.1469-8137.2009.02773.x

Roopa G, Madhusudhan MC, Sunil KCR, Lisa N, Calvin R, Poornima R, Zeinab N, Kini KR, Prakash SH, Geetha N (2015) Identification of Taxol-producing endophytic fungi isolated from *Salacia oblonga* through genomic mining approach. J Genet Eng Biotechnol 13(2):119–127. https://doi.org/10.1016/j.jgeb.2015.09.002

Rosenblueth M, Martı'nez-Romero E (2006) Bacterial endophytes and their interactions with hosts. Mol Plant-Microbe Interact 19:827–837. https://doi.org/10.1094/MPMI-19-0827

Rosier A, Bishnoi U, Lakshmanan V, Sherrier DJ, Bais HP (2016) A perspective on inter-kingdom signaling in plant-beneficial microbe interactions. Plant Mol Biol 90(6):537–548. https://doi.org/10.1007/s11103-016-0433-3

Rouxel T, Grandaubert J, Hane JK et al (2011) Effector diversification within compartments of the *Leptosphaeria maculans* genome affected by repeat induced point mutations. Nat Commun 2(1):1–10. https://doi.org/10.1038/ncomms1189

Rudgers JA, Koslow JM, Clay K (2004) Endophytic fungi alter relationships between diversity and ecosystem properties. Ecol Lett 7(1):42–51. https://doi.org/10.1046/j.1461-0248.2003.00543.x

Rudrappa T, Czymmek KJ, Pare PW, Bais HP (2008) Root-secreted malic acid recruits beneficial soil bacteria. Plant Physiol 148:1547–1556. https://doi.org/10.1104/pp.108.127613

Sahoo S, Sarangi S, Kerry RG (2017) Chapter 19: Bioprospecting of endophytes for agricultural and environmental sustainability. In: Patra JK (ed) Microbial biotechnology, pp 429–458. https://doi.org/10.1007/978-981-10-6847-8_19

Saikkonen K, Faeth SH, Helander M, Sullivan TJ (1998) Fungal endophytes: a continuum of interactions with host plants. Annu Rev Ecol Syst 29(1):319–343. https://doi.org/10.1146/annurev.ecolsys.29.1.319

Saikkonen K, Gundel PE, Helander M (2013) Chemical ecology mediated by fungal endophytes in grasses. J Chem Ecol 39:962–968. https://doi.org/10.1007/s10886-013-0310-3

Saini P, Verma A, Tiwari H, Mishra V, Gautam V (2024) Omics-based approaches in studying fungal endophytes and their associated secondary metabolites. In: Endophytic fungi the hidden sustainable jewels for the pharmaceutical and agricultural industries. Springer, Nature Switzerland, pp 227–224. https://doi.org/10.1007/978-3-031-49112-2_10

Salminen SO, Richmond DS, Grewal SK, Grewal PS (2005) Influence of temperature on alkaloid levels and fall armyworm performance in endophytic tall fescue and perennial ryegrass. Entomol Exp Appl 115:417. https://doi.org/10.1111/j.1570-7458.2005.00303.x

Santangelo JS, Turley NE, Johnson MTJ (2015) Fungal endophytes of *Festuca rubra* increase in frequency following long-term exclusion of rabbits. Botany 93(4):233–241. https://doi.org/10.1139/cjb-2014-0187

Sasse J, Martinoia E, Northen T (2017) Feed your friends: do plant exudates shape the root microbiome? Trends Plant Sci 23(1):25–41. https://doi.org/10.1016/j.tplants.2017.09.003

Saunders M, Kohn LM (2009) Evidence for alteration of fungal endophyte community assembly by host defense compounds. New Phytol 182(1):229–238. https://doi.org/10.1111/j.1469-8137.2008.02746.x

Schardl CL, Liu J-S, White JF Jr, Finkel RA, An Z, Siegel MR (1991) Molecular phylogenetic relationships of nonpathogenic grass mycosymbionts and clavicipitaceous plant pathogens. Plant Syst Evol 178(1):27–41. https://doi.org/10.1007/BF00937980

Schardl CL, Florea S, Pan J, Nagabhyru P, Bec S, Calie PJ (2013a) The epichloae: alkaloid diversity and roles in symbiosis with grasses. Curr Opin Plant Biol 16:480–488. https://doi.org/10.1016/j.pbi.2013.06.012

Schardl CL, Young CA, Hesse U et al (2013b) Plant-symbiotic fungi as chemical engineers: multi-genome analysis of the Clavicipitaceae reveals dynamics of alkaloid loci. PLoS Genet 9(2):e1003323. https://doi.org/10.1371/journal.pgen.1003323

Schmeda-Hirschmann G, Hormazabal E, Astudillo L, Rodriguez J, Theoduloz C (2005) Secondary metabolites from endophytic fungi isolated from the Chilean gymnosperm *Prumnopitys andina* (Lleuque). World J Microbiol Biotechnol 21(1):27–32. https://doi.org/10.1007/s11274-004-1552-6

Schouten A (2016) Mechanisms involved in nematode control by endophytic fungi. In: Leach JE, Lindow S (eds) Annual review of phytopathology, vol 54. Palo Alto, pp 121–142. https://doi.org/10.1146/annurev-phyto-080615-100114

Schueffler A, Anke T (2014) Fungal natural products in research and development. Nat Prod Rep 31:1425–1448. https://doi.org/10.1039/c4np00060a

Schulz B, Boyle C, Draeger S, Römmert A-K, Krohn K (2002) Endophytic fungi: a source of novel biologically active secondary metabolites. Mycol Res 106(9):996–1004. https://doi.org/10.1017/S0953756202006342

Selim KA, Nagia MM, Ghwas DEE (2017) Endophytic fungi are multifunctional biosynthesizers: ecological role an chemical diversity. In: Endophytic fungi: diversity, characterization and biocontrol. Nova Publishers, New York, pp 39–92

Shaffique S, Khan MA, Wani SH, Pande A, Imran M, Kang S-M, Rahim W, Khan SA, Bhatta D, Kwon E-H, Lee I-J (2022) A review on the role of endophytes and plant growth promoting rhizobacteria in mitigating heat stress in plants. Microorganisms 10:1286. https://doi.org/10.3390/microorganisms10071286

Sharma RR, Singh D, Singh R (2009) Biological control of postharvest diseases of fruits and vegetables by microbial antagonists: a review. Biol Control 50:205–221. https://doi.org/10.1016/j.biocontrol.2009.05.001

Shekhawat K, Saad MM, Sheikh A, Mariappan K, Al-Mahmoudi H, Abdulhakim F, Eida AA, Jalal R, Masmoudi K, Hirt H (2021) Root endophyte induced plant thermotolerance by constitutive chromatin modification at heat stress memory gene loci. EMBO Rep 22(3):e51049. https://doi.org/10.15252/embr.202051049

Shen L, Jiao RH, Ye YH, Wang XT, Dr XC, Dr Song YC, Prof Zhu HL, Prof Tan RX (2006a) Absolute configuration of new cytotoxic and other bioactive trichothecene macrolides. Chem Eur J 12:5596. https://doi.org/10.1002/chem.200600084

Shen L, Ye Y-H, Wang X-T, Zhu H-L, Xu C, Song Y-C, Li H, Tan R-X (2006b) Structure and total synthesis of aspernigerin: a novel cytotoxic endophyte metabolite. Chem Eur J 12:4393–4396. https://doi.org/10.1002/chem.200501423

Shen M, Zhao D-K, Qiao Q, Liu L, Wang J-L, Cao G-H, Li T, Zhao Z-W (2015a) Identification of glutathione S-transferase (GST) genes from a dark septate endophytic fungus (*Exophiala pisciphila*) and their expression patterns under varied metals stress. PLoS One 10(4):e0123418. https://doi.org/10.1371/journal.pone.0123418

Shen W, Mao H, Huang Q, Dong J (2015b) Benzenediol lactones: a class of fungal metabolites with diverse structural features and biological activities. Eur J Med Chem 97:747–777. https://doi.org/10.1016/j.ejmech.2014.11.067

Shidore T, Dinse T, Öhrlein J, Becker A, Reinhold-Hurek B (2012) Transcriptomic analysis of responses to exudates reveal genes required for rhizosphere competence of the endophyte *Azoarcus* sp. strain BH72. Environ Microbiol 14:2775–2787. https://doi.org/10.1111/j.1462-2920.2012.02777.x

Shiono Y, Kikuchi M, Koseki T, Murayama T, Kwon E, Aburai N, Kimura K-i (2011) Isopimarane diterpene glycosides, isolated from endophytic fungus *Paraconiothyrium* sp. MY-42. Phytochemistry 72(11–12):1400–1405. https://doi.org/10.1016/j.phytochem.2011.04.016

Shweta S, Zuehlke S, Ramesha BT, Priti V, Kumar PM, Ravikanth G, Spiteller M, Vasudeva R, Uma Shanker R (2010) Endophytic fungal strains of *Fusarium solani*, from *Apodytes dimidiata* E. Mey. ex Arn (Icacinaceae) produce camptothecin, 10-hydroxycamptothecin and 9-methoxy camptothecin. Phytochemistry 71:117–122. https://doi.org/10.1016/j.phytochem.2009.09.030

Shweta S, Gurumurthy BR, Ravikanth G, Ramanan US, Shivanna MB (2013) Endophytic fungi from *Miquelia dentata* Bedd., produce the anticancer alkaloid, camptothecine. Phytomedicine 20(3–4):337–342. https://doi.org/10.1016/j.phymed.2012.11.015

Siegel MR, Latch GCM, Bush LP, Fannin FF, Rowan DD, Tapper BA, Bacon CW, Johnson MC (1990) Fungal endophyte-infected grasses: alkaloid accumulation and aphid response. Chem Ecol 16:3301. https://doi.org/10.1007/BF00982100

Silva GH, Teles HL, Trevisan HC, da Bolzani SV, Young MCM, Pfenning LH, Eberlin MN, Haddad R, Costa-Neto CM, Araújo AR (2005) New bioactive metabolites produced by *Phomopsis cas-*

siae, an endophytic fungus in *Cassia spectabilis*. J Braz Chem Soc 16:1463–1466. https://doi.org/10.1590/S0103-50532005000800029

Slama HB, Chenari Bouket A, Alenezi FN, Khardani A, Luptakova L, Vallat A, Oszako T, Rateb ME, Belbahri L (2021) Olive mill and olive pomace evaporation pond's by-products: toxic level determination and role of indigenous microbiota in toxicity alleviation. Appl Sci 11:5131. https://doi.org/10.3390/app11115131

Soliman SSM, Greenwood JS, Bombarely A, Mueller LA, Tsao R, Mosser DD, Raizada MN (2015) An endophyte constructs fungicide-containing extracellular barriers for its host plant. Curr Biol 25(19):2570–2576. https://doi.org/10.1016/j.cub.2015.08.027

Spiteller P (2015) Chemical ecology of fungi. Nat Prod Rep 32(7):971–993. https://doi.org/10.1039/C4NP00166D

Šraj-Kržič N, Pongrac P, Klemenc M, Kladnick A, Regvar M, Gaberščik A (2006) Mycorrhizal colonisation in plants from intermittent aquatic habitats. Aquat Bot 85:331–336. https://doi.org/10.1016/j.aquabot.2006.07.001

Staniek A, Woerdenbag HJ, Kayser O (2008) Endophytes: exploiting biodiversity for the improvement of natural product-based drug discovery. J Plant Interact 3:75–93. https://doi.org/10.1080/17429140801886293

Stearns JC, Woody OZ, McConkey BJ, Glick BR (2012) Effects of bacterial ACC deaminase on *Brassica napus* gene expression. Mol Plant-Microbe Interact 25:668–676. https://doi.org/10.1094/MPMI-08-11-0213

Stierle A, Strobel G, Stierle D (1993) Taxol and taxane production by *Taxomyces andreanae*, an endophytic fungus of Pacifc yew. Science 260(5105):214–216. https://doi.org/10.1126/science.8097061

Stone JK, Polishook JD, White JF (2004) Endophytic fungi. In: Foster M, Bills G (eds) Biodiversity of fungi. Elsevier/Academic, Burlington, pp 241–270

Strobel GA, Morrison SI, Cassella M (2003) Methods for protection of plants from Oomyocyte pathogens by use of *Serratia marcescens* and isolates. US Patent Appl US2003/0049230 A1

Suryanarayanan TS, Senthilarasu G, Muruganandam V (2000) Endophytic fungi from *Cuscuta reflexa* and its host plants. Fungal Divers 4:117–123

Suryanarayanan TS, Johnson JA (2014) Fungal Endosymbionts of Macroalgae: Need for Enquiries into Diversity and Technological Potential. Journal of Oceanography and Marine Research 2:1. https://doi.org/10.4172/2332-2632.1000119

Sziderics A, Rasche F, Trognitz F, Sessitsch A, Wilhelm E (2007) Bacterial endophytes contribute to abiotic stress adaptation in pepper plants (Capsicum annuum L.). Can J Microbiol 53:1195–1202. https://doi.org/10.1139/W07-082

Tan RX, Zou WX (2001) Endophytes: a rich source of functional metabolites. Nat Prod Rep 18(4):448–459. https://doi.org/10.1039/b100918o

Tawfike AF, Romli M, Clements C, Abbott G, Young L, Schumacher M, Diederich M, Farag M, Edreda-Ebel R (2019) Isolation of anticancer and anti-trypanosome secondary metabolites from the endophytic fungus *Aspergillus flocculus* via bioactivity guided isolation and MS based metabolomics. J Chromatogr B 1106–1107:71–83. https://doi.org/10.1016/j.jchromb.2018.12.032

Teiten MH, Mack F, Debbab A, Aly AH, Dicato M, Proksch P, Diederich M (2013) Anticancer effect of altersolanol A, a metabolite produced by the endophytic fungus *Stemphylium globuliferum*, mediated by its pro-apoptotic and anti-invasive potential via the inhibition of NF-κB activity. Bioorg Med Chem 21(13):3850–3858. https://doi.org/10.1016/j.bmc.2013.04.024

Tejesvi MV, Nalini MS, Basavanna M, Prakash SH, Kini KR, Shetty SH, Subbiah V (2007) New hopes from endophytic fungal secondary metabolites. Bol Soc Quím Méx 1(1):19–26

Terhonen E, Sipari N, Asiegbu FO (2016) Inhibition of phytopathogens by fungal root endophytes of Norway spruce. Biol Control 99:53–63. https://doi.org/10.1016/j.biocontrol.2016.04.006

Thirunavukkarasu N, Suryanarayanan TS, Murali TS, Ravishankar JP, Gummadi SN (2011) L-asparaginase from marine derived fungal endophytes of seaweeds. Mycosphere 2(2):147–155

Tian H, Li XP, Zhao J, Gao HW, Xu QM, Wang JW (2021) Biotransformation of artemisinic acid to bioactive derivatives by endophytic *Penicillium oxalicum* B4 from *Artemisia annua* L. Phytochemistry 185:112682. https://doi.org/10.1016/j.phytochem.2021.112682

Tian Y, Fu X, Zhang G, Zhang R, Kang Z, Gao K, Mendgen K (2022) Mechanisms in growth promoting of cucumber by the endophytic fungus *Chaetomium globosum* strain ND35. J Fungi (Basel) 8(2). https://doi.org/10.3390/jof8020180

Tisserant E, Malbrei ML, Kuo A, Martin F (2013) Genome of an arbuscular mycorrhizal fungus provides insight into the oldest plant symbiosis. Proc Natl Acad Sci USA 110:20117–20122. https://doi.org/10.1073/pnas.1313452110

Tochhawng L, Mishra VK, Passari AK, Singh BP (2019) Endophytic fungi: role in dye decolorization. In: Singh BP (ed) Advances in endophytic fungal research, fungal biology. Springer Nature Switzerland AG. https://doi.org/10.1007/978-3-030-03589-1_1

Toghueo RMK, Boyom FF (2019) Endophytes from ethno-pharmacological plants: sources of novel antioxidants-a systematic review. Biocatal Agric Biotechnol 22:101430. https://doi.org/10.1016/j.bcab.2019.101430

Tohge T, Watanabe M, Hoefgen R, Fernie AR (2013) Shikimate and phenylalanine biosynthesis in the green lineage. Front Plant Sci 4:62. https://doi.org/10.3389/fpls.2013.00062

Toyomasu T, Kagahara T, Okada K, Koga J, Hasegawa M, Mitsuhashi W, Sassa T, Yamane H (2008) Diterpene phytoalexins are biosynthesized in and exuded from the roots of Rice seedlings. Biosci Biotechnol Biochem 72:562–567. https://doi.org/10.1271/bbb.70677

Turbat A, Rakk D, Vigneshwari A, Kocsube S, Thu H, Szepesi A, Bakacsy, Skrbic BD, Jigjiddorj E-A, Vagvolgyi C, Szekers A (2020) Characterization of the plant growth-promoting activities of endophytic fungi isolated from *Sophora favescens*. Microorganisms 8(5):683. https://doi.org/10.3390/microorganisms8050683

Van der Ent S, Van Wees SCM, Pieterse CMJ (2009) Jasmonate signaling in plant interactions with resistance-inducing beneficial microbes. Phytochemistry 70(13–14):1581–1588. https://doi.org/10.1016/j.phytochem.2009.06.009

van Kan JAL (2006) Licensed to kill: the lifestyle of a necrotrophic plant pathogen. Trends Plant Sci 11(5):247–253. https://doi.org/10.1016/j.tplants.2006.03.005

Verma SK, Gond SK, Mishra A, Sharma VK, Kumar J, Singh DK, Kumar A, Kharwar RN (2017) Fungal endophytes representing diverse habitats and their. In: Role in plant protection developments in fungal biology and applied mycology. Springer Nature Singapore, pp 135–157. https://doi.org/10.1007/978-981-10-4768-8_9

Victor C, Moses O, Eze A, Festus O, Charles E (2020) Isolation, identification, and evaluation of biological activities of *Daldinia eschscholtzii*, an endophytic fungus isolated from the leaves of *Musa paradisiaca*. GSC Biol Pharm Sci 12(1):216–228. https://doi.org/10.30574/gscbps.2020.12.1.0225

Waghunde RR, Shelake RM, Pramanik D, Suthar H, Pandey P, Kim J-Y (2021) 2-Application of cereal and vegetable endophytes in plant health management. In: Microbiomes and plant health. Elsevier, pp 29–51. https://doi.org/10.1016/B978-0-12-819715-8.00002-1

Wang Y, Dai C-C (2011) Endophytes: a potential resource for biosynthesis, biotransformation, and biodegradation. Ann Microbiol 61(2):207–215. https://doi.org/10.1007/s13213-010-0120-6

Wang JW, Zhang Z, Tan RX (2001) Stimulation of artemisinin production in *Artemisia annua* hairy roots by the elicitor from the endophytic *Colletotrichum* sp. Biotechnol Lett 23:857–860. https://doi.org/10.1023/A:1010535001943

Wang J, Huang Y, Fang M, Zhang Y, Zheng Z, Zhao Y, Su W (2002) Brefeldin A, a cytotoxin produced by *Paecilomyces* sp. and *aspergillus clavatus* isolated from *Taxus mairei* and *Torreya grandis*. FEMS Immunol Med Microbiol 34(1):51–57. https://doi.org/10.1111/j.1574-695X.2002.tb00602.x

Wang FW, Ye YH, Ding H, Chen Y-X, Tan R-X, Song Y-C (2010) Benzophenones from *Guignardia* sp. IFB-E028, an endophyte on *Hopea hainanensis*. Chem Biodivers 7(1):216–220. https://doi.org/10.1002/cbdv.200800353

Wang Y, Xu L, Ren W, Zhao D, Zhu Y, Wu X (2012) Bioactive metabolites from *Chaetomium globosum* L18, an endophytic fungus in the medicinal plant *Curcuma wenyujin*. Phytomedicine 19(3–4):364–368. https://doi.org/10.1016/j.phymed.2011.10.011

Wang XY, Qin JH, Chen W, Zhou Y, Ren A, Gao Y (2016) Pathogen resistant advantage of endophyte-infected over endophyte-free *Leymus chinensis* is strengthened by pre-drought treatment. Eur J Plant Pathol 144:477–486. https://doi.org/10.1007/s10658-015-0788-3

Wang L, Ren L, Li C, Gao C, Liu X, Wang M, Luo Y (2019) Effects of endophytic fungi diversity in different coniferous species on the colonization of *Sirex noctilio* (hymenoptera: Siricidae). Sci Rep 9(1):1–11. https://doi.org/10.1038/s41598-019-41419-3

Wani ZA, Ashraf N, Mohiuddin T, Riyaz-Ul-Hassan S (2015) Plant-endophyte symbiosis, an ecological perspective. Appl Microbiol Biotechnol 99:2955–2965. https://doi.org/10.1007/s00253-015-6487-3

Weyens N, van der Lelie D, Taghavi S, Vangronsveld J (2009) Phytoremediation: plant–endophyte partnerships take the challenge. Curr Opin Biotechnol 20(2):248–254. https://doi.org/10.1016/j.copbio.2009.02.012

White JF Jr, Torres MS (2010) Is plant endophyte-mediated defensive mutualism the result of oxidative stress protection? Physiol Plant 138(4):440–446. https://doi.org/10.1111/j.1399-3054.2009.01332.x

Wu ZC, Li DL, Chen YC, Xang W-M (2010) A new isofuranonaphthalenone and benzopyrans from the endophytic fungus *Nodulisporium* sp. A4 from *Aquilaria sinensis*. Helv Chim Acta 93(5):920–924. https://doi.org/10.1002/hlca.200900307

Xiao J, Lin LB, Hu JY, Duan D-Z, Shi W, Zhang Q, Han W-B, Wang L, Wang X-L (2018) Pestalustaines A and B, unprecedented sesquiterpene and coumarin derivatives from endophytic fungus *Pestalotiopsis adusta*. Tetrahedron Lett 59(18):1772–1775. https://doi.org/10.1016/j.tetlet.2018.03.078

Xie W, Hao Z, Yu M, Wu Z, Zhao A, Li J, Zhang X, Chen B (2019) Improved phosphorus nutrition by arbuscular mycorrhizal symbiosis as a key factor facilitating glycyrrhizin and liquiritin accumulation in *Glycyrrhiza uralensis*. Plant Soil 439(1–2):243–257. https://doi.org/10.1007/s11104-018-3861-9

Yadav G, Meena M (2021) Bioprospecting of endophytes in medicinal plants of Thar Desert: An attractive resource for biopharmaceuticals. Biotechnol Rep 30:e00629. https://doi.org/10.1016/j.btre.2021.e00629

Yadav G, Meena M (2022) Biological control of plant diseases by endophytes. In: Shah MP, Deka D (eds) Endophytic association: what, why and how (Volume: developments in applied microbiology and biotechnology). Academic/Elsevier, pp 119–135. https://doi.org/10.1016/B978-0-323-91245-7.00007-9

Yadav V, Kumar M, Deep DK, Kumar H, Sharma R, Tripati T, Tuteja N, Saxena AK, Johri AK (2010) A phosphate transporter from the root endophytic fungus *Piriformospora indica* plays a role in phosphate transport to the host plant. J Biol Chem 285(34):26532–26544. https://doi.org/10.1074/jbc.M110.111021

Yadava PK, Bhuyan SK, Bandyopadhyay P (2015) Extraction of proteins for two-dimensional gel electrophoresis and proteomic analysis from an endophytic fungus. Protocol Exchange. https://doi.org/10.1038/protex.2015.084

Yajima A (2014) Recent progress in the chemistry and chemical biology of microbial signaling molecules: quorum-sensing pheromones and microbial hormones. Tetrahedron Lett 55:2773–2780. https://doi.org/10.1016/j.tetlet.2014.03.051

Yan HJ, Li XM, Li CS, Wang BG (2012) Alkaloid and anthraquinone derivatives produced by the marine-derived endophytic fungus *Eurotium rubrum*. Helv Chim Acta 95(1):163–168. https://doi.org/10.1002/hlca.201100255

Yan RM, Zhang ZB, Wang Y, Yang H, Zeng Q, Zhu D (2014) Efficient strategy for maintaining and enhancing the huperzine A production of *Shiraia* sp. Slf14 through inducer elicitation. J Ind Microbiol Biotechnol 41:1175–1179. https://doi.org/10.1007/s10295-014-1461-0

Yang RY, Li CY, Lin YC, Peng G-t, Shi Z-g, Zhou S-n (2006) Lactones from a brown alga endophytic fungus (No. ZZF36) from the South China Sea and their antimicrobial activities. Bioorganic Med Chem Lett 16:4205–4208. https://doi.org/10.1016/j.bmcl.2006.05.081

Yang B, Wang X-M, Ma H-Y, Jia Y, Li X, Dai C-C (2014) Effects of the fungal endophyte *Phomopsis liquidambari* on nitrogen uptake and metabolism in rice. Plant Growth Regul 73:165–179. https://doi.org/10.1007/s10725-013-9878-4

Yang H-R, Yuan J, Liu L-H, Zhang W, Chen F, Dai C-C (2019) Endophytic *Pseudomonas fluorescens* induced sesquiterpenoid accumulation mediated by gibberellic acid and jasmonic acid in *Atractylodes macrocephala* Koidz plantlets. Plant Cell Tissue Organ Cult 138(3):445–457. https://doi.org/10.1007/s11240-019-01640-4

Ye YH, Zhu HL, Song YC, Liu YJ, Tan RX (2005) Structural revision of aspernigrin A, reisolated from *Cladosporium herbarum* IFB-E002. J Nat Prod 68:1106–1108. https://doi.org/10.1021/np050059p

Yuan J, Zhang W, Sun K, Tang MJ, Chen PX, Li X, Dai CC (2019) Comparative transcriptomics and proteomics of *Atractylodes lancea* in response to endophytic fungus *Gilmaniella* sp. AL12 reveals regulation in plant metabolism. Front Microbiol 28(10):1208. https://doi.org/10.3389/fmicb.2019.01208

Yue Q, Miller CJ, White JF, Richardson MD (2000) Isolation and characterization of fungal inhibitors from *Epichloe festucae*. J Agric Food Chem 48:4687–4692. https://doi.org/10.1021/jf990685q

Yung L, Sirguey C, Azou-Barré A, Blaudez D (2021) Natural fungal endophytes from *Noccaea caerulescens* mediate neutral to positive effects on plant biomass, mineral nutrition and Zn phytoextraction. Front Microbiol 12. https://doi.org/10.3389/fmicb.2021.689367

Zakria M, Njoloma J, Saeki Y, Akao S (2007) Colonization and nitrogen-fixing ability of *Herbaspirillum* sp. strain B501 *gfp*1 and assessment of its growth-promoting ability in cultivated rice. Microbes Environ 22:197–206. https://doi.org/10.1264/jsme2.22.197

Zhang JH, Guo SX, Yang JS, Xiao PG (2002) Acta Bot Sin 44:1239. (in Chinese)

Zhang HW, Song YC, Huang WY, Chen JR, Tan RX (2005) Four 6*H*-Dibenzo[*b*,*d*]pyran-6-one derivatives produced by the endophyte *Cephalosporium acremonium* IFB-E007. Helv Chim Acta 88:2861. https://doi.org/10.1002/hlca.200590228

Zhang HW, Song YC, Tan RX (2006) Biology and chemistry of endophytes. Nat Prod Rep 23(5):753–771. https://doi.org/10.1039/b609472b

Zhang CL, Zheng BQ, Lao JP, Mao LJ, Chen SY, Kubicek CP, Lin FC (2008) Clavatol and patulin formation as the antagonistic principle of *Aspergillus clavatonanicus*, an endophytic fungus of *Taxus mairei*. Appl Microbiol Biotechnol 78(5):833–840. https://doi.org/10.1007/s00253-008-1371-z

Zhang XM, Tang DX, Li QQ, Wang BY, Xu ZH, Li WJ, Yu H (2021) Complex microbial communities inhabiting natural *Cordyceps militaris* and the habitat soil and their predicted functions. Antonie van Leeuwenhoek. Int J Gen Mol Microbiol 114:465–477. https://doi.org/10.1007/s10482-021-01534-6

Zhao XM, Wang ZQ, Shu SH, Wang WJ, Xu HJ, Ahn YJ, Wang M et al (2013) Ethanol and methanol can improve Huperzine A production from endophytic *Colletotrichum gloeosporioides* ES026. PLoS One. https://doi.org/10.1371/journal.pone.0061777

Zheng CJ, Shao CL, Guo ZY, Chen JF, Deng DS, Yang KL, Chen YY, Fu XM, Shi ZG, Lin YC, Wang CY (2012) Bioactive hydroanthraquinones and anthraquinone dimers from a soft coralderived *Alternaria* sp. fungus. J Nat Prod 75(2):189–197. https://doi.org/10.1021/np200766d

Zhou J, Diao X, Wang T, Chen G, Lin Q, Tang X, Xu J (2018) Phylogenetic diversity and antioxidant activities of culturable fungal endophytes associated with the mangrove species *Rhizophora stylosa* and *R. mucronata* in the South China Sea. PLoS One 13(6):e0197359. https://doi.org/10.1371/journal.pone.0197359

Zhu XG, Long SP, Ort DR (2010) Improving photosynthetic efficiency for greater yield. Annu Rev Plant Biol 61:235–261. https://doi.org/10.1146/annurev-arplant-042809-112206

Zikmundova M, Drandarov K, Bigler L, Hesse M, Werner C (2002) Biotransformation of 2-benzoxazolinone and 2-hydroxy-1, 4-benzoxazin-3-one by endophytic fungi isolated from

Aphelandra tetragona. Appl Environ Microbiol 68(10):4863–4870. https://doi.org/10.1128/AEM.68.10.4863-4870.2002

Zipfel C, Oldroyd GED (2017) Plant signaling in symbiosis and immunity. Nature 543:328–336. https://doi.org/10.1038/nature22009

Zolla G, Badri DV, Bakker MG, Manter DK, Vivanco JM et al (2013) Soil microbiomes vary in their ability to confer drought tolerance to Arabidopsis. Appl Soil Ecol 68:1–9. https://doi.org/10.1016/j.apsoil.2013.03.007

Zou WX, Meng JC, Lu H, Chen GX, Shi GX, Zhang TY, Tan R (2000) Metabolites of *Colletotrichum gloeosporioides*, an endophytic fungus in *Artemisia mongolica*. J Nat Prod 63:1529. https://doi.org/10.1021/np000204t

Zuccaro A, Lahrmann U, Güldener U, Langen G, Pfiffi S, Biedenkopf D, Wong P, Samans B, Grimm C, Basiewicz M, Murat C, Martin F, Kogel KH (2011) Endophytic life strategies decoded by genome and transcriptome analyses of the mutualistic root symbiont *Piriformospora indica*. PLoS Pathog 7(10):e1002290. https://doi.org/10.1371/journal.ppat.1002290

Fungal Endophytes as Biostimulants to Facilitate Viable Bioprospecting of High-Value Secondary Metabolites

14

Priyanshu Kumar, Bhaskarjyoti Saikia Sonowal, Aina Thakur, Kanika Thakur, Ankush Yadav, Rupesh Bhardwaj, Pritee Kumari, Laishram Amarjit Singh, Mukesh Meena, Prashant Swapnil, Sumit Kumar, and Avinash Marwal

Abstract

Endophytes coexist symbiotically with a wide range of plant species, engaging in complex interactions with both their hosts and the environment. These various microorganisms, which have been found across the plant kingdom, include bacteria, fungi, viruses, and protists. They generate a variety of bioactive compounds that promote plant development and improve tolerance to environmental challenges. They, therefore, have great promise for use as biocontrol and biofertilizers in sustainable agriculture. Furthermore, the production of different industrially useful enzymes and macromolecules by endophytes is being investigated, as is

P. Kumar · B. S. Sonowal · A. Thakur · K. Thakur · A. Yadav · R. Bhardwaj · P. Kumari
L. A. Singh · P. Swapnil (✉)
School of Basic Science, Department of Botany, Central University of Punjab, Bhatinda, Punjab, India
e-mail: prashant.swapnil@cup.edu.in

M. Meena (✉)
Laboratory of Phytopathology and Microbial Biotechnology, Department of Botany, Mohanlal Sukhadia University, Udaipur, Rajasthan, India
e-mail: drmukeshmeena321@mlsu.ac.in

S. Kumar
Department of Plant Pathology, College of Agriculture, Rajmata Vijayaraje Scindia Krishi Vishwa Vidyalaya, Gwalior, Madhya Pradesh, India

A. Marwal
Department of Biotechnology, Vigyan Bhawan-Block B, New Campus, Mohanlal Sukhadia University, Udaipur, Rajasthan, India

M. Meena et al. (eds.), *Metabolic Sustainability of Endophytes*, Microorganisms for Sustainability 52, https://doi.org/10.1007/978-981-96-4004-1_14

their potential for bioremediation and biofuel production. Strategies to improve agricultural productivity and reduce crop yield loss from diverse biotic and abiotic stresses are required for the growing global food demand brought on by a continually expanding human population. Abiotic stressors like salt, heat, drought, heavy metals, and biotic stresses like those brought on by pests, nematodes, and phytopathogens can seriously affect agricultural output and food security. Conventional techniques that rely on artificial chemicals have led to the creation of resistant microorganisms, ecological harm, and unintentional effects on nontarget creatures. Endophytic fungi can be used as a biostimulant to facilitate viable bioprospecting. With an emphasis on bioprospecting, this study intends to explore the current approaches and instruments to clarify the biosynthesis routes of secondary metabolites in endophytic fungi. In addition to discussing future approaches, it will emphasize the reported triumphs and drawbacks of endophytic fungi. Examining the genetic potential of producer strains may become possible if the biosynthetic genes encoding the many secondary compounds extracted from endophytic fungus are identified. This research may yield new and more reasonably priced medications and food additives by improving the synthesis of secondary metabolites through metabolic engineering and discovering novel secondary metabolites.

Keywords

Biostimulants · Endophytic fungi · Secondary metabolites · Drought · Abiotic stress · Biotic stress · Microbiome · Phyto-stimulation · Reactive oxygen species (ROS)

14.1 Introduction

An interesting class of microorganisms known as endophytes has attracted a lot of attention lately because of their special and frequently advantageous relationships with their host plants. The endophytes are microorganisms, mostly bacteria and fungi, that reside inside the tissues of plants without harming the host or producing symptoms of disease (Rodriguez et al. 2008). Numerous studies have been conducted on the diversity and potential uses of these endophytic bacteria, which have been found to live in a variety of plant species, from deserts to tropical rainforests (Strobel 2002; Yadav and Meena 2021). The term endophyte was first defined by Bary (1866) as any organism that grows within plant tissues. Endophytes were defined as asymptomatic microorganisms living inside plants (Carroll 1986) and microorganisms that inhabit internal plant tissues and organs at some point in their lives without causing apparent harm to the host plant (Petrini 1991). The origins of the endophyte study can be found in the 1970s, when researchers started to realize how common these microbes were in plant tissues (Rodriguez et al. 2008). Endophytes differ from other microorganisms associated with plants in a variety of ways. Endophytes, for example, can invade plant tissues' intracellular and intercellular regions, forming close bonds with their hosts (Esmaeel et al. 2018).

Moreover, endophytes are known to generate a wide range of secondary metabolites that may be useful in the sectors of biotechnology, medicine, and agriculture, in addition to facilitating host–microbe interactions (Yadav and Meena 2021). Developments in several scientific fields, including molecular biology, bioinformatics, and microbiology, have made the discovery of endophytes easier. Scholars have utilized an array of methodologies, including both culture-dependent and culture-independent approaches, to separate and describe these elusive microorganisms from various plant species (Deshmukh et al. 2018). The potential of endophytes to be used in a variety of applications, including the synthesis of bioactive chemicals, the improvement of plant stress tolerance, and the creation of new antimicrobial agents, has become more apparent as our understanding of them expands (Yadav and Meena 2021).

14.1.1 Life History Strategies of Endophytic Symbiosis

Different life history strategies highlight the complex and dynamic nature of endophytic symbiosis and its impact on plant health and productivity. Latent endophytes are beneficial microorganisms that live asymptomatic inside plant tissues until they are prompted by external stimuli to manifest their advantageous characteristics. The health and stress tolerance of plants are greatly influenced by these endophytes. Latent endophytic fungi (EFs), for instance, have been demonstrated in tests to improve plant development and provide tolerance to a variety of environmental challenges (Carroll 1986). Quiescent endophytes are those that lie dormant within plant tissues until environmental circumstances favor their growth and activity. Once they become active, these endophytes can support defense systems and enhance general plant health. Studies have shown that dormant endophytic fungi can generate bioactive substances that shield plants from infections and increase their resistance to abiotic stressors (Eid et al. 2019). Mutualistic endophytes are those that form symbiotic relationships with their host plants to offer benefits like improved nutrition absorption, resilience to stress, and defense against infections (Petrini 1991).

In some situations, certain endophytes can behave in a pathogenic manner and harm the host plant. These endophytes are known as pathogenic endophytes. Plant growth and development can be hampered by these pathogenic endophytes, which can result in disease symptoms and decreased agricultural output. Pathogenic endophytic fungi that can harm agricultural systems and cause diseases in plants have been found through research. Some beneficial endophytes, such as *Neotyphodium* and *Epichloë*, give host plants improved nutrient uptake (Schardl et al. 2004), resistance to drought, and anti-herbivore defense. These endophytes improve the host plant's general development and resilience, in addition to shielding it from insects and herbivores. For instance, several *Stagonospora* species (Ernst et al. 2003), *Acremonium strictum*, and *Piriformospora indica* (Waller et al. 2005), all show comparable defensive and growth-promoting properties in their host plants. Protective Endophytes: By generating antibiotics and antifungal substances, endophytes can shield host plants from diseases and pests (Zhang et al. 2006). This defense strengthens the plant's defense mechanisms and lessens the intensity of pathogen attacks. Fungal endophyte colonization in plants can also provide enhanced defense against plant nematodes, although the precise processes behind this antagonistic relationship remain unclear (Bushby et al. 2016). Endophytes differ from other microorganisms associated with plants in a variety of ways. Among these are their capacities to invade plant tissues' intracellular and intercellular regions, enabling them to form close bonds with their hosts (Strobel 2002; Yadav and Meena 2021). Endophytes, also known as the unsung heroes of the plant kingdom, are helpful microorganisms such as fungi and bacteria that coexist peacefully in plant tissues. As microscopic powerhouses that support plant resistance and productivity, these microbial friends are essential for improving plant health and growth. Among the main ways endophytes help plants is by serving as tiny distributors and manufacturers of fertilizers. One such example is the extraordinary capacity of some endophytic bacteria to fix atmospheric nitrogen and transform it into a form that is easily assimilated by the plant. Some are better at solubilizing phosphates, which makes them more available and easier for plant roots to absorb (Dick 2012). Plant growth and development are directly increased by this improved nutrient uptake. Not only do endophytes supply vital nutrients, but they also function as expert pharmacists, creating a wide range of advantageous chemicals. These include phytohormones that are vital for encouraging root elongation, boosting nutrient uptake, and accelerating overall plant growth, such as gibberellins and indole-3-acetic acid.

Endophytes enhance a plant's ability to produce hormones by augmenting the plant's hormone production. In addition to promoting growth, endophytes play a crucial role in plant defense by serving as protectors against various threats. They accomplish this by either inducing the plant's immune system, which primes it for a quicker and more potent defense response, or by creating a plethora of bioactive compounds that can directly inhibit the growth of pathogens (Fadiji and Babalola 2020). The maintenance of plant health and the achievement of maximum yields depend on this improved disease resistance. Endophytes have amazing abilities that provide agriculture with a sustainable answer. Endophytes lessen the need for artificial fertilizers and pesticides by improving nutrient intake, encouraging growth, and increasing disease resistance. This helps the environment and opens the door for an environmental-friendly and sustainable method of farming. As our understanding of the complex realm of plant–microbe interactions deepens, the potential for creating novel approaches to boost plant health, increase agricultural output, and guarantee a more sustainable future for food supply through the use of endophytes becomes increasingly apparent.

14.1.2 Factors Influencing Endophyte Species Richness and Community Assembly

14.1.2.1 At Different Stages of Invasion

Transport, colonization, establishment, and spread are the stages at which different factors influence the richness of endophytes. The initial introduction of endophytes to new settings might be influenced by the way invasive plant species are transported. Endophyte dissemination can occur through vectors, such as wind, water, animals, and human activity (White Jr et al. 2000). The initial pool of endophyte species available for colonization in the new habitat can be ascertained from the source population of the invasive plant. Distinct endophyte communities may be carried by several source populations (Arnold et al. 2003). The degree to which endophytes are unique to a given host plant can have an impact on the success of endophyte colonization. Certain endophytes might favor particular host species, which could have an impact on how well they establish themselves inside the invading plant (Carroll 1988). The variety and makeup of the endophyte community can be impacted by competition between endophyte species for colonization sites within the host plant. The endophytic species that dominate the colonization process may vary depending on competitive interactions (Saikkonen et al. 1998). The host plant's compatibility with the endophytes affects the endophytes' ability to establish themselves within the invading plant host. Successful establishment can be aided by mutualistic relationships, but it can also be hampered by adversarial ones (Rodriguez et al. 2009). Many variables, including pH, moisture content, and soil nutrients, might influence how endophyte communities form in invasive plants (Porras-Alfaro and Bayman 2011). A number of dispersion processes, including seed dissemination, root exudates, and insect vectors, can aid in the spread of endophytes both inside the invasive plant and into neighboring habitats (Vandenkoornhuyse et al. 2015). The expansion of endophyte communities can also be influenced by interactions between native endophytes in the new habitat and among endophyte species within the invasive plant. Endophyte dissemination may be influenced by interactions that are competitive or supportive (Hardoim et al. 2015).

14.1.2.2 At the Transport Stage

The initial stage of an endophyte invasion involves the transportation of a fungal propagule to the host plant. A few elements that impact transportation are air currents, water, soil, and vectors like insects or animals. Endophytic fungi can propagate through wind, rain, or contaminated soil to possible host plants. The step of transport occurs when endophytes relocate from their native habitat to a new setting. Endophytes may be present in seeds and plant materials utilized in agriculture. For instance, endophytic fungi such as *Neotyphodium* may be brought to new areas through international commerce in wheat seeds, which may affect the local plant communities (Schardl et al. 2004). Transported soil for agricultural use may contain endophytes. This soil can affect local soil microbiomes and plant life by introducing unique endophytes to new regions. Endophytes can be present in plants that are carried for forestry or landscaping in nurseries. When nursery-grown trees are planted, for instance, endophytic fungus may colonize new areas. Endophytes can also be transported by wood and logs. Endophytic fungal infestations in wood products can disperse these organisms into unfamiliar habitats, affecting nearby ecosystems (Hulcr and Dunn 2011). Hikers and tourists may unintentionally spread endophytes by wearing contaminated clothes or shoes. Hiking boots, for instance, can spread endophytes from one forest to another. Plant keepsakes that travelers may carry with them have the potential to house endophytes and spread them to other areas. Wind can spread fungal spores over great distances. The endophytic fungus *Cladosporium*, for example, has the ability to spread via the air and colonize new plant hosts that are far from their original position. Endophytes can colonize new geographical regions using atmospheric currents that transfer spores across continents. Fungal spores can spread from plant surfaces into the surrounding area through rain. For instance, endophytes from leaf litter can be transferred to surrounding plants by rain (Levetin 2016). Endophytes can colonize new ecosystems due to their ability to be transported by river and ocean currents. For example, ocean currents can disperse marine endophytes. Endophytes can be carried by animals on their feathers or fur. For example, endophytic fungi can be transported by mammals and birds, exposing them to new plant hosts as they travel across various settings. Animals can consume endophytes and then excrete them, which helps the organisms propagate. The herbivores which graze on plants that have endophytic fungi can spread these fungi through their excrement).

14.1.2.3 At the Colonization Stage

The process of endophytes establishing themselves in a new host or environment is called colonization. Biochemical signaling between the endophyte and host plant is necessary for endophytes to be able to colonize a plant. Auxins, which are produced by endophytes, are signaling molecules that can improve colonization by encouraging plant development and changing the shape of host tissue (Alabid et al. 2019), Recognition mechanisms in host plants have the power to help or impede endophyte colonization. The plants have the ability to identify particular MAMPs, or microbe-associated molecular patterns, which have the potential to elicit immunological reactions and impact the outcome of colonization (Teixeira et al. 2019). Endophyte colonization can be inhibited by antimicrobial chemicals produced by host plants, such as phytoalexins. To establish themselves inside the host, endophytes need to either suppress or resist these chemicals. Certain plants have a defense mechanism called systemic acquired resistance (SAR), which offers persistent protection against a wide range of diseases, including endophytes. For effective colonization, endophytes need a certain range of temperatures and humidity. Endophytic fungi, for example, may prefer colder climates in temperate zones, yet can flourish in warm, humid conditions in tropical woods (Arnold et al. 2003). Extreme temperatures or protracted droughts are examples of harsh environmental circumstances that can prevent endophyte colonization by lowering microbial activity and decreasing the number of suitable host tissues available (Saikkonen et al. 1998). The pH and texture of the soil might affect endophyte colonization. For instance, some endophytic fungi may thrive in acidic soils, while other species may be supported by alkaline soils. Endophyte colonization may be impacted by the availability of nutrients in the soil, such as phosphorus and nitrogen. Rich in nutrients, soils can encourage the growth of plants and create an ideal habitat for the establishment of endophytes (Carroll 1988).

14.1.2.4 At the Establishment Stage

During the establishment phase, endophytes proliferate and keep their populations steady inside the host. Endophytes fight for nutrients and space inside the host with other microbes, such as bacteria and harmful fungi. The variety and makeup of endophyte communities can be influenced by competitive interactions (Arnold et al. 2003). To reduce competition, endophytes may divide resources among the host plants. As an illustration, various endophyte species might use different metabolic pathways or colonize particular organs (Saikkonen et al. 1998). Certain endophytes enhance their establishment by forming mutualistic connections with the host plant or other microbes. To help endophytes survive and thrive, for example, mycorrhizal fungi can promote nutrient exchange (Smith and Read 2008). Endophytes can interact antagonistically with diseases to impede their growth and safeguard the host plant. Endophytic bacteria, for instance, can create antimicrobial substances that inhibit pathogenic fungi (Pieterse et al. 2012). Abiotic stressors, including drought, nutrient shortages, or excessive salt, may make host plants more vulnerable to opportunistic endophyte colonization. By generating substances that reduce stress, endophytes can assist plants in adjusting to external stimuli (Rodriguez et al. 2009). Additionally, plants that are under biotic stress due to herbivory or pathogen attack may be more vulnerable to endophyte colonization. By generating defensive chemicals or fostering systemic resistance, endophytes can increase a plant's tolerance to biotic stresses. Endophytes have the capacity to confer systemic resistance on host plants, hence augmenting their defense mechanisms against herbivores and pathogens. This may foster an atmosphere that is conducive to the establishment of endophytes (Pieterse et al. 2012). Through localized defense responses, such as the generation of reactive oxygen species (ROS) or the activation of defense-related genes, host plants can control the populations of endophytes (Jones and Dangl 2006).

14.1.2.5 At the Spread Stage

The term "spread" describes how endophyte populations grow and multiply inside their host or among other hosts. Endophytes with high spore production have a greater ability to spread. As an example, endophytic fungi such as *Epichloë* species generate a large amount of conidia and ascospores, which aid in their spread (Rodriguez et al. 2009). Endophytes can colonize new hosts and surroundings through the dispersal of spores through a variety of processes, such as wind, water, and animal vectors. Certain endophytes, like hyphal extensions in fungi, are spread by vegetative development. This enables them to spread throughout the host by colonizing surrounding plants or plant tissues (Saikkonen et al. 1998). Clonal propagation is another way that endophytes can proliferate. Aphids, beetles, and mites are examples of insects that can serve as vectors, moving endophytes from one plant to another. The spatial distribution of endophytes is influenced by the feeding and mobility habits of these insects (Clay and Holah 1999). Endophytes are dispersed by plant-eating birds and mammals, either by their droppings or by the spores they carry on their bodies. Along the animals' corridors of migration, this may result in the colonization of new host plants (Verma et al. 2018).

14.2 Fungal Endophytes and Host Physiology

Endophytic fungi (EFs), also known as fungal endophytes, are communities of fungi that form associations either intercellularly or intracellularly with host plant tissues while simultaneously providing benefits to the host and gaining benefits from it (Alam et al. 2021). These fungi establish a mutualistic or symbiotic relationship with the host plant and are categorized into two groups based on this relationship. The Clavicipitaceous group, classified as class I endophytes, inhabits cool-season grasses and infects the ovules of their host plants, subsequently colonizing the rhizomes and branches. In contrast, non-Clavicipitaceous endophytes, which include classes II, III, and IV, enhance drought resistance in their host plants by secreting bioactive compounds. The non-Clavicipitaceous fungus, also known as the non-balansiaceous group, includes class II, III, and IV endophytes. These endophytes, distinguished by their widespread variety and range, are rarely linked with grass hosts. They live in plant tissues in a latent or quiescent state, producing no visible damage to the host. However, when the host plant's chemical composition changes owing to injuries or environmental challenges, these endophytes can penetrate the plant intracellularly via roots, shoots, or rhizomes (Rodriguez et al. 2008). Class II endophytes occupy host tissues via roots, branches, and rhizomes, whereas class III endophytes colonize only through shoots and class IV endophytes exclusively through roots. Class II endophytes transmit both vertically and horizontally, but class III and IV endophytes transmit only horizontally (Rodriguez et al. 2009).

Endosymbiotic microorganisms, most often bacteria and fungus, create mutually beneficial connections with their plant hosts, increasing the overall fitness of the host. These organisms are considered nonpathogenic because they may complete part or all of their life cycle inside the host without having disorderly symptoms (Nair and Padmavathy 2014). The endophytic community is divided into two subgroups: obligatory endophytes and facultative endophytes. Obligate endophytes depend largely on plant metabolism for survival. In contrast, facultative endophytes spend a part of their life cycle outside of the host and are frequently found in the surrounding soil environment, interacting with other plants (Hardoim et al. 2008). Endophytic fungi contribute to plant–microbe interactions in various ways, such as nutrient solubilization in the rhizosphere, promoting plant growth, serving as biocontrol agents against pests and pathogens, inducing systemic resistance to biotic and abiotic stresses, and synthesizing secondary metabolites (Poveda et al. 2020a, b). Fungal secondary metabolites can alter the morphology and physiology of the host plant (Alam et al. 2021). To form symbiotic associations with host plants, endophytic fungi enzymatically degrade the plant's cell wall, facilitating tissue colonization. This involves the production of enzymes like cellulase, laccase, pectinase, and xylanase.

14.2.1 Molecular Interaction of Endophytic Fungi with Host Plants

Endophytic fungi (EFs) engage in mutualistic, commensality, and pathogenic interactions with host plants. In mutualistic symbiosis, both the plants and EFs benefit, with EFs enhancing the plant's stress tolerance and nutrient acquisition. In commensality or latent pathogenic relationships, EFs interact with host plants without necessarily benefiting the plant's physiology and remain harmless without inducing disease symptoms. However, under adverse environmental conditions, these latent fungi can become active, manifesting pathogenic symptoms and potentially leading to the host plant's destruction. Plants in the rhizosphere interact with various microorganisms through pattern-recognition receptors (PRRs), which recognize microbial or pathogen-associated molecular patterns. These PRRs initiate the first layer of plant innate immunity by activating signaling pathways that can inhibit endophytic proliferation, including miRNA-mediated pathways that are part of the plant's defense mechanisms (Plett and Martin 2018).

Extracellular detection of microbial- or pathogen-associated molecular patterns (MAMPs/PAMPs) or damage-associated molecular patterns (DAMPs) begins the first layer of innate immunity via activated defenses, also known as pattern-triggered immunity (Tang et al. 2017). Pathogen-triggered immunity (PTI) is initiated when plant cell wall components, such as oligonucleotides, cellodextrins, cutin monomers, and small peptides, are degraded upon cellular damage. These compounds, released under stress conditions, produce endogenous signals termed DAMPs (damage-associated molecular patterns), which are recognized by pattern-recognition receptors (PRRs). Upon microbial infection, activation of the BIK1 effector kinase within the PRR complex occurs, leading to an elevation in cytosolic calcium (Ca^{2+}) levels mediated by cyclic nucleotide-gated channels (CNGCs). This rise in Ca^{2+} acts as an important signal for initiating pathogen-associated molecular pattern (PAMP) responses applied in plant immunity during PTI. Pathogen-triggered immunity (PTI) and effector-triggered immunity (ETI) are plant defense mechanisms against microbial invasion. PTI responds initially by altering the ionic balance, increasing cytosolic Ca^{2+} and apoplastic reactive oxygen species (ROS), activating the MAPK pathway, and producing nitric oxide (NO). These methods guide the release of phytohormones such as ethylene (ET), jasmonic acid (JA), and salicylic acid (SA). PTI also induces stomatal closure and callose deposition and triggers transcriptional and metabolic changes to improve plant defense. ETI, activated upon successful pathogen infection via effector-triggered susceptibility (ETS), amplifies these responses for robust defense (Nguyen et al. 2021).

In the context of fungal endophytes, many plants have chitin-specific receptors (PR-3) on their surfaces that detect chitin oligomers from fungal cell walls. This recognition activates the plant's defensive systems. Furthermore, plant pattern-triggered immunological signaling pathways such as MIN7 and CAD1 play important roles in controlling the number and support of endophytes in the soil environment

(Chen et al. 2020). These systems help to improve host survival in microorganism-rich environments. In addition to various cellular pathways, plants use extracellular vesicles (EVs) for transporting signaling lipids, proteins, RNA, and metabolites between cells. Research indicates that EVs play a vital role in plant stress responses, exhibiting anti-pathogenic activity by delivering host-derived small interfering RNAs and microRNAs. These molecules can silence fungal genes and dampen stress responses in pathogens, potentially facilitating trans kingdom RNA interference. Conversely, fungal endophytes respond by producing chitin deacetylases upon invading host plants. These enzymes modify chitosan oligomers by deacetylating them, enabling the modified oligomers to evade detection by the host's chitin-specific receptors (Cord Landwehr et al. 2016).

Endophytic bacteria form biofilms to adhere to host plant tissues and communicate, aiding in evading the host's defense systems. They have developed strategies to counteract oxidative stress induced by the host, including reactive oxygen species (ROS) such as superoxide (O_2^-), hydrogen peroxide (H_2O_2), and hydroxyl radicals (OH). Endophytes utilize antioxidant enzymes such as superoxide dismutases (SODs), catalases (CATs), peroxidases (PODs), alkyl hydroperoxide reductases (AhpCs), and glutathione-S-transferases (GSTs) to detoxify ROS, thereby protecting themselves from oxidative damage. This adaptation enables endophytes to establish and sustain symbiotic relationships with host plants. Endophytic microorganisms include genes-producing compounds that inhibit microbe-associated molecular patterns (MAMPs). These MAMPs are frequently recognized by plants, triggering immunological responses. Endophytic communities can reduce MAMP-induced immune responses in host plants while impacting other physiological processes. They activate dormant gene clusters, resulting in the synthesis of new secondary metabolites (SMs), which aid in the defense against harmful microbes. These SMs perform an important function in strengthening the host plant's defense systems against infection. Endophytic fungi also activate the plant's immune system via epigenetic pathways, which promote the manufacture of SMs. This connection causes physiological and phenotypic changes in the host plant, which are regulated by genetic variables, biotic and abiotic stressors, and the activity of the microbiome (Alam et al. 2021). Gene clustering plays a critical role in organizing the synthesis of secondary metabolites, ensuring the preservation of genetic order. Factors such as horizontal gene transfer (HGT), transcription factors, the presence of effector molecules, and gene clustering mechanisms collectively regulate the expression of these genes involved in secondary metabolite production.

14.2.2 Gene Clustering and Horizontal Gene Transfer

The phenomenon known as "gene clustering" describes the grouping of genes that are involved in the manufacture of secondary metabolites, frequently in the vicinity of fungal chromosomes' telomeric or dynamic regions. These gene clusters can serve two primary purposes (Chiang et al. 2015; Henke et al. 2016): either they work together to generate complicated secondary metabolites, or they are situated close to one another for distinct secondary metabolite pathways. Linked genes in higher plant genomes can create transcriptional clusters through the helix-loop-helix domain. As a result, cis-regulatory elements are brought closer together by DNA loops, which raise the local concentration of transcription factors at transcription initiation sites. Protein products may be lost as a result of disturbances in this control, and toxic intermediates may build up in metabolic pathways. Secondary metabolic gene cluster expression depends on chromatin-level gene regulation (Osbourn 2010). Narrow-domain transcription factors (NDTFs) and broad-domain transcription factors (BDTFs) control clustered genes involved in secondary metabolite production. While BDTFs can regulate genes situated elsewhere, NDTFs can only directly regulate genes that are clustered. For instance, by binding to a particular palindromic sequence, the Zn2Cys6 transcription factor non-ribosomal peptide synthetase (NRPS) AflR controls the expression of aflatoxin and sterigmatocystin gene clusters in *Aspergillus* species. Conversely, BDTFs are stimuli-responsive proteins that can affect gene expression in a variety of biological processes, including those that are mediated by ethylene-forming signals (Camehl et al. 2010). Studies have examined the molecular pathways that connect NDTFs and BDTFs, investigating how BDTFs interpret and communicate developmental or environmental cues to NDTFs via chromatin and histone modification pathways or biochemical cascades such as acetylation, phosphorylation, and methylation. These mechanisms trigger the activation of silent genes linked to particular secondary metabolites that are necessary for various growth stages, cellular metabolisms, or environmental circumstances (Fig. 14.1). Furthermore, environmental variables and microbiome activity impact plant phenotypes, resulting in genetic alterations in the endophytic fungi that are connected with the plant (Jia et al. 2016).

Horizontal gene transfer (HGT) is a process where genetic elements are exchanged between separate lineages without involving sexual reproduction. This evolutionary mechanism is particularly notable for effector proteins, which can exploit it to invade, degrade, and manipulate host organisms (Soanes and Richards 2014). HGT facilitates the acquisition of complete metabolic pathways by transferring metabolic clusters from one organism to another. The initiation of gene expression in the recipient organism following HGT is still being studied (Slot and Rokas 2011). For instance, successful functional transfer has been reported for a gene cluster responsible for biosynthesizing sterigmatocystin, a toxic secondary metabolite, from *Aspergillus* species to *Podospora anserina*. Additionally, the presence of specific metabolites like djalonensone in various *Alternaria* fungi and aureonitol in *Chaetomium* sp., along with extracts from *Helichrysum aureonitens*, suggests the role of HGT or genetic recombination during the coevolution of hosts and endophytes. These findings underscore the development of genetic regulatory mechanisms for biosynthesizing secondary metabolites through HGT processes (Aly et al. 2010).

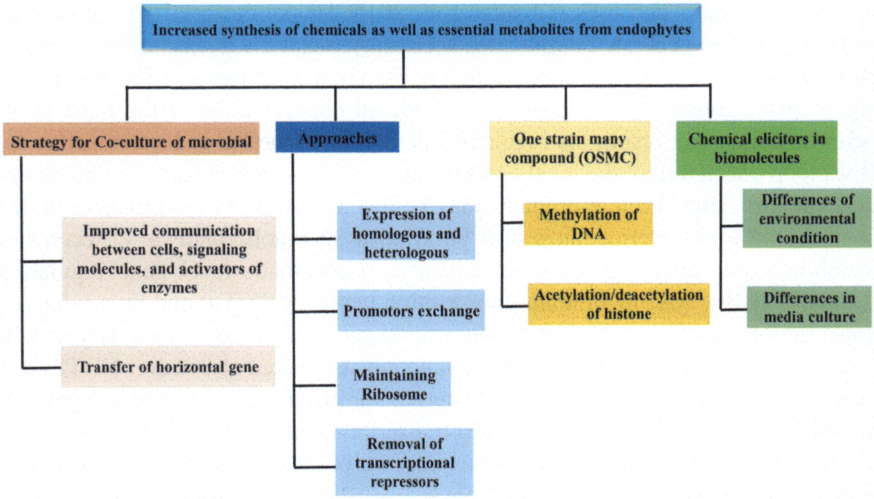

Fig. 14.1 Methods used to increase the synthesis of chemicals and essential metabolites from endophytic fungi (Chettri et al. 2023)

14.3 Types and Role of Secondary Metabolites

Endophytes make microbiomes in plants and contribute to plant development. Endophytic fungi live within plants belonging to Basidiomycota and Ascomycota. The secondary metabolites are of numerous types and vary in every plant. There is a signaling and metabolic interaction between the host plant and the endophytes. New bioactive substances are also synthesized during symbiotic relationships between plants and fungi (Zhao et al. 2010).

Secondary metabolites are organic compounds that are not directly involved in normal growth and development. The bioactive substance responds to various biotic and abiotic stresses. The production of secondary metabolites is a costly process, and the concentration of the production of secondary metabolites is based upon stress situations, while EF is a rich source of SMs. Endophytic fungi have the capacity to produce more bioactive compounds than plants (Jha et al. 2023). The major classes of secondary metabolites are terpenes, phenolics, and N- and S-containing compounds.

14.3.1 Polyketides

Polyketides are the most abundant and well-characterized type of secondary metabolite present in endophytic fungi. The presence of polyketides leads to fungi, which are most commonly and abundantly found in the mycobiome (Muhammad et al. 2024). The biosynthesis of polyketides is the main precursor of enzyme polyketide synthases. Endophytic fungi *Chaetomium globosum* isolated from *Ginkgo bilola* have antifungal activity through the metabolite chaetomugilin A, which acts as a defensive compound (Qin et al. 2009).

14.3.2 Terpenes

Terpenes are the second-largest class found in EFs, which are biosynthesized from the acetyl CoA or glycolytic intermediates by isopentenyl pyrophosphate (IPP), having five major subclasses built up from isoprene units produced by polyporales groups. Defense against pathogens and signaling, communication, pollination attractants, growth promoters, and drug synthesis are the main functions of EFs. They can be isolated from EFs associated with algae, mangroves, and medicinal plants and are generally volatile in nature (Galindo-Solís and Fernández 2022). Endophytic fungi produce diverse bioactive metabolites, including monoterpenes, sesquiterpenes, and diterpenes. Monoterpenes composed of two isoprene units can act as insecticidal, medicinal, and flavoring agents. Sesquiterpenes composed of three isoprene units act as anticancer agents and help in the production of phytohormones (ABA). In fungal endophytes, the most abundant terpene is sesquiterpenoids. They are mostly color and odoriferous compounds to attract organisms (Amirzakariya and Shakeri 2022). Diterpenes composed of four isoprene units act as anticancer agents and antioxidants. Various drugs are obtained from terpenes, which include paclitaxel, vinca alkaloids, fusidic acid, and so on. A tetracyclic diterpenoid paclitaxel, taxol, present in the Taxus trees, has anticancer properties and tumor suppressor activities. They are isolated from *T. brevifolia* (Stierle et al. 1993). It can also isolate various medicinal plants, like moringa, from *Cladosporium oxysporum.* Podophyllotoxin is a carcinogenic compound extracted from the *Podophyllum* roots. *Botryosphaeria laricina* CRS1 from the *Catharanthus roseus* synthesizes vincristine and vinblastine. These indole terpenoids are effective in breast and lung cancers. Camptothecin, with anticancer properties, is extracted from the bark of *Camptotheca acuminate*, commercially from *C. acuminate*, and *Nothapodytes nimmoniana* (Galindo-Solís and Fernández 2022) are isotopomerese inhibitors. An antibiotic, fusidium coccineum from EF *Acremonium pilosum* prevents infections caused by staphylococci (Galindo-Solís and Fernández 2022). Volatile organic compounds (VOCs) synthesized by *Trichoderma* T-51 fungi helps to promote the growth of seedling of *Arabidopsis thaliana* seedling (Galindo-Solís and Fernández 2022).

Steroids are derived from a terpenoid pathway having four fused carbon rings with many oxidation states because of the functional groups attached to the rings, so the structure and function are varied (Theodorakidou et al. 2018). Ergosterol, an antimicrobial compound, can be isolated from *Nodulisporium.* Steroids show moderate antimicrobial activity, so finding an effective drug or pesticide is a challenge (Yu et al. 2010).

14.3.3 Phenols

They are synthesized by the shikimic acid pathway, containing hydroxyl groups with aromatic rings, to make simple phenolic compounds. Phenols are abundantly derived from the endophytic fungi, which are classified into diversed groups. They play important role in the various physiological and defensive mechanisms. They provided protection from ultraviolet (UV) radiations. They help plants grow in poor soil and have defense properties (Singh et al. 2011). Lignins and phenylpropanoids are phenolic compounds having a benzene ring structure. A phenolic compound protects hosts from pathogen attack as it is produced in wounds and lesions and helps in wound healing and adaptability. Extraction of SMs is easily done by wounds and lesions on host plants (Yadav et al. 2020). Coumarin is a phenylpropanoid lactone with a benzopyrone structure that exhibits antioxidant and anti-inflammatory properties. These are generally aromatic compounds. Aflatoxin B1, a derivative of coumarin, has carcinogenic properties isolated from *Aspergillus* spp. Another bioactive compound is scopoletin, derived from *Scopulariopsis brevicaulis*, which has antioxidant and antimicrobial properties. Polyphenol's flavonoids produced by EF polyphenol moieties have a basic skeleton of flavan nucleus including flavones, flavonols, and anthocyanins. These compounds having various applications which are widely used in cosmetics and treating diseases such as Alzheimer's, cancer, and others as they are cost-effective components due to their antioxidant, antimicrobial, antibacterial, anticancer, anti-inflammatory effects (Ullah et al. 2020). An endophytic fungus *Pestalotiopsis adusta* produced two antibiotics pestalachloride A and B, having antifungal activity against plant pathogen (Yu et al. 2010).

14.3.4 Nitrogen-Containing Secondary Metabolites

Alkaloids contain nitrogen compounds, i.e., amines, amide, and indole derivatives. They have antimicrobial activity (Pavithra et al. 2012) Plants have ability to separate from the endophytic fungi. *Clavicipitaceous* fungi are associated with grass synthesize alkaloids which act as insecticidal and herbicide because of their toxicity. (Higginbotham et al. 2013). *Chaetoglobosin* from endophytic fungi *globosum* extracted by host plant *Ginkgo biloba* shows antibacterial activity and has a toxic nature to mammals. *Neonectria ramulariae* endophytic fungi living symbiotically in *Zea mays* synthesized antibiotics pyrrolidines A and B; they having the capacity to protect plant from harmful microbes like *Fusarium* and *Aspergillus flavus* (Yu et al. 2010). Another bioactive property of alkaloids helps to focus on the drug and medicinal use are glycosides, saponins, and other aliphatic compounds. Glycosides are sugar molecule with glycosidic bonds attached with aglycone. Saponins are glycosides consisting of a sugar unit attached to a steroid or triterpenoid which exhibit strong antimicrobial properties and helps to combat microbial pathogens.

14.4 Bioprospecting Strategies for High-Value Secondary Metabolites

14.4.1 Screening Methods for Bioactive Compounds

The beneficial qualities of crude extracts and bioactive substances from endophytic bacteria and fungi have been evaluated using a variety of methodologies. In most modern research, antioxidant capabilities are assessed using the DPPH (2,2-diphenyl-1-picrylhydrazyl) assay. When necessary, NMR (Nuclear magnetic resonance) is added to the GC-MS (Gas chromatography-mass spectrometry), LC-MS (Liquid chromatography-mass spectrometry), and HPLC (High-performance liquid chromatography) techniques used in bioactive component profiling. These methods have completely changed screening procedures, as have high-throughput next-generation sequencing and whole genome sequencing, as well as developments in informatics (Newman 2017). To measure the antibacterial and cytotoxic properties of extracts or compounds, the half-maximal inhibitory concentration (IC_{50}) and minimum inhibitory concentration (MIC) are frequently employed. MIC provides quantitative information on antibiotic effectiveness by measuring the lowest concentration (μg/mL) of an antibiotic that stops bacterial growth that can be seen. MIC offers vital information for choosing the right antibiotics or bioactive substances, improving the effectiveness of treatment and possibly reducing antibiotic resistance (De Azevedo et al. 2018; Michael et al. 2020). A commonly used indicator of medication efficacy, IC_{50} measures the quantity of an inhibitory substance needed to block a biological activity or component by 50%, such as enzymes, cells, or microorganisms. Still, while evaluating IC_{50}, factors like the maximum effect (Emax) or the dose–response curve slope are frequently disregarded (Aykul and Martinez-Hackert 2016). Metrics other than potency should be considered because a medication with a low IC50 may not always produce the desired results in patients, even when given the right dosage. Empirical evidence indicates that Emax is more appropriate at higher drug concentrations, while IC_{50} is better at moderate dosages (Dahlmann 2013). Therefore, finding effective medications to enhance human health management requires merging IC_{50} with other factors. These assays are well documented in the scientific literature and have been widely used in several investigations. These assays provide preliminary knowledge of the potency or activity of the examined substances, providing a framework for additional investigation and utilization (Rajivgandhi et al. 2016).

14.4.2 Distribution of Endophytes in the Tissues of the Host Plant

Endophytes are extensively dispersed throughout the tissues of their host plants, colonizing both aboveground and belowground organs like leaves, flowers, fruits, and seeds, as well as underground ones like roots (Tapfuma et al. 2019). Research has shown that, in comparison to other tissues, the phylloplane, which comprises leaf surfaces, contains a greater concentration of endophytes (Dhayanithy et al.

2019). The nutrient-rich environment of leaf tissues, which is rich in carbon, nitrogen, and phosphorus and supports a diverse microbial community, is probably the cause of this frequency. The host plant's genotype may also have an impact on the endophyte community makeup of the phylloplane (Katoch et al. 2014). Additionally, throughout the year, different amounts of environmental stimuli, including heat, humidity, dust, and human disturbances, are exposed to plant tissues at different times. Though it varies depending on the tissue, this exposure adds to the endophyte richness throughout the plant's life (Harrison and Griffin 2020). For example, the observed diversity of endophytes is influenced by the longer-lasting plant parts, such as bark and stems. Understanding the interactions between endophytes and various host plant tissues is still lacking, despite a wealth of research on endophytes from various sources (Katoch et al. 2017). Long-term coevolutionary constraints that promote a cooperative connection cause bioactive secondary metabolites produced by endophytes to resemble those found in their host plants frequently. This interaction allows for genetic exchange in systemic endophytes, which helps the endophytes to acquire distinctive features (Aboobaker et al. 2019). This coevolutionary pattern was demonstrated, for example, by a study on *Penicillium* sp. isolated from *Murraya paniculata*, which produced the same chemicals as its host plant. Numerous studies were unable to definitively confirm this notion (Rodriguez et al. 2009). Non-systemic endophytes, on the other hand, live in shorter, seasonal cohabitation periods during which they may only share physiological features that are impacted by shifting environmental factors. Because of this, certain endophytes might not produce bioactive substances that are on par with those produced by their host plants. Furthermore, changes in growth characteristics and nutritional composition of the media have a substantial impact on endophytic organisms' metabolic profiles (Mondal and Rai 2019). Different plant tissues have different architectural structures and nutritional compositions, which impact the tissue microbiome's makeup and, in turn, the creation of different bioactive chemicals.

14.4.3 Varieties and Investigation of Bacterial and Fungal Endophytes

The diversity of endophytic fungi found in plants is higher than that of their bacterial counterparts. Fungal endophytes have been investigated more thoroughly than bacterial endophytes, even though Hollis isolated and identified bacterial endophytes from disease-free potato crops as early as 1951 (Latha et al. 2019). In particular, ethnopharmacological plants have been extensively studied for advantageous endophytic bacteria, revealing various species. The paucity of studies on bacterial endophyte assemblages and their bioactive substances, on the other hand, points to a large pool of unidentified beneficial bacteria. The phyla Firmicutes, Proteobacteria, and Actinobacteria are predominant among the reported bacterial endophytes (Wu 2018; Golinska et al. 2015). Notable genera include *Enterobacter*, *Pantoea*, *Pseudomonas*, *and Streptomyces* (all Gram-negative), as well as *Bacillus*, *Microbispora*, *and Paenibacillus* (all Gram-positive). Similarly, many different heterotrophic fungi, mostly belonging to the phylum Ascomycota, colonize the plant species under investigation. *Aspergillus, Cladosporium, Colletotrichum, Curvularia, Diaporthe, Fusarium, Pestalotiopsis, Phoma, and Xylaria* are common genera that are known to produce bioactive chemicals (Palanichamy et al. 2019; Ran et al. 2017).

14.5 The Potential for Effective Secondary Metabolites Bioprospecting from Endophytic Fungi in the Future

Although there are still obstacles to overcome in the sector, a number of new technologies and instruments have the potential to further endophytic fungal bioprospecting initiatives. We will examine these advancements in three main areas in this last section: (1) isolation and cultivation procedures; (2) culture-independent methods; and (3) genome-editing approaches.

14.5.1 Modern Cultivation and Environmental Isolation

Because the endophytes lack the metabolites and cues that the host plant and other members of the microbial community offer, studying them under axenic experimental circumstances is extremely difficult. The majority of endophytic fungi are difficult to cultivate in experiments due to the absence of vital life-supporting components, which makes the isolation procedure more difficult. The ratio of cultivable to non-cultivable fungal endophytes is not currently well researched, although environmental DNA surveys suggest that species diversity is far higher than that of cultivated isolates (Wu et al. 2019). For instance, amplicon sequencing revealed nearly ten times more species in a study on the endophytic fungal community of *Dysphania ambrosioides* than could be isolated through cultivation (Parmar et al. 2018). It is essential to enhance isolation and cultivation protocols in order to expand the number of cultivable species and investigate their potential for biosynthesis. In addition to the technological developments covered in the upcoming sections, tighter cooperation with knowledgeable mycologists and bioprospector training could also aid in resolving the cultivability issue.

14.5.2 The Fungal One-Step Isolation Device (FIND) Is an Innovation in the Fields of Cultivation and Isolation

Rare fungi may now be isolated from both terrestrial and marine samples, thanks to the development of the Fungal One-Step Isolation Device (FIND) (Libor et al. 2019). The isolation chip (iChip), which was first developed to produce bacteria from environmental samples utilizing their naturally occurring nutrients and growth factors, is adapted by this technique. By using hundreds of tiny diffusion chambers in a growing device, each inoculated with a single cell (environmental isolates or created variations of a single strain), FIND and iChip both enable high-throughput isolation of microbes (Berdy et al. 2017). The growth device is subsequently reintroduced into its native habitat, such as by being immersed in seawater or buried in moist soil, in order to promote the diffusion of nutrients and signaling molecules from the surrounding milieu and among microorganisms within the separate chambers. This configuration captures a greater range of biological diversity and allows for the cultivation of new species without the need to optimize artificial growing conditions. A "rhizochip" was created to investigate bacteria in the rhizosphere of the canola plant, despite the fact that the application of this technology for plant-associated fungi has not yet been documented (Gurusinghe et al. 2019). Methods for attaching a FIND to a plant could be devised for the purpose of studying epiphytes, which are organisms that dwell on the surface of plants. It appears difficult to develop a similar mechanism for endophytes, as it would include cutting a hole in a leaf, stem, or bark. Improving the isolation and culture of endophytic fungus has the potential to significantly speed up the search for biosynthetic genes and secondary metabolites.

14.5.3 Microfluidic Devices

Microfluidic platforms have become an effective tool for assessing how plant cells react to various external stimuli, including pathogens and other abiotic or biotic variables like temperature, light, and nutrition (Jiang et al. 2014; Busch et al. 2012). Compared to traditional growing methods, these platforms provide a more precisely controlled environment at a higher throughput. Nevertheless, this method has not yet been widely applied to investigate biotic interactions between plants and their symbionts. Researchers at Argonne National Laboratory, part of the US Department of Energy, made a major discovery. To investigate the relationships between *Populus tremuloides* (aspen tree) and its bacterial root microorganisms, they created a Root Microbiome Interaction (RMI) chip (Noirot-Gros et al. 2020). The RMI chip's microfluidic configuration, combined with cutting-edge live image microscopy, enables real-time, direct visualization of dynamic interactions at the cellular level. However, for now, this technology is limited to observing roots, which is easy enough to do because roots can be immersed and continually perfused to watch biological reactions. There are additional difficulties when studying other plant components because the system has to support both light exposure and gas exchange. Because fungal and bacterial colonization is very different from one another, it is necessary to optimize microfluidic devices in order to research fungal endophytes. Although there has been a discussion of recent developments in microfluidic devices for fungus, we would like to draw attention to a specific chip design (Zhou et al. 2019). This preassembled microfluidic apparatus, which consists of border channels and chambers shaped like the letters "ORNL," was created to investigate binary interactions between different species. Examples include interactions between the fungus *Nicotiana attenuata* seedlings and the plant-growth-promoting bacteria *Mortierella elongata* NVP64+. The microbes were shown to have minimal influence on one another in these binary interactions, despite a small percentage of the bacterial population displaying unique and changed cell morphologies. Plant roots in the coculture of plants and fungi clearly showed signs of endophyte colonization (Millet et al. 2019). Because of their ease of use, these ORNL devices may find widespread usage in the study of endophytic fungi and their biosynthetic gene clusters (BGCs) in the presence of their host plants. This will enable high-spatial and temporal-resolution tracking of the dynamics of these organisms. It is possible to conceive a setting comparable to the one used to investigate plant development from a single protoplast in microfluidic devices, with the goal of examining the reciprocal influence of microbiota and plants (Sakai et al. 2019). This method could make secondary metabolite detection and downstream metabolomics analysis easier for the aerial sections of plants as opposed to the roots.

14.5.4 Progress in Culture-Independent Methods for Microbial Community Analysis

Natural product bioprospecting is moving away from conventional studies that concentrate on the biosynthetic potential of isolated organisms and towards more comprehensive methods. These include analyzing individual organisms without prior cultivation (single-cell omics) or analyzing microbial communities as a whole (meta-omics). Although meta-omics methodologies have been thoroughly reviewed in recent reviews, we will focus on advancements related to fungi (Van Der Hooft et al. 2020). As single-cell analysis of fungus is still in its infancy, we shall discuss more general viewpoints regarding its future directions.

14.6 Study of Fungal Communities via Meta-Omics

As sequencing prices have fallen and computer analysis pipelines have advanced, shotgun metagenomics has emerged as a critical tool for understanding microbial communities, especially for species that are not cultivable (Pérez-Cobas et al. 2020). Shotgun metagenomics can provide high-quality assemblies of complete chromosomes, entire genomes of microbial communities, or even biosynthetic gene clusters (BGCs) with enough sequencing depth (Youngblut et al. 2020; Sugimoto et al. 2019). However, a lot of the shotgun metagenomics initiatives that are underway now mostly concentrate on bacteria, ignoring fungal DNA in the process of preparing samples. This restriction limits the number of useful datasets that are available to researchers who are studying fungal BGCs (Ye et al. 2019). However, more recently, computer processes like Find Fungi have been developed with the purpose of identifying fungal genes or contigs in metagenomic datasets, especially those related to animal microbiomes. Furthermore, the performance of a number of tools that were initially created for more general use but may be used to identify bacterial and eukaryotic contigs in metagenomic samples has been compared. Notably, because of its strong performance in cross-validating hits and determining the last common ancestor of query sequences, an extension of the MMSeqs2 platform has attracted much attention in the field of environmental genomics (Segawa et al. 2020; Santos-Júnior et al. 2019). Contamination of reference genomes with incorrectly allocated contigs is a frequent problem in metagenomics studies. By creating a tool to remove these contaminants, the creators of MMSeqs2 have solved this problem and improved the accuracy and dependability of database entries. It is anticipated that these developments in reference databases and methods would greatly improve the study of both new and old metagenomic datasets, possibly revealing a multitude of fascinating fungal contigs for further investigation (Mirdita et al. 2020). Technologies such as Oxford Nanopore and Pacific Biosciences' long-read sequencing are beginning to show promise as metagenomic library tools. These technologies are excellent at retrieving functional genomic units, complete chromosomes, and entire genomes from bacteria and fungus while having greater error rates, particularly in Oxford Nanopore sequencing (McKenzie et al. 2020; Moss et al. 2020). More recently, real-time selective sequencing has made it possible to manipulate sequencing runs to enrich or deplete particular sequences. This strategy might make it easier to retrieve less abundant species in a microbiome by suppressing the sequencing of extremely abundant species (Payne et al. 2020). In addition to these methods of culture-independent genomics, meta-transcriptomics and meta-metabolomics provide a thorough understanding of microbial communities. They make it easier to analyze biochemical profiles and gene expression levels, giving information on the metabolic processes that occur over particular time periods. Because of the large volume of data collected, integrating metagenomics and metabolomics datasets presents issues that require standardizing and improving data structures for efficient tool interoperability (Van Der Hooft et al. 2020). There are not many studies using meta-omics to look into endophytic fungus yet. Nevertheless, a recent study that correlated the microbiome's diversity with metabolite profiles combined holobiont metabolomics with amplicon sequencing of endophytic fungi in distinct plant sections of *Ephedra sinica* (Zhang et al. 2020). Future research is anticipated to uncover new biosynthetic gene clusters (BGCs) and secondary metabolites due to the high-throughput nature of culture-independent approaches and their capacity to examine uncultivable organisms.

14.7 Single-Cell Omics for Investigating Environmental Microbes

Single-cell omics seek information from individual cells of animals or microbial cells within a population. This method works especially well for researching organisms that are hard to cultivate or that are rare in a community. Single-cell genomics is one of the most sophisticated omics methods available, capable of generating nearly whole genome assemblies of ambient bacteria (Dam et al. 2020; Pachiadaki et al. 2019). In single-cell genomics, cells are isolated by sorting them using flow cytometry, and then multiple displacement amplification and sequencing are performed. This approach has considerable promise for expanding the genomic research of unculturable organisms, even though full genome coverage is rarely accomplished. In contrast to shotgun metagenomics, single-cell methods appear to be able to retrieve a greater number of genomes from bacterial environmental samples (Woyke et al. 2017). Nevertheless, there are difficulties in using single-cell genomics with endophytic fungus, especially when preparing single cells. Spores, which are easily extracted from cultivable fungi, have been the subject of most single-cell genomics studies conducted on fungi. Alternative methodologies must be developed because typical single-cell genomic approaches are impractical for non-cultivable species, such as endophytic fungi (Montoliu-Nerin et al. 2020). The most promising approaches, according to recent developments in plant single-cell genomics, probably entail nucleus isolation and/or the creation of protoplasts by cell wall digestion, which requires the digestion of both fungal and host plant cells. Following the preparation of a cell or nuclei solution, sorting can be done by hand or via fluorescence-activated cell sorting (FACS). The number of collected cells or nuclei and the availability of appropriate markers for staining needed for FACS analysis are two considerations that influence the decision between manual sorting and FACS (Luo et al. 2020). Manual sorting is currently more common than FACS in the plant research community, and it may be useful when examining endophytic fungi. However, since flow cytometry has been successfully used to separate filamentous fungi, this technique may be used for high-throughput sorting of fungal endophytes. Multimodal single-cell omics techniques, which integrate genomic, transcriptomic, and proteomic data from each isolated cell, have been made possible by advancements in animal single-cell technologies (Ma et al. 2020; Stuart and Satija 2019). It is predicted that these workflows will include further physiological observations prior to carrying out these in-depth studies. Similar developments in the field of investigating environmental microbial isolates, such as endophytic fungi, are anticipated in the near future.

14.8 Genome Editing of Novel Fungal Isolates

A major difficulty in filamentous fungus is the experimental validation of biosynthetic gene clusters (BGCs) in their native hosts. Several methodologies for genome editing have been thoroughly reviewed in recent studies (Mei et al. 2019). Here, we highlight recent advancements in filamentous fungal genome editing using CRISPR/Cas9, which show potential for use in organisms lacking complex genetic tools. In order to produce the active Cas9/guide RNA complex in vivo, the *Cas9* gene and guide DNA were traditionally delivered by expression cassettes in CRISPR/Cas9 techniques in filamentous fungi (Abdallah et al. 2017). For newly isolated strains, it is frequently impossible to obtain appropriate DNA shuttles, selective markers, and functioning promoter and terminator sequences in the original host, all of which are necessary for this method. A different approach that involves preparing the Cas9-gRNA ribonucleoprotein complex in vitro and transforming it into fungal protoplasts has surfaced recently (Wang et al. 2018). Although this method has not been used on endophytic fungi yet, it can potentially improve bioprospecting efforts in these organisms with scarce molecular resources. In addition to enabling the creation of gene editing tools, CRISPR/Cas9 systems also aid in developing gene expression tools for biotechnological uses, such as activation and interference systems that control gene expression (Zheng et al. 2019). The technique known as CRISPR/Cas9 can streamline the process of experimentally validating BGCs, hasten the creation of expression systems for biotechnological procedures involving endophytic fungi, and create new opportunities for genetic modification inside these species.

14.9 Fungal Endophytes as Biostimulants

Initially, it was assumed that secondary metabolites were not essential for growth and were generally for plant defense, i.e., host plant resistance to pathogens. However, further identification and research concluded that secondary metabolites have a tremendous role in plant development (Fig. 14.2). The chemical ecological concept explains the origin of secondary metabolites for defense purposes, the attraction of pollinators, and some metabolic functions. Secondary metabolites produced as a result of symbiosis between endophytic fungi and their host plants help both the fungi and their hosts survive in harsh conditions, providing mutual benefits. However, in limited resource conditions, these fungi may compete with other organisms by producing allelopathic chemicals. These bioactive compounds, which are typically low-molecular-weight secondary metabolites, play a key role in their

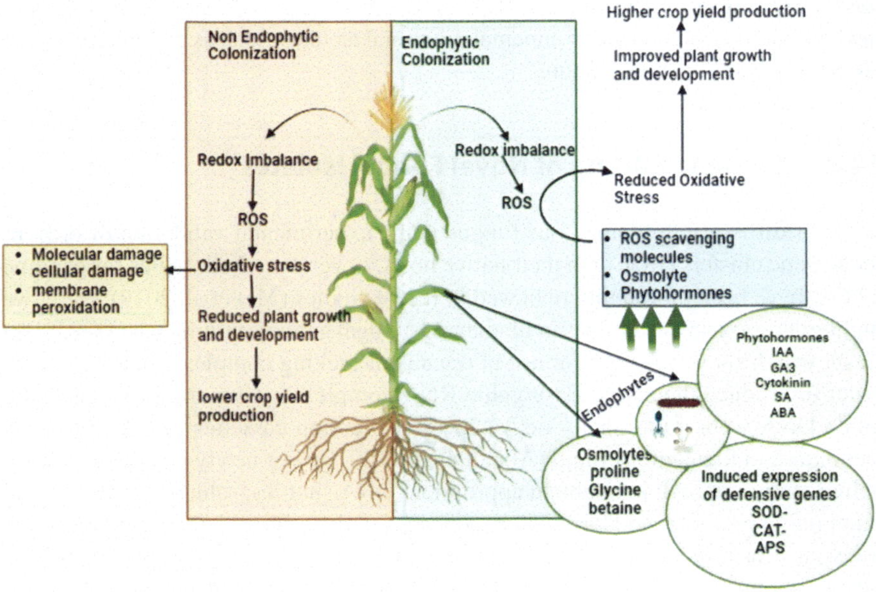

Fig. 14.2 Strategies incorporated by endophytes in agricultural plants to cope with the stress of salinity and drought (Verma et al. 2021)

adaptation and survival (Fox and Howlett 2008). A number of bioactive compounds are synthesized by the endophytic fungi, which are useful for various purposes including for human use. Drugs are one of the important compounds extracted from the EFs; beside that, there are multiple roles including the following.

14.9.1 Plant Protection

Fungus endophytes produce secondary metabolites that are somehow beneficial to plants. These also inhibit many phytopathogens. The importance of secondary metabolites produced by endophytic fungi against various abiotic and biotic stress is highlighted in Table 14.1. These bioactive substances are anticancer, antibiotics, insecticidal and produce allelopathic chemicals (Alam et al. 2021). Alkaloids prevent the plants from the insect and herbivores (Verma et al. 2009).

14.9.2 Ecological Interaction and Communication to Compete and Survive

Secondary metabolites help in competing with another organism and they can harm other organisms, helping the host organism to survive in adverse conditions. Plant pathogens are highly susceptible against these endophytic fungi secondary metabolites (Williams et al. 1989). Also in adverse conditions they can provide tolerance to plants. For example, in experiment observed that *Penicillium variabile* an EF produces *hydroperoxin oxylipin* (helping in decrease mycotoxins), a secondary metabolite that has allelopathic behavior to the plant pathogen *Fusarium oxysporum* (Combès et al. 2012). Diversification in microbiomes is huge; the communities remain communicated with each other as it means they coexist in nature. SMs are used as signal molecules, and the coexistence leads to synthesization of more secondary metabolites production.

14.9.3 Stress Response and Virulence

Secondary metabolites prevent various biotic and abiotic stress conditions. Certain SMs have antimicrobial activities which prevent pathogens and herbivores attack. Abiotic stress such as drought, heavy metals, and salinity can be tolerated by hormone regulation, ROS tolerance, and increased membrane stability. SMs produced by fungal endophytes give protection against microbes present in the external environment. Fungal endophytes are predominantly observed in symbiotic associations with plants (Macheleidt et al. 2016). For example, *Fusarium graminearum* forms a symbiotic relationship with *Triticum*, where the compound trichothecene deoxynivalenol functions as a virulence factor (Kazan et al. 2012).

Table 14.1 Important crops and their beneficial endophytic fungi

S. No.	Agricultural crops	Beneficial endophytic fungi	Specific functions	References
1.	Rice (*Oryza sativa*)	*Phoma glomerata, Penicillium simplicissimum, Galactomyces geotrichum, Fusarium oxysporum, Phoma* sp., *Aspergillus ustus*	Growth-promoting factors, salt tolerance, improves mineral nutrition and quality especially under low nitrogen content in soil, ameliorates crop production	Tang et al. (2022), Potshangbam et al. (2017)
2.	Wheat (*Triticum aestivum*)	*G. etunicatum, G. intraradices, G. fasciculatum, T. atroviride, Glomus spp., Trichoderma atroviride, Alternaria alternata*	Improves seed germination, shoot and root growth, chlorophyll content, alleviates drought tolerance, ameliorates crop yield	Qiang et al. (2019), Colla et al. (2015)
3.	Barley (*Hordeum vulgare*)	*Serendipita indica, Fusarium culmorum, Tricholderma harzianum, Paenibacillus polmyxa*	Enhances the nutrient uptake, increases stress tolerance and plant growth, resistance to plant, biocontrol agent, produces antibiotics and plant growth hormones	White Jr et al. (2000)
4.	Maize (*Zea mays*)	*Fusarium sacchari, Gibberella intermedia, Trichoderma atroviride, Aspergillus awamori, Metarhizium robertsii*	Plant growth-promoting factors, induces disease resistance by improving plant immune response, protects against black cutworm larvae, improves crop yield	Ahmad et al. (2020), Mehmood et al. (2019), Contreras-Cornejo et al. (2018))
5.	Sorghum (*Solanum tuberosum*)	*Gibberella fujikuroi, Fusarium oxysporum, Trichoderma atroviride, Aspergillus awamori, Sarocladium zeae*	Plant growth-promoting factors, induces disease resistance by improving plant immune response, protects against black cutworm larvae, improves crop yield	Ahmad et al. (2020), Mehmood et al. (2019), Potshangbam et al. (2017)

(continued)

Table 14.1 (continued)

S. No.	Agricultural crops	Beneficial endophytic fungi	Specific functions	References
6.	Citrus	*Alternaria alternata*, *Alternaria citri*, *Alternaria rosae*, *Alternaria* sp., *Aspergillus* sp., *Colleotrichum karstii*, *Diaporthe eres*, *Piriformospora indica*, *Cladosporium* sp., *Pseudozyma* sp., *Meyerozyma* sp.	Acts as a biocontrol agent against pathogenic bacteria and fungal species, improves soil quality by enhancing total phosphatase activity, better fruit quality due to enhanced macronutrient content	Cheng et al. (2022), Sadeghi et al. (2019)
7.	Banana	*Fusarium* sp., *Phoma* sp., *Nigrospora* sp., *Penicillium* sp., *Colleotichum* sp., *Piriformospora indica*	Tolerance to cold stress, resistance to pathogenic fungi, protects from nematode infection, increases productivity	Kisaakye et al. (2023), Li et al. (2021), Zakaria and Aziz (2018)
8.	Soybean (*Glycine max*)	*Penicillium minioluteum*, *Porostereum spadiceum*, *Rhizopus oryzae*, *Paecilomyces formosus*	Protects from abiotic salinity and thermal stress, promotes photosynthetic activity, improves plant growth by increased macronutrient uptake, disease resistance	Bilal et al. (2020), Ismail et al. (2020), Bajaj et al. (2018), Hamayun et al. (2017), Khan et al. (2012)
9.	Tomato (*Solanum lycopersicum*)	*G. intraradices*, *T. atroviride*, *Fusarium solani*, *Pochinia* sp., *Pythium* sp., *Piriformospora* sp.	Acts as biofertilizers by improving nitrogen content in the soil, increases biomass production, counteracts bacterial pathogens	Sinno et al. (2020), Pappas et al. (2018) and Colla et al. (2015)
10.	Potato (*Solanum tuberosum*)	*Rhizophagus irregularis*, *Epicoccum nigrum*, *Curvularia lunata*	Mitigates oxidative stress in roots and shoots during growth, acts as a biocontrol agent against blackleg disease, antimicrobial properties	Deja-Sikora et al. (2020), Bagy et al. (2019)
11.	Sunflower (*Helianthus annuus*)	*Penicillium citrinum*, *Talaromyces assiutensis*, *Aspergillus terreus*, *Rhizophus oryzae*, *Brassica napus*, *Piriformospora indica*	Alleviates thermal stress, disease resistance, biocontrol agent against pathogenic fungi, mitigates cadmium toxicity, improves chlorophyll content under stress	Farhat et al. (2023), Ismail et al. (2020), Shahabivand et al. (2017), Waqas et al. (2015)

(continued)

Table 14.1 (continued)

S. No.	Agricultural crops	Beneficial endophytic fungi	Specific functions	References
12.	Cotton (*Gossypium herbaceum*)	*Penicillium simplicissimum, Leptosphaeria* sp., *Talaromyces flavus, Acremonium* sp., *Beauveria bassiana, Purpureocillium lilacinum*	Protects against cotton wilt caused by *Verticillium* sp., promotes plant growth, protects against aphid infestations, controls crop infestation	Tang et al. (2022), Potshangbam et al. (2017), Ahmad et al. (2020)

14.10 Applications of Secondary Metabolites Produced by Endophytic Fungus

The secondary metabolites synthesized by EFs are various beneficial chemicals that have a role in the pharma companies like medicine, agriculture, cosmetics and fragrance, and many others (Hyde and Soytong 2008). Endophytes are accessible for various secondary metabolites, which can be beneficial for the commercial and industrial levels.

Pharmacological properties: These are treasures of bioactive compounds for medicine and pharmaceuticals. Fungal endophytes are an abundance of bioactive compounds, which are beneficial in pharmaceuticals. Antibacterial, anticancer and antioxidants properties of secondary metabolites of EFs are helpful in drug manufacturing.

Antimicrobial activity: Pathogenic bacteria can be prevented by various alkaloids, peptides, phenols, steroids having antimicrobial activity. *Aspergillus* is derived from medicinal plant *Callistemon subulatus* having antimicrobial activity by the secretion of compounds like isoshamixanthone.

Anticancer activity: Many compounds derived from EFs are cytotoxic. From the roots, needles and barks of *Taxus brevifolia* a little amount of paclitaxel is synthesized which is helpful in treating lung, ovarian, breast, and prostate cancers (Yu et al. 2010). Taxol (paclitaxel) a diterpenoid found in *Taxomyces andreanae*, extracted from the pacific yew plant, an anticancer drug is being used in the treatment of ovarian and breast cancers. Camptothecin produced by *Aspergillus niger* is effective against cancer by leading to replication errors in cancer cells.

Antioxidant properties: These are used for treating ROS-linked diseases and for antiaging. Antioxidant properties present in Pertacin compounds are used for antifungal activities. *Penicillenols* from *Penicillium notanum* is another example.

Antibiotic properties: Bioactive compounds have antibiotic properties and help in improving plant health. DAPG synthesized by *pseudomonas* spp., a phenolic antibiotic, helps against soil-borne disease. An antibacterial activity against Gram-negative bacteria from the altersetin from the *Alternaria* spp.

Siderophore: A chelating agent siderophore produced from fungus increases the bioavailability of iron in plants.

Antidiabetic properties: Many endophytic fungi produce metabolites which can be a source for antidiabetic drugs. In the leaves of *Combretum dolichopetalum*, the fungus *Nigrospora oryzae* in its purified state is low in the sugar level (Singh and Kumar 2023). Endophytic fungi extracted from *Rauvolfia* also have these properties.

Many compounds have various properties. Podophyllotoxin is a sinopodophyllum plant-based bioactive aryltetralin compound having tremendous properties like antiviral, anticancer, antioxidant, antibacterial, and antirheumatic properties (Zhao et al. 2010).

Agriculture: The secondary metabolite makes plants tolerant to biotic and abiotic stresses and resistance against the pathogens. They can be used as biopesticide in organic farming. EFs also help ericoid plants to survive in the harsh environment. Endophytes incorporated into different plant parts have the ability to fix nitrogen. Alkaloids produce by EFs are herbivory and act as insecticidal (Sharma et al. 2021). These can act as biocontrol agents and natural pesticides. Antifungal properties of *Epichloe typhina* symbiotically associated with the perennial grass *Phleum pratense* against *Cladosporium herbarum* (Sharma et al. 2021). Some EFs help plant growth by uptaking nutrients. In some endophytic fungi, a phosphate-solubilizing bioactive compound improves soil by degrading phosphate and secretes growth promoters. Their production can be seen in *Hymenoscyphus ericae* (Sharma et al. 2021). Soil contaminated with heavy metals can be tolerated by these endophytic secretions (Sharma et al. 2021). These can help in improving soil by bioremediation. Volatile compound from *Muscodor albus* of the host plant *Cinnamomum zeylanicum* have mycofumigation properties (Sharma et al. 2021). Penicillium from host plant *Triticum aestivum* having phosphate-solubilizing properties act as biofertilizer. Siderophores with antibacterial activity are synthesized by *Pharotheca* associated with *Pinus* and *Fusarium* associated with *Rhododendron* (Sharma et al. 2021).

Industrial applications: Several industrial products, such as, lipases, pectinase, cellulases, and many bio-pigments, are obtained through bioactive metabolites produced by endophytic fungi, which can be utilized for biofuel production. Biodegradation of polyurethane by EF *Pestalotiopsis* microspore from host Ecuadorian Amazonian plant (Sharma et al. 2021). Biofuel production as mycodiesel from *Gliocladium roseum* from host plant *Eucryphia cordifolia* (Strobel et al. 2008).

Cosmetics and fragrance: The fungal endophytes bioactive compounds have antioxidant, anti-inflammatory properties and so help in the cosmetics used in skin and hair care. Antioxidant property protecting from the oxidative stress, which has antioxidant and antiaging formula. Antimicrobial compounds are used for increasing the shelf life of cosmetic products. Melanin production is also hit by certain metabolites which have skin lighting effects. Some metabolites are volatile in nature which can be used as base perfumes. Methyl eugenol is derived from *Alternaria* having a refreshing aroma.

Food industry: The bioactive compounds of endophytic fungi can be used as pigments, preservatives, and flavors. The nontoxic flavoring agent and synthetic color synthesized by these can be used as an alternate for synthetic ones. They help in increasing the shelf life of processed food by inhibiting pathogens. Fatty acids produced by EFs having oxidative stability protect food oxidation and rancid for some time. Edible pigments synthesized by fungal endophyte *Phoma* sp. from host plant *Clerodendrum longifolia* helps in the food industry (Sharma et al. 2021). Chitosans from *A. flavus* and *C. cladosporioides* are used as food preservative. EFs that are coated in the fruits and vegetables act as antioxidant and antimicrobial, enhancing the shelf life.

14.11 Role of Fungal Endophytes in Mitigating Biotic Stress

Biotic stress, brought on by pathogenic microorganisms, is a major obstacle to healthy plant growth and has detrimental impacts worldwide (Chaudhary et al. 2022). Significant biotic stresses, such as bacteria, pests, and nematodes, cause a rise in reactive oxygen species (ROS), which affects a plant's physiological and molecular functions. In the end, this lowers crop output and may even kill plants (Savary et al. 2019). Because they invade plant tissues without harming them, fungi called endophytes present a viable answer to the problems associated with conventional farming. These endophytes help their host plants become more resilient to biotic stress by producing phytohormones, enhancing nutrient uptake, and reducing pathogen-caused damage. These include the creation of lytic enzymes, mycoparasitism, the manufacturing of secondary metabolites, and the activation of induced systemic resistance (Fig. 14.3).

14.11.1 Against Fungal Pathogens

Plant diseases caused by fungal pathogens significantly negatively influence agriculture, lowering crop quality and productivity (Akram et al. 2023). In the end, these phytopathogenic fungi damage host plants by producing toxins essential to disease development (Peng et al. 2021). Many of these poisons are low-molecular-weight secondary metabolites that can cause wilting, necrotic spots, stunted growth, chlorosis, and even plant death. Even in extremely small amounts, these toxins have the ability to interfere with normal physiological processes in plants (Verma et al. 2022). Despite being widely employed to manage pathogenic fungi, chemical fungicides harm the environment. Fungal endophytes can greatly lessen the effects of fungal illnesses and the harmful effects of chemical fungicides on the environment (Cheong et al. 2017). These endophytes use a variety of strategies, including the production of bioactive metabolites, hormone signaling molecules, and the activation of defense-related gene expression, to suppress fungal infections and promote plant resistance (Adeleke et al. 2022). It is commonly known that antimicrobial compounds generated by fungal endophytes as secondary metabolites can inhibit

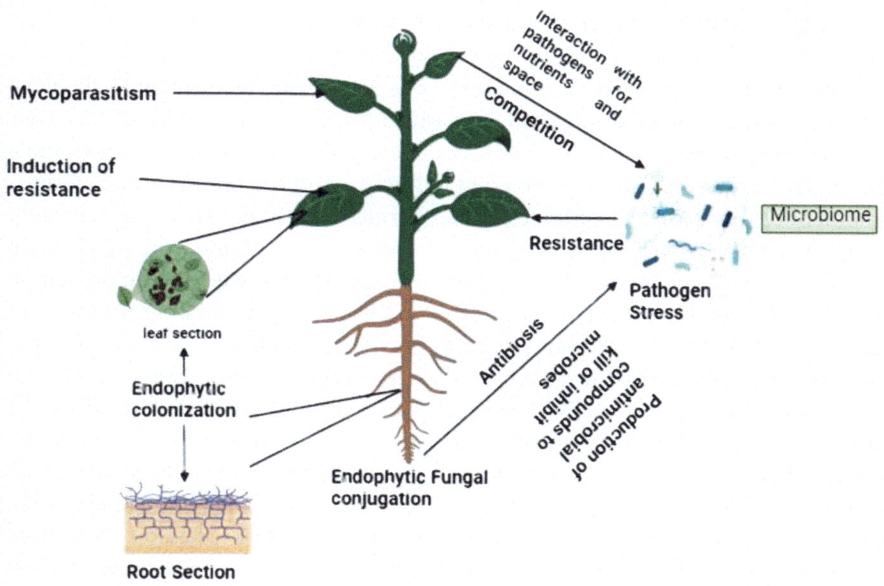

Fig. 14.3 Different mechanisms developed by endophytic fungi to inhibit potential plant infections (Verma et al. 2021)

the growth and development of pathogenic fungi, consequently decreasing their virulence and prevalence in the host plant (Gupta et al. 2022). Certain fungal endophytes can regulate host plants by producing antimicrobial chemicals that prevent fungal pathogens from vying for nutrients and space or through mycoparasitic activity, in which one fungus parasitizes another (Grabka et al. 2022). Furthermore, by inducing a systemic reaction, fungal endophytes frequently colonize host plants in a systemic manner, strengthening the plant's defenses against fungal infections. However, depending on the particular endophyte, host plant, and pathogen species involved, the impact of fungal endophytes on fungal infections might differ greatly. These complex relationships could have an adverse effect on the host plant's physiology and immune system, either directly or indirectly (Khare et al. 2018). For example, it has been demonstrated that *Hypoxylon rubiginosum* reduces the symptoms of dieback disease in *Fraxinus excelsior* seedlings by generating antifungal metabolites that inhibit *Hymenoscyphus fraxineus*, such as phomopsidin and its novel derivative, 10-hydroxyphomopsidin. Endophytic isolates such as *A. insulicola* and *A. melleus* showed strong antifungal activity against *Pythium aphanidermatum* in in-vitro dual-culture tests. They caused abnormal mycelial growth and damaged the hyphal walls, significantly suppressing *Pythium*'s growth. Additionally, they showed notably higher activity of cellulase and β-glucanase (Karunasinghe et al. 2020). Future studies might use fungal endophytes to improve crop resilience and productivity by increasing agricultural plants' resistance to phytopathogen infections.

14.11.2 Against Bacterial and Viral Pathogens

Reduced yields and disease control expenses can result in substantial financial losses caused by bacterial infections in various plant species (Sundin et al. 2016). Certain bacteria-borne infections can completely destroy crops and cause great harm. The use of disease-resistant plant cultivars, crop rotation, and the administration of copper-based bactericides and antibiotics are common management techniques to control bacterial plant diseases. Sustainable agricultural techniques might lessen their need for chemical treatments for bacterial illnesses by utilizing endophytes as biocontrol agents (Poveda et al. 2020a, b). By generating a variety of antibacterial secondary metabolites, including acetic acid, acetol, aliphatic compounds, alkaloids, hexanoic acid, peptides, phenylpropanoids, polyketides, and terpenoids, which target various bacterial pathogens and lower the prevalence of disease, fungal endophytes, in particular, are known to improve crop health and protect plants from bacterial infections (Grabka et al. 2022). In *Anthurium andraeanum* plants, the root endophytic *Piriformospora indica* inoculation stimulates the host's antioxidant activities and pathogenesis-related gene expression, protecting the root system, fostering growth, and conferring disease resistance against *Ralstonia solanacearum* infection (Lin et al. 2019). Via alterations in the activities of PAL (Phenylalanine ammonia-lyase), POD, β-1,3-glucanase, and total phenols linked to plant defense systems, potential endophytic *Fusarium solani and Trichoderma asperellum* have been discovered to produce resistance against *R. solanacearum*, which causes wilt disease in *Capsicum annuum* (Irawati et al. 2020). Even though the susceptible host-plant-associated endophytic fungus cannot directly combat *R. solanacearum*, the lipid-rich endometabolites extracted from the vertically transmitted endophytic *Penicillium* sp. strain showed the ability to attenuate the virulence factors of phytopathogenic *R. solanacearum*. Fusarium-lateritium-secreted protein (FlSp1), a novel protein secreted by the endophytic *F. lateritium*, has been found to function as an immune-triggering effector in *Nicotiana benthamiana*, where it upregulates ROS and manipulates the plant immune system to inhibit colonization and enhance resistance against *R. solanacearum* (Huang et al. 2023).

Plant viral infections can lower agricultural yields and quality and, in certain situations, potentially kill infected plants, resulting in large financial losses. Consequently, controlling these infections is necessary to safeguard crops and provide food security. Utilizing virus-resistant crop types, crop rotation, controlling vector hosts, steering clear of contaminated plant materials, and good sanitation are some strategies for controlling plant viral infections (Tatineni and Hein 2023). Certain fungal endophytes with entomopathogenic activity against insect vectors may indirectly inhibit viral disease transmission. For example, *Zea mays* plants have been discovered to be colonized by the fungal endophytes *Metarhizium anisopliae* and *Trichoderma harzianum*. This has been shown to lessen the severity of sugarcane mosaic virus illness and lower virus titers, which has protective implications (Kiarie et al. 2020). Furthermore, at low concentrations, plant immune inducers derived from the endophytic fungus *Paecilomyces variotii* have demonstrated high effectiveness in improving *Nicotiana benthamiana* resistance to the Potato X virus through the positive regulation of salicylic acid signaling pathways and RNA silencing. Additionally, a substantial source of secondary metabolites with antiviral qualities that fight viral illnesses can be found in fungal endophytes. As such, endophytic fungal priming of plants provides an eco-friendly substitute approach to viral disease management (Lacerda et al. 2022).

14.11.3 Role of Endophytic Fungi Against Pests and Parasitic Nematodes

Insects and mites that feed on plant tissues and spread plant viruses are examples of plant pests that can seriously harm agricultural crops. Effective pest control techniques such as chemical pesticides, pest-resistant crop types, and integrated pest management are crucial to reduce insect damage and preserve crop health and overall production (Deguine et al. 2021; Yadav and Meena 2022). Most terrestrial plants are home to various possible fungal endophytes, which can alter plant defense mechanisms and influence insect behavior and community structures either directly or indirectly. Fungal endophytes are important for both plant defense against various pest species and pest control. Certain fungal endophytes generate bioactive compounds or secondary metabolites that have insecticidal effects and are poisonous or repulsive to pests (Shah et al. 2019). By competing with pests for nutrients and space, these endophytes might reduce the resources pests require for survival and reproduction, hence indirectly decreasing pest populations (Chandran et al. 2021; Sinno et al. 2021). Additionally, by triggering plant metabolic pathways that result in chemicals poisonous to pests, fungal endophytes might strengthen the host plant's defenses against pests. Furthermore, the endophytic fungal colonization of plants has a major impact on the volatile compounds released by plants that are caused by pests. These volatile compounds have the ability to repel pests and draw in their natural enemies, which helps to improve plant resistance to pests that are significant to agronomy. When its culture filtrates and spores were applied, *Trichoderma hamatum* was found to demonstrate entomopathogenic activity against the larvae of *Spodoptera littoralis* (Lana et al. 2023). This was mostly because the fungus produced the siderophore rhizoferrin, which was responsible for its insecticidal activity. *Beauveria bassiana* and *Beauveria varroae*, two fungal endophytes, not only colonized various parts of *Beta vulgaris* plants (leaves, stems, and roots) following foliar spraying or seed inoculation without appearing to have any negative effects, but they also boosted plant growth and decreased *S. littoralis* survival (Darsouei et al. 2024). This demonstrates how applying fungal endophytes to pest management techniques may provide sustainable and environment-friendly ways to lessen pest damage and improve crop protection. Nematodes parasitic on plants are a worldwide agricultural hazard, resulting in large-scale financial losses, especially in tropical and subtropical areas. Although most species of these nematodes mainly target plant roots, they can also feed on seeds, stems, leaves, flowers, and other plant components. They cause harm to the root system and make it easier for secondary diseases to enter by creating feeding sites. Agricultural difficulties can be exacerbated by plant-parasitic nematodes, which can also serve as carriers of several plant viruses that can cause illnesses and even plant mortality (Phani et al. 2021). Chemical nematodes are commonly employed in traditional nematode control strategies. However, these substances may have unforeseen consequences for ecosystems and helpful soil microbial populations. There is much interest in investigating different approaches to managing nematodes because of ecological and environmental concerns. Because they can stop nematodes from growing and spreading, fungi endophytes have become possible biocontrol agents against nematodes (Miao et al. 2018). Through various strategies, such as parasitism, nematode paralysis, antibiosis, space competition, and the release of lytic enzymes, fungal endophytes can reduce plant-parasitic nematode infestations (Poveda et al. 2020a, b). Nematodes and their eggs can be physically trapped with hyphal loops, or they can parasitize nematode eggs and larvae and create nematocidal chemicals. Bioactive compounds secreted by certain endophytic fungi influence the colonization of plants by nematodes, either directly or indirectly, through the surrounding soil (Le et al. 2016; Meena et al. 2022). Furthermore, fungal endophytes initiate both induced and acquired systemic resistance, which aid in the movement of chemical defense compounds within the plant and encourage the synthesis of hormones like strigolactones, jasmonic acid, and salicylic acid, which helps the plant develop resistance against nematodes (Poveda et al. 2020a, b). *Meloidogyne graminicola* juveniles completely died when exposed to an endophytic *Aspergillus niger* associated with galls, both in vitro and in vivo. In addition to reducing egg hatching and inducing defense responses in *Oryza sativa* plants, this fungus demonstrated ovicidal characteristics and provided indirect protection against *M. graminicola* infection. In order to manage nematodes in agricultural plants, utilizing fungal endophytes as a seed treatment or root inoculation is a viable approach (Dutta et al. 2023). To optimize their potential for nematode control and their efficacy as biocontrol agents against a wide variety of nematodes, it is imperative to comprehend the bioactive compounds generated by these endophytes.

14.12 Role of Fungal Endophytes in Reducing Abiotic Stress

Drought, salt, heat, and heavy metals are examples of abiotic stresses, which are major environmental problems that can negatively influence agricultural production and cause yield losses in excess of 50% globally. Stresses like these can permanently harm a plant, leading to stunted development, severe damage, or even death (Murali et al. 2021a). They also prevent growth and metabolism, lowering plant biomass and active ingredients. Numerous studies have shown that fungal endophytes increase plants' resistance to various abiotic stressors (Table 14.2).

Table 14.2 As fungal endophytes that prevent diseases in their host plants, fungal endophytes are essential for biocontrol. By the synthesis of antimicrobial chemicals that impede the growth of pathogens, among other processes, they contribute to the defense of plants against diseases

S. No.	Endophytic Fungi	Host	Function of host plant	References
	Drought stress			
1.	*P. minioluteum*	*Chenopodium quinoa*	Positive plant–symbiont interactions were demonstrated by colonized plants' increased resistance, improved root development, and changed radicle growth	González-Teuber et al. (2017)
2.	*Darksidea strain, Knufia* sp., *and Leptosphaeria* sp.	*Ammopiptanthus mongolicus*	Plant biomass increased as a result of the endophyte's symbiotic connection, which differed depending on the strain of the plant	Li et al. (2018)
3.	*Embellisia chlamydospora, Knufia* sp., *Leptosphaeria* sp., *and Phialophora* sp.	*Hedysarum scoparium*	Plant roots that were successfully colonized created a mutually advantageous relationship with their host plants, which resulted in a rise in nutritional content, improved antioxidant enzyme activities, and overall biomass of the plants	Li et al. (2019)
4.	*Acrocalymma vagum*	*Ormosia hosiei*	Enhanced pigment content, transpiration rate, stomatal conductance, and net photosynthetic rate, decreased intracellular CO_2 concentration, and preserved the integrity of cell membranes, mitochondria, and chloroplasts	Liu and Wei (2021)
5.	*P. indica*	*H. vulgare*	Plant biomass grew as a result of colonization, and proteins essential for photosynthesis, signal transduction, ROS scavenging, and plant defense responses accumulated	Ghabooli et al. (2013)

(continued)

Table 14.2 (continued)

S. No.	Endophytic Fungi	Host	Function of host plant	References
6.	*P. indica*	*H. vulgare*	Colonized plant roots displayed altered amino acid metabolism in the host, as well as increased activity in the electron transfer chain and photosystem, as well as an accumulation of proteins linked to primary metabolism, energy regulation, photorespiration, autophagy, and transporters	Ghaffari et al. (2019)
7.	*Z. erostrata*	*S. lycopersicum* and *T. aestivum*	In the rhizosphere, extensively developed melanized mycelium showed improved nutrient mineralization and water uptake, enhanced plant biomass production, stimulated proline buildup, and decreased levels of lipid peroxide. It also showed higher resistance to drought	Miranda et al. (2023)
8.	*Neotyphodium coenophialum*	*Lolium arundinaceum*	Increased amounts of free fructose, glucose, trehalose, glutamic acid, proline, and sugar alcohols in plants, as well as fungal metabolites such alkaloids, mannitol, and loline. Resulted in noticeably higher tillering and survival of re-watered plants	Nagabhyru et al. (2013)

(continued)

Table 14.2 (continued)

S. No.	Endophytic Fungi	Host	Function of host plant	References
9.	Ascomycota sp. and *Cladosporium cladosporioides*	*N. benthamiana*	Plants that were colonized exhibited improved resistance, postponed the wilting of shoot tips, elevated relative moisture content, biomass, proline, soluble protein, soluble sugar, and antioxidant enzyme activity (PPO, POD, and CAT); additionally, they displayed decreased production of reactive oxygen species and electrical conductivity, and upregulated genes linked to drought defense	Dastogeer et al. (2018)
10.	*Nectria haematococca*	*S. lycopersicum*	Boosted proline accumulation in shoots, produced tolerance to drought stress, and markedly improved plant growth metrics	Valli and Muthukumar (2018)
11.	*A. vagum*, *F. acuminatum* and *Paraboeremia putaminum*	*Glycyrrhiza uralensis*	Raised plant biomass and glycyrrhizin concentration, improved plant root development, nutrient absorption, photosynthetic and antioxidant characteristics, colonized plants and created strain-dependent symbiosis, and altered the soil microbiota	He et al. (2019)
12.	*A. chlamydospora* and *Preussia terricola*	*G. uralensis*	Colonized plant roots, leading to higher levels of accessible nitrogen, soil organic matter, and glycyrrhizic acid concentration, as well as increased total plant biomass and root biomass	He et al. (2021)
13.	*Neocamarosporium* sp. and *Periconia macrospinosa*	*C. sativus* and *S. lycopersicum*	Increased proline content, development of the plants, and activity of antioxidant enzymes	

(continued)

Table 14.2 (continued)

S. No.	Endophytic Fungi	Host	Function of host plant	References
14.	*A. aculeatus*, *Meyerozyma guilliermondi* and *Microdochium majus*	*Moringa oleifera*	Increased stomatal aperture and less water potential loss were the results of enhanced plant growth attributes, total chlorophyll and carotenoid levels, primary and secondary metabolite production, decreased abscisic acid levels, increased activity of antioxidant enzymes like CAT, APX, and overall antioxidant capacity, and decreased ROS production. Additionally, upregulated gene expression of *MolAPX*, *MolHSF3*, and *MolHSF19* was observed	Moghaddam et al. (2021)
	Stress			
1.	*Neocamarosporium* sp. and *P. macrospinosa*	*C. sativus* and *S. lycopersicum*	Increased levels of proline, chlorophyll, and antioxidant enzyme activity in plants	Moghaddam et al. (2021)
2.	*Y. lipolytica*	*Z. mays*	Improvement in the properties of plant growth, such as higher levels of carotenoid and chlorophyll, less electrolyte leakage, more water content in seedlings compared to their parent plants, lower levels of internal abscisic acid, and higher levels of endogenous indole acetic acid. Proline, CAT, and POD production were also efficiently controlled	Jan et al. (2019)

(continued)

Table 14.2 (continued)

S. No.	Endophytic Fungi	Host	Function of host plant	References
3.	*Bipolaris* sp.	*G. max*	Plants that produced more organic acids, like indole acetic acid and gibberellins, were able to withstand salt stress better, and their length, weight, and chlorophyll content, all increased along with their salicylic acid levels. Endogenous abscisic acid decreased as a result, while antioxidant and oxidative stress indicators such as PPO, POD, superoxide anion, and malondialdehyde increased. Furthermore, while the expression levels of GmFDL19, GmNARK, and GmSIN1 declined, plant resistance to NaCl stress increased	Lubna et al. (2022)
4.	*A. terreus*	*O. sativa* and *Z. mays*	The biomass of the plants, their relative water content, photochemical efficiency, and oxidative balance, all increased significantly. The content of gibberellic acid was higher, photosynthesis and antioxidant defense systems were activated, and oxidative damage indicators including malondialdehyde and hydrogen peroxide were downregulated. There was also a positive interaction between microorganisms and plants	Siddiqui et al. (2022)

(continued)

Table 14.2 (continued)

S. No.	Endophytic Fungi	Host	Function of host plant	References
5.	*Paecilomyces formosus*	*C. sativus*	The synthesis of gibberellins and indole acetic acid promoted growth features and enhanced shoot length. It also lessened the negative effects of salt stress and led to the buildup of proline and antioxidants. In addition, endogenous abscisic acid levels were decreased, membrane damage and electrolyte leakage were minimized, and water potential was preserved	Khan et al. (2012)
6.	*F. verticillioides*	*G. max*	Gibberellins were created, seed germination and plant growth were enhanced, and plant length and fresh weight were markedly improved. It decreased lipid peroxidation, successfully lessened the harmful effects of salt stress, and increased protein content and the activity of antioxidant enzymes like POD, CAT, and SOD. It also revealed higher salicylic acid concentration and decreased abscisic acid levels	Radhakrishnan et al. (2013)
7.	*Stemphylium lycopersici*	*Z. mays*	It raised the levels of phenolics, flavonoids, and indole acetic acid, lowered the levels of malondialdehyde, and improved the activity of antioxidant enzymes such APX and CAT. It also increased the contents of $Mg2+$, $K+$, $Ca2+$, P, and N while decreasing the levels of $Na+$ and $Cl-$ ions and the ratios of $Na+/K+$ and $Na+/Ca2+$	Ali et al. (2022)

(continued)

Table 14.2 (continued)

S. No.	Endophytic Fungi	Host	Function of host plant	References
8.	*P. indica*	*H. vulgare*	It enhanced the potassium/sodium ratio in the foliar solution, controlled ion accumulation, promoted the growth and biomass of shoots in plants, and promoted the accumulation of proteins related to photosynthesis, energy production, signal transduction, protein translation and degradation, and antioxidant defense	Alikhani et al. (2013)
9.	*P. indica*	*H. vulgare*	It assisted in the isolation of metabolites, ions, and genes that were variably regulated to determine stress tolerance	Ghaffari et al. (2016)
10.	*B. bassiana*	*S. tuberosum*	It raised the number of stolons, elevated the activity of antioxidant enzymes such as SOD and POD, accelerated plant development, and reduced the adverse effects of salt stress. It also accumulated free proline	Tomilova et al. (2023)
11.	*Sordariomycetes* sp. and *Melanconiella elegans*	*Vigna unguiculata*	It enhanced stomatal conductance, plant growth parameters, and colonization rate	Farias et al. (2019)
12.	*C. globosum* and *Microsphaeropsis arundinis*	*T. aestivum*	It effectively colonized the plant, encouraged plant growth, and increased the biomass and germination rate of the seeds	Nguyen et al. (2021)

(continued)

Table 14.2 (continued)

S. No.	Endophytic Fungi	Host	Function of host plant	References
	Heat stress			
1.	*A. flavus*	*H. annuus* and *G. max*	Secondary metabolites were generated, resulting in elevated concentrations of flavonoids, phenolics, salicylic acid, and indole acetic acid. Additionally, it caused plants to have increased levels of proline and abscisic acid and lower levels of flavonoids, phenols, AAO (ascorbate oxidase), and CAT	Ismail et al. (2019)
2.	*Thermomyces lanuginosus*	*Cullen plicata*	It showed efficacious growth-promoting activity, enhanced plant survival, and higher levels of ascorbic acid, flavonoids, and total carbohydrates. It also increased the concentrations of antioxidant enzymes as CAT, POD, and PAL	Ali et al. (2019)
3.	*A. niger*	*G. max* and *H. annuus*	In contrast to lowering lipid peroxidation and ROS concentration, it raised plant biomass, height, and chlorophyll levels. Furthermore, it increased proline and phenolic levels, lowered the content of abscisic acid, and improved ROS scavenging activities like GR, CAT, AAO, POD, and SOD	Ismail et al. (2020)

(continued)

Table 14.2 (continued)

S. No.	Endophytic Fungi	Host	Function of host plant	References
4.	*A. japonicus*	*G. max* and *H. annuus*	It showed elevated levels of salicylic acid, phenolics, flavonoids, and indole acetic acid, which enhanced plant biomass and lessened the impacts of heat stress. Additionally, it controlled the actions of abscisic acid, CAT, and AAO. Moreover, it improved the seedlings' nutritional quality, which included phenolics, flavonoids, lipids, proteins, and soluble carbohydrates	Ismail et al. (2018)
	Cold stress			
1.	*Fusarium* sp. and *Pyrenophora* sp	*B. oleracea*	It enhanced plant development and resilience to cold	Poveda et al. (2020a)
2.	*P. indica*	*A. thaliana*	Cold stress response genes, including *WRKY*, *ERF*, *bHLH*, *HSF*, *MYB*, and NAC transcription factors, were expressed more frequently as a result	Jiang et al. (2021)
	Heavy metal stress			
1.	*Exophiala pisciphila*	*Z. mays*	By colonizing plant roots and markedly enhancing plant growth, antioxidants, and antioxidant enzyme activity, it demonstrated improved tolerance to cadmium stress. It transferred subcellular cadmium to the cell wall, changed the chemical form of metals into an inactive state, and improved cadmium tolerance by bioaugmentation	Wang et al. (2016)

(continued)

Table 14.2 (continued)

S. No.	Endophytic Fungi	Host	Function of host plant	References
2.	*Gaeumannomyces cylindrosporus*	*Z. mays*	Through colonizing plant roots and increasing plant biomass, height, and basal diameter, it demonstrated enhanced tolerance to lead stress. Additionally, it changed the lead accumulation and translocation processes in plants and increased photosynthetic efficiency	Ban et al. (2017)
3.	*Phialophora mustea*	*L. esculentum*	It improved plant growth and showed enhanced tolerance to zinc and cadmium stress by colonizing plant roots. It decreased metal absorption and buildup, improved resistance to cadmium and zinc stress, and raised the activity of antioxidant enzymes including SOD and POD. It also reduced the amount of malondialdehyde in leaves and lessened the harm caused by membrane lipid peroxidation	Zhu et al. (2018)
4.	*P. indica*	*N. tabacum*	It improved the plants' ability to withstand cadmium stress, which resulted in a rise in cadmium accumulation in the roots and a fall in the accumulation in the leaves. Additionally, it increased glutathione levels and POD activity. Moreover, it markedly increased the expression of GS, POD, and proteins linked to photosynthesis	Su et al. (2021)

(continued)

Table 14.2 (continued)

S. No.	Endophytic Fungi	Host	Function of host plant	References
5.	*Paecilomyces lilacinus*	*S. lycopersicum*	By enhancing plant growth, weight, sugar, flavonoids, phenols, indole acetic acid, proline, protein content, and relative water content, it improved plant tolerance to cobalt and lead stress. It also reduced the harm that lead and cobalt stress produced	Musa et al. (2023)

14.12.1 Drought Stress

Drought is a major abiotic stressor that reduces crop output in agriculture and directly interferes with developing nations' efforts to meet their targets for food security. Plant life is negatively impacted by inadequate water availability because it can lead to osmotic stress, hinder photosynthesis, reduce nutrient uptake, and increase reactive oxygen species (ROS) (Gowtham et al. 2021; Kumar et al. 2022). Plants that are stressed by drought display a range of physiological, biochemical, and molecular responses, including stomatal closure, decreased rates of photosynthesis and transpiration, the creation of stress hormones, the activation of the antioxidant defense system, and an increase in reactive oxygen species (ROS) (Murali et al. 2021b). It is crucial to create safe management plans for sustainable crop production to lessen drought stress's impact. Studies have demonstrated that fungal endophyte inoculation considerably mitigates the deleterious effects of drought stress on plants. Greater drought tolerance results from the favorable plant–symbiont association, improving transpiration, nutritional content, pigments, photosynthesis, stomatal conductance, total biomass, leaf shape and anatomical structure, and antioxidant enzyme activities (González-Teuber et al. 2017). *P. indica* and *Zopfiella erostrata* are two examples of fungal endophytes that are essential for increasing plant resistance to drought stress. *Plantago indica* colonizes the roots of *Hordeum vulgare*, enhancing the electron transfer chain and photosystem activity, increasing the accumulation of protein, and impacting many metabolic pathways such as transport, energy regulation, and amino acid metabolism. Similarly, *Zopfiella erostrata* forms melanized mycelium in the rhizosphere by deeply colonizing the roots of *S. lycopersicum and T. aestivum* (Miranda et al. 2023). This reduces lipid peroxide buildup in drought and increases plant biomass output, nutrient mineralization, and water intake. The efficiency of these fungal endophytes varies according to the individual strains, host plants, and climatic factors; this emphasizes how crucial host–fungus compatibility is for providing plants with drought tolerance.

14.12.2 Salt Stress

Especially in arid areas, salt stress is a major threat to plant growth and productivity worldwide (Hasanuzzaman and Fujita 2022). Extended exposure causes the creation of reactive oxygen species (ROS), which raises intracellular osmotic pressure and hazardous levels of sodium (Na^+). These levels damage plant metabolism, membranes, proteins, nucleic acids, and the K^+/Na^+ balance (Jan et al. 2019). Effective solutions for reducing salt stress are essential for optimal plant development and sustainable agriculture. By regulating defense genes, endogenous hormones, and antioxidant systems that lower ROS levels, fungus endophytes improve plant tolerance to salt (Lubna et al. 2022). By building up antioxidants and proline, controlling phytohormones, erecting physical barriers to prevent salt uptake in roots, and preserving plant water potential and ionic balance, these endophytes, through mutualistic interactions and phytohormone secretion, enhance plant growth and mitigate the effects of salt stress (Khan et al. 2012; Ali et al. 2022). The mutualistic root endophyte *P. indica* dramatically increased *H. vulgare* biomass and growth under salt stress by triggering systemic reactions that changed ion concentrations and proteome patterns (Alikhani et al. 2013). Because *Yarrowia lipolytica* secretes exogenous indole acetic acid, regulates endogenous abscisic acid and indole acetic acid levels, and produces proline, POD, and CAT in plants, it has been shown to increase plant development in saline environments through endophytic interaction with *Z. mays* (Jan et al. 2019). Furthermore, the entomopathogenic endophyte *B. bassiana* boosted plant development in *Solanum tuberosum* by reducing the effects of salt stress, most likely by enhancing the accumulation of free proline and activating antioxidant enzymes in plant tissues (Tomilova et al. 2023). Future studies may use fungal endophytes to improve the salt stress tolerance of agricultural plants.

14.12.3 Heat and Cold Stress

Elevated temperatures cause heat stress, a major abiotic factor that affects plants by interfering with normal physiological processes like reduced activity of mitochondrial and chloroplast enzymes, increased protein denaturation, changed membrane fluidity, increased ROS levels, and impaired photosystem II-mediated electron transport (Ismail et al. 2019). The way that different species and developmental stages react to heat stress in plants varies. By controlling the physiological, biochemical, and metabolic processes of plants, fungi endophytes are essential in reducing these impacts. When compared to untreated plants, the application of fungal endophytes counteracts heat stress by maintaining maximal photosystem II performance, improving water usage efficiency and photosynthesis, encouraging root growth, and accelerating the accumulation of osmolytes and antioxidant enzymes (Ali et al. 2018; Ismail et al. 2020). Fungal endophytes significantly improve plant resistance to heat stress and prevent environmental stress.

Due to excessive malondialdehyde and hydrogen peroxide accumulation brought on by low temperatures that cause cold stress, cell membrane lipid peroxidation can rise and could cause plant mortality (Valizadeh-Kamran et al. 2017). By lowering hydrogen peroxide and malondialdehyde levels, fungus endophyte colonization helps to mitigate the harm that membrane lipid peroxidation under cold stress causes to leaf cells (Jiang et al. 2021). Root endophyte *P. indica* efficiently increases host plant resistance to cold stress by controlling genes related to metabolic pathways, phytohormone signaling, and lipid and inorganic ion transport. *P. indica* increases the cold resistance of *M. acuminata* seedlings by boosting antioxidant capacity, increasing the accumulation of soluble sugars, and regulating the expression of genes that respond to cold, such as CSD1C, Why 1, HOS1, and CBF7-1 in leaves. This study shows that fungi endophytes may help crop plants become more resilient to cold stress (Li et al. 2021).

14.12.4 Heavy Metal Stress

Aquatic life and ecosystems are seriously threatened by heavy metal poisoning in developing nations. Due to the overuse of chemicals and the landfilling of municipal and industrial effluents, industrial waste and effluents are frequently dumped straight into rivers and build up on soil surfaces, which negatively affects agricultural areas (Sharma and Kumar 2021). These persistent and nonbiodegradable contaminants undergo chemical changes that allow them to remain in the soil environment. By accumulating in soil and irrigation water, heavy metals can damage plants and pose health concerns to animals and people up the food chain. Elevated quantities of metals such as lead, nickel, cobalt, zinc, copper, and cadmium are hazardous to plants and other natural systems (Liu et al. 2023). Thus, in order to protect human health and the health of ecosystems, effective steps must be taken to remove and remediate heavy metal contamination. By removing heavy metals from the soil and neutralizing their detrimental effects on plants, fungi endophytes establish vital symbiotic partnerships with plants that boost antioxidant enzyme activity (Wang et al. 2016). For example, *P. indica* enhanced *N. tabacum*'s ability to withstand cadmium stress by upregulating the expression of proteins involved in photosynthesis, and glutathione (GSH), and peroxidase (POD) activity, thereby mitigating the oxidative damage brought on by cadmium. Inoculating *Paecilomyces lilacinus* as an endophyte results in a considerable production of indole acetic acid and secondary metabolites that stimulate the growth of *S. lycopersicum* and lessen the harmful effects of heavy metals like cobalt and lead. Therefore, by aiding in the biosorption of heavy metals, fungal endophytes lessen the harmful effects of heavy metal toxicity (Musa et al. 2023). This eco-friendly bioremediation method promotes a more robust agricultural ecology by addressing heavy metal toxicity.

14.13 Future Perspectives

Researchers are increasingly concentrating on environment-friendly methods for sustainable food production in response to the mounting problems of pollution, overuse of agrochemicals, population expansion, food imbalance, and disappearing fertile land. The capacity of fungi endophytes to mitigate biotic and abiotic challenges and foster mutualistic relationships that improve plant development has garnered them a great deal of interest in the field of agriculture. Finding novel fungal endophytes that can flourish in challenging environments and have a variety of host relationships is essential. One potential way to lessen dependency on agrochemicals is to use these endophytes as biofertilizers, either alone or in groups, as part of integrated stress management techniques. In order to create improved biocontrol products that benefit the ecology and farmers alike, consortia with prospective fungal endophyte strains from nearby agricultural farms can be established. This is an environmentally responsible and crucial step towards sustainability. By using cutting-edge biotechnological and bioinformatic tools like RNA interference, the CRISPR-Cas system, metabolomics, and next-generation sequencing, extensive studies of the endophytic fungal community can unveil the intricate relationships between fungal endophytes and their hosts as well as the molecular patterns involved in stress tolerance mechanisms. To secure a safer future and enable the development of organic food crops, powerful bio-inocula based on advanced investigations into particular host–fungal endophyte connections are needed. Fungal endophytes can be added to cropping systems through foliar applications, seed treatments, or other means to increase agricultural productivity and produce beneficial environmental effects. It is difficult to comprehend the possible negative impacts of fungal endophytes on host plants, especially the possibility of toxins entering the food chain. It is imperative to research the possibility of fungal endophytes transforming into latent pathogens upon introduction to various crop plants. Depending on the physiological and anatomical characteristics of the plant species as well as signal transduction during colonization, some endophytic fungi that are advantageous to one species (host) may be pathogenic to others (nonhost), with various degrees of virulence. Furthermore, during growth cycles and in response to shifting environmental conditions, beneficial endophytic fungi may cause morphological, physiological, biochemical, and molecular alterations in plants that could result in their transformation into latent pathogens. Fungal endophytes have been reported to have the ability to become pathogenic and infect both the host and other plants under the right circumstances. Determining the ways in which recently introduced endophytes interact with native plant endobiomes presents another challenge. This matters because pure cultures of fungal endophytes rarely yield antimicrobials; instead, they usually only do so when other endophytes are present. On the other hand, some endophytes that are introduced have the potential to infect new host plants with disease or completely replace native endophytes during colonization. Because there is currently inadequate knowledge on the field performance of included endophytes and how climatic changes may affect their functioning, more research is necessary. The widespread commercial application of fungal endophytes is hindered by a number of issues, such as an incomplete understanding of particular plant–endophyte interactions, pathogen-plant defense mechanisms, physiological and chemical barriers, gene expression, and defense signaling pathways in agriculture. Furthermore, it is extremely difficult to register fungal bioagent products prior to commercialization because the approval procedure entails complex licensing laws and regulatory requirements. Famous scientists are making important contributions in spite of these obstacles, and they have high hopes for creating new, commercially successful fungal endophyte formulations and inoculations. The goal of these developments is to make climate-smart and sustainable agriculture methods possible.

14.14 Conclusion

For almost a century, the industrial manufacturing of pharmaceuticals and nutraceuticals has relied on fungi, which are abundant producers of secondary metabolites. Bioprospectors have spent the past three decades studying endophytic fungi, a very varied but mysterious species that lives within plant tissues. Even though it is often difficult to grow these organisms and keep their metabolite production steady, a number of phenotypic studies have shown the identification of new metabolites. To rationally modify or remodel these pathways for commercial use, however, would require identifying the underlying biosynthetic pathways, which is where the true issue resides. Only nine examples of endophyte secondary metabolites with definitively characterized biosynthetic gene clusters (BGCs) were found in our extensive literature review. Genome mining and BGC identification are difficult because of the small number of whole genome sequences of endophytic fungi and the frequently low predictive value of known BGCs for fungal BGCs. Rule-independent methods for predicting BGCs and sophisticated procedures for cultivation and experimental verification are essential for this endeavor. Furthermore, the increasing amount of biochemical research will improve training datasets for existing and upcoming machine learning-based systems. Genetic and genomic tool developments have greatly aided and will continue to aid in the bioprospecting of natural products. To summarize, the utilization of fungal endophytes presents a viable and efficient approach to alleviating crop losses while diminishing the adverse effects linked to traditional chemical techniques. By addressing the problems caused by biotic and abiotic stressors in plants, this method helps reduce the number of chemical pesticides and fertilizers used in agriculture. An important biological tool for promoting both commercially and environmentally sustainable farming methods is fungi endophytes. They guarantee long-term food security, improve soil health, and increase plant resilience. Because they encourage plant growth, yield economically valuable enzymes for biofuels, and produce nanomaterials and bioactive compounds for the food, textile, and cosmetic sectors, endophytes hold great promise for sustainable agricultural techniques. They are useful in biomedicine because they also create molecules with medicinal significance. All plant species' phyllosphere and rhizosphere have yielded endophytes and biotechnological developments have made it easier to isolate, screen, and identify a wide range of uncultivated endophytes. To improve this understanding and make better use of endophytes, a deeper comprehension of the mechanisms behind their various applications is necessary. Identifying novel endophytic microorganisms, their niche-specific functionality, and the molecular biology of their active metabolites should be the main areas of study in the future. Furthermore, comprehending how endophytes alter the expression of genes in their plant hosts may open the door to long-term uses in a range of biotechnology fields.

Acknowledgments The authors would like to extend their sincere appreciation to the funding agency Science and Engineering Research Board (SERB), State University Research Excellence (SURE) & Empowerment and Equity Opportunities for Excellence in Science, New Delhi, India. All the authors acknowledge their host institute for infrastructure support.

Conflict of Interest The authors declare no conflict of interest.

Funding This research was supported by funding agency Science and Engineering Research Board (SERB), State University Research Excellence (SURE) (File Number: SUR/2022/005216) and Empowerment and Equity Opportunities for Excellence in Science (File Number: EEQ/2022/000455), New Delhi, India. The research was also supported by the Ministry of Education and SPD-RUSA Rajasthan through the RUSA-2.0 project and the funding agency DST-FIST (GP193).

Data Availability Not applicable.

References

Abdallah QA, Ge W, Fortwendel JR (2017) A simple and universal system for gene manipulation in *Aspergillus fumigatus*: *in vitro*-assembled cas9-guide RNA ribonucleoproteins coupled with microhomology repair templates. MSphere 2(6):e00446-17

Aboobaker Z, Viljoen A, Chen W, Crous P, Maharaj V, Van Vuuren S (2019) Endophytic fungi isolated from *Pelargonium sidoides* DC: antimicrobial interaction and isolation of a bioactive compound. S Afr J Bot 122:535–542

Adeleke BS, Ayilara MS, Akinola SA, Babalola OO (2022) Biocontrol mechanisms of endophytic fungi. Egypt J Biol Pest Control 32(1):46

Ahmad I, Ahmad B, Boote K, Hoogenboom G (2020) Adaptation strategies for maize production under climate change for semi-arid environments. Eur J Agron 115:126040

Akram S, Ahmed A, He P, He P, Liu Y, Munir S, He Y (2023) Uniting the role of endophytic fungi against plant pathogens and their interaction. J Fungi 9(1):72

Alabid I, Glaeser SP, Kogel KH (2019) Endofungal bacteria increase fitness of their host fungi and impact their association with crop plants. Curr Issues Mol Biol 30(1):59–74

Alam B, Lǐ J, Gě Q, Khan MA, Gōng J, Mehmood S, Yuán Y, Gǒng W (2021) Endophytic fungi: from symbiosis to secondary metabolite communications or vice versa? Front Plant Sci 12:791033

Ali AH, Abdelrahman M, Radwan U, El-Zayat S, El-Sayed MA (2018) Effect of *Thermomyces* fungal endophyte isolated from extreme hot desert-adapted plant on heat stress tolerance of cucumber. Appl Soil Ecol 124:155–162

Ali AH, Radwan U, El-Zayat S, El-Sayed MA (2019) The role of the endophytic fungus, *Thermomyces lanuginosus*, on mitigation of heat stress to its host desert plant *Cullen plicata*. Biol Futura 70(1):1–7

Ali R, Gul H, Rauf M, Arif M, Hamayun M, Husna Khilji SA, Ud-Din A, Sajid ZA, Lee I (2022) Growth-promoting endophytic fungus (*Stemphylium lycopersici*) ameliorates salt stress tolerance in maize by balancing ionic and metabolic status. Front Plant Sci 13:890565

Alikhani M, Khatabi B, Sepehri M, Nekouei MK, Mardi M, Salekdeh GH (2013) A proteomics approach to study the molecular basis of enhanced salt tolerance in barley (*Hordeum vulgare* L.) conferred by the root mutualistic fungus *Piriformospora indica*. Mol BioSyst 9(6):1498

Aly AH, Debbab A, Kjer J, Proksch P (2010) Fungal endophytes from higher plants: a prolific source of phytochemicals and other bioactive natural products. Fungal Divers 41(1):1–16

Amirzakariya BZ, Shakeri A (2022) Bioactive terpenoids derived from plant endophytic fungi: an updated review (2011–2020). Phytochemistry 197:113130

Arnold AE, Mejía LC, Kyllo D, Rojas EI, Maynard Z, Robbins N, Herre EA (2003) Fungal endophytes limit pathogen damage in a tropical tree. Proc Natl Acad Sci USA 100(26):15649–15654

Aykul S, Martinez-Hackert E (2016) Determination of half-maximal inhibitory concentration using biosensor-based protein interaction analysis. Anal Biochem 508:97–103

Ban Y, Xu Z, Yang Y, Zhang H, Chen H, Tang M (2017) Effect of dark septate endophytic fungus *Gaeumannomyces cylindrosporus* on plant growth, photosynthesis and Pb tolerance of maize (*Zea mays* L.). Pedosphere 27(2):283–292

Bagy HMK, Hassan EA, Nafady NA, Dawood MF (2019) Efficacy of arbuscular mycorrhizal fungi and endophytic strain *Epicoccum nigrum* ASU11 as biocontrol agents against blackleg disease of potato caused by bacterial strain *Pectobacterium carotovora* subsp. atrosepticum PHY7. Biol Control 134:103–113

Bajaj R, Huang Y, Gebrechristos S, Mikolajczyk B, Brown H, Prasad R, Varma A, Bushley KE (2018) Transcriptional responses of soybean roots to colonization with the root endophytic fungus *Piriformospora indica* reveals altered phenylpropanoid and secondary metabolism. Sci Rep 8(1):10227

Berdy B, Spoering AL, Ling LL, Epstein SS (2017) *In situ* cultivation of previously uncultivable microorganisms using the ichip. Nat Protoc 12(10):2232–2242

Bilal S, Shahzad R, Imran M, Jan R, Kim KM, Lee IJ (2020) Synergistic association of endophytic fungi enhances *Glycine max* L. resilience to combined abiotic stresses: heavy metals, high temperature and drought stress. Ind Crop Prod 143:111931

Busch W, Moore BT, Martsberger B, Mace DL, Twigg RW, Jung J, Pruteanu-Malinici I, Kennedy SJ, Fricke GK, Clark RL, Ohler U, Benfey PN (2012) A microfluidic device and computational platform for high-throughput live imaging of gene expression. Nat Methods 9(11):1101–1106

Busby PE, Ridout M, Newcombe G (2016) Fungal endophytes: modifiers of plant disease. Plant Mol Biol 90:645–655

Camehl I, Sherameti I, Venus Y, Bethke G, Varma A, Lee J, Oelmüller R (2010) Ethylene signalling and ethylene-targeted transcription factors are required to balance beneficial and nonbeneficial traits in the symbiosis between the endophytic fungus *Piriformospora indica* and *Arabidopsis thaliana*. New Phytol 185(4):1062–1073

Carroll G (1986) The biology of endophytism in plants with particular reference to woody perennials. In: Fokkema NJ, Van Den Heuvek J (eds) Microbiology of the phyllosphere. Cambridge University Press, pp 205–222

Carroll G (1988) Fungal endophytes in stems and leaves: from latent pathogen to mutualistic symbiont. Ecology 69(1):2–9

Chandran H, Meena M, Swapnil P (2021) Plant growth-promoting rhizobacteria as a green alternative for sustainable agriculture. Sustainability 13(19):10986. https://doi.org/10.3390/su131910986

Chaudhary P, Agri U, Chaudhary A, Kumar A, Kumar G (2022) Endophytes and their potential in biotic stress management and crop production. Front Microbiol 13:933017

Chen T, Nomura K, Wang X, Sohrabi R, Xu J, Yao L, Paasch BC, Ma L, Kremer J, Cheng Y, Zhang L, Wang N, Wang E, Xin X, He SY (2020) A plant genetic network for preventing dysbiosis in the phyllosphere. Nature 580(7805):653–657

Cheng XF, Xie MM, Li Y, Liu BY, Liu CY, Wu QS, Kuča K (2022) Effects of field inoculation with arbuscular mycorrhizal fungi and endophytic fungi on fruit quality and soil properties of Newhall navel orange. Appl Soil Ecol 170:104308

Cheong SL, Cheow YL, Ting ASY (2017) Characterizing antagonistic activities and host compatibility (via simple endophyte-calli test) of endophytes as biocontrol agents of *Ganoderma boninense*. Biol Control 105:86–92

Chettri D, Rathod J, Verma AK, Ghosh, Verma AK (2023) Bioprospecting potentials of endophytic microbes in the sustainable biotechnological applications. Symbiosis 90(2):127–150

Chiang Y, Ahuja M, Oakley CE, Entwistle R, Asokan A, Zutz C, Wang CCC, Oakley BR (2015) Development of genetic dereplication strains in *Aspergillus nidulans* results in the discovery of aspercryptin. Angew Chem 55(5):1662–1665

Clay K, Holah J (1999) Fungal endophyte symbiosis and plant diversity in successional fields. Science 285(5434):1742–1744

Colla G, Rouphael Y, Di Mattia E, El-Nakhel C, Cardarelli M (2015) Co-inoculation of *Glomus intraradices* and *Trichoderma atroviride* acts as a biostimulant to promote growth, yield and nutrient uptake of vegetable crops. J Sci Food Agric 95:1706–1715

Combès A, Ndoye I, Bance C, Bruzaud J, Djediat C, Dupont J, Nay B, Prado S (2012) Chemical communication between the endophytic fungus *Paraconiothyrium variabile* and the phytopathogen *Fusarium oxysporum*. PLoS One 7(10):e47313

Contreras-Cornejo HA, Macías-Rodríguez L, Del-Val E, Larsen J (2018) The root endophytic fungus *Trichoderma atroviride* induces foliar herbivory resistance in maize plants. Appl Soil Ecol 124:45–53

Cord Landwehr S, Melcher RLJ, Kolkenbrock S, Moerschbacher BM (2016) A chitin deacetylase from the endophytic fungus *Pestalotiopsis* sp. efficiently inactivates the elicitor activity of chitin oligomers in rice cells. Sci Rep 6(1):38018

Dahlmann HA (2013) Spotlight. Chem Res Toxicol 26(12):1776–1777

Dam HT, Vollmers J, Sobol MS, Cabezas A, Kaster A (2020) Targeted cell sorting combined with single cell genomics captures low abundant microbial dark matter with higher sensitivity than metagenomics. Front Microbiol 11:01377

Darsouei R, Karimi J, Stelinski LL (2024) Endophytic colonization of sugar beet by *Beauveria varroae* and *Beauveria bassiana* reduces performance and host preference in army worm, *Spodoptera littoralis*. Crop Protect 175:106441

Dastogeer KM, Li H, Sivasithamparam K, Jones MG, Wylie SJ (2018) Fungal endophytes and a virus confer drought tolerance to *Nicotiana benthamiana* plants through modulating osmolytes, antioxidant enzymes and expression of host drought responsive genes. Environ Exp Bot 149:95–108

De Azevedo Silva F, Liotti RG, De Araújo Boleti AP, De Melo RÉ, Passos MBS, Santos ELD, Sampaio OM, Januário AH, Branco CLB, Da Silva GF, De Mendonça EAF, Soares MA (2018) Diversity of cultivable fungal endophytes in *Paullinia cupana* (Mart.) Ducke and bioactivity of their secondary metabolites. PLoS One 13(4):e0195874

Deguine J, Aubertot J, Flor RJ, Lescourret F, Wyckhuys KA, Ratnadass A (2021) Integrated pest management: good intentions, hard realities. A review. Agron Sustain Dev 41(3):38

Deja-Sikora E, Kowalczyk A, Trejgell A, Szmidt-Jaworska A, Baum C, Mercy L, Hrynkiewicz K (2020) Arbuscular mycorrhiza changes the impact of potato virus Y on growth and stress tolerance of *Solanum tuberosum* L. in vitro. Front Microbiol 10:2971

Deshmukh SK, Gupta MK, Prakash V, Saxena S (2018) Endophytic fungi: a source of potential antifungal compounds. J Fungi (Basel) 4(3):77–77

Dhayanithy G, Subban K, Chelliah J (2019) Diversity and biological activities of endophytic fungi associated with *Catharanthus roseus*. BMC Microbiol 19(1):22

Dick RP (2012) Manipulation of beneficial microorganisms in crop rhizospheres. In: Microbial ecology in sustainable agroecosystems. CRC Press, Taylor & Francis, pp 41–66

Dutta S, Mondal S, Hazra A, Ghosh S, Panja B, Chakrabarti M, Mukherjee A (2023) Evaluation of root-gall associated fungal endophytes for the control of *Meloidogyne graminicola* infecting rice. Biol Control 186:105341

Eid AM, Salim SS, Hassan SD, Ismail MA, Fouda A (2019) Role of endophytes in plant health and abiotic stress management. In: Kumar V, Prasad R, Kumar M, Choudhary DK (eds) Microbiome in plant health and disease: challenges and opportunities. Springer, Singapore, pp 119–144

Ernst M, Mendgen KW, Wirsel SGR (2003) Endophytic fungal mutualists: seed-borne *Stagonospora* spp. enhance reed biomass production in axenic microcosms. Mol Plant-Microbe Interact 16(7):580–587

Esmaeel Q, Miotto LCV, Rondeau M, Leclère V, Clément C, Jacquard C, Sanchez L, Barka EA (2018) *Paraburkholderia phytofirmans* PsJN-plants interaction: from perception to the induced mechanisms. Front Media 9:2093

Fadiji AE, Babalola OO (2020) Elucidating mechanisms of endophytes used in plant protection and other bioactivities with multifunctional prospects. Front Bioeng Biotechnol 8:467

Farias GC, Nunes KG, Soares MA, De Siqueira KA, Lima WC, Neves ALR, De Lacerda CF, Filho EG (2019) Dark septate endophytic fungi mitigate the effects of salt stress on cowpea plants. Braz J Microbiol 51(1):243–253

Farhat H, Urooj F, Irfan M, Sohail N, Majeed S, Ullah S, Shafique HA (2023) Biological control potential of endophytic fungi with amelioration of systemic resistance in sunflower and GC–MS metabolic profiling of *Talaromyces assiutensis*. Curr Microbiol 80:61

Fox EM, Howlett BJ (2008) Secondary metabolism: regulation and role in fungal biology. Curr Opin Microbiol 11(6):481–487

Galindo-Solís JM, Fernández FJ (2022) Endophytic fungal terpenoids: natural role and bioactivities natural role and bioactivities. Microorganisms 10(2):339

Ghabooli M, Khatabi B, Ahmadi FS, Sepehri M, Mirzaei M, Amirkhani A, Jorrín-Novo JV, Salekdeh GH (2013) Proteomics study reveals the molecular mechanisms underlying water stress tolerance induced by *Piriformospora indica* in barley. J Proteomics 94:289–301

Ghaffari MR, Ghabooli M, Khatabi B, Hajirezaei MR, Schweizer P, Salekdeh GH (2016) Metabolic and transcriptional response of central metabolism affected by root endophytic fungus *Piriformospora indica* under salinity in barley. Plant Mol Biol 90(6):699–717

Ghaffari MR, Mirzaei M, Ghabooli M, Khatabi B, Wu Y, Zabet-Moghaddam M, Mohammadi-Nejad G, Haynes PA, Hajirezaei MR, Sepehri M, Salekdeh GH (2019) Root endophytic fungus *Piriformospora indica* improves drought stress adaptation in barley by metabolic and proteomic reprogramming. Environ Exp Bot 157:197–210

Golinska P, Wypij M, Agarkar G, Rathod D, Dahm H, Rai M (2015) Endophytic actinobacteria of medicinal plants: diversity and bioactivity. Antonie Van Leeuwenhoek 108(2):267–289

González-Teuber M, Urzúa A, Plaza P, Bascuñán-Godoy L (2017) Effects of root endophytic fungi on response of *Chenopodium quinoa* to drought stress. Plant Ecol 219(3):231–240

Gowtham H, Duraivadivel P, Ayusman S, Sayani D, Gholap S, Niranjana S, Hariprasad P (2021) ABA analogue produced by *Bacillus marisflavi* modulates the physiological response of *Brassica juncea* L. under drought stress. Appl Soil Ecol 159:103845

Grabka R, D'Entremont TW, Adams SJ, Walker AK, Tanney JB, Abbasi PA, Ali S (2022) Fungal endophytes and their role in agricultural plant protection against pests and pathogens. Plan Theory 11(3):384

Gupta S, Choudhary M, Singh B, Singh R, Dhar MK, Kaul S (2022) Diversity and biological activity of fungal endophytes of *Zingiber officinale* Rosc. with emphasis on *Aspergillus terreus* as a biocontrol agent of its leaf spot. Biocatal Agric Biotechnol 39:102234

Gurusinghe S, Brooks TL, Barrow RA, Zhu X, Thotagamuwa A, Dennis PG, Gupta VVSR, Vanniasinkam T, Weston LA (2019) Technologies for the selection, culture and metabolic profiling of unique rhizosphere microorganisms for natural product discovery. Molecules 24(10):1955

Hamayun M, Hussain A, Khan SA, Kim HY, Khan AL, Waqas M, Irshad M, Iqbal A, Rehman G, Jan S, Lee IJ (2017) Gibberellins producing endophytic fungus *Porostereum spadiceum* AGH786 rescues growth of salt affected soybean. Front Microbiol 8:686

Hardoim PR, Van Overbeek LS, Van Elsas JD (2008) Properties of bacterial endophytes and their proposed role in plant growth. Trends Microbiol 16(10):463–471

Hardoim PR, Van Overbeek LS, Berg G, Pirttilä AM, Compant S, Campisano A, Döring M, Sessitsch A (2015) The hidden world within plants: ecological and evolutionary considerations for defining functioning of microbial endophytes. Microbiol Mol Biol Rev 79(3):293–320

Harrison JG, Griffin EA (2020) The diversity and distribution of endophytes across biomes, plant phylogeny and host tissues: how far have we come and where do we go from here? Environ Microbiol 22(6):2107–2123

Hasanuzzaman M, Fujita M (2022) Plant responses and tolerance to salt stress: physiological and molecular interventions. Int J Mol Sci 23(9):4810

He C, Wang W, Hou J (2019) Plant growth and soil microbial impacts of enhancing licorice with inoculating dark septate endophytes under drought stress. Front Microbiol 10:2277

He C, Wang W, Hou J, Li X (2021) Dark septate endophytes isolated from wild licorice roots grown in the desert regions of Northwest China enhance the growth of host plants under water deficit stress. Front Microbiol 12:522449

Henke MT, Soukup AA, Goering AW, McClure RA, Thomson RJ, Keller NP, Kelleher NL (2016) New aspercryptins, lipopeptide natural products, revealed by HDAC inhibition in *Aspergillus nidulans*. ACS Chem Biol 11(8):2117–2123

Higginbotham SJ, Arnold AE, Ibañez A, Spadafora C, Coley PD, Kursar TA (2013) Bioactivity of fungal endophytes as a function of endophyte taxonomy and the taxonomy and distribution of their host plants. PLoS One 8(9):e73192

Huang J, He Z, Wang J, Zha X, Xiao Q, Liu G, Li KJ (2023) A novel effector FlSp1 inhibits the colonization of endophytic *Fusarium lateritium* and increases the resistance to *Ralstonia solanacearum* in tobacco. J Fungi 9(5):519

Hulcr J, Dunn RR (2011) The sudden emergence of pathogenicity in insect–fungus symbioses threatens naive forest ecosystems. Proc Royal Soc B: Biol Sci 278:2866–2873

Hyde K, Soytong K (2008) The fungal endophyte dilemma. Fungal Divers 33:163–173

Irawati AFC, Mutaqin KH, Suhartono MT, Widodo W (2020) The effect of application endophytic fungus *Trichoderma* spp. and *Fusarium* spp. to control bacterial wilt in chilli pepper. Walailak J Sci Technol 17(6):559–569

Ismail N, Hamayun M, Hussain A, Iqbal A, Khan SA, Lee I (2018) Endophytic fungus *Aspergillus japonicus* mediates host plant growth under normal and heat stress conditions. Biomed Res Int 2018:1–11

Ismail N, Hamayun M, Hussain A, Khan SA, Iqbal A, Lee I (2019) *Aspergillus flavus* promoted the growth of soybean and sunflower seedlings at elevated temperature. Biomed Res Int 2019:1–13

Ismail N, Hamayun M, Hussain A, Iqbal A, Khan SA, Lee I (2020) *Aspergillus niger* boosted heat stress tolerance in sunflower and soybean via regulating their metabolic and antioxidant system. J Plant Interact 15(1):223–232

Jan FG, Hamayun M, Hussain A, Jan G, Iqbal A, Khan A, Lee I (2019) An endophytic isolate of the fungus *Yarrowia lipolytica* produces metabolites that ameliorate the negative impact of salt stress on the physiology of maize. BMC Microbiol 19(1):3

Jha P, Kaur T, Chhabra I, Panja A, Paul S, Kumar V, Malik T (2023) Endophytic fungi: hidden treasure chest of antimicrobial metabolites interrelationship of endophytes and metabolites. Front Microbiol 14:1227830

Jia M, Chen L, Xin H, Zheng C, Rahman K, Han T, Qin L (2016) A friendly relationship between endophytic fungi and medicinal plants: a systematic review. Front Microbiol 7:906

Jiang H, Xu Z, Aluru MR, Dong L (2014) Plant chip for high-throughput phenotyping of *Arabidopsis*. Lab Chip 14(7):1281–1293

Jiang W, Pan R, Buitrago S, Wu C, Abdelaziz ME, Oelmüller R, Zhang W (2021) Transcriptome analysis of *Arabidopsis* reveals freezing-tolerance related genes induced by root endophytic fungus *Piriformospora indica*. Physiol Mol Biol Plants 27(2):189–201

Karunasinghe TG, Maharachchikumbura SSN, Velazhahan R, Al-Sadi AM (2020) Antagonistic activity of endophytic and rhizosphere fungi isolated from sea purslane (*Sesuvium portulacastrum*) against Pythium damping-off of cucumber. Plant Dis 104(8):2158–2167

Katoch M, Singh G, Sharma S, Gupta N, Sangwan PL, Saxena AK (2014) Cytotoxic and antimicrobial activities of endophytic fungi isolated from *Bacopa monnieri* (L.) Pennell (Scrophulariaceae). BMC Complement Altern Med 14(1):52

Katoch M, Singh A, Singh G, Wazir P, Kumar R (2017) Phylogeny, antimicrobial, antioxidant and enzyme-producing potential of fungal endophytes found in *Viola odorata*. Ann Microbiol 67(8):529–540

Kazan K, Gardiner DM, Manners JM (2012) On the trail of a cereal killer: recent advances in *Fusarium graminearum* pathogenomics and host resistance. Mol Plant Pathol 13(4):399–413

Khan AL, Hamayun M, Kang S, Kim Y, Jung H, Lee J, Lee I (2012) Endophytic fungal association via gibberellins and indole acetic acid can improve plant growth under abiotic stress: an example of *Paecilomyces formosus* LHL10. BMC Microbiol 12(1):3

Khare E, Mishra J, Arora NK (2018) Multifaceted interactions between endophytes and plant: developments and prospects. Front Microbiol 9:2732

Kiarie S, Nyasani JO, Gohole LS, Maniania NK, Subramanian S (2020) Impact of fungal endophyte colonization of maize (*Zea mays* L.) on induced resistance to thrips- and aphid-transmitted viruses. Plants 9(4):416

Kisaakye J, Fourie H, Coyne D, Cortada L, Khamis FM, Subramanian S, Masinde S, Haukeland S (2023) Endophytic fungi improve management of the burrowing nematode in banana (*Musa spp.*) through enhanced expression of defence-related genes. Nematology 25(4):427–442

Kumar R, Swapnil P, Meena M, Selpair S, Yadav BG (2022) Plant growth-promoting rhizobacteria (PGPR): approaches to alleviate abiotic stresses for enhancement of growth and development of medicinal plants. Sustain For 14(23):15514

Lacerda ÍCDS, Polonio JC, Golias HC (2022) Endophytic fungi as a source of antiviral compounds—a review. Chem Biodivers 19(6):e202100971

Lana M, Simón O, Velasco P, Rodríguez VM, Caballero P, Poveda J (2023) First study on the root endophytic fungus *Trichoderma hamatum* as an entomopathogen: development of a fungal bioinsecticide against cotton leafworm (*Spodoptera littoralis*). Microbiol Res 270:127334

Latha P, Karthikeyan M, Rajeswari E (2019) Endophytic bacteria: prospects and applications for the plant disease management. In: Plant health under biotic stress. Springer, Singapore, pp 1–50

Le H, Padgham J, Hagemann M, Sikora R, Schouten A (2016) Developmental and behavioural effects of the endophytic *Fusarium moniliforme* Fe14 towards *Meloidogyne graminicola* in rice. Ann Appl Biol 169(1):134–143

Levetin E (2016) Aerobiology of agricultural pathogens. In: Manual of environmental microbiology, 4 ed. ASM Press, pp 1–20. https://doi.org/10.1128/9781555818821.ch3.2.8

Li X, He X, Hou L, Ren Y, Wang S, Su F (2018) Dark septate endophytes isolated from a xerophyte plant promote the growth of *Ammopiptanthus mongolicus* under drought condition. Sci Rep 8(1):7896

Li X, He X, Zhou Y, Hou Y, Zuo Y (2019) Effects of dark septate endophytes on the performance of *Hedysarum scoparium* under water deficit stress. Front Plant Sci 10:903

Li D, Bodjrenou DM, Zhang S, Wang B, Pan H, Yeh K, Lai Z, Cheng C (2021) The endophytic fungus *Piriformospora indica* reprograms banana to cold resistance. Int J Mol Sci 22(9):4973

Libor B, Harms H, Kehraus S, Egereva E, Crüsemann M, König GM (2019) Isolation of fungi using the diffusion chamber device FIND technology. Beilstein J Org Chem 15:2191–2203

Lin H, Xiong J, Zhou H, Chen C, Lin F, Xu X, Oelmüller R, Xu W, Yeh K (2019) Growth promotion and disease resistance induced in *Anthurium* colonized by the beneficial root endophyte *Piriformospora indica*. BMC Plant Biol 19(1):40

Liu Y, Wei X (2021) Dark septate endophyte improves the drought-stress resistance of *Ormosia hosiei* seedlings by altering leaf morphology and photosynthetic characteristics. Plant Ecol 222(7):761–771

Liu Z, Bai Y, Gao J, Li J (2023) Driving factors on accumulation of cadmium, lead, copper, zinc in agricultural soil and products of the North China Plain. Sci Rep 13:7429. https://doi.org/10.1038/s41598-023-34688-6

Lubna N, Khan MA, Asaf S, Jan R, Waqas M, Kim K, Lee I (2022) Endophytic fungus *Bipolaris* sp. CSL-1 induces salt tolerance in *Glycine max* L. via modulating its endogenous hormones, antioxidative system and gene expression. J Plant Interact 17(1):319–332

Luo C, Fernie AR, Yan J (2020) Single-cell genomics and epigenomics: technologies and applications in plants. Trends Plant Sci 25(10):1030–1040

Ma A, McDermaid A, Xu J, Chang Y, Ma Q (2020) Integrative methods and practical challenges for single-cell multi-omics. Trends Biotechnol 38(9):1007–1022

Macheleidt J, Mattern DJ, Fischer J, Netzker T, Weber J, Schroeckh V, Valiante V, Brakhage AA (2016) Regulation and role of fungal secondary metabolites. Annu Rev Genet 50:371–392

McKenzie SK, Walston RF, Allen JL (2020) Complete, high-quality genomes from long-read metagenomic sequencing of two wolf lichen thalli reveals enigmatic genome architecture. Genomics 112(5):3150–3156

Meena M, Yadav G, Sonigra P, Nagda A, Mehta T, Swapnil P, Harish, Marwal A (2022) Role of elicitors to initiate the induction of systemic resistance in plants to biotic stress. Plant Stress 5:100103

Mehmood A, Hussain A, Irshad M, Hamayun M, Iqbal A, Khan N (2019) In vitro production of IAA by endophytic fungus *Aspergillus awamori* and its growth promoting activities in *Zea mays*. Symbiosis 77:225–235

Mei Y, Zhu Y, Huang P, Yang Q, Dai C (2019) Strategies for gene disruption and expression in filamentous fungi. Appl Microbiol Biotechnol 103(15):6041–6059

Miao G, Han J, Zhang K, Wang S, Wang C (2018) Protection of melon against Fusarium wilt-root knot nematode complex by endophytic fungi *Penicillium brefeldianum* HS-1. Symbiosis 77(1):83–89

Michael A, Kelman T, Pitesky M (2020) Overview of quantitative methodologies to understand antimicrobial resistance via minimum inhibitory concentration. Animals 10(8):1405

Millet LJ, Aufrecht J, Labbé J, Uehling J, Vilgalys R, Estes ML, Guennoc CM, Deveau A, Olsson S, Bonito G, Doktycz MJ, Retterer ST (2019) Increasing access to microfluidics for studying fungi and other branched biological structures. Fungal Biol Biotechnol 6(1):8

Miranda V, Silva-Castro GA, Ruiz-Lozano JM, Fracchia S, García-Romera I (2023) Fungal endophytes enhance wheat and tomato drought tolerance in terms of plant growth and biochemical parameters. J Fungi 9(3):384

Mirdita M, Steinegger M, Breitwieser F, Söding J, Karin EL (2020) Fast and sensitive taxonomic assignment to metagenomic contigs. bioRxiv 37(18):3029–3031

Moghaddam MSH, Safaie N, Soltani J, Hagh-Doust N (2021) Desert-adapted fungal endophytes induce salinity and drought stress resistance in model crops. Plant Physiol Biochem 160:225–238

Mondal S, Rai VR (2019) Molecular profiling of endophytic *Streptomyces cavourensis* MH16 inhabiting *Millingtonia hortensis* Linn. and influence of different culture media on biosynthesis of antimicrobial metabolites. Sci Nat 106(9–10):51

Montoliu-Nerin M, Sánchez-García M, Bergin C, Grabherr M, Ellis B, Kutschera VE, Kierczak M, Johannesson H, Rosling A (2020) Building *de novo* reference genome assemblies of complex eukaryotic microorganisms from single nuclei. Sci Rep 10(1):1303

Moss EL, Maghini DG, Bhatt AS (2020) Complete, closed bacterial genomes from microbiomes using nanopore sequencing. Nat Biotechnol 38(6):701–707

Muhammad M, Basit A, Ali K, Ahmad H, Li WJ, Khan A, Mohamed HI (2024) A review on endophytic fungi: a potent reservoir of bioactive metabolites with special emphasis on blight disease management. Arch Microbiol 206(3):129

Murali M, Gowtham H, Singh SB, Shilpa N, Aiyaz M, Niranjana S, Amruthesh K (2021a) Bioprospecting of ACC deaminase producing Rhizobacteria towards sustainable agriculture: a special emphasis on abiotic stress in plants. Appl Soil Ecol 168:104142

Murali M, Singh SB, Gowtham H, Shilpa N, Prasad M, Aiyaz M, Amruthesh K (2021b) Induction of drought tolerance in *Pennisetum glaucum* by ACC deaminase producing PGPR-*Bacillus amyloliquefaciens* through antioxidant defense system. Microbiol Res 253:126891

Musa M, Jan FG, Hamayun M, Jan G, Khan SA, Rehman G, Ali S, Lee I (2023) An endophytic fungal isolate *Paecilomyces lilacinus* produces bioactive secondary metabolites and promotes growth of *Solanum lycopersicum* under heavy metal stress. Agronomy 13(3):883

Nagabhyru P, Dinkins RD, Wood CL, Bacon CW, Schardl CL (2013) Tall fescue endophyte effects on tolerance to water-deficit stress. BMC Plant Biol 13(1):127

Nair DN, Padmavathy SJTSWJ (2014) Impact of endophytic microorganisms on plants, environment and humans. Sci World J 2014:250693

Newman D (2017) Screening and identification of novel biologically active natural compounds. F1000Research 6:783

Nguyen Q, Iswanto ABB, Son GH, Kim SH (2021) Recent advances in effector-triggered immunity in plants: new pieces in the puzzle create a different paradigm. Int J Mol Sci 22(9):4709

Noirot-Gros M, Shinde S, Akins C, Johnson J, Zerbs S, Wilton R, Kemner K, Noirot P, Babnigg G (2020) Functional imaging of microbial interactions with tree roots using a microfluidics setup. Front Plant Sci 11:408

Osbourn A (2010) Gene clusters for secondary metabolic pathways: an emerging theme in plant biology. Plant Physiol 154(2):531–535

Pachiadaki MG, Brown JM, Brown J, Bezuidt O, Berube PM, Biller SJ, Poulton NJ, Burkart MD, La Clair JJ, Chisholm SW, Stepanauskas R (2019) Charting the complexity of the marine microbiome through single-cell genomics. Cell 179(7):1623–1635

Palanichamy P, Kannan S, Murugan D, Alagusundaram P, Marudhamuthu M (2019) Purification, crystallization and anticancer activity evaluation of the compound alternariol methyl ether from endophytic fungi *Alternaria alternata*. J Appl Microbiol 127(5):1468–1478

Pappas ML, Liapoura M, Papantoniou D, Avramidou M, Kavroulakis N, Weinhold A, Broufas GD, Papadopoulou KK (2018) The beneficial endophytic fungus *Fusarium solani* strain K alters tomato responses against spider mites to the benefit of the plant. Front Plant Sci 9:1603

Parmar S, Li Q, Wu Y, Li X, Yan J, Sharma VK, Wei Y, Li H (2018) Endophytic fungal community of *Dysphania ambrosioides* from two heavy metal-contaminated sites: evaluated by culture-dependent and culture-independent approaches. Microb Biotechnol 11(6):1170–1183

Pavithra N, Sathish L, Ananda K (2012) Antimicrobial and enzyme activity of endophytic fungi isolated from tulsi. J Pharm Biomed Sci (JPBMS) 16:2014

Payne A, Holmes N, Clarke T, Munro R, Debebe BJ, Loose M (2020) Readfish enables targeted nanopore sequencing of gigabase-sized genomes. Nat Biotechnol 39(4):442–450

Peng Y, Li SJ, Yan J, Tang Y, Cheng JP, Gao AJ, Yao X, Ruan JJ, Xu BL (2021) Research progress on phytopathogenic fungi and their role as biocontrol agents. Front Microbiol 12:670135

Pérez-Cobas AE, Gomez-Valero L, Buchrieser C (2020) Metagenomic approaches in microbial ecology: an update on whole-genome and marker gene sequencing analyses. Microbial. Genomics 6(8):mgen000409

Petrini O (1991) Fungal endophytes of tree leaves. In: Andrews JH, Hirano SS (eds) Microbial ecology of leaves. Springer, New York, pp 179–197

Phani V, Khan MR, Dutta TK (2021) Plant-parasitic nematodes as a potential threat to protected agriculture: current status and management options. Crop Prot 144:105573

Pieterse CM, Van Der Does D, Zamioudis C, Leon-Reyes A, Van Wees SC (2012) Hormonal modulation of plant immunity. Annu Rev Cell Dev Biol 28(1):489–521

Plett JM, Martin FM (2018) Know your enemy, embrace your friend: using omics to understand how plants respond differently to pathogenic and mutualistic: microorganisms. Plant J 93(4):729–746

Porras-Alfaro A, Bayman P (2011) Hidden fungi, emergent properties: endophytes and microbiomes. Annu Rev Phytopathol 49(1):291–315

Potshangbam M, Devi SI, Sahoo D, Strobel GA (2017) Functional characterization of endophytic fungal community associated with *Oryza sativa* L. and *Zea mays* L. Front Microbiol 8:325

Poveda J, Abril-Urias P, Escobar C (2020a) Biological control of plant-parasitic nematodes by filamentous fungi inducers of resistance: *Trichoderma*, mycorrhizal and endophytic fungi. Front Microbiol 11:992

Poveda J, Zabalgogeazcoa L, Soengas P, Rodríguez VM, Cartea ME, Abilleira R, Velasco P (2020b) *Brassica oleracea* var. *acephala* (kale) improvement by biological activity of root endophytic fungi. Sci Rep 10(1):981507

Qin JC, Zhang YM, Gao JM, Bai MS, Yang SX, Laatsch H, Zhang AL (2009) Bioactive metabolites produced by *Chaetomium globosum*, an endophytic fungus isolated from Ginkgo biloba. Bioorg Med Chem Lett 19(6):1572–1574

Qiang X, Ding J, Lin W, Li Q, Xu C, Zheng Q, Li Y (2019) Alleviation of the detrimental effect of water deficit on wheat (*Triticum aestivum* L.) growth by an indole acetic acid-producing endophytic fungus. Plant Soil 439:373–391

Radhakrishnan R, Khan AL, Lee I (2013) Endophytic fungal pre-treatments of seeds alleviates salinity stress effects in soybean plants. J Microbiol 51(6):850–857

Rajivgandhi G, Vijayan R, Kannan M, Santhanakrishnan M, Manoharan N (2016) Molecular characterization and antibacterial effect of endophytic actinomycetes *Nocardiopsis* sp. GRG1 (KT235640) from brown algae against MDR strains of uropathogens. Bioact Mater 1(2):140–150

Ran X, Zhang G, Li S, Wang J (2017) Characterization and antitumor activity of camptothecin from endophytic fungus *Fusarium solani* isolated from *Camptotheca acuminate*. Afr Health Sci 17(2):566

Rodriguez RJ, Henson JM, Volkenburgh EV, Hoy MS, Wright L, Beckwith F, Kim Y, Redman RS (2008) Stress tolerance in plants via habitat-adapted symbiosis. ISME J 2(4):404–416

Rodriguez RJ, White JF Jr, Arnold AE, Redman RS (2009) Fungal endophytes: diversity and functional roles. New Phytol 182(2):314–330

Sadeghi F, Samsampour D, Seyahooei MA, Bagheri A, Soltani J (2019) Diversity and spatiotemporal distribution of fungal endophytes associated with *Citrus reticulata* cv. Siyahoo. Curr Microbiol 76:279–289

Saikkonen K, Faeth SH, Helander M, Sullivan TJ (1998) Fungal endophytes: a continuum of interactions with host plants. Annu Rev Ecol Syst 29(1):319–343

Sakai K, Charlot F, Saux TL, Bonhomme S, Nogué F, Palauqui JJ, Fattaccioli J (2019) Design of a comprehensive microfluidic and microscopic toolbox for the ultra-wide spatio-temporal study of plant protoplasts development and physiology. Plant Methods 15(1):79

Santos-Júnior CD, Sarmento H, de Miranda FP, Henrique-Silva F, Logares R (2019) Uncovering the gene machinery of the Amazon river microbiome to degrade rainforest organic matter. Microbiome 8:151

Savary S, Willocquet L, Pethybridge SJ, Esker P, McRoberts N, Nelson A (2019) The global burden of pathogens and pests on major food crops. Nat Ecol Evol 3(3):430–439

Schardl CL, Leuchtmann A, Spiering MJ (2004) Symbioses of grasses with seedborne fungal endophytes. Annu Rev Plant Biol 55(1):315–340

Segawa T, Takeuchi N, Mori H, Rathnayake RMLD, Li Z, Akiyoshi A, Satoh H, Ishii S (2020) Redox stratification within cryoconite granules influences the nitrogen cycle on glaciers. FEMS Microbiol Ecol 96(11):fiaa199

Shahabivand S, Parvaneh A, Aliloo AA (2017) Root endophytic fungus *Piriformospora indica* affected growth, cadmium partitioning and chlorophyll fluorescence of sunflower under cadmium toxicity. Ecotoxicol Environ Saf 145:496–502

Shah S, Sharma A, Srivastava M (2019) Ecological significance of endophytes in the plant-insect microbial interaction: a review. Indian J Entomol 81(4):747

Sharma P, Kumar S (2021) Bioremediation of heavy metals from industrial effluents by endophytes and their metabolic activity: recent advances. Bioresour Technol 339:125589

Sharma H, Rai AK, Dahiya D, Chettri R, Nigam PS (2021) Exploring endophytes for *in vitro* synthesis of bioactive compounds similar to metabolites produced *in vivo* by host plants. AIMS Microbiol 7(2):175–199

Siddiqui ZS, Wei X, Umar M, Abideen Z, Zulfiqar F, Chen J, Hanif A, Dawar S, Dias DA, Yasmeen R (2022) Scrutinizing the application of saline endophyte to enhance salt tolerance in rice and maize plants. Front Plant Sci 12:770084

Singh VK, Kumar A (2023) Secondary metabolites from endophytic fungi: production, methods of analysis, and diverse pharmaceutical potential. Symbiosis 90(2):111–125

Singh LP, Singh Gill S, Tuteja N (2011) Unraveling the role of fungal symbionts in plant abiotic stress tolerance. Plant Signal Behav 6(2):175–191

Sinno M, Ranesi M, Gioia L, d'Errico G, Woo SL (2020) Endophytic fungi of tomato and their potential applications for crop improvement. Agriculture 10:587

Sinno M, Ranesi M, Di Lelio I, Iacomino G, Becchimanzi A, Barra E, Molisso D, Pennacchio F, Digilio MC, Vitale S, Turrà D, Harizanova V, Lorito M, Woo SL (2021) Selection of endophytic *Beauveria bassiana* as a dual biocontrol agent of tomato pathogens and pests. Pathogens 10(10):1242

Smith SE, Read DJ (2010) Mycorrhizal symbiosis, 3 ed. Academic press, Elsevier, New York

Soanes D, Richards TA (2014) Horizontal gene transfer in eukaryotic plant pathogens. Ann Rev Phytopathol 52(1):583–614

Stierle A, Strobel G, Stierle D (1993) Taxol and taxane production by *Taxomyces andreanae*, an endophytic fungus of Pacific yew. Science 260(5105):214–216

Strobel GA (2002) Rainforest endophytes and bioactive products. Crit Rev Biotechnol 22(4):315–333

Strobel GA, Knighton B, Kluck K, Ren Y, Livinghouse T, Griffin M, Spakowicz D, Sears J (2008) The production of myco-diesel hydrocarbons and their derivatives by the endophytic fungus *Gliocladium roseum* (NRRL 50072). Microbiology 154(Pt 11):3319–3328

Stuart T, Satija R (2019) Integrative single-cell analysis. Nat Rev Genet 20(5):257–272

Su Z, Zeng Y, Li X, Perumal AB, Zhu J, Lu X, Dai M, Liu X, Lin F (2021) The endophytic fungus *Piriformospora Indica* assisted alleviation of cadmium in tobacco. J Fungi 7(8):675

Sugimoto Y, Camacho FR, Wang S, Chankhamjon P, Odabas A, Biswas A, Jeffrey PD, Donia MS (2019) A metagenomic strategy for harnessing the chemical repertoire of the human microbiome. Science 366(6471):eaax9176

Sundin GW, Castiblanco LF, Yuan X, Zeng Q, Yang C (2016) Bacterial disease management: challenges, experience, innovation and future prospects. Mol Plant Pathol 17(9):1506–1518

Tang D, Wang G, Zhou JM (2017) Receptor kinases in plant-pathogen interactions: more than pattern recognition. Plant Cell 29:618–637

Tang MJ, Lu F, Yang Y, Sun K, Zhu Q, Xu FJ, Zhang W, Dai CC (2022) Benefits of endophytic fungus *Phomopsis liquidambaris* inoculation for improving mineral nutrition, quality, and yield of rice grains under low nitrogen and phosphorus condition. J Plant Growth Regul 41:2499–2513

Tapfuma KI, Uche-Okereafor N, Sebola TE, Hussan R, Mekuto L, Makatini MM, Green E, Mavumengwana V (2019) Cytotoxic activity of crude extracts from *Datura stramonium's* fungal endophytes against A549 lung carcinoma and UMG87 glioblastoma cell lines and LC-QTOF-MS/MS based metabolite profiling. BMC Complement Altern Med 19(1):330

Tatineni S, Hein GL (2023) Plant viruses of agricultural importance: current and future perspectives of virus disease management strategies. Phytopathology 113(2):117–141

Teixeira PJ, Colaianni NR, Fitzpatrick CR, Dangl JL (2019) Beyond pathogens: microbiota interactions with the plant immune system. Curr Opin Microbiol 1(49):7–17

Theodorakidou M, Nikola OA, Lambrou G (2018) The multiple roles of steroids and anabolic steroids and its relations to cardiovasular and musculoskeletal pathology: a brief review. J Res Pract Musculoskeletal Syst 02:22–30

Tomilova OG, Kryukova NA, Efimova MV, Kolomeichuk LV, Kovtun IS, Glupov VV (2023) The endophytic entomopathogenic fungus *Beauveria bassiana* alleviates adverse effects of salt stress in potato plants. Horticulturae 9(10):1140

Ullah A, Munir S, Badshah SL, Khan N, Ghani L, Poulson BG, Emwas AH, Jaremko M (2020) Important flavonoids and their role as a therapeutic agent. Molecules 25(22):5243

Waqas M, Khan AL, Hamayun M, Shahzad R, Kang SM, Kim JG, Lee IJ (2015) Endophytic fungi promote plant growth and mitigate the adverse effects of stem rot: an example of *Penicillium citrinum* and *Aspergillus terreus*. J Plant Interact 10:280–287

Valizadeh-Kamran R, Toorchi M, Mogadam M, Mohammadi H, Pessarakli M (2017) Effects of freeze and cold stress on certain physiological and biochemical traits in sensitive and tolerant barley (*Hordeum vulgare*) genotypes. J Plant Nutr 41(1):102–111

Valli PPS, Muthukumar T (2018) Dark septate root endophytic fungus *Nectria haematococca* improves tomato growth under water limiting conditions. Indian J Microbiol 58(4):489–495

Van Der Hooft JJJ, Mohimani H, Bauermeister A, Dorrestein PC, Duncan KR, Medema MH (2020) Linking genomics and metabolomics to chart specialized metabolic diversity. Chem Soc Rev 49(11):3297–3314

Vandenkoornhuyse P, Quaiser A, Duhamel M, Le Van A, Dufresne A (2015) The importance of the microbiome of the plant holobiont. New Phytol 206(4):1196–1206

Verma VC, Kharwar RN, Strobel GA (2009) Chemical and functional diversity of natural products from plant associated endophytic fungi. Nat Prod Commun 4(11):1511–1532

Verma S, Kingsley K, Bergen M, Kowalski K, White J (2018) Fungal disease prevention in seedlings of rice (*Oryza sativa*) and other grasses by growth-promoting seed-associated endophytic bacteria from invasive *Phragmites australis*. Microorganisms 6(1):21

Verma H, Kumar D, Kumar V, Kumari M, Singh SK, Sharma VK, Droby S, Santoyo G, White JF, Kumar A (2021) The potential application of endophytes in management of stress from drought and salinity in crop plants. Microorganisms 9(8):1729

Verma A, Shameem N, Jatav HS, Sathyanarayana E, Parray JA, Poczai P, Sayyed RZ (2022) Fungal endophytes to combat biotic and abiotic stresses for climate-smart and sustainable agriculture. Front Plant Sci 13:953836

Waller F, Achatz B, Baltruschat H, Fodor J, Becker K, Fischer M, Heier T, Hückelhoven R, Neumann C, Von Wettstein D, Franken P, Kogel K (2005) The endophytic fungus *Piriformospora indica* reprograms barley to salt-stress tolerance, disease resistance, and higher yield. Proc Natl Acad Sci USA 102(38):13386–13391

Wang J, Li T, Liu G, Smith JM, Zhao Z (2016) Unraveling the role of dark septate endophyte (DSE) colonizing maize (*Zea mays*) under cadmium stress: physiological, cytological and genic aspects. Sci Rep 6(1):22028

Wang Q, Cobine PA, Coleman JJ (2018) Efficient genome editing in *Fusarium oxysporum* based on CRISPR/Cas9 ribonucleoprotein complexes. Fungal Genet Biol 117:21–29

White JF Jr, Bacon CW, Stone JK (2000) An overview of endophytic microbes: endophytism defined. In: Microbial endophytes. CRC Press, Taylor & Francis, pp 17–44

Williams DH, Stone MJ, Hauck PR, Rahman SK (1989) Why are secondary metabolites (natural products) biosynthesized? J Nat Prod 52(6):1189–1208

Woyke T, Doud DFR, Schulz F (2017) The trajectory of microbial single-cell sequencing. Nat Methods 14(11):1045–1054

Wu B (2018) Secondary metabolites of endophytic fungus *Aspergillus* TQ67 from *Suaeda glauca*. Chin Pharm J 24:504–507

Wu B, Hussain M, Zhang W, Stadler M, Liu X, Xiang M (2019) Current insights into fungal species diversity and perspective on naming the environmental DNA sequences of fungi. Mycology 10(3):127–140

Yadav G, Meena M (2021) Bioprospecting of endophytes in medicinal plants of Thar Desert: an attractive resource for biopharmaceuticals. Biotechnol Rep 30:e00629

Yadav G, Meena M (2022) Biological control of plant diseases by endophytes. In: Shah MP, Deka D (eds) Endophytic association: what, why and how (Volume: developments in applied microbiology and biotechnology). Academic Press, Elsevier, pp 119–135. https://doi.org/10.1016/B978-0-323-91245-7.00007-9

Yadav V, Wang Z, Wei C, Amo A, Ahmed B, Yang X, Zhang X (2020) Phenylpropanoid pathway engineering: an emerging approach towards plant defense. Pathogens 9(4):312

Ye SH, Siddle KJ, Park DJ, Sabeti PC (2019) Benchmarking metagenomics tools for taxonomic classification. Cell 178(4):779–794

Youngblut ND, De La Cuesta-Zuluaga J, Reischer GH, Dauser S, Schuster N, Walzer C, Stalder G, Farnleitner AH, Ley RE (2020) Large-scale metagenome assembly reveals novel animal-associated microbial genomes, biosynthetic gene clusters, and other genetic diversity. MSystems 5(6). https://doi.org/10.1128/msystems.01045-20

Yu H, Zhang L, Li L, Zheng C, Guo L, Li W, Sun P, Qin L (2010) Recent developments and future prospects of antimicrobial metabolites produced by endophytes. Microbiol Res 165(6):437–449

Zakaria L, Aziz WNW (2018) Molecular identification of endophytic fungi from banana leaves (*Musa* spp.). Trop Life Sci Res 29:201

Zhang HW, Song YC, Tan RX (2006) Biology and chemistry of endophytes. Nat Prod Rep 23(5):753

Zhang Q, Xue X, Miao S, Cui J, Qin X (2020) Differential relationship of fungal endophytic communities and metabolic profiling in the stems and roots of *Ephedra sinica* based on metagenomics and metabolomics. Symbiosis 81(2):115–125

Zhao J, Zhou L, Wang L, Shan L, Zhong L, Liu X, Gao X (2010) Endophytic fungi for producing bioactive compounds originally from their host plants. In: Current research, technology and education topics in applied microbiology and microbial biotechnology, 1st edn. Formatex Research Center, Badajoz, Spain pp 567–576

Zheng Y, Su T, Qi Q (2019) Microbial CRISPRI and CRISPRA systems for metabolic engineering. Biotechnol Bioprocess Eng 24(4):579–591

Zhou W, Le J, Chen Y, Cai Y, Hong Z, Chai Y (2019) Recent advances in microfluidic devices for bacteria and fungus research. TrAC: Trends Anal Chem 112:175–195

Zhu L, Li T, Wang C, Zhang X, Xu L, Xu R, Zhao Z (2018) The effects of dark septate endophyte (DSE) inoculation on tomato seedlings under Zn and Cd stress. Environ Sci Pollut Res Int 25(35):35232–35241